INTERNATIONAL ASTRONOMICAL UNION

UNION ASTRONOMIQUE INTERNATIONALE

GALACTIC BULGES

PROCEEDINGS OF THE 153TH SYMPOSIUM OF THE INTERNATIONAL ASTRONOMICAL UNION, HELD IN GHENT, BELGIUM, AUGUST 17-22, 1992

EDITED BY

HERWIG DEJONGHE

Astronomical Observatory, University of Ghent, Belgium

and

HARM J. HABING

Sterrewacht Leiden, The Netherlands

KLUWER ACADEMIC PUBLISHERS

DORDRECHT / BOSTON / LONDON

Library of Congress Cataloging-in-Publication Data

International Astronomical Union. Symposium (153rd : 1992 : Ghent, Belgium)
Galactic bulges : proceedings of the 153rd Symposium of the International Astronomical Union held in Ghent, Belgium, August 17-22, 1992 / edited by Herwig Dejonghe and Harm J. Habing.
p. cm.
Includes index.
ISBN 0-7923-2424-2 (alk. paper)
1. Galactic bulges--Congresses. I. Dejonghe, Herwig. II. Habing, H. J. (Harm Jan), 1937- . III. Title.
QB857.5.B84I58 1992
523.1'12--dc20 93-27853

ISBN 0-7923-2424-2

Published on behalf of
the International Astronomical Union
by
Kluwer Academic Publishers, P.O. Box 17, 3300 AA Dordrecht, The Netherlands.

Kluwer Academic Publishers incorporates
the publishing programmes of
D. Reidel, Martinus Nijhoff, Dr W. Junk and MTP Press.

Sold and distributed in the U.S.A. and Canada
by Kluwer Academic Publishers,
101 Philip Drive, Norwell, MA 02061, U.S.A.

In all other countries, sold and distributed
by Kluwer Academic Publishers Group,
P.O. Box 322, 3300 AH Dordrecht, The Netherlands.

NOTICE

Due to an unfortunate mistake, the copyright lines appearing at the bottom of the opening page of each contribution in this volume are incorrect. The copyright of each and every contribution in this volume is held by the International Astronomical Union and not by Kluwer Academic Publishers.

Printed on acid-free paper

Printed in the Netherlands

TABLE OF CONTENTS

CONTRIBUTED PAPERS

INDICES

PREFACE

In the early summer of '89 a very informal meeting on the bulge of our Galaxy was held in Leiden. During that meeting Michael Rich proposed to hold a more properly organised symposium on "Galactic Bulges" in a few years time. After some discussion a Scientific Organising Committee was founded and after some manoeuvring a chairman was chosen, a local organiser was assigned and two editors were given instructions. A good thing about the location of the meeting was that Ghent is a very beautiful city and had never before hosted an IAU symposium. It could be that this, plus the fact that he is a very keen amateur astronomer led H.M. the King of Belgium to offer his patronage to the meeting – an offer that we gratefully and – we hope – gracefully accepted.

The meeting took place at a resort some 15 km outside Ghent. Most participants were housed on the premises – a very convenient situation. This feeling of togetherness made up for the small shortcomings of the lecture room, which is normally used as a sports hall. The weather was fair, except on the day of the barbecue when pouring rain forced us to go inside. We want to thank the local organisers and all their helpers for the enormous amount of work they put into the organisation, the success of which can be measured by the importance of the only major disaster: on the first day no poster would stick onto the wall. We especially would like to thank the two conference secretaries, Nicole Baeck and Monique Schaetsaert. Thanks are also due to P. Batsleer who edited the book of abstracts and to M. Winsall for his help in transcribing the tapes containing the discussions.

As for these proceedings: we thank all invited speakers for finally sending in their reviews – we could not help but notice that the reviewers from the other side of the ocean obeyed the deadline more strictly than those from this side. Unfortunately, two reviewers missed all deadlines – and these proceedings. Scepticism made us decide not to schedule a separate session on the possible triaxiality of the bulge – or in other words, about the bar. The subject index to these proceedings shows that the participants corrected automatically for this mistake. We printed a few Russian contributions, although the authors were not able to participate in the meeting; we felt justified by the very special circumstances of these authors. Ideally all contributed papers should be of equal length. However, although all scientists are equal, it happened that some contributions are more equally long than others. A novelty was provided by the portraits of speakers drawn during their presentation by L. Debrouwere, who agreed to work under the poor lighting conditions of a lecture room. The results are true caricatures: the victims refuse to recognize what all others do.

For the cover we selected a diagram discussed by Harding and Morrison in their contribution. Large data bases have always been valued in astronomical research; so have data of high quality. New is the possibility to obtain large data bases with entries of high quality. Old problems suddenly take on a new appearance; read the discussion of the diagram on the cover and of other similar diagrams elsewhere in this volume.

We hope that these proceedings carry with them the atmosphere of an enjoyable and fruitful meeting, and that they will be of use for some time in the future.

Ghent and Leiden, April 15, 1993

H.Dejonghe and H.J.Habing

Scientific Organizing Committee:

H.J. Habing (the Netherlands, Chair), C. Chiosi (Italy), H. Dejonghe (Belgium), M. Feast (South Africa), K. Freeman (Australia), G. Gilmore (United Kingdom), F. Hartwick (Canada), J. Melnick (ESO), I. Novikov (Russia), M. Rich (USA).

Local Organizing Committee:

H. Dejonghe (Chair), N. Baeck, M. Luwel, M. Schaetsaert, P. Smeyers, H. Steyaert, B. Van der Walle, F. Verheest.

Sponsoring IAU Commissions:

28 (Galaxies), 33 (Structure and dynamics of the galactic system), 35 (Stellar constitution).

This symposium would not have been possible without the support of:

the International Astronomical Union, het Nationaal Fonds voor Wetenschappelijk Onderzoek (Belgium), the City Council of Ghent, the Province of East Flanders, the University of Ghent, the University of Leiden, de Limburgse Volkssterrenwacht, Kluwer Academic Publishers, AGFA-GEVAERT, Kredietbank, Digital Equipment Corporation, Prof. Dr. H. Steyaert.

List of Participants

Nobuo ARIMOTO
Institut für
Theoretische Astrophysik der
Universität Heidelberg
Im Neuenheimer Feld 561
D-6900 Heidelberg
GERMANY
e47@dhdwiz1.bitnet

Richard ARNOLD
Sterrewacht Leiden
Postbus 9513
NL–2300 RA Leiden
THE NETHERLANDS
arnold@strw.leidenuniv.nl

Marc BALCELLS
Obs del Roque de los Muchachos
Apartado 321
E–38700 Santa Cruz de la Palma
Tenerife, Canarias, SPAIN
balcells@rug.nl

Beatriz BARBUY
Universidade de São Paulo
Instituto Astronomico e Geofisico
Depto. Astronomia
C.P. 9638
São Paulo 01065, BRAZIL
barbuy@iag.usp.ansp.br

Peter BATSLEER
Sterrenkundig Observatorium
Krijgslaan 281 (S9)
B–9000 Gent, BELGIUM
peter@astro.rug.ac.be

William BAUM
University of Washington
Astronomy Department
2124 NE Park Road
Seattle, WA 98195, U.S.A.
baum@phast.phys.washington.edu

Giuseppe BERTIN
Scuola Normale Superiore
Piazza dei Cavalieri 7
I–56126 Pisa, ITALY
bertin@vaxsns.infn.it

Francesco BERTOLA
Università di Padova
Dipartimento di Astronomia
Vicolo dell'Osservatorio, 5
I–35122 Padova, ITALY
bertola@astrpd.infn.it

Joris BLOMMAERT
Sterrewacht Leiden
Postbus 9513
NL–2300 RA Leiden
THE NETHERLANDS
blommaert@hlerul51

Xavier CALBET
Instituto de Astrofisica de Canarias
E–38200 La Laguna (Tenerife), SPAIN
fax : 22-605210

Tapan Kumar CHATTERJEE
Universitate A. Puebla
Facultad de Ciencias, F.M.
Apartado Postal 1316
Puebla 72000, MEXICO

Martin COHEN
University of California
Radio Astronomy Laboratory
Campbell Hall 601
Berkeley, CA 94720, U.S.A.
cohen@bkyast.berkeley.edu

Rafael CUBARSI
Universitat Politècnica de Catalunya
Dept. Matemàtica aplicada i Telematica
P.O. Box 30002
E-08080 Barcelona, SPAIN
fax : 34-3-4016801

Herwig DEJONGHE
Sterrenkundig Observatorium
Krijgslaan 281 (S9)
B–9000 Gent, BELGIUM
dejonghe@astro.rug.ac.be

Ronaldo DE SOUZA
Universidade de São Paulo
IAG-USP
Depto. Astronomia
C.P. 9638
Sao Paulo 01065, BRAZIL
ronaldo@iag.usp.ansp.br

Tim DE ZEEUW
Sterrewacht - Huygens Laboratorium
Postbus 9513
NL–2300 RA Leiden
THE NETHERLANDS
tim@rulhsy.leidenuniv.nl

Sandra DOS ANJOS
Universidade de São Paulo
Instituto Astronomico e Geofisico
Depto. Astronomia
C.P. 9638
Sao Paulo 01065, BRAZIL
sandra%iagusp@brfapesp.bitnet

Neil EVANS
Institute of Astronomy
Madingley Road
Cambridge CB3 OHA, U.K.
nwe10@uk.ac.cam.cus

Michael FEAST
Institute of Astronomy
Madingley Road
Cambridge CB3 OHA, U.K.
mfeast@ast-star.cam.ac.uk

Henry FERGUSON
Institute of Astronomy
Madingley Road
Cambridge CB3 OHA, U.K.
hcf@ast-star.cam.ac.uk

Marijn FRANX
Harvard–Smithsonian
Center for Astrophysics
60 Garden Street
Cambridge, MA 02138, U.S.A.
franx@cfa0.harvard.edu

Ken FREEMAN
Mt.Stromlo and Siding Springs
Observatories
Private Bag, Weston Creek Post Office
Canberra, ACT 2611, AUSTRALIA
kcf@mso.anu.edu.au

Alexei FRIDMAN
Institute of Astronomy
Russian Academy of Sciences
48 Pyatnitskaya Street, 109017 Moscow
RUSSIA
iaas@node.ias.msk.su

Ortwin GERHARD
Landessternwarte Königstuhl
D–6900 Heidelberg, GERMANY
c68@dhdurz1

Ian GLASS
South African Astronomical Observatory
P.O. Box 9
Observatory 7935
SOUTH AFRICA
isg@saao.ac.za

Martin GROENEWEGEN
Universiteit van Amsterdam
Sterrenkundig Instituut
Kruislaan 403
NL–1098 SJ Amsterdam
THE NETHERLANDS
groen@astro.uva.nl

Harm HABING
Sterrewacht - Huygens Laboratorium
Postbus 9513
NL–2300 RA Leiden
THE NETHERLANDS
habing@rulh11.leidenuniv.nl

Paul HARDING
KPNO/NOAO
950 North Cherry Avenue
Tucson, AZ 85719, U.S.A.
pharding@noao.edu

Hashima HASAN
Space Telescope Science Institute
3700 San Martin Drive
Baltimore, MD 21237, U.S.A.
hasan@stsci

Manuel HERNANDEZ-PAJARES
Universitat Politècnica de Catalunya
Dept. Matemàtica aplicada i Telematica
P.O. Box 30002
E–08080 Barcelona, SPAIN
matmhp@mat.upc.es

Jean HEYVAERTS
Observatoire de Paris–Meudon, DAEC
F–92195 Meudon, FRANCE
melama::heyvaerts

Christopher HUNTER
Florida State University
Mathematics Department
Tallahassee, FL 32306, U.S.A.
hunter@math.fsu.cdu

Rodrigo IBATA
Institute of Astronomy
Madingley Road
Cambridge CB3 OHA, U.K.

Hideyuki IZUMIURA
Tokyo Gakugei University
Dept. of Astronomy and Earth Science
4-1-1 Nukuikita-Machi
Koganei-shi, Tokyo 184, JAPAN
izumiura@clezio.u-gakugei.ac.jp

Frank JANSEN
Space Science Dept. of ESA
ESTEC
Postbus 299
NL–2200 AG Noordwijk
THE NETHERLANDS
fax : +31-1719-84698

Stephen KENT
Fermilab / MS 127
P.O. Box 500
Batavia, IL 60510, U.S.A.
skent@fnal.bitnet

Ivan KING
University of California
Astronomy Department
Berkeley, CA 94720, U.S.A.
king@bkyast.berkeley.edu

John KORMENDY
University of Hawaii at Manoa
Institute for Astronomy
2680 Woodlawn Drive
Honolulu, HA 96822, U.S.A.
kormendy@uhifa.ifa.hawaii.edu

Koen KUIJKEN
Harvard-Smithsonian Center for Astroph
60 Garden Street
Mail Stop 20
Cambridge, MA 02138, U.S.A.
kuyken@cfa.harvard.edu

Dirk LAURENT
Sterrenkundig Observatorium
Krijgslaan 281 (S9)
B–9000 Gent, BELGIUM
dirk@astro.rug.ac.be

Terence MAHONEY
Instituto de Astrofisica de Canarias
E–38200 La Laguna (Tenerife)
SPAIN
fax : 22-605210

M. L. MALAGNINI SICURANZA
Università degli Studi di Trieste
Dipartimento di Astronomia
Via Tiepolo 11
I–34131 Trieste, ITALY
malagnini@astrts.astro.it

Louis MARTINET
Geneva Observatory
Ch. des Maillettes 51
CH–1290 Sauverny
SWITZERLAND
martinet@cgeuge54.bitnet

Dante MINNITI
University of Arizona
Steward Observatory
Tucson, AZ 85721, U.S.A.
dante@as.arizona.edu

Edmundo MORENO
UNAM – Instituto de Astronomia
Apartado Postal 70-264
Ciudad Universitaria
Mexico D.F. 04510, MEXICO
fax : 5-548-3712

Heather MORRISON
KPNO/NOAO
950 North Cherry Avenue
Tucson, AZ 85719, U.S.A.
heather@noao.edu

Yuen Keong NG
Sterrewacht Leiden
Postbus 9513
NL–2300 RA Leiden
THE NETHERLANDS
ng@rulhl1.leidenuniv.nl

Ken'ichi NOMOTO
University of Tokyo
Fac. Science - Dept. of Astronomy
Bunkyo-ku, Tokyo 113, JAPAN
nomoto@apsun1.astron.s.u–
tokyo.ac.jp

Colin NORMAN
Space Telescope Science Institute
3700 San Martin Drive
Baltimore, MD 21218, U.S.A.
norman@stsci.edu

Leonid OSSIPKOV
University of St. Petersburg
Institute of Comput.Math.
Bibliotechnaya pl 2
198904 Peterhoff, RUSSIA
fax : 7-812-4286649
kvk@astr.lgu.spb.su

Irina PETROVSKAYA
University of St. Petersburg
Astronomical Institute
Bibliotechnaya pl 2
198904 Peterhoff, RUSSIA
fax : 7-812-4286649
kvk@astr.lgu.spb.su

Daniel PFENNIGER
Geneva Observatory
Ch. des Maillettes 51
CH-1290 Sauverny
SWITZERLAND
pfenniger@cgeuge54.bitnet

Valerij POLYACHENKO
Institute of Astronomy
Russian Academy of Sciences
48 Pyatnitskaya Street
109017 Moscow, RUSSIA
iaas@node.ias.msk.su

George PRESTON
Observatories of the Carnegie
Institution of Washington
813 Santa Barbara Street
Pasadena, CA 91101, U.S.A.
gwp@ociw.caltex.edu

Alvio RENZINI
Dipartimento di Astronomia
Cp 596
I-40100 Bologna, ITALY

Mike RICH
Columbia University
Department of Astronomy
Box 42, Pupin Hall
538 West 120th Street
New York, NY 10027, U.S.A.
rmr@cuthry.bitnet

Hans Walter RIX
Institute for Advanced Study
181 Von Neumann Drive
Princeton, NJ 08540, U.S.A.
rix@guinness.ias.edu

Annie ROBIN
Observatoire de Besançon
B.P. 1615
F–25010 Besançon Cédex, FRANCE
robin@frobes51

Alex RUELAS
UNAM – Instituto de Astronomia
Apartado Postal 70-264
Ciudad Universitaria
Mexico D.F. 04510, MEXICO
rarm@alfa.astroscu.unam.mx

Prasenjit SAHA
University of Toronto
CITA, Mc Lennan Labs
60 St George Street
Toronto, Ontario M5S 1A7
CANADA
saha@cita.utoronto.ca

Jaume SANZ-SUBIRANA
Universitat Politècnica de Catalunya
Dept. Matemàtica aplicada i Telematica
P.O. Box 30002
E–08080 Barcelona, SPAIN
fax : 34-3-4016801

Cecilia SCORZA
Landessternwarte Königstuhl
D–6900 Heidelberg
GERMANY
y92@dhdurz1

Jerry SELLWOOD
Rutgers University
Dept. of Physics and Astronomy
P.O. Box 849
Piscataway, NJ 08855, U.S.A.
sellwood@physics.rutgers.edu

Maartje SEVENSTER
Sterrewacht - Huygens Laboratorium
Postbus 9513
NL–2300 RA Leiden
THE NETHERLANDS
habing@rulh11.leidenuniv.nl

Olga SIL'CHENKO
Sternberg Astronomical Institute
University Avenue 13
Moscow 119899, RUSSIA
snn@sai.msk.su

Lorant SJOUWERMAN
Sterrewacht - Huygens Laboratorium
Postbus 9513
NL–2300 RA Leiden
THE NETHERLANDS
lorant@rulh11.leidenuniv.nl

Grazyna STASINSKA
Observatoire de Paris–Meudon, DAEC
F–92195 Meudon Cédex, FRANCE
grazyna@frmeu51

Thomas STATLER
Dept. of Physics and Astronomy
CB 3255 Phillips Hall
University of North Carolina
Chapel Hill, NC 27599, U.S.A.
tss@physics.unc.edu

Ewa SZUSZKIEWICZ
Queen Mary and Westfield College
School of Mathematical Sciences
Astronomy Unit
University of London
Mile End Road
London E1 4NS, U.K.
ewa@maths.qmw.ac.uk

Donald TERNDRUP
Ohio State University
Dept. of Astronomy
174 West 18th Avenue
Columbus, OH 43210-1106, U.S.A.
terndrup@baade.mps.ohio-state.edu

Frank THEEDE
Institut für Reine und
Angewandte Kernphysik
Christian-Albrechts-Universität zu Kiel
Olshausenstrasse 40
Otto Hahn Platz 1, D–2300 Kiel
GERMANY
pke97c@rz.uni-kiel.dbp.de

Joachim TRUMPER
Max Planck Inst. für Physik und Astroph
Giessenbachstrasse
D–8046 Garching bei München
GERMANY
fax : 089-32-993569

Neil TYSON
Princeton University
Department of Astrophysical Sciences
Peyton Hall
Princeton, NJ 08544, U.S.A.
ndt@astro.princeton.edu

Stephane UDRY
Geneva Observatory
Ch. des Maillettes 51
CH–1290 Sauverny
SWITZERLAND
udry@scsun.unige.ch

Bartel VAN DE WALLE
Sterrenkundig Observatorium
Krijgslaan 281 (S9)
B–9000 Gent, BELGIUM
bartel@astro.rug.ac.be

Antonia VARELA PEREZ
Instituto de Astrofysica de Canarias
E–38200 La Laguna, Tenerife
SPAIN
fax : 22-605210

Paul VAUTERIN
Sterrenkundig Observatorium
Krijgslaan 281 (S9)
B–9000 Gent, BELGIUM
mark@astro.rug.ac.be

Ted VON HIPPEL
Institute of Astronomy
Madingley Road
Cambridge CB3 OHA, U.K.
ted@mail.ast.cam.ac.uk

Keiichi WADA
Hokkaida University
Faculty of Science
Department of Physics
Sapporo 060, JAPAN
wada@astro1.phys.hokudai.ac.jp

Wilfred WALSH
Australia Telescope National Facility
P.O. Box 76
Epping, NSW 2121
AUSTRALIA
wwalsh@atnf.csiro.au

Nicholas WALTON
University College London
Department of Physics and Astronomy
Gowerstreet
London WCIE 6BT,VIC, U.K.
naw@uk.ac.ucl.star

Patricia WHITELOCK
South African Astronomical Observatory
P.O. Box 9
Observatory 7935
SOUTH AFRICA
paw@saao.ac.za

Albert WHITFORD
Lick Observatory
University of California
Santa Cruz, CA 95064, U.S.A.
board@helios.ucsc.edu

Heinz WIECHEN
Theoretische Physik IV
Ruhr-Universität Bochum
Postfach 102148
D–4630 Bochum, GERMANY
hw@tp4.ruhr-uni-bochum.de

Anders WINNBERG
Onsala Space Observatory
S–43900 Onsala, SWEDEN
anders@oden.oso.chalmers.se

Mark WINSALL
Sterrenkundig Observatorium
Krijgslaan 281 (S9)
B–9000 Gent, BELGIUM
mark@astro.rug.ac.be

Joan WROBEL
NRAO, P.O. BOX 0
Socorro, NM 87801, U.S.A.
jwrobel@nrao.edu

Monique WYBO
Sterrenkundig Observatorium
Krijgslaan 281 (S9)
B–9000 Gent, BELGIUM
monique@astro.rug.ac.be

Dennis ZARITSKY
Carnegie Observatories
813 Santa Barbara Street
Pasadena, CA 91101, U.S.A.
dennis@lynx.ociw.edu

Werner ZEILINGER
European Southern Observatory
Karl Schwarzschild Strasse 2
D–8046 Garching bei München
GERMANY
wzeil@eso.org

Harald ZIEGLER
Theoretische Physik IV
Ruhr-Universität Bochum
Postfach 102148
D–4630 Bochum, GERMANY
hz@tp4.ruhr-uni-bochum.de

1. D. Laurent
2. R. Ibata
3. P. Saha
4. Y.K. Ng
5. H. Izumiura
6. N. Arimoto
7. P. Batsleer
8. K. Wada
9. A.M. Varela Perez
10. T. Mahoney
11. H. Dejonghe
12. A. Whitford
13. E. Szuszkiewicz
14. S. Dos Anjos
15. K. Chatterjee
16. D. Zaritsky
17. A. Ruelas
18. P. Harding
19. A. Robin
20. T. Von Hippel
21. L. Ossipkov
22. I. King
23. H. Hasan
24. E. Moreno
25. M. Cohen
26. I. Petrovskaya
27. M. Malagnini Sicuranza
28. P. Whitelock
29. J. Wrobel
30. D. Minniti
31. J. Trumper
32. R. De Souza
33. J. Heyvaerts
34. O. Sil'chenko
35. G. Stasinska
36. H. Morrison
37. K. Nomoto
38. R. Arnold
39. H. Ferguson
40. C. Hunter
41. M. Wybo
42. H. Habing
43. L. Sjouwerman
44. A. Fridman
45. X. Calbet
46. F. Jansen
47. A. Renzini
48. F. Bertola
49. T. de Zeeuw
50. D. Pfenniger
51. V. Polyachenko
52. I. Glass
53. W. Walsh
54. A. Winnberg
55. H.W. Rix
56. D. Terndrup
57. M. Franx
58. N. Tyson
59. H. Wiechen
60. S. Kent
61. M. Groenewegen
62. G. Preston
63. F. Theede
64. M. Feast
65. G. Bertin
66. K. Kuijken
67. T. Statler
68. N. Evans
69. W. Baum
70. L. Martinet
71. M. Rich
72. M. Winsall
73. J. Kormendy
74. P. Vauterin
75. N. Walton
76. J. Blommaert
77. M. Balcells
78. J. Sellwood
79. S. Udry
80. C. Norman

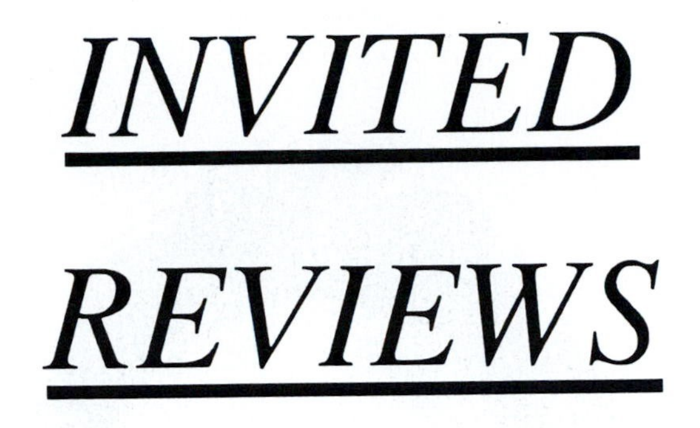
INVITED
REVIEWS

Ivan KING

REVIEW OF THE GALACTIC BULGE

IVAN R. KING
Astronomy Department, University of California, Berkeley, CA 94720, U.S.A.

For sins that I have been unable to identify, I have been asked to give a review talk on the Galactic bulge. This is a subject to which I have not been very close in recent years, so that I have been able to approach it with a fairly open mind, though I must acknowledge a lot of generous help from Mike Rich, Don Terndrup, and Jay Frogel. But any errors, and especially any prejudices, are purely my own. And in my reading I have undoubtedly missed some important papers, but I hope that their authors will forgive me.

This is a field in which the activity has accelerated in recent years, and I think that it is completely appropriate to have another symposium on the subject only 2 1/2 years after the Workshop in La Serena. I am struck that nearly all the papers that I have read in preparing this summary are dated 1989 or later. In fact I rather suspect—and hope—that the subject will look rather different five days from now, after all the material that is presented at this meeting.

1. Historical Background

I have always thought that any subject that involves stellar populations profits from a historical approach, or at least from a historical introduction. People had seen bulges in the middle of galaxies for a long time, but the study of bulges really begins with the detection of red giants at the edge of the M31 bulge by Baade, nearly half a century ago (Baade 1944). This was of course the famous paper in which he invented the concept of stellar populations. Since he saw globular-cluster-like red giants in Andromeda, Baade identified the center of Andromeda with what he called Population II. A few years later he found the RR Lyrae stars in the region that is now known fondly as Baade's Window, and the identification with his Population II became even tighter. But we know now that it was wrong, and that the dominant population, both in the M31 bulge and in Baade's Window, is an old metal-rich population.

I would like to use this as an example of why, after nearly fifty years of further development, we ought to give up completely the use of the terms "Population I" and "Population II." Those terms conjure up all the errors of the past. We should be describing populations as old or young, and metal-rich or metal-poor, not with the simplistic numbers I and II. (There, I promised you some prejudice, didn't I?) The detailed facts, of course, are even worse. We are going to hear talk at this meeting of oxygen-to-iron ratios, nitrogen enhancement, a continuum of ages, and all sorts of things, until the jungle of populations becomes quite impenetrable. Mind you, I am not suggesting that we recite an entire pedigree every time we refer to a population; at our present stage of knowledge, age and metallicity seem to me adequate labels. But my basic point is that population labels should be descriptive, not just a pair of oversimplified pigeonholes labeled I and II.

H. Dejonghe and H. J. Habing (eds.), Galactic Bulges, 3–20.

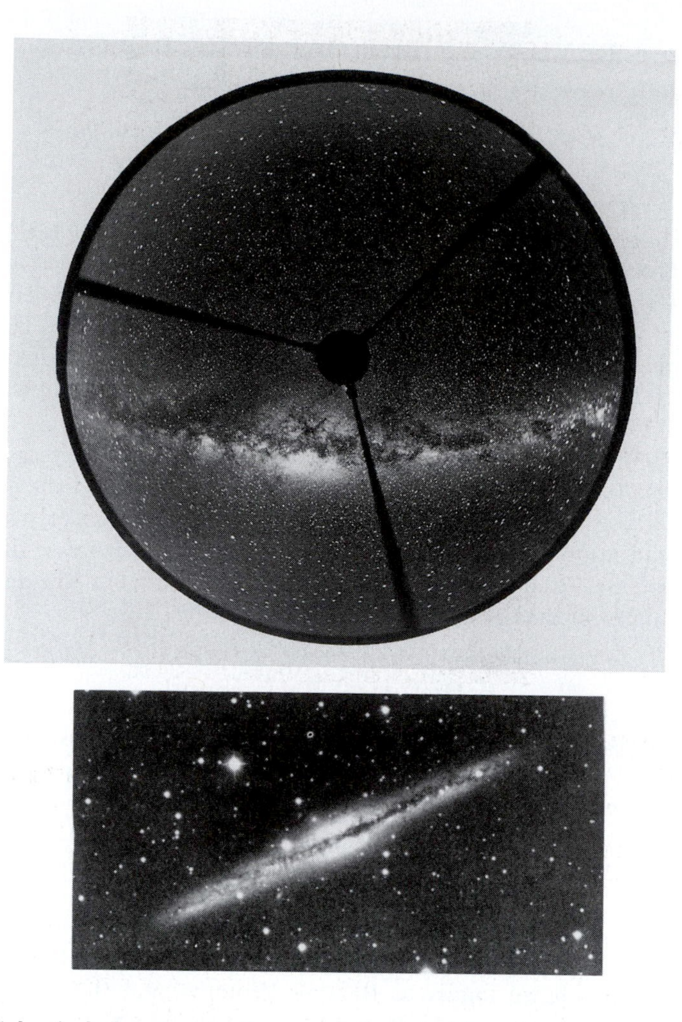

Fig. 1. The Galactic bulge seen in the near infrared with a "fisheye" camera, compared with the edge-on galaxy NGC 891.

But back to bulges, more specifically. The populations of bulges got straightened out much better in the 1950's. First Baum and Schwarzschild (1955) showed that the bulge of Andromeda had far too high a surface brightness for its paltry sprinkling of red giants. The light had to be coming from some other component. Then the situation became a lot clearer during that wonderful summer that Bill Morgan spent going through Nick Mayall's spectra at Lick Observatory. The spectra of bulges turned out to be like those of solar-type stars (Morgan and Mayall 1957). (And, by the way, during this same orgy of classification Morgan [1956] discovered the strong-lined globular clusters.) And after that it went even farther: by the mid-1960's Spinrad (1966; also Spinrad and Taylor 1967) was talking about super-metal-rich bulges. And that gets us onto the modern track.

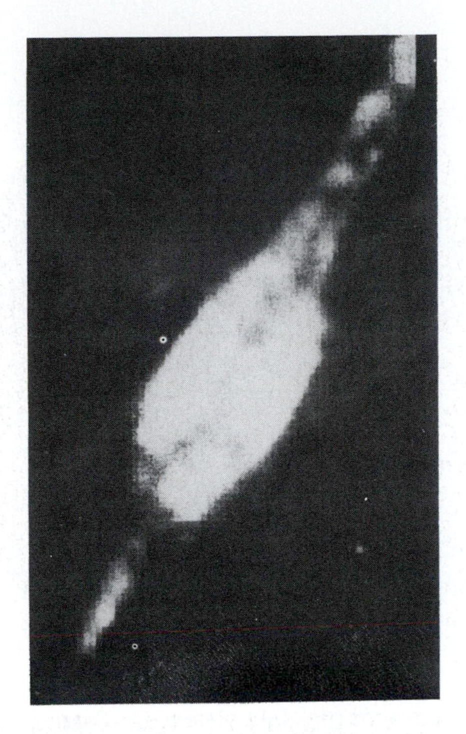

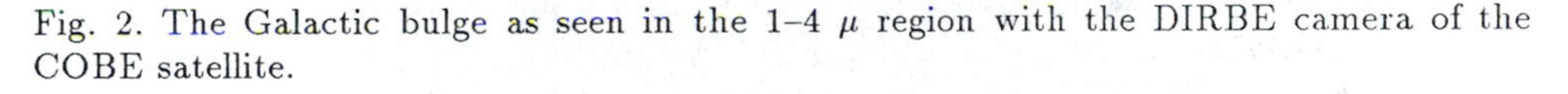

Fig. 2. The Galactic bulge as seen in the 1–4 μ region with the DIRBE camera of the COBE satellite.

It is easy to observe the bulges of other galaxies, but our own bulge is much more obscure. (That was an intentional pun.) As you all know, there are 30 magnitudes of visual absorption between us and the Galactic center, and one has to go to fairly long wavelengths to see through it. But fortunately our bulge *does* bulge out; and a little bit off the plane, shorter wavelengths penetrate. The bulge was first detected by Stebbins and Whitford (1947), at a wavelength of 1 micron. (I am happy to note that 45 years later Albert Whitford is still working actively on the bulge, and is present at this meeting.) A few years later the Henyey–Greenstein camera was able to photograph the edges of the bulge on infrared film. What I will show you instead (Fig. 1) is a more modern picture, made by the University of Bochum camera; the comparison galaxy is NGC 891. Here we can see the bulge peering out on both sides of the obscuring matter, but notice how much less obscured it is on the southern side, where our "windows" are.

Finally, in the modern era we have spacecraft. The COBE DIRBE map (Fig. 2) is probably the best delineation of the red giant stars of the bulge that we have.

At longer wavelengths, IRAS is very informative. The IRAS survey map, in which all the bands are color-coded and mixed, does not show the bulge; but the situation changes completely if we single out the IRAS point sources, as shown in the 12-micron map (Fig. 3). These are dominated by stars with circumstellar

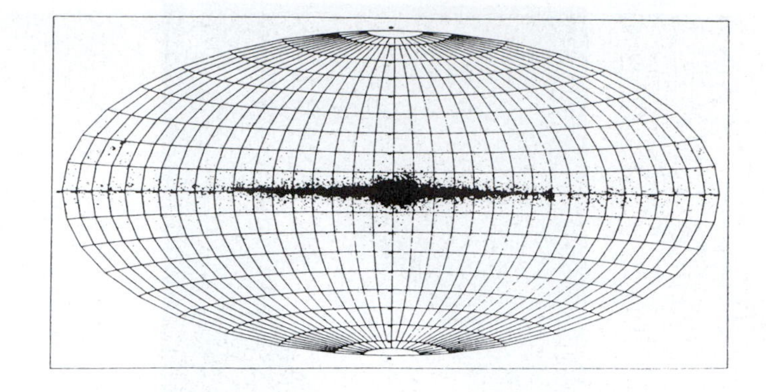

Fig. 3. The IRAS point-source map, which in the bulge is dominated by emission from the dust shells surrounding OH/IR stars.

dust shells—principally the OH/IR stars—and we see the bulge quite clearly. But I will not emphasize this map now, because the stars that it shows belong to a later discussion of the populations in the bulge.

The COBE map (Fig. 2) shows the potential of the near infrared for penetrating the absorption between us and the Galactic center. Fortunately this is not a venture that has to be done completely from space. We have a good ground-based window at 2 microns that has already produced useful information. But now this so-called K band has been revolutionized by the appearance of array detectors. Figure 4 (Gatley *et al.* 1989) is a K-band montage of rather less than a square degree around the Galactic center, which I imagine is the bright spot in the lower middle. You can see by the patchiness that absorption troubles have by no means disappeared, but my uninformed hope is that we will eventually be able to get around them by using something like a JHK two-color diagram in a way similar to what we do with the UBV two-color diagram in the visible. This is an exciting new prospect, and it is just beginning to be exploited, although there do appear to be difficulties in the interpretation (Davidge 1991). The general subject of K-band imaging will be the topic of the following paper, by Glass.

While I am on the topic of instrumental developments, there is another one that is very important for this field. Until recently, nearly all spectra were taken one at a time. But now nearly every major observatory has, or is developing, a multi-object spectrograph. It will be of tremendous value to studies of the Galactic bulge to be able to get radial velocities, spectral classes, and chemical abundances in great numbers—and, in fact, we will hear of such studies during this symposium.

And there are other new contributions from space, too. Bill Baum will be telling us about Hubble Space Telescope observations of Baade's Window, and I am pleased to say that late tomorrow evening HST will be pointed at Baade's Window again, imaging with the Faint Object Camera, which has four times the resolving power of the material that Bill will be talking about (but unfortunately over a much smaller

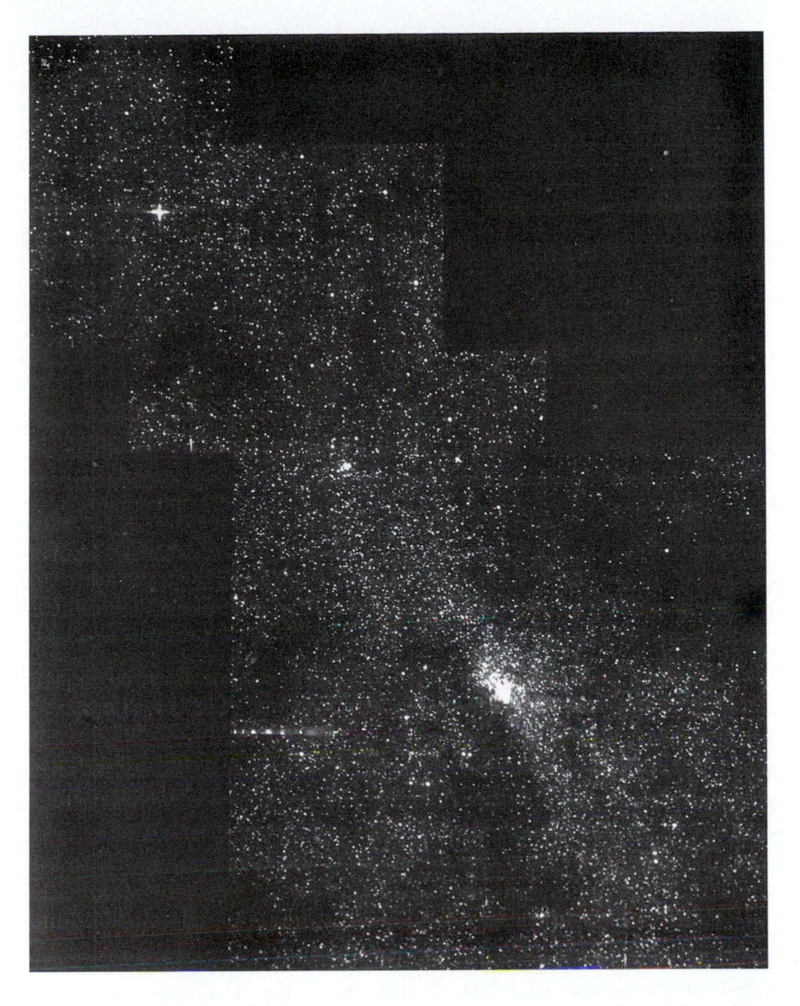

Fig. 4. Mosaic of K-band images, showing a region around the Galactic center.

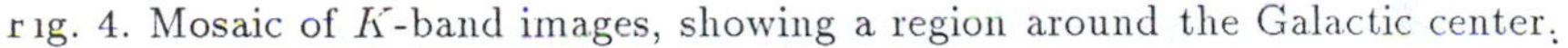

field)[1].

2. The Problems of the Bulge

But this is enough general introduction; now I want to talk about the characteristics of the bulge itself: what we know and what we don't know, and what we are likely to be discussing in the next few days. I will have occasion, of course, to refer to a lot of specific papers, but I am not attempting to give comprehensive literature references, partly because other speakers will give detailed lists in their fields, and partly because I hope and expect that much of the work to which I refer will be superseded by presentations given at this meeting.

[1] The images were taken on 18 August 1992 and appear to be quite satisfactory.

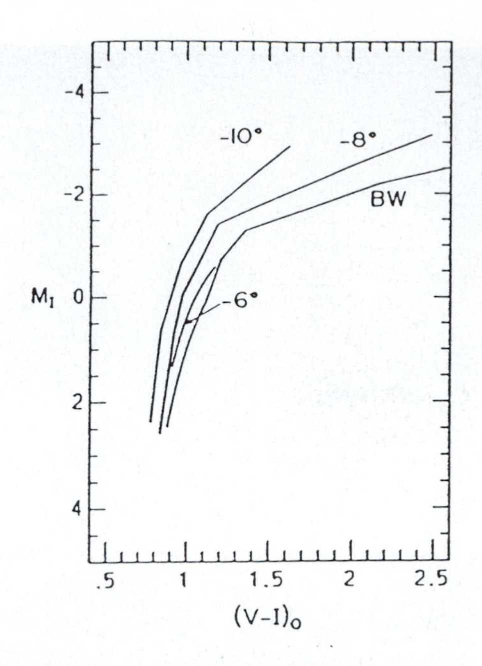

Fig. 5. Schematic giant branches derived by Terndrup (1988) in Baade's Window and in other fields farther from the Galactic center.

I should mention, by the way, that I am going to confine my attention to the smooth stellar bulge, which after all is what we can compare with other galaxies. That means that I am not going to discuss the disorderly conduct that is going on in the Galactic center, where we have ionized gas, supernova remnants, and maybe even a mini-AGN. That is galaxy-nucleus stuff; I don't want to mix it up with the bulge.

The problems of the bulge fall into four general areas: what is there, how it is distributed, how it moves, and how it got that way. I don't mean that these are separate areas; they are very much interrelated. But at least we can distinguish separate areas of fact—except for the last one, "how it got that way," which still has a lot more fancy in it than fact. (Sorry; more prejudice. But I *am* glad to see that Colin Norman has labeled his "formation" talk with the word "speculation.")

3. Populations

First, the populations in the bulge. (Notice that I use the plural.) Again, we can subdivide this into three questions: what kind of stars are we able to observe, what chemical abundances do they have, and what are their ages?

3.1. Population Tracers

One approach is just to see what we see—to make a color–magnitude diagram. Obviously for this we want a relatively unobscured region, and the favorite has

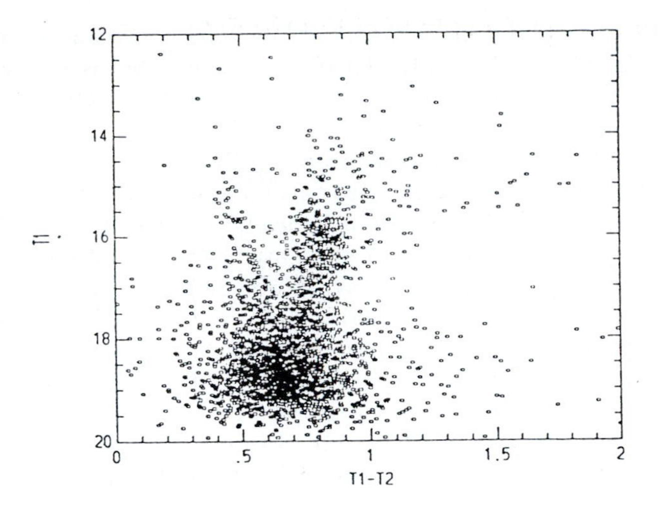

Fig. 6. A color-magnitude diagram in Baade's Window, in the Washington photometric system (Geisler and Friel 1992).

been Baade's Window. Here are some examples. Figure 5 shows Terndrup's (1988) schematic rendition of the giant branches in Baade's Window and in some other fields. It really is schematic, though. In Figure 6 are some actual data, from Geisler and Friel (1992), in the Washington photometric system. You can see that there is a lot of scatter, and a lot of interference from field stars, particularly from the main-sequence disk stars of the foreground. And at a not-very-faint magnitude, crowding gets you. That is why we are so interested in using HST in Baade's Window.

And I should mention that interesting CMD work can be done in other fields (see, especially, Tyson [1991 and in this volume], and two poster papers by Harding and Morrison in this volume), and in other bands (Davidge 1991).

Then there are particular types of stars. The M giants are especially useful, because they are luminous, spectroscopically easy to recognize, and particularly easy to observe at the longer wavelengths. They are also a population indicator, because metal-poor populations don't make late-type gM's. Victor Blanco and his associates have done a lot with M giants. (For a summary, with detailed references, see McCarthy and Blanco 1990.)

A related problem is the search for carbon stars. The fact that there are so few C's in the bulge, compared with, say, the Magellanic Clouds, is another indication that we are dealing with a metal-rich population. (The work on C stars is also summarized by McCarthy and Blanco 1990.)

Another type of M stars is the Miras, the long-period pulsators. They are not regular M giants, though. They sit at the tip of the asymptotic giant branch; so this is a pinpointed evolutionary stage. Also, each Mira has a period; and this is another population indicator, although I am not sure that it is a totally calibrated pointer. For further details about Miras, see Whitelock in this volume.

As a Mira evolves farther up the AGB, it develops a shell of dust and gas, and becomes an OH/IR star. As this name applies, these stars have radio emission, and they also show up as point sources in the IRAS survey. In both cases we get a view that is free of interstellar absorption. For further details about OH/IR stars, see both Habing and Dejonghe in this volume.

In the next stage of evolution, the star throws off a gas shell and evolves rapidly to the left in the HR diagram. It gets so hot that it sets the shell into luminescence as a planetary nebula. (I might remark parenthetically that a year ago I didn't know a thing about planetaries, but then I looked at my first ultraviolet HST image of the center of M31, and it was full of post-asymptotic-giant-branch stars, which are the nuclei of planetaries. I had to learn about them in a hurry. In a similar way I took on the preparation of this talk as an exercise to get me ready to do something intelligent with my HST observations of Baade's Window. In the bookstore on my campus they sell a sign that says, "Four years ago I couldn't even spell inginere, but now I are one." My position is a little bit like that.) A fair amount has been done about planetaries in the Galactic bulge; for further details, see Stasinska in this volume.

But we can go no further with this evolutionary track. The next stage is to become a white dwarf, and they are too faint to see in the bulge.

The stellar types that I have mentioned so far are all characteristic of a particular stage of evolution, and in some cases of a particular kind of population. So it is appropriate that I end the list with the RR Lyrae stars. Betty Blanco has done a great deal of work on RR Lyraes in the bulge (Blanco 1984, 1992), but unfortunately she is not here. But we are happy to have with us George Preston, who is a patriarch of the field, and he will tell us about RR Lyraes.

3.2. Abundances

These are the objects whose observational status we will be discussing. Now for the basic questions about them. First there is the nature of the populations in the bulge. This raises two kinds of questions: abundance and age. Let us look first at the abundances. It is immediately obvious that there is a large range, when we see RR Lyraes that are a signature of a metal-poor population and late M giants that occur only in a metal-rich population. But to study the *distribution* of abundances, we can use neither of these types, because each of them is so biased toward one end of the distribution. (Notice that the Miras and the OH/IR stars may also be biased in this way.) The one type of luminous star that occurs in populations of all abundances is the K giants.

The abundance distribution of a sample of K giants in Baade's Window was studied spectroscopically by Rich (1988), and more recently Geisler and Friel (1992) have studied a much larger sample, using Washington-system photometry. In Figure 7 is shown their distribution of [Fe/H]; I call attention to three striking characteristics: it has a big spread, the mean is greater than zero, and the extreme high is around +1.

Another feature of Fig. 7 that is noteworthy is that the horizontal scale at the left is linear, unlike the customary logarithmic [Fe/H]. The quantities shown are therefore just what is wanted by those who model scenarios of element enrichment,

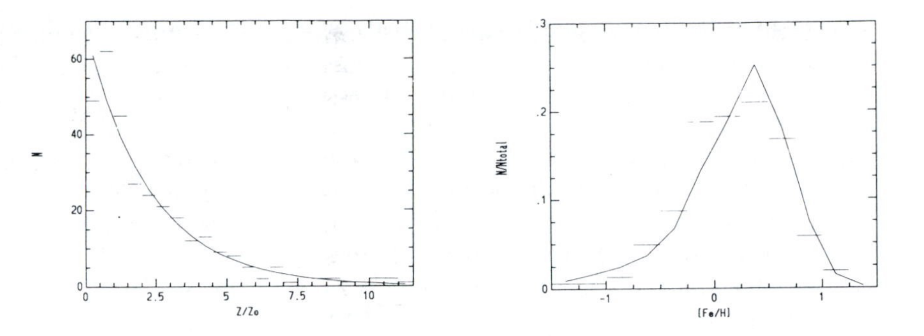

Fig. 7. The distribution of abundances in Baade's Window (Geisler and Friel 1992), on linear and logarithmic scales. The horizontal lines represent their histograms; the continuous lines are from theoretical models.

because they need to deal in actual amounts rather than ratios.

Geisler and Friel made careful efforts to avoid biasing their selection of K giants, but even here there is one bias that still seems to remain. Although they appear to have produced an unbiased abundance distribution of K giants, this is still one step removed from the basic answer. The remaining question to ask is, what is the ratio, at each abundance, of the number of K giants to the number of middle-main-sequence stars? It is the latter that we should take as the indicator of true numbers. This is an important additional question, yet I am not aware of its ever being treated.

We should remember also that there are more quantities to be concerned about than just [Fe/H]. At the Santa Cruz globular-cluster workshop a month ago, Frogel (1993) showed some infrared color-color diagrams in which the bulge stars didn't match up with any other population. In fact, there were other mismatches too; for example, in these diagrams the field population of the halo doesn't match any type of globular cluster. I don't really know what to make of facts like these, but I think that they may be trying to tell us something quite important. Try the following: because different elements affect various parts of the spectrum in different ways, these differences in two-color diagrams are telling us that the details of the enrichment process depend on the environment. If that is true, it is a very important statement.

And if the foregoing is really so, it constitutes an argument against the formation of the bulge by mergers. We cannot produce a unique population by merely mixing ordinary ones.

Uniqueness also makes one think of the RR Lyrae stars in Baade's Window. (See the discussion by Blanco 1984.) For a given shape and amplitude of light curve, they have shorter periods than any other RR Lyraes known. It would be fascinating to know what RR Lyraes are like even closer to the Galactic center. Yet strangely, in a field only a degree and a half farther from the center (Blanco 1992) the RR Lyraes have the periods of an ordinary Oosterhoff Type I globular cluster.

An interesting discussion of the ability of various populations to produce RR

Lyrae stars is given by Renzini and Greggio (1990), in a paper to which I shall soon have occasion to refer again.

The RR Lyraes in Baade's Window offer an excellent example, by the way, of population selection in the distribution of abundances. Walker and Terndrup (1991) find, from ΔS values, that their [Fe/H] distribution peaks sharply around -1, quite unlike that of the K giants. But this is surely a result of the fact that RR Lyrae stars are made much more readily in the low-metallicity part of the population. This bias is also illustrated by the fact that RR Lyraes are not found in the more metal-rich globular clusters. They exist in the field, though; putting these two facts together, it is easy to conclude that their number per main-sequence star must be very low in metal-rich populations.

The low abundances found by Ratag *et al.* (1992) for planetary nebulae in the bulge may have a similar explanation. The central stars of planetaries are post-asymptotic-giant-branch stars, and these may be lacking in old populations of high metal abundance.

The details of chemical abundances in the Galactic bulge certainly pose a problem. For the discussion of a possible scenario, see the bold approach by Matteucci and Broccato (1990). But in any case, this is a problem that is going to be with us for a long time.

3.3. Ages in the Galactic Bulge

Now for the problem of ages. To start with, it is obvious that in the bulge we are dealing with an old population. But there are two important age questions about this population: *how* old is it? and is any of it appreciably younger than the rest?

I would like to look at the second question first. The strongest evidence for younger stars in the bulge is the excessively bright upper limit of the asymptotic giant branch, which in our familiar solar-metal-abundance population would be a sure sign of more-massive, younger stars. But Frogel (1990) argues strongly that this luminosity might instead be an effect of the supermetallicity that we know exists in the bulge. This is a very uncertain area of questions, however. The best theoretical exploration that I know of (Renzini and Greggio 1990) emphasizes the extent to which late evolution for metal-rich stars depends on the rate of helium production during metal enrichment. (For an example, see Figure 8.) This is a quantity that we know very little about—except that it is surely not zero and is likely to be at least 1 or 2. The only direct evidence I know to cite is that the Sun's helium content has a ΔY of 0.04–0.05 above the primeval value, along with a ΔZ of 0.02. But evolutionary tracks change with Z, so who knows how $\Delta Y/\Delta Z$ changes with Z itself? For further details I refer you directly to their paper, and of course to whatever Renzini has to say later in this volume.

As for the ages of the oldest stars in the bulge, I just don't know. We have very little direct evidence. For this reason I am looking forward very much to Bill Baum's account of the work of HST's WFPC team in Baade's Window—as I am looking forward to my own observations there. *If* we can do accurate enough photometry, correct reliably for interstellar absorption, and trust the theoretical isochrones that we fit to, then the main-sequence turnoff can give us an age (or a distribution of

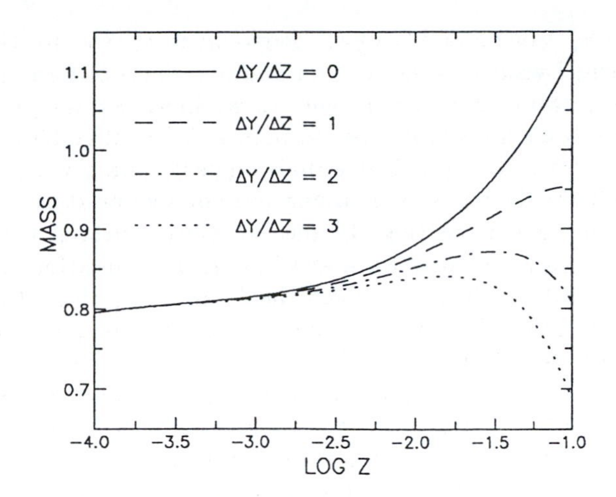

Fig. 8. The mass of stars evolving off the main sequence at age 15 Gyr, as a function of metal abundance and helium production rate (Renzini and Greggio 1990). Note how rapidly the different curves diverge at metal abundances above solar ($\log Z = -1.7$).

ages).

The suggestion has been made on other grounds, however, that the bulge is the oldest population in the Galaxy. Young-Wook Lee (1992) believes his theories of horizontal branch structure and evolution well enough that he claims to be able to age-date the RR Lyrae stars in Baade's Window. His ages are relative rather than absolute, but he asserts that the bulge is a gigayear older than the oldest halo population. I am not really competent to judge Lee's arguments, but Renzini's discussion of the uncertainties in helium production makes me skeptical of them.

Before I leave the problem of population mixtures in the bulge, let me make one firm point. Dynamically, there *must* be a mixture. Because of the very behavior of potential wells, the halo stars must have their highest absolute density at the Galactic center (even though they surely constitute only a small fraction of the population there), so they must be a contributor to the bulge. This is less obvious for the population of the thin disk, where rotation-supported circular orbits do allow the possibility of a hole in the center. Such a hole does seem unlikely, however, in view of the results presented later in this volume by Whitford, who points out the existence of a set of OH/IR stars in a flattened, rapidly rotating disk close to the Galactic center.

4. Relation of the Bulge to Other Components

Perhaps the most intriguing question about the bulge population is, what other component(s) of the Milky Way is it related to? Is it the center of the disk (or perhaps of the thick disk), or of the halo, or is it a completely independent component of the Milky Way? For that matter, we know that there are strong population gradients within the bulge itself; is the bulge a single component or a mixture of several?

In this connection, I want to speak out on another abuse of terminology that I think creates a great deal of unnecessary confusion, and sometimes even error, in the attempt to understand the overall structure of the Milky Way. This is the use of the pernicious term "spheroid" to mix, in a single unhappy pot, the metal-rich bulge and the metal-poor halo. This is very common terminology (and can undoubtedly be found in numerous papers in this volume, to whose authors I apologize for criticizing them so), but it is truly an abomination. It is used in of one of the most popular stellar-distribution models for the Galaxy, and in the most widely read book on galactic astronomy. What it does is to tempt the unwitting reader to equate the bulge with the center of the halo and then to slip even more unwittingly into thinking of the bulge as metal-poor. It is possible that they *might* be part of the same population, with a continuous transition from one to the other—although I rather doubt it. But the burden of proof is on the facts, and is not to be avoided by naïvely using the same name for both. Please, let us relegate the term "spheroid" to describing geometrical shapes of astronomical bodies; it is indeed useful for that. But in the context of components of the Galaxy it simply creates a mess and should be abolished. Just because the halo and the bulge both have a spheroidal shape, that does not make them population siblings.

Please excuse that diatribe, but I think that the subject of it is important. Back now to what other components the bulge is related to. At one time I used to feel very comfortable with the idea that a bulge is simply what a stellar disk likes to do dynamically at its center. There might have been some dynamical virtue in the idea, but it fails completely to address the population difference. There is clearly a thin-disk population represented at the Galactic center, and the scandalous goings on in the innermost 20 parsecs of the plane do not accord at all with the serene distributions in an old bulge.

Perhaps a similar dynamical idea could be used to connect the bulge with the thick disk. Again there are population differences, but we know that population gradients exist in the bulge. The principal difficulty is that on the dynamical side this hypothesis is merely hand-waving. I know of no dynamical analysis or modeling that either supports or refutes it. But there are other more direct difficulties. We know very little about the structure of the thick disk, except vertically in the solar neighborhood. Of course the suggestion has been made that the thick disk is part of the same system as the disk globular clusters, in which case we do know something about its density distribution. But this connection raises the population difficulty that the system of disk globulars shows no tendency for metallicities to increase inward.

Allowing for population gradients does indeed open the question of whether the bulge could be the inward continuation of the halo, with the population gradient simply getting very strong in this innermost region that we call the bulge. I see two strong arguments against this. One is that the halo does not have a strong metallicity gradient at all, and perhaps has none at all when the disk globulars are properly separated from the halo globulars. The other is that the halo has much less flattening than the bulge, so that tying the two systems into one would require a rather unlikely change of ellipticity with radius. (The same statement can alternatively be couched in terms of amount of rotation.)

All in all, I have to admit that I have no idea whether the bulge has an intimate relationship with any other component of the Milky Way. For an interesting discussion of many of the facts and arguments, see Carney *et al.* (1990).

5. Density Distribution

Now into another major area, the density distribution in the bulge. Let us look first at the radial distribution. The very center was first mapped by Becklin and Neugebauer (1968). At 2.2 μ they found a central spike that was not resolved by their smallest aperture of 5 arcsec, and outside it a dropoff that went spatially as the -1.8 power of distance. They mapped out to 10 arcmin. A source of information at greater radii is the 2.4-μ survey of Kent *et al.* (1991). Their vertical profile suggests that the Becklin–Neugebauer trend continues out to about 1°, beyond which the distribution becomes exponential, with a scale height of about 375 pc. Another study was done by Harmon and Gilmore (1988), who looked at the distribution of IRAS point sources at 12 μ. They were unable to sample the innermost 4° because of IRAS confusion, but outside that radius their distribution agrees quite well with an exponential of 375 pc scale height.

Here again we have to take note of some uncertainties. The K-band studies refer to red giant stars in general, whereas the IRAS data sketch the distribution of OH/IR stars. Population gradients could lead to systematic differences. And the 12-μ data are absorption-free, whereas the K band is subject to an absorption of 2 1/2 to 3 magnitudes, which is spatially variable. Surely the mapping of the absorption in the surroundings of the Galactic center is one of our urgent needs.

There are other components that behave quite differently, however. The late M giants, studied by Victor Blanco (1988), drop off extremely steeply. This is not really a density gradient, however; it is a population gradient. Its most plausible interpretation is that only very metal-rich stars *can* be late M giants, and as the metallicity falls off away from the center these stars no longer occur. In fact, contemplating this phenomenon leads me to wonder whether we have any information at all on the density distribution in the bulge that is free from bias by population selection. Which brings me to an even more fundamental question: what do we really mean by the bulge, anyway? Rather than trying to answer that, I leave it for everyone to contemplate.

The abundance gradient is clearly important, but we are not in good agreement about it. In Figure 9 I have reproduced two versions of it. Frogel (1990), using infrared indices, finds a steady dropoff. Tyson (1991, p. 183), who used Washington photometry, interprets his results as showing a level abundance out to 10° latitude, followed by a sharp drop. We just don't have our facts straight yet.

Another question is the flattening of the bulge. There have been various answers about this, but I think that the best of them is the COBE map Fig. 2, which shows the bulge as rather flattened and quite boxy. This map refers to the 1–4 μ range, so it is red-giant light that we are looking at. Perhaps here again we need to worry a little about what a representative population is.

Finally, there is the question of the symmetries of the bulge. The evidence is

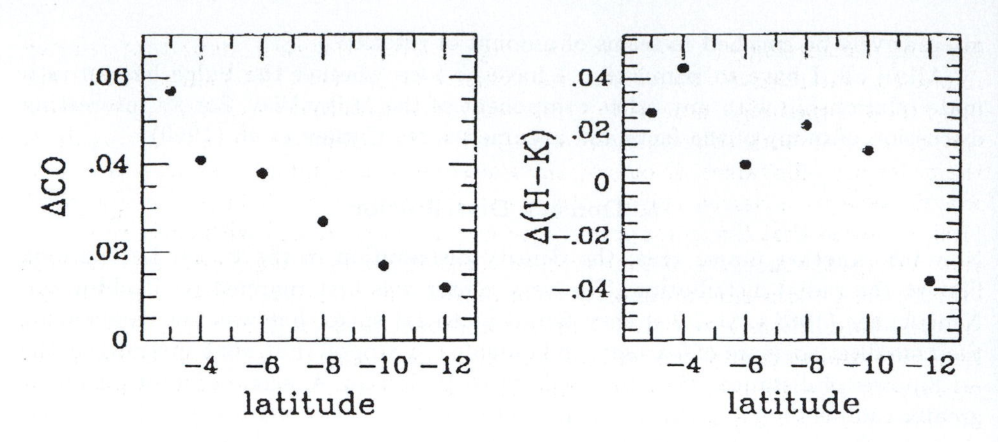

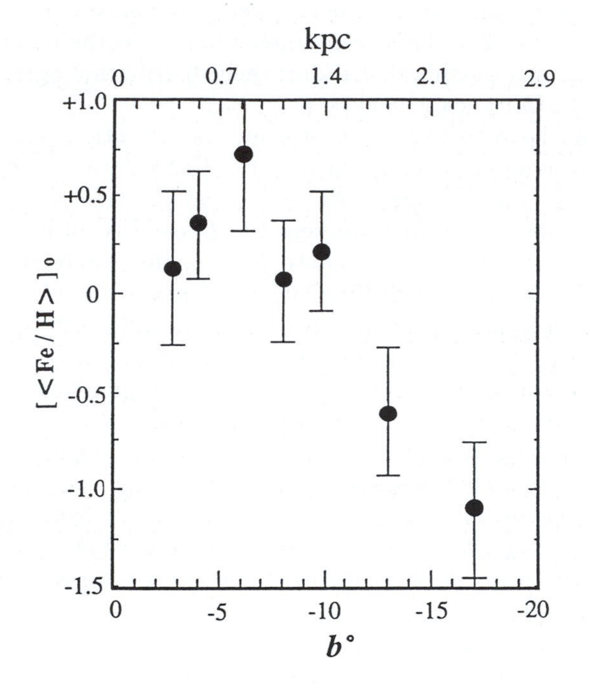

Fig. 9. Dependence of metal abundance on Galactic latitude, according to Frogel (top, 1990) and Tyson (bottom, 1991).

becoming increasingly strong that the bulge is triaxial, with a shape that is often referred to alternatively as a bar. First, Blitz and Spergel (1991) showed that the 2.4-μ map of the bulge indicated in many separate ways that the bulge has a barred shape, inclined both to the line of sight and to the Galactic plane—on a scale of the order of a kiloparsec. (Note, by the way, that this, if interpreted as a true bar, would make us a barred spiral with a peculiarly short bar.) Then Binney *et al.* (1991) showed that the gas motions were very much in accord with such a picture. But perhaps the most compelling evidence is the claim by Whitelock and Catchpole (1992), using the period-luminosity-color relation developed by Feast *et al.* (1990), that the bulge Miras at positive Galactic longitude are closer than those at negative longitude. But strangely, the stellar kinematics do not show any sign of the influence of a bar. I hope that de Zeeuw's paper in the present volume will shed more light on these perplexing questions.

It should be noted in this connection that there is no *a priori* reason to expect the Galactic bulge to be axisymmetric. We know, after all, that giant elliptical galaxies are triaxial, and the isophote twist in the M31 bulge shows that it is triaxial too.

(And note also that triaxiality could defeat any attempt to find the distance to the Galactic center by comparing radial velocities with proper motions.)

One final point about the shape of the bulge: Kent (1992) has done a quite nice dynamical model of the bulge, but it is axisymmetric. I hope that it can now be modified in some way to take into account the apparent asymmetry.

6. Kinematics of the Bulge

I now turn to the kinematics of the bulge, with a little dynamics thrown in. The basic dynamical fact to seize upon is the virial theorem, which, in simple language says, "You've got to have motion in order to resist gravitation." The motion is of course a mixture of rotation and random velocity dispersion. In the z direction the only support is velocity dispersion, so the vertical extent of the bulge tells us to expect a sizable dispersion. This is also borne out by the flattening of the bulge, where rotation and velocity dispersion combine to draw the bulge out farther in its equatorial plane. In fact, one can easily show, from the tensor virial theorem, that for a bulge the shape of ours one should expect comparable levels of rotation and velocity dispersion—and that is just about what we see. And to pursue this line of reasoning a bit further, if you want a bar, you've got to have more velocity dispersion in the direction of the long axis.

There have been many radial-velocity studies of various types of object in the bulge. K giants have been studied by Minniti *et al.* (1992), M giants by Walker *et al.* (1990), Miras by Catchpole (1990) and by Menzies (1990), OH/IR stars by Lindkvist *et al.* (1989) and by Le Poole and Habing (1990), planetary nebulae by Kinman *et al.* (1989), and main-sequence stars (perhaps at too high a latitude really to be called bulge) by Harding (1990). And no doubt this enumeration has missed some studies.

I will not go into these studies individually, as their results are rather similar. We tend to see linear rotation curves, typically with a slope of about 12 km/sec per degree of longitude. Typical velocity dispersions are 100–120 km/sec. But there is a tendency for the higher-metal-abundance types to rotate a little faster, and be a little more flattened to the plane. In this connection, it is time to repeat here my caution about the inhomogeneity of the K giants. In any kinematical study they should be divided into abundance groups. Something similar applies to the Miras and their various successor types.

In addition one other interesting study should be mentioned, although it is not clear that it refers to the bulge rather than to the center of the disk. McGinn *et al.* (1989) and Sellgren *et al.* (1989) discuss radial velocities of stars within 100 arcsec of the Galactic center. They find a rapid rotation, with the mean velocity changing by more than 100 km/sec in a few parsecs. This would argue for a disky population. On the other hand, their velocity dispersion of 100 km/sec would argue for a bulgy population. What are we to make of this?

There is one new development that I find quite exciting. That is the entry of proper motions into the field. Spaenhauer, Jones, and Whitford (1992) have worked with repeats of Baade's original plates and have measured proper motions in Baade's Window. Their accuracy is good enough for the velocity dispersions to be highly significant. Not that these motions are simple to interpret, however. In the latitude direction we see only the effect of velocity dispersion, whereas the proper motions in longitude also include the effect of rotation. The near and far sides of the bulge go in opposite directions; we of course can't tell them apart, so rotation has the effect of making the dispersion in proper motions larger in the longitude component. This effect does show up, and as one might expect, it is greater for the metal-rich stars than for the metal-poor stars.

7. The Origin of the Bulge

This is a subject that I won't even try to touch on. I will just refer you to Colin Norman's paper in this volume.

8. How Typical is Our Bulge?

Whether our bulge is typical of those of other galaxies is almost impossible to answer (although Frogel 1990 assembles a very useful summary of the facts). The problem is that we can see the bulges of many other galaxies so clearly, whereas we have to look at our own bulge through all the intervening muck. In many ways what we can see of it does seem typical, but there is one way in which it differs very much from the bulge of M31. The latter looks like a purely old population, whereas we get all sorts of young-population signals from a thin layer around the very center of the Milky Way. No doubt there will be more to say on this general question at the next symposium that is held on the fascinating problems of the Galactic bulge.

References

Baade, W.: 1944, *Astrophys. J.* 100, 137
Baum, W. A., and Schwarzschild, M.: 1955, *Astron. J.* 60, 247

Becklin, E. E., and Neugebauer, G.: 1968, *Astrophys. J.* **151**, 145

Binney, J., Gerhard, O. E., Stark, A. A., Bally, J., and Uchida, K. I.: 1991, *Monthly Notices Roy. Astron. Soc.* **252**, 210

Blanco, B. M.: 1984, *Astron. J.* **89**, 12

Blanco, B. M.: 1992, *Astron. J.* **103**, 1872

Blanco, V. M.: 1988, *Astron. J.* **95**, 1400

Blitz, L., and Spergel, D. N.: 1991, *Astrophys. J.* **379**, 631

Carney, B. W., Latham, D. W., and Laird, J. B.: 1990, in Jarvis, B. J., and Terndrup, D. M., eds., *Bulges of Galaxies*, Garching:ESO, p. 127

Catchpole, R. M.: 1990, in Jarvis, B. J., and Terndrup, D. M., eds., *Bulges of Galaxies*, Garching:ESO, p. 111

Davidge, T.: 1991, *Astrophys. J.* **380**, 116

Feast, M. W., Glass, I. S., Whitelock, P. A., and Catchpole, R. M.: 1990, *Monthly Notices Roy. Astron. Soc.* **241**, 375

Frogel, J. A.: 1990, in Jarvis, B. J., and Terndrup, D. M., eds., *Bulges of Galaxies*, Garching:ESO, p. 177

Frogel, J. A.: 1993, in Smith, G., and Brodie, J., eds., *The Globular Cluster–Galaxy Connection*, San Francisco:ASP, p. 000

Gatley, I., Joyce, R., Fowler, A., DePoy, D., and Probst, R.: 1989, in Morris, M. ed., *The Galactic Center* (I.A.U. Symposium No. 136), Dordrecht:Kluwer, p. 361

Geisler, D., and Friel, E. D.: 1992, *Astron. J.* **104**, 128

Harding, P.: 1990, in Jarvis, B. J., and Terndrup, D. M., eds., *Bulges of Galaxies*, Garching:ESO, p. 105

Harmon, R., and Gilmore, G.: 1988, *Monthly Notices Roy. Astron. Soc.* **235**, 1025

Kent, S. M.: 1992, *Astrophys. J.* **387**, 181

Kent, S. M., Dame, T. M., and Fazio, G.: 1991, *Astrophys. J.* **378**, 131

Kinman, T. D., Feast, M. W., and Lasker, B. M.: 1989, *Astron. J.* **95**, 804

Lee, Y.-W.: 1992, *Publ. Astron. Soc. Pacific* **104**, 798

Le Poole, R. S., and Habing, H. J.: 1990, in Jarvis, B. J., and Terndrup, D. M., eds., *Bulges of Galaxies*, Garching:ESO, p. 33

Lindkvist, M., Winnberg, A., Habing, H. J., Matthews, H. G., and Olnon, F. M.: 1989, in Morris, M. ed., *The Galactic Center* (I.A.U. Symposium No. 136), Dordrecht:Kluwer, p. 503

Matteucci, F., and Broccato, E.: 1990, *Astrophys. J.* **365**, 539

McCarthy, M. F., and Blanco, V. M.: 1990, in Jarvis, B. J., and Terndrup, D. M., eds., *Bulges of Galaxies*, Garching:ESO, p. 11

McGinn, M. T., Sellgren, K., Becklin, E. E., and Hall, D. N. B.: 1989, *Astrophys. J.* **338**, 824

Menzies, J. W.: 1990, in Jarvis, B. J., and Terndrup, D. M., eds., *Bulges of Galaxies*, Garching:ESO, p. 115

Minniti, D., White, S. D. M., Olszewski, E. W., and Hill, J. M.: 1992, *Astrophys. J. (Letters)* **393**, L47

Morgan, W. W.: 1956, *Publ. Astron. Soc. Pacific* **68**, 509

Morgan, W. W., and Mayall, N. U.: 1957, *Publ. Astron. Soc. Pacific* **69**, 291

Ratag, M. A., Pottasch, S. R., Dennefeld, M., and Menzies, J.: 1992, *Astron. Astrophys.* **255**, 255

Renzini, A., and Greggio, L.: 1990, in Jarvis, B. J., and Terndrup, D. M., eds., *Bulges of Galaxies*, Garching:ESO, p. 47

Rich, R. M.: 1988, *Astron. J.* **95**, 828

Sellgren, K., McGinn, M. T., Becklin, E. E., and Hall, D. N. B.: 1989, *Astrophys. J.* **359**, 112

Spaenhauer, A., Jones, B. F., and Whitford, A. E.: 1992, *Astron. J.* **103**, 297

Spinrad, H.: 1966, *Publ. Astron. Soc. Pacific* **78**, 367

Spinrad, H., and Taylor, B. J.: 1967, *Astron. J.* **72**, 320

Stebbins, J., and Whitford, A. E.: 1947, *Astrophys. J.* **106**, 235

Terndrup, D. M.: 1988, *Astron. J.* **96**, 884

Tyson, N. D.: 1991, thesis, Columbia Univ.

Tyson, N. D., and Rich, R. M.: 1990, in Jarvis, B. J., and Terndrup, D. M., eds., *Bulges of Galaxies*, Garching:ESO, p. 119

Walker, A., Sharples, R., and Cropper, M.: 1990, in Jarvis, B. J., and Terndrup, D. M., eds., *Bulges of Galaxies*, Garching:ESO, p. 99

Walker, A. R., and Terndrup, D. M.: 1991, *Astrophys. J.* **378**, 119
Whitelock, P. A., and Catchpole, R. M.: 1992, in Blitz, L., ed., *The Center, Bulge, and Disk of the Milky Way*, Dordrecht:Kluwer, p. 103

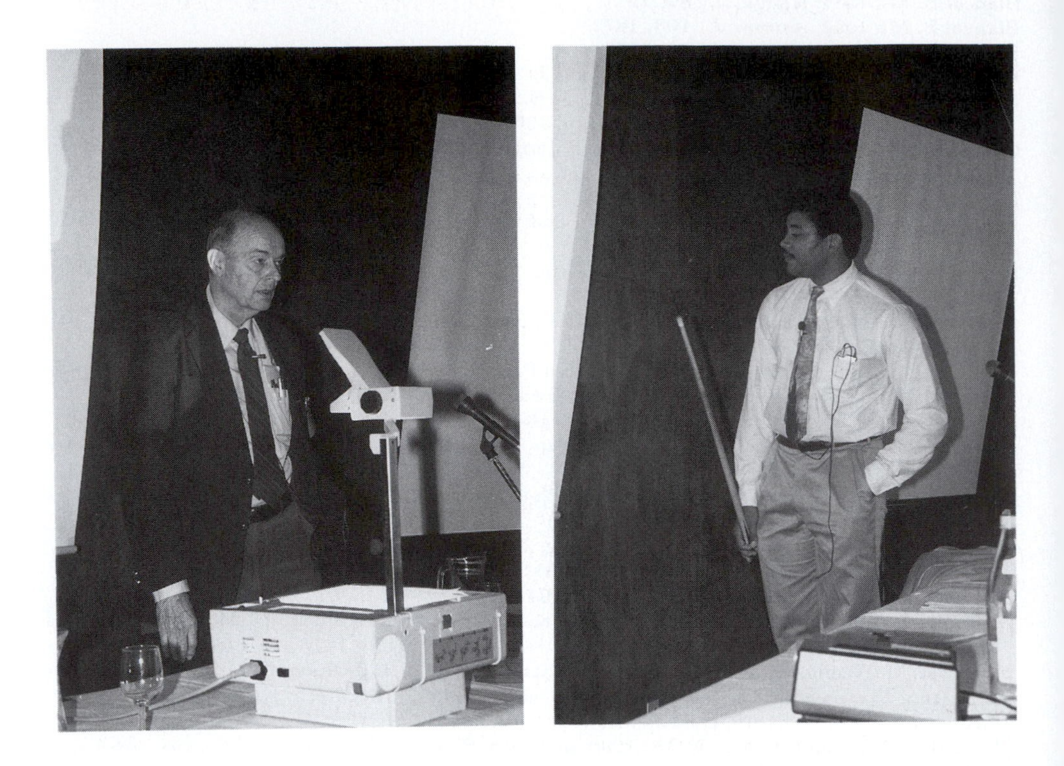

W. Baum and N. Tyson, each during his talk

GENERAL SURVEYS OF THE BULGE IN THE INFRARED

I.S. GLASS

South African Astronomical Observatory, PO Box 9, Observatory 7935, South Africa

October 16, 1992

Abstract. The Bulge has been mapped in the infrared both coarsely from balloon platforms and in detail from the IRAS satellite and the ground. The information revealed by each type of survey is discussed. Particular attention is given to surveys of the Sgr I clear field, its content of long-period variables and their implications for distance determination and the high end of the Bulge luminosity function.

Key words: Bulge – surveys – Sgr I

1. Introduction

Infrared surveys are particularly important for studies of the Bulge because they are relatively unaffected by the absorption due to the interstellar dust which occupies much of the galactic plane.

More than half of the starlight component is emitted beyond 1μm. Photospheres remain the main source of radiation to the end of the K band (2.4μm), but at longer wavelengths circum- and inter-stellar matter become dominant. By 12μm, the shortest IRAS wavelength, photospheric radiation is almost insignificant and even apparently stellar sources are often cool dust shells.

Photography stops at a wavelength of about 1.1μm, beyond which electronic detectors are used. Infrared area detectors have small formats at present but are increasing in size; for example at the moment it is possible to work out to 2.2μm (K band) with 256 × 256 pixels using the NICMOS3 HgCdTe arrays. InSb arrays of similar size, sensitive up to the M band (4.8μm), are also starting to become available. PtSi arrays 512 square have been used out to the K band and others up to 1024 square are under development. Although the PtSi arrays have generally fewer defective pixels than other types, they have only about 5% quantum efficiency and around 70% fill factor. Nevertheless, their large formats show promise for large-scale survey work.

From the ground, surveys of extended sources at wavelengths longer than the K band are not possible due to the very high background flux per unit area which overwhelms ordinary area detectors. For *point* sources it is however becoming feasible to examine larger and larger areas with detectors connected to fast processors which co-add many exposures per second in real time. Observations from space thus have clear advantages for large scale survey work and mapping of extended objects, and are of course a necessity at wavelengths where the earth's atmosphere is opaque.

2. Infrared Extinction

Preliminary to the general discussion, it is important to emphasize the behaviour of interstellar extinction in the infrared. The following is a table in general use at

H. Dejonghe and H. J. Habing (eds.), Galactic Bulges, 21–37.

SAAO, based on the famous van de Hulst (1949) No 15 curve. Some data from Rieke & Lebofsky (1985) are also shown.

The Rieke & Lebofsky law, which was based mainly on observations of heavily obscured high-luminosity stars in the direction of the Galactic Centre, differs from the van de Hulst one quite significantly, ie from 15% higher at J to 60% higher at L.

A parameterization of the extinction law of the form $\tau = k\ \lambda^{-1.85 \pm 0.05}$, based on comparison of observed to calculated line ratios in an obscured HII region, gives extinction values from the R (0.64μm) to M (4.8μm) bands about 10% lower than the van de Hulst ones (Landini *et al*, 1984).

TABLE I
Interstellar extinction for common photometric bands.

Band	λ	A_λ	A_λ(R & L)
U	0.36	1.56	
B	0.44	1.33	
V	0.55	1.00	1.00
R_C	0.64	0.78	
I_C	0.79	0.59	
J	1.25	0.245	0.282
H	1.65	0.142	0.175
K	2.2	0.081	0.112
L	3.5	0.036	0.058
M	4.8		0.023
N	10		0.052

The extinction law has been investigated experimentally at SAAO by Glass (unpub.) for A stars and Laney & Stobie (in prep.) for Cepheids. For example, in the near infrared, the ratio of the colour excess E_{B-V} to E_{J-H} and E_{H-K} are found to agree with the van de Hulst law to within a few percent. However, E_{B-V}/E_{V-K} is only known to about 1% which means that the ratio of A_K to A_V is known to 10%. At longer wavelengths the determination of extinction becomes more difficult because the values are so small and "infrared excesses" become more common. It is always important to remember that different groups use infrared filters which differ slightly in effective wavelength from one to another and that slightly different ratios will therefore apply.

3. Low-resolution studies of the Bulge

Absolutely calibrated large-scale observations of the Bulge are difficult from the ground because of atmospheric absorption and emission, which complicate the determination of background levels. Apart from ordinary thermal radiation which becomes prominent around 2.4μm, there is rapidly variable emission from OH high (~90 km) in the atmosphere (airglow). For this reason, such work has traditionally been carried out from balloon-borne platforms in carefully chosen bands and more recently from space vehicles such as the Shuttle and COBE.

The existing surveys are summarized in the Table II

TABLE II
Large area maps of the bulge at low resolution.

Group	Resolution	$\lambda(\mu m)$	$\Delta\lambda$	Range in l	Reference
Kyoto	1° sq	2.4	.09	−10° to +30°	Okuda *et al* (1977), Maihara *et al* (1978)
Nagoya	2° sq	2.4	.1	−10° to +30°	Ito *et al* (1977)
MPIA	2° diam	2.4	.15	0° to +10°	Hofman *et al* (1977)
		3.4	.2		
MPIA	1° sq	2.45	.09	−7° to +7°	Hofman *et al* (1978)
Nagoya++	1.7° sq	2.4	.09	−70° to +50°	Hayakawa *et al* (1981)
		3.4	.14		
Kyoto	0.6° sq	2.38	.08	−15° to +33°	Oda *et al* (1979)
Nagoya++	0.5° diam	2.4	.09	−12° to +12°	Matsumoto *et al* (1982)
	0.7° diam	3.4	.14		
Kyoto	0.4° diam	2.4	.08	−5° to + 5°	Hiromoto *et al* (1984)
Harv-Smith	~1°	2.4	1.3	−8° to +100°	Kent *et al* (1992)
COBE	0.7° sq	JKLM		−96° to +96°	NASA photograph (1990)

The last map is presented as a three-colour photograph combining the J (1.25), K (2.2) and L (3.4μm) images obtained by the DIRBE (Diffuse Infrared Background Experiment) on the COBE (Cosmic Background Explorer). No quantitative results of this experiment have been released at the time of writing.

3.1. Interpretation

The Nagoya group (Matsumoto *et al*, 1982) data at 0.5° resolution are the most refined of the large-scale surveys and they have been analysed by Blitz & Spergel (1991) who searched for a bar which appears from dynamical considerations to be affecting the gas kinematics in the central few kpc of the Galaxy. The somewhat bar-shaped bulge revealed by COBE and the Nagoya data have been identified with the hypothesized bar and analysed for non-axisymmetric properties in the hope of revealing its orientation to the line of sight. It is assumed that at galactic latitudes beyond ± 3° the data are unaffected by reddening. They placed the data in 1° bins and subtracted each bin at "negative" longitude from its counterpart at positive longitude to form a difference map which clearly shows an excess at positive longitudes, i.e. the near end of the bar is at positive longitudes and appears wider. Further, the bar appears to be tilted so that there is slightly more flux at negative b and positive l with less at positive b and "negative" l.

To be certain that the measured effects are not consequences of uneven extinction, they appeal to the Milky Way CO maps of Dame *et al* (1987) and the well-known correlation of dust and gas distributions. If anything, there is slightly more CO at positive than negative longitudes, mainly in a narrow plume, which would work against rather than for the longitudinal asymmetry. In general, the maximum visual extinction in the region examined is estimated to be less than 1.3 mag, corresponding to less than 0.13 mag at 2.4μm. The asymmetry noted is of order 15%-40%, considerably more than any possible differential extinction could account for.

We may compare the above results with those of Binney *et al* (1991) who considered a particle in a bar-like potential arising from a prolate ellipsoid of axis ratio 0.75 and mass density varying as $r^{-1.75}$. Its orbits were compared with the parallelogram-like ^{12}CO (longitude, velocity) diagram for $|b| < 0.1°$. They concluded that the bar's axis should be at about 16° to the line of sight and co-rotation would occur at 2.4 kpc. The pattern speed of the bar is 81 km s^{-1} kpc^{-1} from the dynamics of the gas which is similar to the mean velocities of planetary nebulae as a function of galactic longitude (Kinman *et al*, 1988). Mira variables at b$\sim$ 7° (Menzies, 1990) also rotate at these speeds.

4. Surveys with high resolution

The IRAS satellite as is well known, surveyed the Bulge with sufficient resolution to pick up individual objects.

However, it is also known that the high source density in the galactic plane led to incomplete surveys, or 'shadows', in the regions l=10° to 30°, b=0° to 5° and l=330° to 350°, b=0° to −5°, especially at the longer wavelengths. Fig VIII.D.2 of the *IRAS Explanatory Supplement* (Beichman *et al*, 1988) shows that possibly about 1/3 of the 12 micron sources within 1° of the plane, in the longitude range 10° to 30°, were not detected. The effects of crowding and saturation are worse nearer to the Galactic Centre and become more severe at longer wavelengths.

It was quickly realised by Habing (1985) that the distribution of IRAS sources with 12μm fluxes less than 5 Jy and comparable 25μm fluxes very much resembles the view of many edge-on spiral galaxies, showing a small bulge and a disc. Feast (1985) showed that many of the IRAS sources in the Sgr I and NGC6522 fields of the Bulge could be identified with known mira variables. Criteria were devised by van der Veen & Habing (1988) in terms of the [12]−[25], [25]−[60] two colour diagram (Fig 1) in order to separate objects according to degree of dominance by circumstellar shells, interpreted by them in evolutionary terms. The shells associated with O-rich variables have temperatures of around 200-300K.

The spatial distribution of sources in the category IIIb (thick dust shells) is particularly concentrated towards the Bulge. Thirty-seven IIIb bulge candidates were observed a number of times in the KLMN and the ESO narrow N band filters (2.2 to 10μm) and most were shown to have silicate absorption features, indicative of optically thick dust shells as expected (van der Veen & Habing, 1990). Periods of 500 to 1990 days were derived for these stars. Some of the longer period stars were undersampled in phase and have been since been revised to have shorter periods (van der Veen & Habing, in preparation). More than 50% of the sample are identified with OH/IR stars from the te Lintel Hekkert *et al* (1990) survey. Blommaert *et al*

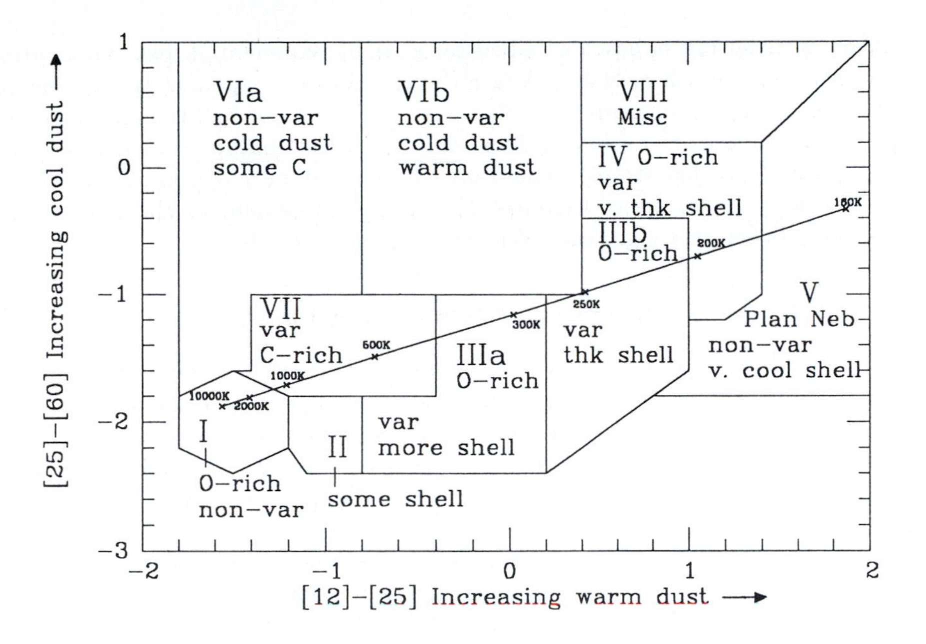

Fig. 1. Regions in the IRAS two-colour diagram which denote various types of stars with dusty envelopes (van der Veen & Habing, 1988).

(1992) report on a similar survey starting from OH/IR stars from a field within 30′ of the Centre which appears to show that they are closely related to the van der Veen & Habing stars and also that their bolometric magnitudess are below the (extrapolated) LMC P-L relation for M-type miras.

4.1. IRAS and the Bar

More evidence for the bar-like structure of the Bulge has come from analysis of IRAS data. Early work on the distribution of Bulge sources by Harmon & Gilmore (1988) showed a slight asymmetry. Nakada *et al* (1991) analysed a sample of sources away from the Plane, having a narrow range of colours on the border of IIIa and IIIb and assuming $M_{Bol} \sim$ constant. They found evidence that the sources in the first quadrant are nearer the sun.

Almost simultaneously, Weinberg (1992a) made a similar analysis by selecting variable sources away from the Plane and having [12]−[25] colours in the reddest quartile (presumably mostly of type IIIb), assuming constant M_{Bol} and calculating photometric distances using M_{Bol} estimated for each star from its IRAS fluxes. A map of source numbers with inferred position in the galactic plane for $F_{12} > 2$ Jy just reveals the Galactic Centre with a bar in the 1st quadrant. Going to $F_{12} > 1$ Jy begins to show the other end of the bar. The semimajor axis of the bar is 5 Kpc and the tilt to the line of sight 36°.

In a more recent paper, Weinberg (1992b) has carried his analysis of the spatial distribution of the variable $F_{12} > 2$ Jy sources somewhat further and finds that the entire distribution is best explained by a single exponential disk distribution with a = 4.5 kpc and h $\sim$ 800pc. Any $r^{1/4}$-type Bulge component is quite small. Weinberg's choice of minimum 12μm flux is such that about half the IRAS sources in the Sgr I window are excluded. Thus only the most luminous variables of the Bulge contribute to his sample. These are the longest period sources, which are likely to belong to the most flattened (old disc) component of the long period variable population (see e.g. Feast, Whitelock & Sharples, 1992).

4.2. Near-Infrared Surveys and their follow-up

The nearest infrared surveys, conducted by Blanco and collaborators in the I band ($\sim$ 0.8μm), are capable of showing Bulge stars in the "clear" windows of low extinction. However, as pointed out by Frogel (1988), the brightest stars at I are not necessarily the most luminous bolometrically because of line blanketing effects. Photographic surveys are however very useful in that large areas can be covered in a short time.

Objective prism surveys, also by Blanco and collaborators, have been used for spectral classification of Bulge stars in the clear windows on a large scale. Some confusion has entered the literature due to the fact that the Blanco classifications were not exactly according to the Morgan-Keenan standard system, as pointed out by Terndrup *et al* (1990). In effect, the M stars of subtypes between M0 and M7 were classified too late by one to two subtypes.

Frogel & Whitford (1987) showed that Bulge stars have bluer J-K and H-K colours than local giants of the same spectral class, and this conclusion remains in spite of the revision of the objective prism spectral types. However one should remember that for stars of high metallicity, the bands used for classification purposes are more pronounced, independent of luminosity and temperature, and thus to some extent spectral type becomes confusing, especially when discussing luminosities.

Independently of spectral type, the mean locus in the J$-$H, H$-$K diagram shows a progression from left to right as one goes from globular clusters to local giants to Bulge giants (Frogel, Terndrup, Blanco & Whitford, 1990). This is a powerful indicator of metallicity differences, although there are some "local" giants at a mean distance of 500 pc from the Plane with colours similar to those in the Bulge (Feast, Whitelock & Carter, 1990).

In the colour-magnitude diagram, an interesting characteristic is that the luminosities of Bulge M stars extend to half a magnitude greater than those in globular clusters. Frogel (1988) points out that the K, J-K diagram does not change significantly with latitude, and this suggests that the effect noted in the J-H, H-K diagram occurs mainly in the H band. J-K is a good indicator of temperature, independent of metallicity (Terndrup, Frogel & Whitford, 1991).

It is often remarked that the Bulge giant branch is quite broad and therefore indicative of a spread in metallicity. Little account seems to have been taken of the spread caused by the different distances of Bulge stars in the line of sight, which will of course become larger as distance from the Centre increases. Indications from mira work are that a spread of magnitudes should be expected, having a FWHM of about 0.5 to 0.7 mag in the Sgr I field at $\sim$3° from the Centre. The line of sight depth will also be affected by the barred nature of the Bulge.

The choice of late-type M-stars by Frogel and co-workers for discussion of metallicity questions may give a distorted view of the population as a whole. It is well-known that a population that is metal-rich will allow evolution to later spectral types than one which is metal-poor. Thus, in a population containing stars which cover a large metallicity range, it is likely that the late M stars will be part of the metal rich fraction of the general distribution. K giants may be more representative of average metallicity, since all stars go through a K giant phase. A sample of 88 Bulge giants observed by Rich (1988) shows about that about 50% exceed solar metal abundance and 20% are super metal rich.

4.3. K-band Surveys

K-band-infrared surveys provide a more objective way of picking out the true high-luminosity late-type stars from a population because of the monotonic relation between K mag and m_{Bol} for giants (excepting those possessing very thick shells).

So far, K surveys have covered relatively small areas. The largest JHK survey of the bulge that is sensitive enough to show individual stars remains that of the SAAO group, covering an area of $2^{\circ} \times 1^{\circ}$ in RA and Dec. It stretches from just west of the G.C. to about 1.5° east of it. Qualitatively (Glass, Catchpole & Whitelock, 1987) it shows that even at 2.2μm wavelength there remain considerable patches of extinction ($A_V > 60$ mag). This extinction is strongly correlated with low-velocity molecular gas, presumed to be in circular orbits in the foreground. The vast column densities of high velocity gas revealed by the CO surveys are not correlated with readily visible features in the extinction and if there is dust associated with this material in the usual proportions its location must be behind the Centre.

One of the unexpected results of the SAAO survey was a very marked increase in the number of stars of high luminosity ($5 < K < 6$ or $M_{Bol} \sim -5$) towards the Centre (Fig 2). These have a flattened distribution lying along the longitude axis, but are very much peaked towards the Centre even along that axis. Haller & Rieke (1989) have independently found a population of luminous M stars within a few arcmin of the Galactic Centre. These objects do not appear to vary and are believed to owe their high luminosity to relative youth. Evidence for recent star formation near the Centre and mostly close to the Galactic Plane has been presented by Moneti, Glass & Moorwood (1992). The IRAS maps of unresolved continuum emission (presumably due to dust) presented by Cox & Laureijs (1989) also suggest the possibility of ongoing star formation.

The quantitative results of the SAAO survey are presented by Catchpole, Whitelock & Glass (1990). Stars from sub-fields of 400″ by 400″ were plotted on colour-magnitude (H-K, K) diagrams. Assuming that most of the stars were late-type giants and that they possessed the colours of giants from other fields, the average reddening for each sub-field could be found and the true stellar distribution in the plane of the sky at K could be determined.

A composite H and K′(2.15μm) picture of the central part (3.6° E-W by 3.6° N-S) of the galaxy has been made by combining 35 fields observed with the 512 x 512 PtSi-based infrared camera of the National Observatory of Japan by Munetaka

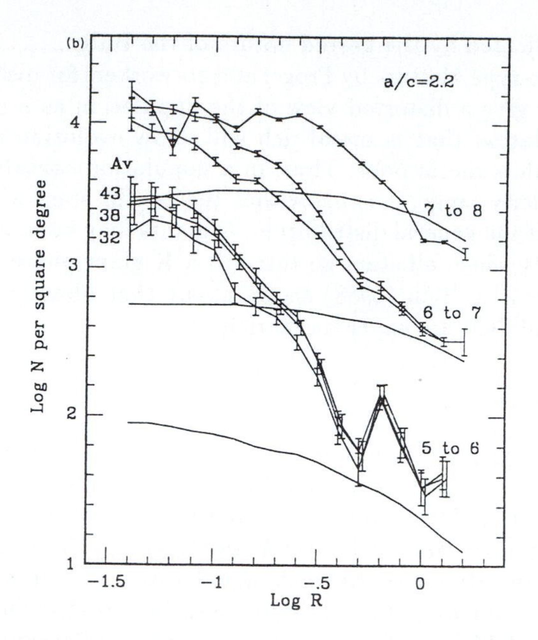

Fig. 2. No of stars per square degree in various (dereddened) K magnitude intervals shown as a function of log R (degrees from the Centre) for elliptical annuli having a/c = 2.2, corrected for crowding, interstellar absorbtion and contamination by Disc stars. The continuous curves show the Disc numbers that have been subtracted. The A_V numbers represent the A_V out to which counts have been made. For the $7 < K < 8$ curve, the upper limit of A_V was 32 mag. From Catchpole, Whitelock & Glass (1990)

Ueno (private communication) of the University of Tokyo. The resolution is about 10 arcsec. Each field was exposed twice for 240s with a small shift between exposures. The limiting mags are 12.0 at K′ and 13.0 at H.

4.3.1. The Sgr I field

The Sgr I field, lying at l=1.4°, b=−2.6°, is one of those clear fields surveyed at B,V & I for long-period variables by Lloyd Evans (1976) and others. A portion (205 arcmin^2) of this field (Catchpole, Glass & Whitelock, unpublished) was scanned in J,H & K at the time of the 1° by 2° survey referred to above (see Fig 3). Sixteen mira variables were identified in the scanned area, of which five of the reddest and brightest were detected by IRAS. From their agreement with the P-L-C relation these miras are clearly part of the Bulge and not of the foreground. The non-miras cannot be so easily placed as to distance. Some objects to the blue of the giant branch are probably foreground. Walker, Sharples & Cropper (1990) found from a kinematic survey of 225 stars in the NGC 6522 window that the brightest stars at I in their sample have systematically lower dispersion than the others, suggesting that they are foreground (disc) objects.

The 19 IRAS sources in the whole 1200 arcmin^2 Sgr I field of Lloyd Evans

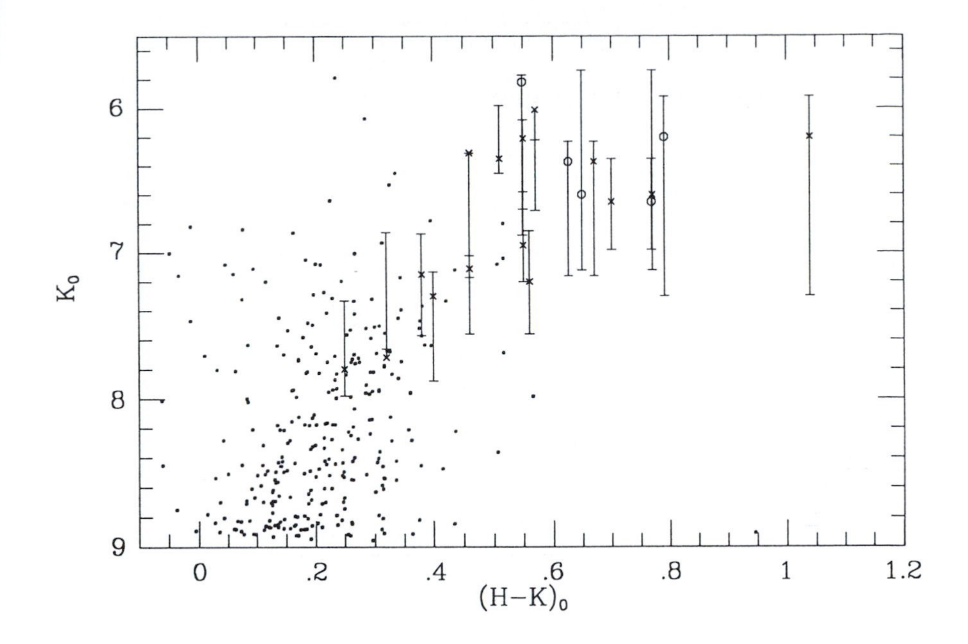

Fig. 3. K_0 vs $(H-K)_0$ for a 205 arcmin^2 portion of the Sgr I clear field from the SAAO JHK survey of the central Bulge. Mira variables which were found by Lloyd Evans (1976) or previous workers are marked with crosses. The bars represent the magnitude limits of the variables as observed during a programme of photometry of individual stars (Glass, Whitelock, Catchpole & Feast, in preparation). Colours should be accurate to about ±0.1 mag.

(1976) have been found to be mira or mira-like variables with only two exceptions. A programme of repeated observations of these and other known mira variables, expanding on the photometry by Glass & Feast (1982), has been completed recently (Glass, Whitelock, Catchpole & Feast, in preparation) to determine better periods, magnitudes and colours. The Sgr I miras fall slightly below the P-L relation (see Fig 4) but follow closely the period-luminosity-colour relation established for Large Magellanic Cloud M-type miras by Feast *et al* (1989).

The miras, because of their adherence to a period-luminosity relation, are clearly identifiable as members of the Bulge, with a few possible exceptions. This implies that their true luminosity function can be established, without fear of contamination by foreground stars.

The distance to the Centre can be obtained from the photometrically determined distances of individual mira variables (Glass & Feast, 1982). We are now using the LMC P-L-C relation to repeat the determination. A prelimary analysis of the improved data, based on modelling the mira distribution as a function of distance from the Centre, an ellipsoidal Bulge, an assumed power-law stellar density distribution and a particular value for the combined intrinsic and observational scatter has been performed (Fig 5). The distance itself is not particularly sensitive to the modelling parameters. (Discussed further by Whitelock, these proceedings.)

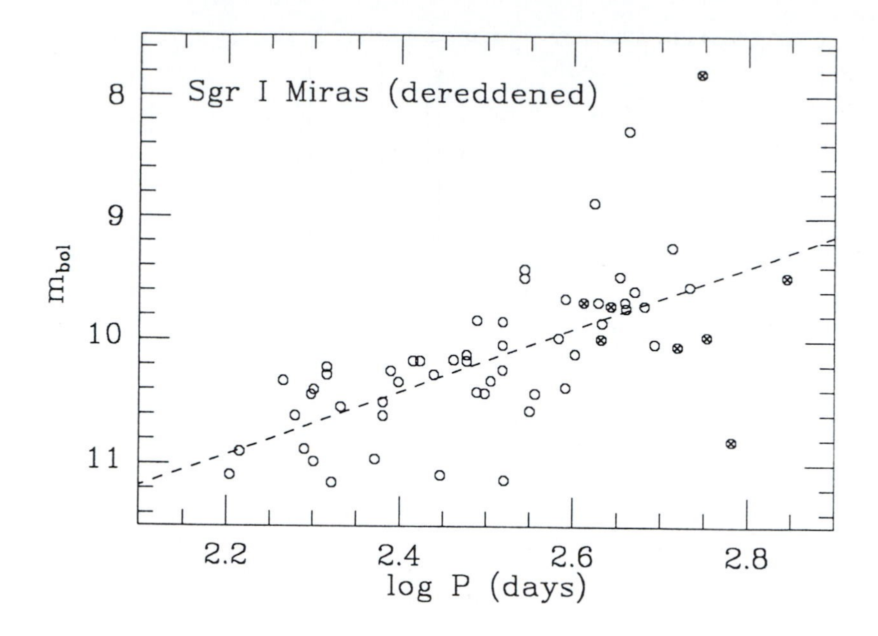

Fig. 4. The average bolometric mags of miras in the Sgr I field as a function of period. (from Glass, Whitelock, Catchpole & Feast, in preparation). The best fit P-L relation, omitting the two (apparently) most luminous sources, is shown, viz $m_{Bol} = -2.54 \log P + 16.53$. The scatter in this diagram is partly due to the depth of the field in the line of sight, the colour dependence of M_{Bol}, the observational errors and the actual intrinsic scatter. (The filled circles make use of IRAS data)

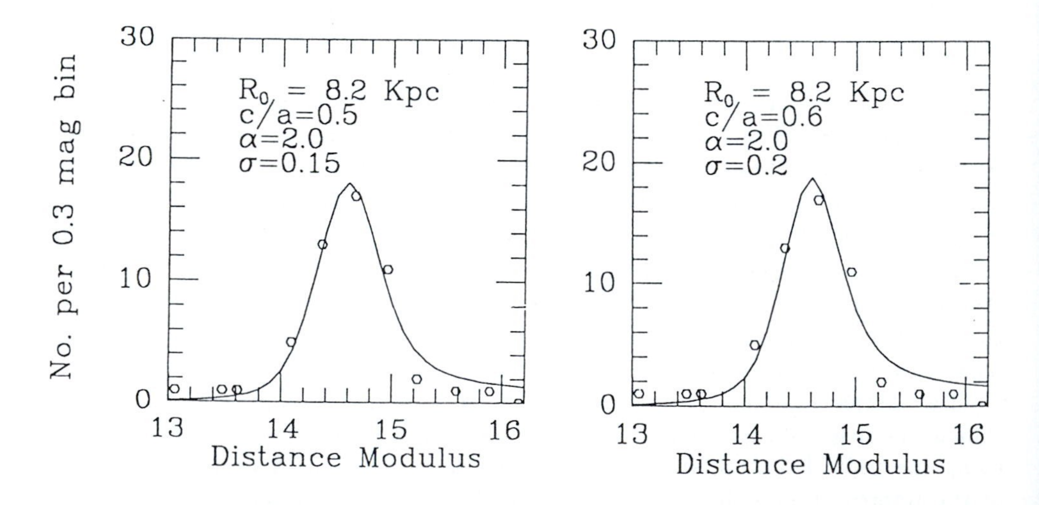

Fig. 5. Two examples of the fit to the distance distribution of the Sgr I miras from their M_{Bol} determined from the P-L-C relationship. c/a is the axial ratio of the model assumed for the ellipsoidal Bulge and α is the exponent of the power law density distribution assumed for mira variables. σ is the assumed combined intrinsic and observational scatter.

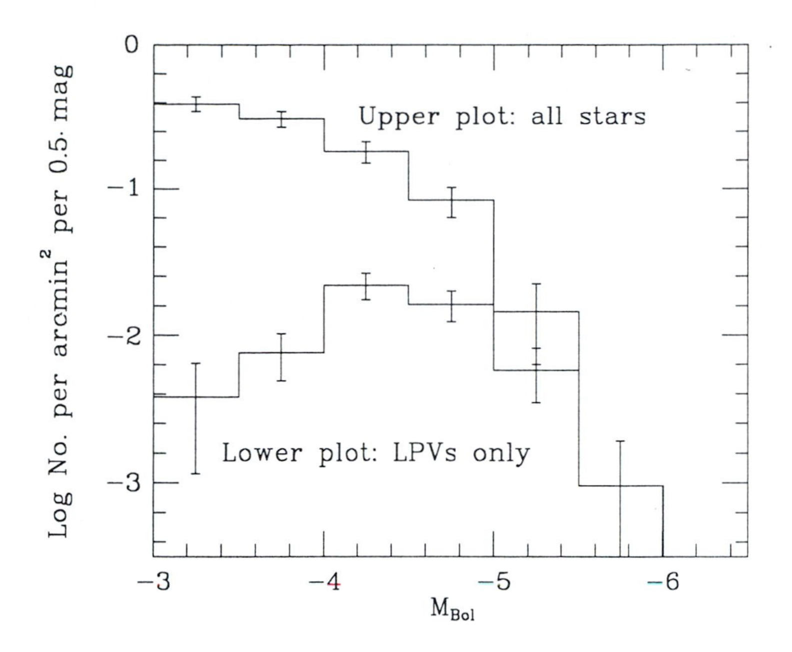

Fig. 6. Apparent luminosity function (not corrected for distance spread) for the Sgr I field based on multi-phase photometry of miras and the K scans mentioned in the text. The bolometric magnitudes of the non-miras have been formed simply by adding 3.1 to their de-reddened K mags. The tail-end of the luminosity function is comparable with that for NGC 6522 found by Frogel & Whitford (1987). Although Frogel & Whitford's cut-off M_{Bol} is usually given as ~ -4.9 (if the distance modulus = 14.6), there are a few LPVs in each field which extend the distributions to an average M_{Bol} of ~ -5.5. The two apparently most luminous stars of Fig 4 have been omitted as they probably lie in the foreground.

The luminosity function of the mira variables in the Sgr I field can be determined quite accurately because the distances of the miras are known and we do not need to subtract the numbers of foreground stars according to some model (Fig 6).

It is clearly interesting to know how mira variables can be detected most efficiently in the Bulge. Fig. 7 shows the success of the various techniques. The blue survey was conducted by Oosterhoff & Ponsen (1968), that in the V and I band by Lloyd Evans (1976) and the IRAS results are taken from the Point Source Catalog, regardless of detection quality (Glass, 1986). Clearly none of these surveys were complete for all period ranges. The Lloyd Evans technique (Lloyd Evans, private communication) probably yields completeness to about 300 days. There is a possibility that some of the longer-period variables were not detected due to faintness at I and diminished IRAS sensitivity near the Galactic Plane. It appears that a 2.2μm survey of rather modest depth but large area would be ideal for this sort of work. It would offer the possibility of mapping the reddening from mira colours in considerable spatial detail. The relationship between mira kinematics and period for local miras and the period distribution of miras in globular clusters also suggests a strong connection between age (or perhaps metallicity) and period for miras which

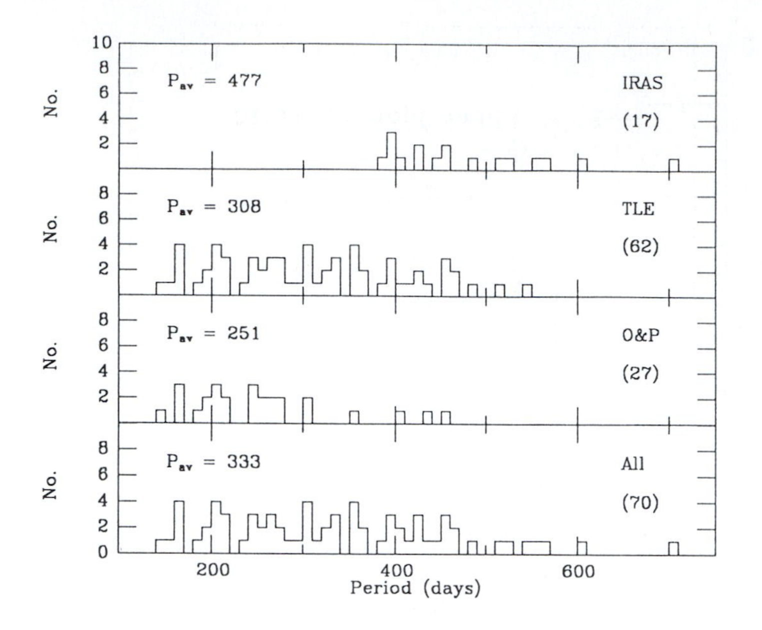

Fig. 7. Histograms of the miras in the Sgr I field according to method of detection.

could be of use in exploring the formation of the Bulge.

The average J-H, H-K colours of the Sgr I miras follow those of the solar neighbourhood rather than the LMC, at least for periods of up to about 300 days. Beyond 300 days, they are distinctly redder than the other samples, as previously noted by Wood & Bessell (1983). If these results are interpreted in terms of metallicity, we can say that the short period miras of the Bulge and solar neighbourhood and Galactic Centre are similar to each other but more metal rich than their LMC counterparts, while the longer period miras in the Bulge are more metal rich than those known in the solar neighbourhood and are not represented at all in the LMC. However, the solar neighbourhood sample is mainly based on *visual and blue* surveys and our knowledge of the long-period end is thus distorted. The one population difference that is certain is the existence of carbon miras locally and in the LMC and their complete absence in the Bulge.

4.3.2. The NGC 6522 field

Part of the NGC 6522 field (201 arcmin^2) has been scanned at 2.2μm by Ruelas-Mayorga & Teague (1992). This was extended to fainter magnitudes by scanning a smaller area with a larger telescope. The brightest 165 objects found in the scans were re-observed in a programme of discrete photometry. From this work they constructed a Cumulative Counts Function (integrated K luminosity function) for the NGC 6522 window. In an application of the NICMOS3 near-infrared array, Davidge (1991) has examined a 1.6 arcmin^2 area of the NGC 6522 field. This work is compared to that of Ruelas-Mayorga & Teague (1992) in Fig.9. The agreement

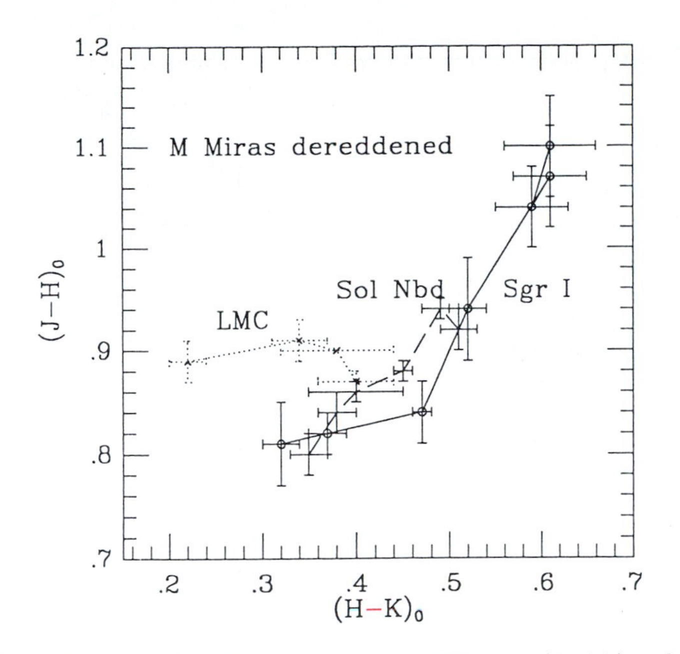

Fig. 8. Summary J-H, H-K diagram for LMC, local (O-rich) and Sgr I miras.

is fairly satisfactory. This field and also the Sgr II field have been surveyed for mira variables at I by Lloyd Evans (1976). Many of the miras in NGC 6522 were observed repeatedly by Glass & Feast (1982) and once by Wood & Bessell (1983).

4.3.3. Other fields that have been surveyed

Ruelas-Mayorga & Teague (1992, in press) have made scans of three small fields (56 to 185 arcmin^2) along the latitude axis of the galaxy, at latitudes of 3.5°, 4° and 5°.

4.3.4. Outer Bulge

Multi-phase photometry of IRAS sources in two fields lying at $7° < |b| < 8°$, $345° < l < 15°$, to determine periods and average magnitudes, has been reported by Whitelock, Feast & Catchpole (1991). One of the fields partially overlaps Plaut field No 3, at l=0°, b=−10°, which has been surveyed using B and R photographic plates for miras and other variable stars by Wesselink (1987). Although our knowledge of mira variables at intermediate periods is likely to be uncertain in this field, there is a strong suggestion that the numbers of very long-period variables that IRAS detects compared to those of short period is greater in the Sgr I field than in the −10° degree field (see contribution by P.A. Whitelock), again a suggestion of a metallicity gradient in the mira population. The SAAO photometric observations of IRAS sources have turned up very few miras of period exceeding 700 days, although other authors claim to have found longer period sources on the basis of

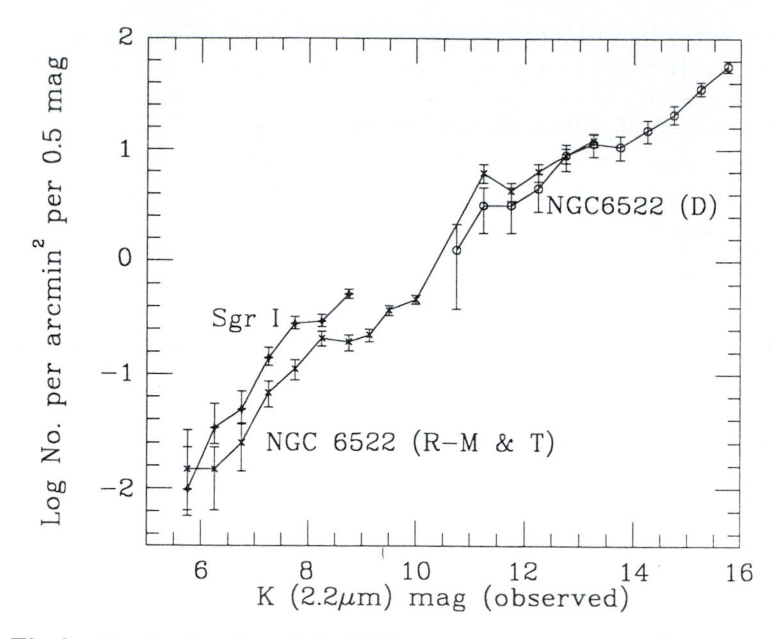

Fig. 9. The luminosity function of the NGC 6522 field at K as determined from data obtained by Ruelas-Mayorga & Teague (1992) and Davidge (1992). The luminosity function from the unpublished SAAO scans of Sgr I is also shown. All data are presented as observed, ie without any correction for interstellar absorption (which increases K by ~0.15). About 30% of the stars brighter than K=6.7 are non-miras and thus possibly foreground).

sparser observations. Only in two cases has direct evidence been found for an incorrect period (Whitelock *et al*, 1991), but the suspicion remains that many of the long periods claimed or hypothesized by interpreters of IRAS data, for example Harmon & Gilmore (1988), are not realistic. An analysis of the bolometric mags derived from this programme by Whitelock & Catchpole (1992) and using the P-L relationship to obtain distances, shows that there is a marked longitude asymmetry which can be modelled by a prolate spheroidal stellar distribution having $x_0 = 760$ pc, $y_0 = z_0 = 190$ pc with an angle of inclination to the line of sight of 45°, lending further weight to the bar hypothesis.

5. Extension to other galaxies

Surveys of the bulge of M31, which might be expected to be similar to that of the Milky Way, have been conducted by Rich & Mould (1991) and Davies, Frogel & Terndrup (1991). The first of these papers finds that the luminosity function in M31 extends considerably beyond Frogel & Whitford's cut off, by almost 1 mag. However, taking the distance modulus to the Centre as 14.6, the difference is hardly significant and amounts to an argument about small-number statistics. The Davies *et al* paper finds a similar result, but the suggestion is made that the M31 bulge sample is contaminated by luminous disc stars from a younger population. Studies of the luminosity function in other galaxies less similar to ours show more convincing evidence for extensive populations of higher luminosity stars than are met with in the Bulge. For example, Freedman (1992) finds a highly luminous group of late-type stars (including many non-variables) in the dwarf elliptical M32, which she suggests are from an intermediate age population. Similarly, and perhaps more expectedly, Mould (1992) finds intermediate age populations of high luminosity in the Magellanic Clouds.

6. Acknowledgements

I thank R.M. Catchpole, M.W. Feast & P.A. Whitelock for permission to quote unpublished material from joint programmes concerning the Sgr I region. P.A. Whitelock and T. Lloyd Evans are thanked for critical readings of a draft of this article.

References

Beichman, C.A., Neugebauer, G., Habing, H.J., Clegg, P.E. & Chester, T.J., 1988. *IRAS Explanatory Supplement*, NASA RP-1190, Washington, D.C.
Binney, J., Gerhard, O.E., Stark, A.A., Bally, J. & Uchida, K.I., 1991. *MNRAS*, **252**, 210.
Blitz, L. & Spergel, D.N., 1991. *Ap J*, **379**, 631.
Blommaert, J.A.D.L., van Langevelde, H.J., Habing, H.J., van der Veen, W.E.C.J. & Epchtein, N., 1992. In Warner, B., (ed) *Variable Stars & Galaxies; Symposium in Honour of M.W. Feast*, A.S.P. Conf. Ser., San Franscisco.
Catchpole, R.M., Whitelock, P.A. & Glass, I.S., 1990. *MNRAS*, **247**, 479.
Cox, P. & Laureijs, R., 1989. In Morris, M., (ed) *IAU Symp. 136, The Centre of the Galaxy*, Reidel, Dordrecht.
Dame, T.M., Ungerechts, H., Cohen, R.S., de Geus, E.J., Grenier, I.A., May, J., Murphy, D.C., Nyman, L.-A. & Thaddeus, P., 1987. *Ap J*, **322**, 706.
Davidge, T.J., 1991. *Ap J*, **380**, 116.
Davies, R.L., Frogel, J.A. & Terndrup, D.M., 1991. *AJ*, **102**, 1729.
Feast, M.W., 1985. In Israel, F.P., (ed) *Light on Dark Matter*, Reidel, Dordrecht.
Feast, M.W., Glass, I.S., Whitelock, P.A. & Catchpole, R.M., 1989. *MNRAS*, **241**, 375.
Feast, M.W., Whitelock, P.A., & Carter, B.S., 1990. *MNRAS*, **247**, 227.
Feast, M.W., Whitelock, P.A. & Sharples, R., 1992. In Barbuy, B. & Renzini, A., (eds) *IAU Symposium 149: The Stellar Populations of Galaxies*, Reidel, Dordrecht.
Freedman, W.L., 1992. *AJ*, (in press).
Frogel, J.A., 1988. *Ann Rev A Ap*, **26**, 51.
Frogel, J.A., Terndrup, D.M., Blanco, V.M. & Whitford, A.E., 1990. *Ap J*, **353**, 494.
Frogel, J.A. & Whitford, A.E., 1987. *Ap J*, **320**, 199.
Glass, I.S., 1986. *MNRAS*, **221**, 879.
Glass, I.S., Catchpole, R.M. & Whitelock, P.A., 1987. *MNRAS*, **227**, 373.
Glass, I.S. & Feast, M.W., 1982. *MNRAS*, **198**, 199.
Habing, H.J., 1985. In Israel, F.P., (ed) *Light on Dark Matter*, Reidel, Dordrecht.
Habing, H.J., Olnon, F.M., Chester, T., Gillett, F., Rowan-Robinson, M. & Neugebauer, G., 1985. *A Ap*, **152**, L1.
Haller, J.W. & Rieke, M.J., 1989. In Morris, M., (ed) *IAU Symp. 136, The Centre of the Galaxy*, Reidel, Dordrecht.
Harmon, R. & Gilmore, G., 1988. *MNRAS*, **235**, 1025.
Hayakawa, S., Matsumoto, T., Murakami, H., Uyama, K., Thomas, J.A. & Yamagami, T., 1981, *A Ap*, **100**, 116.
Hiromoto, N., Maihara, T., Mizutani, K., Takami, H., Shibai, H., & Okuda, H., 1984. *A Ap*, **139**, 309.
Hofman, W., Lemke, D. & Thum, C., 1977. *A Ap*, **57**, 111.
Hofman, W., Lemke, D. & Frey, A., 1978. *A Ap*, **70**, 427.

Ito, K., Matsumoto, T. & Uyama, K., 1977. *Nature*, **265**, 517.
Kent, S.M., Mink, D., Fazio, G., Koch, D., Melnick, G., Tardiff, A. & Maxson, C., 1992. *Ap J Suppl Ser*, **78**, 403.
te Lintel Hekkert, P., Caswell, J. & Habing, H.J., *A Ap*, 1990.
Lloyd Evans, T., 1976. *MNRAS*, **174**, 169.
Kinman, T.D., Feast, M.W. & Lasker, B.M., 1988. *AJ*, **95**, 804.
Landini, M., Natta, A., Oliva, E., Salinari, E. & Moorwood, A.F.M., 1984. *A Ap*, **134**, 284.
Maihara, T., Oda, N., Sugiyama, T & Okuda, H., 1978. *PASJ*, **30**, 1.
Matsumoto, T., Hayakawa, S., Koizumi, H., Murakami, H., Uyama, K., Yamagami, T. & Thomas, J.A., 1982. In Riegler, G.R. & Blandford, R.D., (eds) *The Galactic Center*, AIP Conference Ser. 83, Amer. Inst. Phys., New York, USA.
Menzies, J.W., 1990. In Jarvis, B.J & Terndrup, D.M., *Bulges of Galaxies*, ESO Cnf & Workshop Proc No 35.
Moneti, A., Glass, I.S. & Moorwood, A.F.M., 1992. *MNRAS*, **258**, 705.
Mould, J.R., 1992. In Barbuy, B. & Renzini, A., (eds) *IAU Symposium 149: The Stellar Populations of Galaxies*, Reidel, Dordrecht.
Nakada,, Y., Deguchi, S., Hashimoto, O., Izumiura, H., Onaka, T., Sekiguchi, K. & Yamamura, I., 1991, *Nature*, **353**, 140.
Oda, N., Maihara, T., Sugiyama, T. & Okuda, H. 1979. *A Ap*, **72**, 309.
Okuda, , H., Maihara, T., Oda, N. & Sugiyama, T., 1977. *Nature*, **265**, 515.
Oosterhoff, P. Th. & Ponsen, J., 1968. *BAN Suppl Ser*, **1**, 397.
Rieke, G.H. & Lebofsky, M.J., 1985. *Ap J.*, **288**, 618.
Rich, R.M., 1988. *AJ*, **95**, 828.
Rich, R.M. & Mould, J.R., 1991. *AJ*, **101**, 1286.
Ruelas-Mayorga, R.A. & Teague, P.F., 1992. *A Ap Suppl.*, **93**, 61.
Terndrup, D.M., Frogel, J.A. & Whitford, A.E., 1990. *Ap J*, **357**, 453.
Terndrup, D.M., Frogel, J.A. & Whitford, A.E., 1991. *Ap J*, **378**, 742.
van de Hulst, H.C., 1949. *Rech astr Obs Utrecht*, **11**, pt 2.
van der Veen, W.E.C.J. & Habing, H.J., 1988. *A Ap*, **194**, 125.
van der Veen, W.E.C.J. & Habing, H.J., 1990. *A Ap* **231**, 404.
Walker, A., Sharples, R. & Cropper, M., 1990. In Jarvis, B.J. & Terndrup, D.M., (eds) *Bulges of Galaxies*, ESO Cnf & Workshop Proc No 35.
Weinberg, M.D., 1992a. *Ap J*, **384**, 81.
Weinberg, M.D., 1992b. *Ap J*, **392**, L67.
Wesselink, Th., 1987. PhD Thesis, Nijmegen.
Whitelock, P.A. & Catchpole, R.M., 1992. In Blitz, L., (ed) *Large Scale Distribution of Gas and Dust in the Galaxy*, Kluwer, Dordrecht.
Whitelock, P.A., Feast, M.W. & Catchpole, R.M., 1991. *MNRAS*, **248**, 276.
Wood, P.R. & Bessell, M.S., 1983. *Ap J*, **265**, 748.

DISCUSSION

Sellwood: I wonder what you mean by the Bulge. In particular, your conception that there seems to be a bar at the centre of the Galaxy. But you would not accept Weinberg's point of view, that there was nothing else. So what is it that you mean by the Bulge? Is it the bar, or is it something with additional components to that?

Glass: I would say that it is essentially a conglomeration that is dominated by the bar, but I don't except Weinberg's point of view that there is nothing else.

Sellwood: So the bar is a large part of it, but not all of it?

Glass: I should imagine that it's a matter of which population you're looking at.

Feast: I think that the point of Weinberg is, that he comes to the conclusion that there is nothing that shows an $r^{1/4}$ distribution. But I think that's not new, there is practically nothing that shows an $r^{1/4}$ and that has been known for some while. If you say "You have to have an $r^{1/4}$ distribution in order to have a bulge", then there's no bulge. But whatever the group of stars is, whether you think of it as a concentration of the disc or a concentration of the spheroid, there is in fact a large concentration of objects there and that in a way is the important thing for many studies. The rest tends to be rather a question of terminology.

Rich: The M31 luminosity function of field four: we in fact do confirm your Sagittarius field, we found exactly the cut-off at -5. We've now extended the study to several other fields, as I will show later. We can disprove any evidence of disk contamination, there is a sharp cut-off of the luminosity function, just as you see. As we go in closer it goes up to -5.5 or so.

King: With regard to Sellwood's question: I think part of the answer is in the distinction I urge people to make about what population you're selecting your stars from. The metal rich population can't make Miras because it doesn't make an asymptotic giant branch.

At registration time: M. Schaetsaert, N. Baeck, J. Sellwood, G. Bertin, S. Kent, H. Hasan

Ian Glass
'92

LONG-PERIOD VARIABLES AND CARBON STARS IN THE GALACTIC BULGE

PATRICIA WHITELOCK

South African Astronomical Observatory,
P O Box 9, Observatory, Cape 7935, South Africa

October 12, 1992

Abstract. The review covers the properties of red variables in globular clusters and the Galactic Bulge. Details are given of our current understanding of the Mira evolutionary phase. There is evidence that Miras in the LMC and the Bulge occupy different parts of the instability strip but obey the same PLC relation. The Bulge contains at least 2×10^4 Miras of which 100 or so have luminosities in excess of $M_{bol} = -5$ mag. The Mira phase lasts more than 10^5 yr. These objects originate from stars with a wide range of metallicity, but it is currently unclear if the most metal-rich stars reach the top of the AGB to become Miras. Preliminary data suggest that the distribution of the Miras along the minor axis of the Bulge is different from that of the late-M stars but similar to the 2.4 μm luminosity.

Our knowledge of the Bulge carbon stars is briefly reviewed. It is suggested that, by analogy with the carbon-rich dwarfs, these stars are probably best understood as the products of binary evolution.

Key words: Galactic Bulge – Mira Variables – LPVs – IRAS sources – Carbon Stars – Metallicity Gradients

1. Introduction

Studies of variable stars have made a substantial contribution to our understanding of the Galactic Bulge, and show promise of doing the same for the bulges of other galaxies. Detailed investigations of the red variables, the typical flux distribution of which peaks longward of 1 μm, had to await the development of infrared techniques and surveys (see Glass's contribution to these proceedings). In particular, the wide variety of studies which followed the IRAS survey have made a major contribution to the field. This review describes our current understanding of low-mass red variables as derived from studies of globular clusters and the Galactic Bulge. Some emphasis is given to the Mira variables which, although they are the most luminous objects in the Bulge, are far from being fully understood. The review covers recent work on the period-luminosity-colour (PLC) relation which was discovered for Miras in the LMC but is proving crucial to our understanding of Miras in the Bulge. It also appears that the distribution of Miras in the Bulge follows the bulk luminosity rather more closely than do other groups of Bulge stars, enhancing their potential as tracers of galactic structure. The recent review by Whitelock (1990) gives more background to the subject while that by Whitelock & Feast (1992) concentrates on the evolutionary connection between Miras and planetary nebulae. The papers by Habing and Dejonghe in these proceedings describe the OH/IR sources which are a particularly interesting subgroup of the Mira variables.

The final section of the review considers the enigmatic Bulge carbon stars. This is a topic which the organizers asked me to cover, although the carbon stars are not obviously related to the red variables.

H. Dejonghe and H. J. Habing (eds.), Galactic Bulges, 39–54.

2. Red Variables in the Bulge

The early surveys of the Bulge, such as those by Gaposchkin, Oosterhoff and Plaut, used blue photographic plates and concentrated on finding RR Lyrae stars. They also noted the presence of numerous short period (mostly P< 300 day) red variables, including many Mira and semi-regular (SR) variables. This supported early ideas that the Bulge variable star population was similar to that found in globular clusters. More recent surveys using red sensitive plates, such as those by Terzan and collaborators (e.g. Terzan & Ounnas 1988), have identified larger numbers of LPVs but are far from being complete. Lloyd Evans (1976) provided the only survey specifically aimed at finding red variables and determining their periods. He used near-infrared (I) and V plates and found 121 large amplitude red variables in the Sgr I, Sgr II and NGC 6522 Baade windows. This is the most complete survey to date for Mira variables in the Bulge, although others are now in progress. It is not however complete for the longer period, dust-enshrouded, variables with energy distributions which peak in the infrared (see e.g. Glass 1986). The IRAS satellite was particularly sensitive at detecting this kind of star as recent surveys have shown (e.g., Feast 1986, Whitelock *et al.* 1986, te Lintel Hekkert 1990, Whitelock *et al.* 1991–hereafter WFC, Blommaert 1992).

3. Red Variables in Globular Clusters

Studies of globular clusters have provided much of our basic understanding of red variables. The most luminous stars in globular clusters are all variable and the most luminous of the variables are the Miras (see Fig 1 of Feast & Whitelock 1987). These all have bolometric luminosities $M_{bol} < -3.5$ mag, the approximate luminosity at which the core helium-flash occurs. For this reason it is clear that they are on the asymptotic giant branch (AGB); furthermore there are no stars brighter than the Miras and we therefore presume that they are the last, reasonably long lived, phase of AGB evolution. Our current picture of the Bulge is consistent with this, although there is much greater ambiguity attached to the brightness of individual stars due to the line-of-sight depth of the Bulge.

The SR variables within a particular cluster obey a period luminosity (PL) relation, the slope of which is that predicted for an evolutionary track (Feast 1989). It thus appears that SR variables evolve into Miras. The combined Miras from all clusters containing such variables, obey a very tight PL relation (Menzies & Whitelock 1985; Feast 1984) which is too steep to be understood as an evolutionary track. For this and other reasons the Mira PL relation is interpreted as the locus of end points of AGB evolution for stars of different initial mass (Feast & Whitelock 1987). We might expect the SRs in the Bulge to be similar to those in globular clusters but rather little is known about these stars beyond the fact that they are numerous.

Figure 1 illustrates the PL relation for Galactic Miras with independent distances. The distances of the globular cluster Miras were calculated assuming that the magnitude of their horizontal branch, $M_V(HB)$, is given by: $M_V(HB) = 0.15[Fe/H]+0.73$ (Walker 1992), except for ω Cen where $M_V(HB) = 0.25$ (Dickens

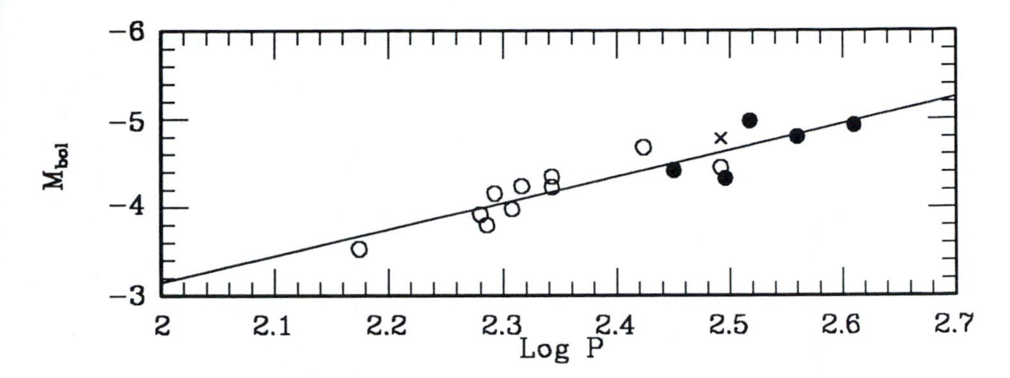

Fig. 1. The PL relation for Galactic Miras. Globular cluster photometry (open circles) is from Menzies & Whitelock (1985) with distances derived as described in the text. The photometry for the other stars comes from Catchpole *et al.* (1979) and unpublished SAAO observations. Distances were derived from Robertson & Feast (1981) for those with spectroscopic parallaxes (closed circles) and from the trigonometric parallax for R Leo (cross, Gatewood 1992). The line follows the LMC PL relation assuming that $(m\text{-}M)_0=18.5$ mag; it is not fitted to the points.

1989) was used. The bolometric magnitudes were calculated from infrared photometry as described by Menzies & Whitelock (1985). The line has not been fitted to the points, it is simply the PL relation derived for Miras in the LMC (Feast *et al.* 1989) on the assumption that the distance modulus to the LMC is 18.5. As discussed by Walker (1992) $(m\text{-}M)_o=18.5$ for the LMC is consistent with values derived from the Cepheid variables, main-sequence fitting to clusters and the ring around SN 1987A. It is clear that, within the illustrated period range, the same PL relation applies to Miras in the LMC, the Galactic globular clusters and the solar neighbourhood.

Within the globular clusters, Mira variables are found only in the most metal-rich clusters. Some of the metal deficient clusters contain SRd variables with luminosities which suggest that they are near the top of the GB or AGB. Like Miras they show emission lines at certain phases of their variability cycle. They have earlier spectral types, lower amplitudes and shorter periods than do the Miras; differences which can be understood as a consequence of their lower metallicities. These SRd variables can be viewed as a metal-deficient analogue of the Miras. The Mira and SRd pulsation periods are a function of the metallicity of the parent cluster (Feast 1981). The longest period Mira thought to be a cluster member has a period of about 300 day and is associated with a cluster of near solar metallicity (NGC 5927). Unfortunately once we leave the clusters there is no way of determining the metallicity of a Mira. The distribution of periods among Bulge Miras covers a larger range than the clusters extending to 700 day or more. There is as yet no strong evidence that the short-period Bulge Miras differ from those found in globular clusters. The longer period objects we presume have originated from progenitors which are more metal rich than those found in the clusters, although

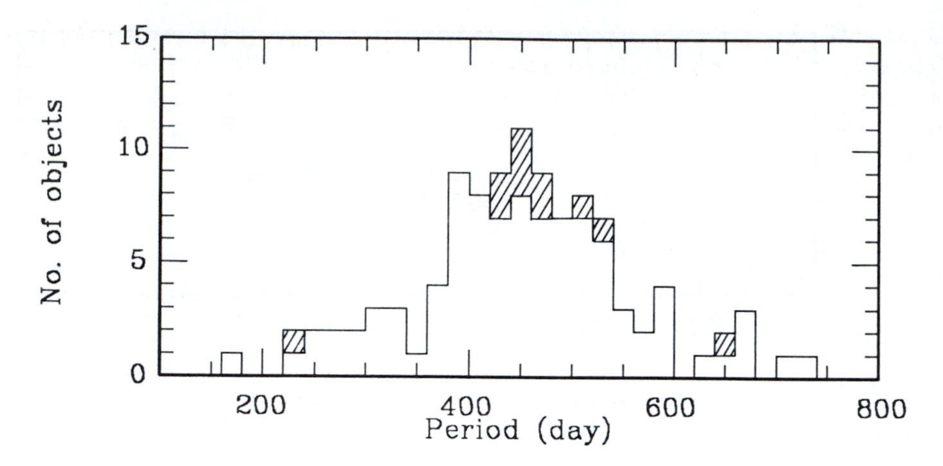

Fig. 2. Histogram of the periods of the IRAS Miras discussed by WFC. The shaded area represents those stars which are within 5.5 kpc of the sun.

the longest period objects may originate from merged binaries (Renzini & Greggio 1990), and are therefore not necessarily very metal rich.

4. Mira Variables in the Galactic Bulge

WFC published the results of a survey for Miras from two strips across the Bulge covering galactic latitudes of $b = +7^\circ$ to $+8^\circ$ and $b = -7^\circ$ to -8°. The sources were selected from the IRAS point source catalogue (1985) as having colours indicative of high mass-loss rates. Periods were determined for 104 of the 113 Miras discovered and Fig 2 illustrates the distribution of these periods. Miras closer than 5.5 kpc were classified as foreground to the Bulge and indicated as such on the histogram; about 13% of the IRAS Miras fell into this category. It is notable that the period distribution of these stars covers the same range as that of the Bulge Miras. It is possible that these Miras represent an extension of the Bulge population into the solar neighbourhood. They may be related to the stars discussed by Grenon (1990) which he suggests formed in the Bulge with orbits which bring them into the solar neighbourhood.

4.1. Gradients within the Bulge

The WFC survey covered only Miras with high mass-loss rates. It is of considerable interest to obtain a complete sample of Miras in a number of well defined regions of the Bulge in order to examine gradients as a function of position. With the intention of isolating a complete sample of Miras at $l \sim 0^\circ$ and $b \sim -7^\circ.5$, all of the IRAS sources in the 3.8 square degree area where the -7° to -8° strip overlaps the Baade/Plaut field number 3 were examined (Whitelock, Catchpole & Marang in prep.). JHKL photometry was obtained for all the 12μm IRAS sources in this region which had not already been examined by WFC. Of the 89 IRAS sources observed 47 have the colours of Miras and periods were determined for 40 of these. These periods cover approximately the same range as those measured by WFC but are somewhat lower on average; very few Miras had periods less than 350 day. The mass-loss rates for these stars are discussed by Whitelock & Feast (1992). Within the same 3.8 square degrees there were 45 Miras discovered by Plaut (1971) (see also Wesselink 1987). Most of these had periods less than 300 day. Only 4 of the IRAS Miras were in Plaut's list and 2 of these were clearly foreground stars. This lack of overlap between the Plaut and IRAS samples indicates that the sample is incomplete. The period distributions of the various samples suggests that the missing stars will be mostly in the 300 to 400 day period range.

There is considerable evidence that a metallicity gradient exists in the Bulge (e.g., van den Bergh & Herbst 1974, Terndrup 1988, Frogel *et al.* 1990). The gradient is in the sense that there is a greater concentration of metal-rich stars towards the Galactic Centre, although the slope and extent of this gradient is still subject to some uncertainty. It is interesting to examine what the Miras tell us about such gradients. Unfortunately, there is, as yet, no reliable way to establish the metallicities of Miras directly. We would expect long-period Miras to have evolved either from an old metal-rich population or from any intermediate age population which may be present. Terndrup's (1988) result on the main-sequence turn-off seems to rule out a large intermediate age population. So in the first instance it seems reasonable to assume that Miras in the Bulge with periods of over $\sim$ 400 day are old but probably have metallicities above the solar value. As stated above there are reasons to think that variables in the long-period tail of the period distribution may be in part the result of binary mergers and are therefore not necessarily very metal rich.

Figure 3 illustrates histograms of the period distribution for Miras in three Bulge fields at different galactic latitudes. Note that the distributions have not been normalized and that the high latitude field covers a larger area (3.9 square degrees) than do the Baade window fields (0.33 square degrees). All three histograms show a large range of periods from around 150 day to over 600 day. This is consistent with our views of a range of metallicities within the Bulge population. We also know from WFC that there are Miras with periods in the range 700-800 day present in other parts of the $7^\circ < |b| < 8^\circ$ field but that they are rare.

As stated above the survey of the $b = -7^\circ.5$ field is probably incomplete for periods between 300 and 400 day. For this reason the only reliable way to examine differences between the distributions is to take the ratio, $\Re$, of the number of short period ($200 \leq P \leq 300$) to the number of long period ($400 \leq P \leq 600$) Miras. The following values were derived: Sgr I $\Re = 1.1 \pm 0.3$; NGC 6522 $\Re = 1.8 \pm 0.7$; $b = -7^\circ.5$ $\Re = 3.9 \pm 1.5$; where the errors are calculated assuming $\sqrt{N}$ statistics. There thus appears to be a difference between the inner and outer fields in the sense that the inner field has proportionally almost four times as many long-period Miras as does the outer field. It must however be remembered that there are differences in the way that Miras were discovered in the various fields and that this conclusion may have to be modified when any systematic effects are taken into account. Note in particular that Fig 3 includes Miras which are not strictly part of the Bulge but which lie in the foreground. This will be a slightly more serious contamination of

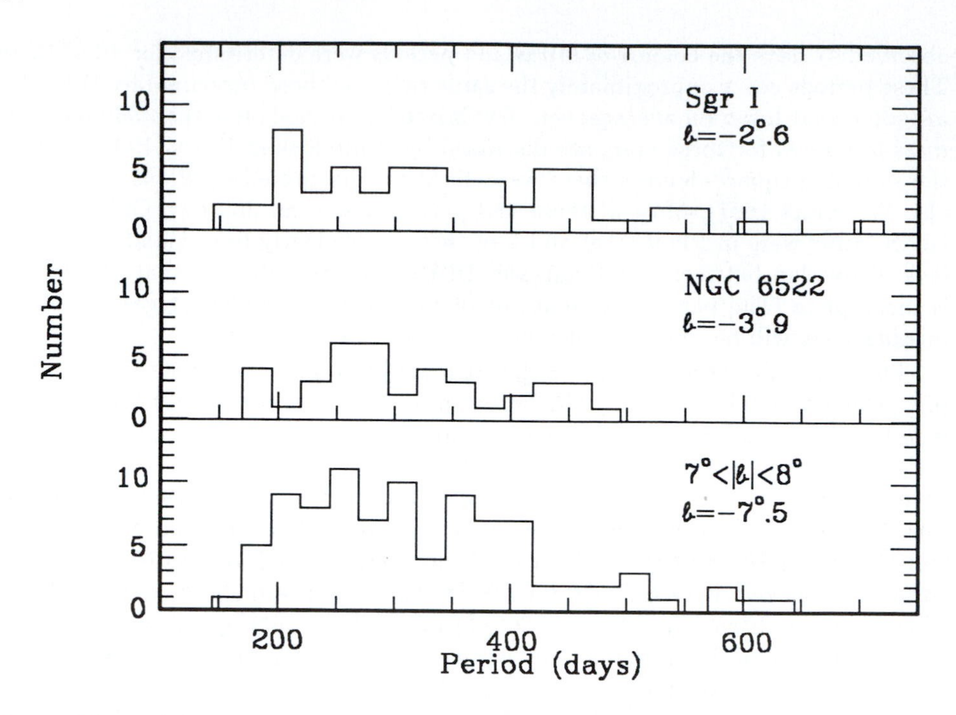

Fig. 3. The period distribution of Miras in the Bulge. The periods for the Miras are from Glass *et al.* (in prep.) for Sgr I, Lloyd Evans (1976) for NGC 6522 and from Whitelock *et al.* (in prep.) for $b = -7°$.

the outer field than the inner ones, but correcting for this effect should not grossly affect the result. There are also Miras in the $b = -7°.5$ field for which the periods have not been determined.

It is also interesting to note that the median of the distribution for the NGC 6522 field is at about 300 day. Rich (1988) finds that the median metallicity for K giants in the same window is at [Fe/H]~ 0. Thus *if* all of the K stars evolve into Miras and *if* stars of different metallicity spend proportionally the same time as K giants and as Miras, then a 300 day Bulge Mira must have [Fe/H]~ 0. This would be consistent with what is known from Miras in globular clusters. Note however that Miras are absent from the metal deficient clusters.

Blanco (1988) has pointed out differences in the radial gradient of early-M and of late-M stars. Figure 4 compares the radial gradient of the Miras with that of the late-M stars (M7 and later from Blanco 1988). The M stars have been corrected for foreground stars, the Miras have not. As discussed above, all three fields surveyed for Miras are probably incomplete, but the outermost field ($b = -7°.5$) is likely to be less complete than the other two. The two solid lines on Fig 4 are exponentials with scale heights of $2°.7$, the value determined by Kent *et al.* (1991) for the 2.4 μm

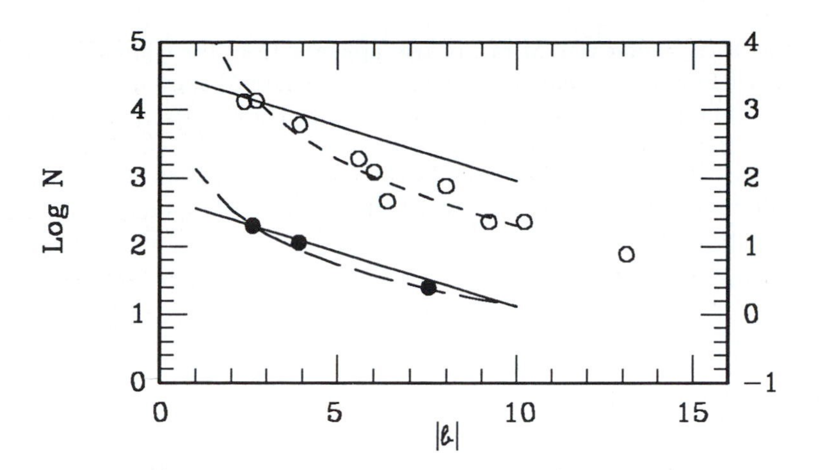

Fig. 4. The number of Miras (closed circles, left axis) compared to the number of late-M stars (open circles, right axis) per square degree as a function of galactic latitude, b. The solid lines are drawn at the slope of the model from Kent *et al.* (1991) for the logarithmic 2.4 μm surface luminosity. The broken lines are the power-laws discussed in the text. All of the lines are normalized to the measurements for the Sgr I window.

radial surface brightness within 10° of the Galactic Center. They are normalized to the number of Miras and M stars at $|b| \sim 2^{\circ}.7$, respectively. This comparison of surface brightness and surface number density implies an assumption that the luminosity function remains constant over the range of latitude considered. Such an assumption, although not strictly justified, may be a reasonable first approximation. The difference between the distribution of M stars and that of the 2.4 μm surface brightness has been been commented on by Tyson (1992). Figure 4 illustrates two points of particular interest. First, the trend shown by the Miras is indicative of the behaviour of the general Bulge luminosity, and it seems likely that the same will be true in extragalactic bulges. Secondly, the late-M stars have a steeper radial gradient than do the Miras.

The radial gradient in volume number density, N, is sometimes expressed as a power law of the distance from the Galactic Centre, R, so that $N \propto R^{-\alpha}$. The equivalent expression for the surface number density, ρ, as illustrated in Fig 4, is then $\rho \propto R^{1-\alpha}$ over the range of latitude under consideration. For late M stars $\alpha \sim 4.2$ (Blanco & Terndrup 1989) as is illustrated by the line of short dashes in Fig 4. A value of $\alpha \sim 3$ provides a good fit to the Mira densities as is shown by the line of long dashes. Note however that a preliminary analysis of the line-of-sight distribution of the Miras in Sgr I requires $\alpha \sim 2$ (Whitelock & Feast 1992 Fig 2 and Glass *et al.* in prep.). The difference between the radial gradient discussed above and the line-of-sight distribution may be a consequence of the triaxial nature of the Bulge (see §4.2).

The difference in the radial distribution of Miras and late-M stars is puzzling, since it is generally assumed that the M stars, whether they are on the giant branch

(GB) or the AGB, evolve into Miras. It also appears, from the comparison with the 2.4 μm surface luminosity, that the behaviour of the late-M stars is not indicative of the luminous red, presumably K and early-M, Bulge stars Possibly the very late-M stars (i.e. the most metal rich ones) take some alternative evolutionary track and do not become AGB stars in the normal way. The late evolution of metal rich stars has been discussed in detail by Greggio & Renzini (1990) (see also Horch *et al.* 1992) in an attempt to explain the ultraviolet excess observed in super-metal-rich elliptical galaxies. It appears that under certain conditions metal rich stars may not undergo AGB evolution. It is not yet clear if we should expect this kind of scenario to apply to stars in the Galactic Bulge. Two alternative explanations of the different distributions are that the Mira descendants of metal-rich M giants have much shorter lifetimes than their globular cluster counterparts or that the late-M giants evolve into long-period Miras and that these have a much steeper density gradient than do short period ones. Better statistics for the Miras are necessary before progress can be made with this aspect of the problem.

In looking at the broad question of density gradients in the Bulge it is interesting to note that the RR Lyrae variables show an even steeper gradient than the late-M stars between $|b| \sim 3^\circ.9$ and $|b| \sim 8^\circ$ as can be seen from Fig 1 in Feast *et al.* (1992). Note that the point representing the RR Lyraes at $|b| \sim 3^\circ.9$ on this diagram has not been corrected for completeness and the true slope is therefore steeper than it appears. This is in contrast to the gradient of RR Lyraes at higher latitudes which is shallower than that of the M stars. In addition it is now known (Walker & Terndrup 1991, Wesselink 1987) that the RR Lyrae variables at $b \sim 3^\circ.9$ are more metal rich than those at higher latitudes. Therefore the metal-rich RR Lyraes (those with [Fe/H]~ -1) must have an extremely steep density gradient. This suggests that the interpretation of gradients in the Bulge may not be entirely trivial.

4.2. Structure of the Bulge

Mira variables have a considerable, but largely unrealized, potential for the study of Galactic structure. This is because the distances to individual stars can be determined by use of the PL or PLC relations. In practice of course determining periods and mean luminosities for large numbers of variables is a time consuming activity. Whitelock & Catchpole (1992) (see also Whitelock 1992) demonstrated the usefulness of this approach in examining the structure of the Bulge. They looked at the distribution of distance moduli for IRAS Miras from WFC, separating them into two groups depending on whether they were in the first ($l > 0$) or fourth ($l < 0$) Galactic quadrant. The distributions are illustrated in Fig 5a together with the best fitting model of a triaxial ellipsoidal Bulge. It is clear from this that the peak of the distribution at negative longitudes is further from us than the peak on the other side. Figure 5b shows the longitude distribution for the same sources and model. The model illustrated is of the form:

$$N \propto exp\left[-\left(\frac{x}{x_o}\right)^2 - \left(\frac{y}{y_o}\right)^2 - \left(\frac{z}{z_o}\right)^2\right]^{1/2}$$

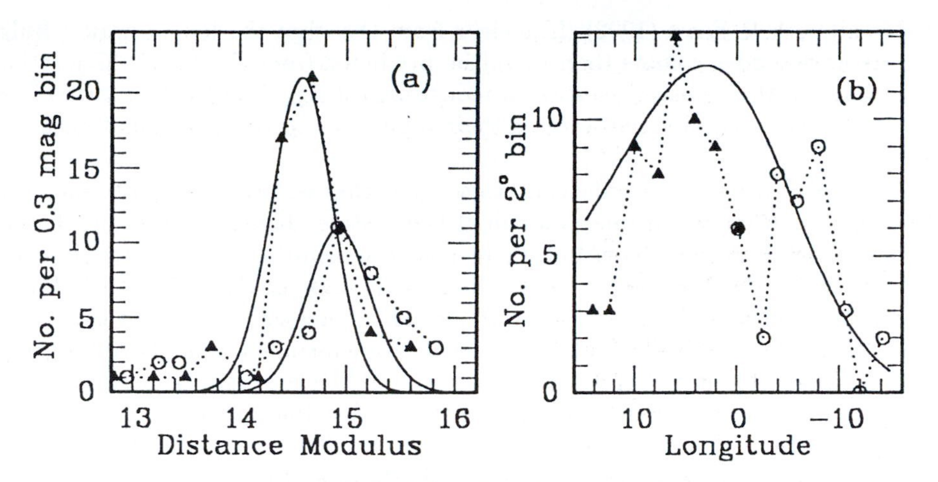

Fig. 5. (a) The distribution of IRAS Miras in distance modulus. The open circles and closed triangles represent those with negative and positive longitudes respectively. (b) The latitude distribution of the same sources. The solid lines represent one of the best fitting models, as discussed in the text.

and the best fitting scale lengths are x_0=700pc, $y_0 = z_0 - 175$pc and the x axis is inclined at an angle θ=45° to the line of sight. A rather large distance to the Galactic Centre, R_0=9.1 kpc, is required for this fit, because the distances to the Miras were obtained from the PL rather than the PLC relation, the latter not being applicable to Miras with thick dust shells. Because rather few Miras were used to define this model and because of possible systematic biases in the IRAS data-base, the exact parameters of the fit should not be over interpreted. In particular, if the differences in the distributions of Miras along the minor axis of the Bulge and in the line of sight, which were discussed in §4.1, are a consequence of Bulge triaxiality then the angle θ must be less than 45° (note that Binney *et al.* (1991) deduce $\theta = 16 \pm 2°$ on dynamical grounds). The result does however illustrate the potential of Mira variables as detailed probes of structure within the Bulge and suggests that a detailed study of a large number of stars could be very rewarding. The derived asymmetry for the Bulge is at least qualitatively consistent with other evidence for triaxiality as summarized by de Zeeuw and by Gerhard in these proceedings. Menzies (1990) measured radial velocities for a number of the visually brighter Miras from WFC. His results show the expected high velocity dispersion but also clearly illustrate the rotation of the Bulge.

4.3. The Period-luminosity-colour Relation

A preliminary analysis of multi-phase IR observations of Miras in the Sgr I window ($l = 1°.4, b = -2°.6$) suggests that these stars diverge somewhat from the LMC PL and period-colour (PC) relations (Glass, Whitelock, Feast & Catchpole in prep.). A comparison of the (J-K) colours of the Bulge and LMC Miras is shown in Fig

1 of Whitelock & Feast (1992). It is clear from this that the longer period Bulge sources have redder colours than would be predicted from the LMC PC relation. Note that the Miras with circumstellar reddening ($(J\text{-}K)_o \gtrsim 4.0$) were omitted from consideration in the first instance as their photospheric colours cannot easily be deduced.

The Galactic Centre is sufficiently close to us that we observe a finite spread in the line of sight for the distance moduli of Bulge stars. This spread means that we cannot assume that any given Bulge star is at the distance of the centre. In order to compare the luminosities of Bulge Miras with those in the LMC the following procedure was adopted. The LMC PL relations, expressed in terms of bolometric and K (2.2 μm) magnitude (Feast *et al.* 1989), were assumed to apply to the Bulge Miras and used to derive individual distance moduli. The resulting distributions of distance moduli can be fitted with models for the distribution of stars in the Bulge from which the distance to the centre can be derived. The result is R_o=9.1 kpc from the bolometric PL and R_o=8.6 kpc from the K PL (note that although the absolute error on these distances is at least $\pm$0.5 kpc, the relative error is only $\pm$0.1 kpc). These values not only disagree, they are significantly further than the currently accepted distance of $R_o = 7.8 \pm 0.8$ kpc (Feast 1987). This strongly suggests that the Bulge Miras are fainter than we would expect from the LMC PL relations. However, if the distance moduli are derived via the LMC PLC relations then both of the resulting distributions can be fitted by models with $R_o = 8.2$ kpc. Further work is in progress but this seems to confirm the existence of the PLC relation and imply that the Bulge Miras, or at least the longer period ones, occupy a different part of the instability strip from the LMC Miras, probably due to metallicity effects (see Wood *et al.* 1991).

The implications of this are quite significant (see also §4.4). It is clear that in principle the PLC relation should be used in preference to the PL relation for determining distances to Miras. However, in practice the PL relation seems quite adequate for most environments (see Fig 1). It is probable that difficulties with the PL relation will only manifest themselves in very metal-rich environments. The distances to the long-period (P>600 day) Galactic OH/IR stars are rather poorly known (van Langevelde *et al.* 1990) but on average their luminosities fall below an extrapolation of the LMC PL relation (WFC), as do those of the Bulge Miras. This is in contrast to the LMC OH/IR sources (Wood *et al.* 1991) which lie on an extrapolation of the Mira PL relation. It is possible therefore that some of the Galactic OH/IR sources may be less luminous than their LMC counterparts. In view of this the procedure used by WFC to derive a Mira PL relation, by combining data on LMC Miras and Galactic OH/IR sources (their equation 2), may not be strictly valid. However, much more accurate distances are required for the Galactic OH/IR sources before firm conclusions can be drawn on this point.

Blommaert *et al.* (1992) have made a detailed study of a few of the OH/IR stars within 30 arcmin of the Galactic Centre, and pointed out that they are fainter than would be predicted from the LMC PL relation (see their Fig 3). These are particularly valuable observations because the concentration of stars near the centre is such that it is possible to assume that almost all the sources are at the distance of the centre. There are distinct difficulties in doing photometry of these stars due

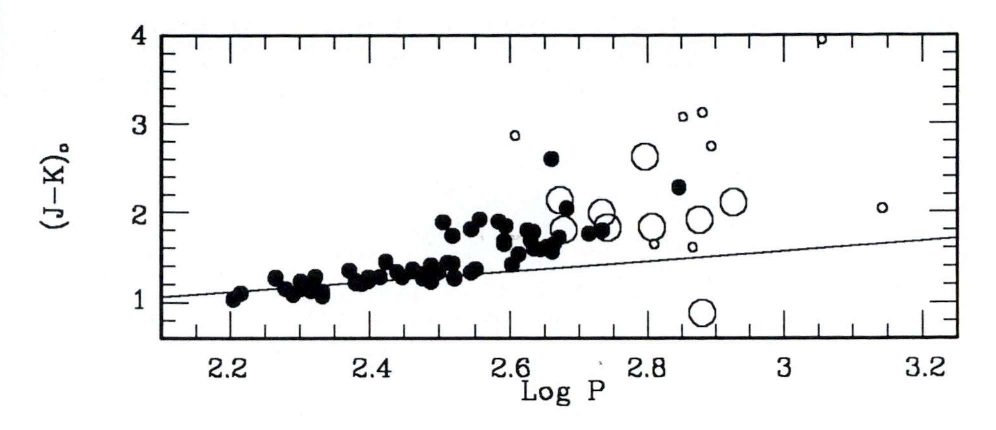

Fig. 6. A PC diagram comparing the observed reddening-corrected colours of the Sgr I Miras (closed circles, Glass *et al.* in prep.) with those calculated using the LMC PLC relation for the Galactic Centre OH/IR sources (large open circles for those with accurate periods and luminosities, small circles for those with uncertain periods and/or luminosities) from Blommaert *et al.* (1992). The line is the PC relation for LMC Miras from Feast *et al.* (1989).

to the very crowded nature of the fields and to the high, but uncertain, interstellar extinction corrections. Unfortunately it is not possible to apply the PLC relations directly to these sources as they have thick dust shells so that their (J-K) colours are strongly influenced by circumstellar extinction. However, we can reverse the problem and assume that they are at the distance of the Galactic Centre ($R_o = 8.0$ kpc) and use the PLC relation to calculate their photospheric (J-K). This has been done and in Fig 6 the result is compared to the measured colours of the Sgr I Miras. Those OH/IR sources with accurate luminosities and periods follow the trend established by the slightly shorter period Sgr I Miras (with one exception which might be closer than the centre, or in a different evolutionary state from the others). This suggests that these OH/IR sources might also follow the same PLC relation as the LMC Miras.

4.4. Numbers and Luminosities of Miras in the Bulge

It is possible to combine the data from the WFC survey with the estimate of the total number of Miras in the 3.8 square degree overlap with the Baade/Plaut field (described at the beginning of §4.1) to calculate the total number of Miras in the Bulge, bearing in mind that the survey was probably incomplete for Miras with periods in the 300 to 400 day range. There were 95 Miras in 3.8 square degrees seven of which were in the WFC survey which had a total of 113 high-mass-loss Miras from two one-degree wide strips across the Bulge. Therefore these two strips will contain a total of approximately $(113/7) \times 95$ (–15% foreground) = 1300 Miras of all kinds. Assuming that the density distribution of Miras in the Bulge falls off as

R^{-2} then the total number of Miras within a radius of 2 kpc from the centre must be at least 2×10^4. If the Mira population is actually more centrally concentrated than these assumptions imply then the total number will be larger.

The lifetime of a Bulge Mira may be estimated via Renzini & Buzzoni's (1986) fuel consumption theorem, $N_j = B(t)L_T t_j$. This relates the number of objects, N_j, in a specific evolutionary phase with the duration of that phase, t_j, via the total luminosity of the population, L_T, and the stellar death rate per unit luminosity, $B(t)$. Assuming $B(t) \sim 2 \times 10^{-11}\ L_{\odot} yr^{-1}$ and $L_T \sim 10^{10}\ L_{\odot}$ (Renzini & Greggio 1990) we find a Mira lifetime of at least 10^5 yr. This can be compared with other estimates: Renzini & Greggio found 2.5×10^5 yr for the globular cluster Miras using the same expression – these are short period objects (P< 300day); Jura & Kleinmann (1992) have recently estimated 2×10^5 yr for Miras with intermediate periods (300-400 day) in the solar neighbourhood. Although the Bulge Miras show a large range of periods, from 100 to perhaps 800 day, the bulk of them have periods less than 400 day.

There is a good deal of uncertainty surrounding the luminosity of the brightest stars in the Galactic and other bulges (see e.g. the contributions to this meeting by Rich and by Renzini). The most luminous stars are assumed to be representative of the last significant stage of star formation, although it is not entirely clear how chemical composition affects the chronology. Provided an old or intermediate age population is sufficiently large that it includes stars in the short lived Mira phase then these stars will be the most luminous of that population. In view of the above findings it is clear that luminosities of Bulge Miras which were determined via the PL relation will have been overestimated. Extrapolating from WFC and more recent work in the $b = -7^{\circ}.5$ field it is possible to estimate that the Bulge will contain very roughly 100 stars with luminosities greater than $M_{bol} = -5.0$ mag. As already stated these may be the result of binary mergers rather than the evolution of stars with $M_i > 1.2 M_{\odot}$. However this estimate does not take into account the luminous stars which have a highly flattened distribution around the Galactic Centre (Catchpole *et al.* 1990, Feast 1989 and Whitford these proceedings) and must therefore be regarded as a lower limit. Evidence for luminous stars near the Galactic Centre and in the bulges of other galaxies is discussed in Rich's contribution to these proceedings.

Whitelock & Feast (1992) compare the relative numbers and ages of Miras and planetary nebulae in the Bulge. They conclude that Miras outnumber planetary nebulae by about 10 to 1, although there is a high level of uncertainty in the assumptions made about both types of object. They also conclude that there are no compelling reasons to doubt that most, if not all, Miras evolve into planetary nebulae. However, the large number of planetary nebulae that seem to have evolved from binary systems (e.g. Livio 1992) may cast doubt on the idea that most planetary nebulae have evolved through a Mira phase.

5. The Bulge Carbon-Stars

Azzopardi *et al.* (1991) have published coordinates and finding charts for the carbon stars in the Bulge. Their astrophysical properties have been discussed by Tyson

& Rich (1991) and Westerlund *et al.* (1991) and, except where noted, it is upon these two papers that the following summary is based. One of the most notable characteristics of these Bulge carbon stars is their low luminosity ($M_{bol} \gtrsim -2.5$ mag). They therefore cannot be in a similar evolutionary state to the stars discussed above. They could only be near the end of their AGB evolution if they are going to evolve off the AGB prematurely, to become post early-AGB stars, but according to Greggio & Renzini (1990) such stars will be nitrogen-rich and carbon-deficient. They are certainly not near the top of any normal AGB, where the Miras are, nor can they be on the thermal pulsing part of the AGB. That being the case their carbon enrichment cannot result from dredge-up following a thermal pulse, as is thought to occur for the luminous carbon stars found in the Magellanic Clouds and elsewhere. The Bulge carbon stars are very rare, being outnumbered several hundred to one by M stars, so they must represent either a short-lived phenomenon or one which occurs only in a small fraction of the population, or both. These stars are presumed to be metal-rich because of their strong absorbtion features, in particular Na D and CN bands.

Azzopardi *et al.* (1988) suggested the Bulge carbon stars might have become enriched in carbon through binary mass-transfer, presumably by analogy with the barium and CH stars. However, the discovery that many of the Bulge carbon stars have enhanced ^{13}C but no s-process enhancement led both Tyson & Rich and Westerlund *et al.* to reject the binary hypothesis and to suggest instead that the Bulge stars may be closely related to the R-type carbon stars. These are not generally thought to be binaries and their carbon enrichment is as yet unexplained although much of the literature favours a suggestion made by Dominy (1984) that the enrichment could be associated with mixing during the core-He flash.

It seems to me that recent work on dwarf carbon stars is highly relevant to this problem. There are four known dwarf carbon stars (Green *et al.* 1991). Limited spectroscopic studies have shown that at least two and possibly three of these have enhanced ^{13}C without any obvious overabundance of the s-process elements (Lloyd Evans 1992 & refs therein). If these stars are indeed dwarfs then any explanation for the anomalous abundances involving the core-He flash is out of the question and binary interaction would seem to offer the only viable solution. It therefore seems reasonable to suggest that both the dwarf- and the Bulge carbon stars must originate from binary interactions. More work is required to confirm the apparent similarity of the abundance peculiarities found in the dwarfs and the Bulge carbon stars.

Acknowledgements

I am grateful to Michael Feast and John Menzies for helpful discussion. I would also like to thank Robin Catchpole, Michael Feast and Ian Glass for permission to quote work in preparation.

References

Azzopardi, M., Lequeux, J. & Rebeirot, E.: 1988, *Astron. Astrophys.*, **202**, L27

Azzopardi, M., Lequeux, J., Rebeirot, E. & Westerlund, B. E.: 1991, *Astron. Astrophys. Suppl.*, **88**, 265
Binney, J., Gerhard, O. E., Stark, A. A., Bally, J. & Uchida, K. I.: 1991, *Mon. Not. R. astr. Soc.*, **252**, 210
Blanco, V. M.: 1988, *Astronom. J.*, **95**, 1400
Blanco, V. M. & Terndrup, D. M.: 1989, *Astronom. J.*, **98**, 843
Blommaert, J. A. D. L.: 1992, *PhD thesis*, Rijksuniversiteit te Leiden
Blommaert, J. A. D. L., van Langevelde, H. J., Habing, H. J., van der Veen, W. E. C. J. & Epchtein, N.: 1992, in *Variable Stars and Galaxies*, ASP conf. ser. 30, ed., B. Warner, in press
Catchpole, R. M., Robertson, B. S. C., Lloyd Evans, T. H. H., Feast, M. W., Glass, I. S. & Carter, B. S.: 1979, *SAAO circ.*, **1**, 61
Catchpole, R. M., Whitelock, P. A. & Glass, I. S.: 1990, *Mon. Not. R. astr. Soc.*, **247**, 479
Dickens, R. J.: 1989, in *The Use of Pulsating Stars in Fundamental Problems of Astronomy*, IAU Coll 111, ed., E. G. Schmidt, Cambridge Univ. Press, p. 141
Dominy, J. F.: 1984, *Astrophys. J. Suppl.*, **55**, 27
Feast, M. W.: 1981, in *Physical Processes in Red Giants*, eds., I. Iben & A. Renzini, Reidel, p. 193
Feast, M. W.: 1984, *Mon. Not. R. astr. Soc.*, **211**, 51P
Feast, M. W.: 1986, in *Light on Dark Matter*, ed., F. P. Israel, Reidel, p. 339
Feast, M. W.: 1987, in *The Galaxy*, eds., G. Gilmore & B. Carswell, Reidel, p. 1
Feast, M. W.: 1989, in *The Use of Pulsating Stars in Fundamental Problems of Astronomy*, IAU Coll 111, ed., E. G. Schmidt, Cambridge Univ. Press, p. 205
Feast, M. W. & Whitelock, P. A.: 1987, in *Late Stages of Stellar Evolution*, eds., S. Kwok & S. R. Pottasch, Reidel, p. 33
Feast, M. W., Whitelock, P. A. & Sharples, R.: 1992, in *The Stellar Populations of Galaxies*, IAU Sym. 149, eds., B. Barbuy & A. Renzini, Kluwer, p. 77
Feast, M. W., Glass, I. S., Whitelock, P. A. & Catchpole, R. M.: 1989, *Mon. Not. R. astr. Soc.*, **241**, 375
Frogel, J. A., Terndrup, D., Blanco, V. M. & Whitford, A. E.: 1990, *Astrophys. J.*, **353**, 494
Gatewood, G.: 1992, *Publ. Astronom. Soc. Pacif.*, **104**, 23
Glass, I. S.: 1986, *Mon. Not. R. astr. Soc.*, **221**, 879
Green, P. J., Margon, B. & MacConnell, D. J.: 1991, *Astrophys. J.*, **380**, L31
Greggio, L. & Renzini, A.: 1990, *Astrophys. J.*, **364**, 35
Grenon, M.: 1990, in *Bulges of Galaxies*, ESO-CTIO Workshop, eds., B. J. Jarvis & D. M. Terndrup, ESO, p. 143
Horch, E., Demarque, P. & Pinsonneault, M.: 1992, *Astrophys. J.*, **388**, L53
IRAS Point Source Catalogue: 1985, US Government Publication Office
Jura, M. & Kleinmann, S G.: 1992, *Astrophys. J. Suppl.*, **79**, 105
Kent, S. M., Dame, T. M. & Fazio, G.: 1991, *Astrophys. J.*, **378**, 131
Livio, M.: 1992, in *Planetary Nebulae*, IAU Sym. 155, eds., R. Weinberger & A. Acker, Kluwer, in press
Lloyd Evans, T.: 1976, *Mon. Not. R. astr. Soc.*, **174**, 169
Lloyd Evans, T.: 1992, *Mon. Not. R. astr. Soc.*, submitted
Menzies, J. W.: 1990, in *Bulges of Galaxies*, ESO-CTIO Workshop, eds., B. J. Jarvis & D. M. Terndrup, ESO, p. 115
Menzies, J. W. & Whitelock, P. A.: 1985, *Mon. Not. R. astr. Soc.*, **212**, 783
Plaut, L.: 1971, *Astron. Astrophys. Suppl.*, **4**, 75
Renzini, A. & Buzzoni, A.: 1986, in *Spectral Evolution of Galaxies*, eds., C. Chiosi & A. Renzini, Reidel, p. 195
Renzini, A. & Greggio, L.: 1990, in *Bulges of Galaxies*, ESO-CTIO Workshop, eds., B. J. Jarvis & D. M. Terndrup, ESO, p. 47
Rich, R. M.: 1988, *Astronom. J.*, **95**, 828
Robertson, B. S. C. & Feast, M. W.: 1981, *Mon. Not. R. astr. Soc.*, **196**, 111
te Lintel Hekkert, P.: 1990, *PhD thesis*, Rijksuniversiteit te Leiden
Terndrup, D. M.: 1988, *Astronom. J.*, **96**, 884
Terzan, A. & Ounnas, Ch.: 1988, *Astron. Astrophys. Suppl.*, **76**, 205
Tyson, N. D.: 1992, in *Variable Stars and Galaxies*, ASP conf. ser. 30, ed., B. Warner, in press
Tyson, N. D. & Rich, R. M.: 1991, *Astrophys. J.*, **367**, 547

van den Bergh, S. & Herbst, E.: 1974, *Astronom. J.*, **79**, 603
van Langevelde, H. J., van der Heiden, R., van Schooneveld, C.: 1990, *Astron. Astrophys.*, **239**, 193
Walker, A. R.: 1992, *Astrophys. J.*, **390**, L81
Walker, A. R, & Terndrup, D. M.: 1991, *Astrophys. J.*, **378**, 119.
Wesselink, Th.: 1987, PhD thesis, Nijmegen
Westerlund, B. E., Lequeux, J., Azzopardi, M. & Rebeirot, E.: 1991, *Astron. Astrophys.*, **244**, 367
Whitelock, P. A.: 1990, in *Confrontation Between Stellar Pulsation and Evolution*, eds., C. Cacciari & G. Clementini, ASP Conf. Ser. 11, p. 365
Whitelock, P. A.: 1992, in *Variable Stars and Galaxies*, ASP conf. ser. 30, ed., B. Warner, in press
Whitelock, P. A. & Catchpole, R. M.: 1992, in *The Center, Bulge and Disk of the Milky Way*, Ed L. Blitz, Kluwer, in press
Whitelock, P. A. & Feast, P. A.: 1992, in *Planetary Nebulae*, IAU Sym. 155, eds., R. Weinberger & A. Acker, Kluwer, in press
Whitelock, P. A., Feast, M. W. & Catchpole, R. M.: 1986, *Mon. Not. R. astr. Soc.*, **221**, 1
Whitelock, P. A., Feast, M. W. & Catchpole, R. M.: 1991, *Mon. Not. R. astr. Soc.*, **248**, 276, (WFC)
Wood, P. R., Moore, G. K. G. & Hughes, S. M. G.: 1991, in *The Magellanic Clouds, IAU Sym. 148*, ed(s)., R. Haynes & D. Milne, Kluwer, 259

DISCUSSION

Terndrup: With reference to the RR-Lyraes, Walker and I have indeed shown that the RR-Lyraes are formed only by the most extreme metal poor tail of the K-giants. Have you looked into whether the presence of somewhat younger stars, which would also produce more RR-Lyraes in the instability strip, account for that steeper gradient in the RR-Lyra density distribution.

Whitelock: No I haven't looked into that at all. But obviously you have to look at the total picture. If you need more massive stars to explain the RR-Lyraes, then you will, presumably, end up with more massive stars becoming M-giants and Miras.

Blommaert: The Miras of 200 to 300 days period, in the Palomar-Groningen field number 3 have, I believe, mass losses of $10^{-6} M_{\odot}$/yr and the life time would be around 10^5yr you say, so that would mean that they lose $10^{-1} M_{\odot}$ in total?

Whitelock: I don't think there is any reason to suspect that the mass loss remains constant through the lifetime of a Mira. I think that there is every reason to believe that the mass-loss rate must increase during the Mira evolution. That is consistent with the fact that you find a very large range in mass loss at a given period for the Miras.

Blommaert: But these IRAS sources can be interpreted as an evolutionary stage after the Miras, because they have larger mass losses and higher mass loss rates.

Whitelock: We do have objects in different evolutionary stages and it is true that the ones in the most advanced evolutionary stage you expect to have the highest mass-loss rate. But this is complicated by the fact that they are thermally pulsing (i.e. undergoing helium shell flashes) and the mass-loss rate will fluctuate with the pulses. But then, we've also got a range of initial masses and the evidence, as far as the kinematics of the solar neighborhood Miras are concerned, is that Miras of different period must have different initial masses because they have quite different kinematics. Although we are probably not dealing with a large range of initial masses in the Bulge, perhaps just a few tenths of a solar mass.

Tyson: If you do appeal to binarity to explain the bulge carbon stars, then you change the focus of the problem, because you then upset what I had always interpreted to be very well understood evolutionary behavior of the secondary star that would be transferring matter.

Whitelock: I won't argue with that statement, but we also have to explain these dwarf carbon stars. They seem to have the same characteristics and unless our complete understanding of stellar evolution is wrong, then you can't get dwarf carbon stars unless they are in binary systems.

During a coffee break

Patricia WHITELOCK

Donald TERNDRUP

OH/IR STARS AS TRACERS OF GALACTIC POPULATIONS

H.J.HABING
Sterrewacht Leiden

Abstract. I discuss observable quantities of OH/IR stars and their interpretation. The strong 1612 MHz maser is found in about 1/2 of all the stars that reach the AGB and then have the following properties: they are oxygen–rich, of high–luminosity $(3,000 \rightarrow 30,000 L_{\odot})$, vary in luminosity with a long period (P>500 d), lose mass rapidly $(\dot{M} > 10^{-6} M_{\odot}/\mathrm{yr})$ and have a metallicity high enough to form sufficient amounts of dust (the metallicity of the LMC is probably the lowest value sufficient to produce OH/IR stars). OH/IR stars can be several Gyr old. There are at least three different populations each characterized by different galactic orbits. (1) OH/IR stars in the galactic disk on almost circular orbits; (2) OH/IR stars in the galactic bulge on orbits of lower angular momentum; (3) OH/IR stars within 150 pc from the galactic center forming a rapidly rotating system.

Key words: OH/IR stars

1. The transparancy of interstellar dust

In Figure 1 I show an ESO photograph of an area of 60 degrees by 40 degrees around the center of our Galaxy- it contains most of the galactic bulge. The image is made up by stars and by interstellar matter and it is *not* nicely ellipsoidal. Clearly the picture is affected by thc irregular distribution of the interstellar matter causing an intermingling of dark and bright areas. Drawn onto this figure are the many small regions where in recent times different observers have studied the stellar population of the bulge. The rectangle at the center represents the area covered by an infrared study by Catchpole et al. (1990; see Glass' review in these proceedings). Look at the distribution of these small regions: it shows the difficulty of penetrating, at visual wavelengths, the galactic center, that "heart of darkness". Studies of the stellar population in small windows as Sgr I and Baade's window are invaluable; yet there are too few windows to give us a convincing picture of how the properties of the stellar population vary over the bulge- interpolation bctween the various windows leaves too much uncertainty.

Dust absorption hides the inner Galaxy at all UV, visual and at the shortest infrared wavelengths; for wavelengths longer than a few microns the galactic plane is essentially transparent (also for hard X-radiation; much of what I will say is therefore also valid for X-ray sources). If we want to study the stellar population *everywhere* in the galactic bulge we have to study it where we can see the stars: in the infrared, at radio wavelengths, in X- rays. What populations of stars are there available?

2. OH/IR stars : observable quantities, results, interpretation

In this review I discuss stars that emit practically all their energy in the middle-infrared, and that have a remarkable, strong maser line of hydroxyl, OH, at 1612 MHz or 18 cm: the OH/IR stars. In addition the stars can be detected in a few other maser lines and in thermally excited molecular lines (especially CO and SiO).

H. Dejonghe and H. J. Habing (eds.), Galactic Bulges, 57–70.

Fig. 1. An ESO photograph (The Messenger, 1986, 46, 14-15) of the galactic bulge. The axes show the galactic coordinate system (l,b), the outermost tickmarks indicate 5°. The dashed square indicates the Palomar Groningen Field 3, and two of Baade's windows are indicated by filled circles. Terndrup's (1988) deep CCD photometry fields are indicated by asterisks; the open circles are Blanco and Terndrup's (1989) survey fields for late-type giants. The solid rectangle at the center is the field studied by Catchpole et al. (1990). (Courtesy: ESO and J. Blommaert)

I will discuss (1) stellar parameters that can be measured, (2) the nature of the stars (structure, luminosity) and (3) the question what galactic populations they represent (or equivalently: age and atomic composition). Readers who want a more elaborate introduction are referred to the proceedings of conferences dedicated to these and similar stars (e.g. Mennessier and Omont, 1990). In the next review Dejonghe discusses kinematical and dynamical conclusions about the bulge that are based on the observations of OH/IR stars.

2.1. Observable quantities

In Figure 2a I show the (infrared) spectrum of one of the brightest OH/IR stars, in figure 2b its 1612 MHz spectrum and in figure 2c the CO(2-1) emission line. The infrared spectrum (figure 2a) is explained very well by spherically symmetric outflow of gas and dust around a late type giant: an elaborate modelling technique has been developped over the last 15 years (e.g. David and Papoular, 1992; Justtanont and Tielens, 1992). Figure 3 shows a collection of model fits to spectra for some M-type Miras and OH/IR stars. The models form a one dimensional series, with the optical depth in the silicate absorption band at 9.7μm increasing from the upper left to the lower right plot (Bedijn, 1987). From the observed spectrum one thus derives the value of the optical depth of the envelope, and m_{bol}, the apparent bolometric magnitude, by integration over all wavelengths (or $L_*/4\pi d^2$, where L_* is the stellar luminosity and d the distance).

The models just mentioned give a good representation of the spectra as observed. However, the calculations are based on an assumed density distribution and this distribution is not necessarily in agreement with the dynamical effects of radiation pressure: the photons transfer momentum to the dust particles and this momentum is transmitted by friction very efficiently to the gas. Recently Tignon, Tielens and I have made model calculations that simultaneously take into account the radiative transfer and the radiation pressure. Our specific aim was to study how the outflow velocity, v_{out} depends on parameters such as the luminosity, L_*, the mass loss rate, $\dot{M}$, and the dust-to- gas ratio, δ. Our main conclusion is that $v_{out} \propto L_*^{\frac{1}{3}} \delta^{\frac{1}{2}}$ and that v_{out} is affected by $\dot{M}$ only when $\dot{M} < 10^{-6}\ M_\odot/$ yr. This result agrees with the systematic properties measured for v_{out} (see below). It then follows that an estimate of δ may be obtained once v_{out} and L_* have been measured. In oxygen–rich stars the dust consists of silicates and silicium is not created by fusion inside AGB stars, thus the value of δ is likely to represent the initial abundance of silicium. δ is a measure of the metallicity of the OH/IR star.
Here it is necessary to point out a persistent misunderstanding, or a plain error, that continues to appear in the litterature: The outflow of matter represents a flow of momentum equal to $\dot{M} v_{out}$; the radiation emitted by the star carries with it a momentum flow equal to L_*/c; let β be the ratio between the two. Now it is often stated that $\beta \leq 1$: the momentum flow of the photons has to exceed that of the gas "because the shell is radiation-driven". This conclusion is wrong: if the photons get reflected often enough β might reach the value ∞! That in circumstellar shells β may easily be larger than 1 has been argued in detail, and convincingly, by Gail and Sedlmayer (1986) and Lefèvre (1989), but statements to the contrary can be found in publications as recent as 1992.

The maser line profile in figure 2b is quite characteristic for 1612 MHz OH/IR stars: two narrow peaks separated by many times the peak width; a sharp increase in flux at the outside of the peaks, but on the inner side a much more gentle slope. The line shape is quite well explained, qualitatively and quantitatively (Spaans and van Langevelde, 1992) by a spherical shell of hydroxyl molecules that are radiatively pumped by infrared emission from the center. Each of the two peaks defines a Doppler shifted velocity; the average of these two velocities is the stellar radial velocity, v_*; the difference equals twice the expansion velocity of the shell, v_{out}. The total line flux is another parameter that can be well measured. Theory and observations lead to the insight (see e.g. Elitzur, 1992) that the 1612 MHz maser is pumped by OH line photons at 35 and at 53 μm, and it is predicted that the number of maser photons is a certain fraction of the number of pump photons. In recent years it has been shown that this is indeed the case; the efficiency of the conversion of pump into maser photons is a strong function of the (dust) optical

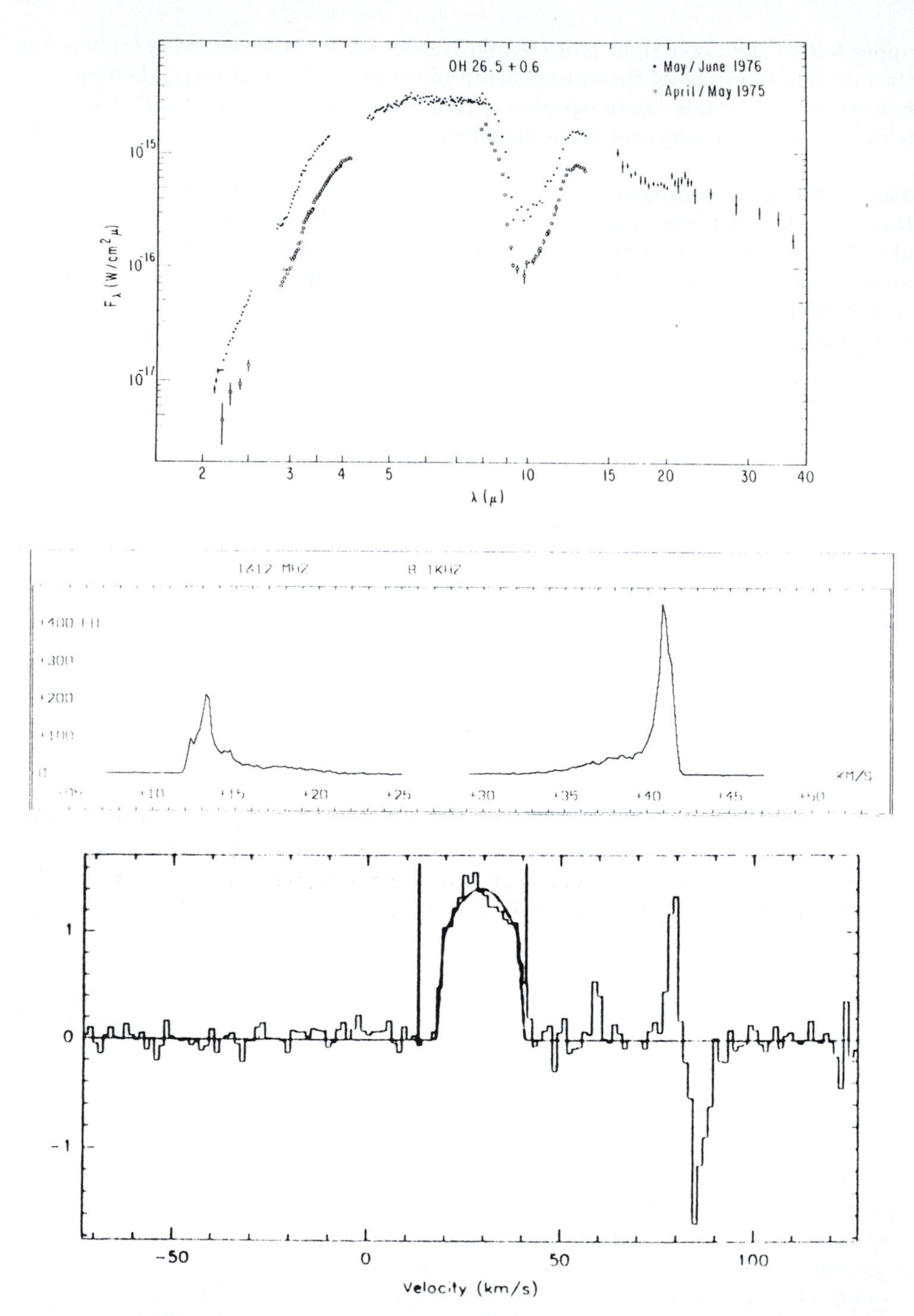

Fig. 2. The OH/IR star OH26.5+0.6 or AFGL 2205. Figure 2a gives the infrared spectrum (Forrest et al., 1978); Figure 2b the 1612 MHz OH maser line (Andersson et al., 1974) and Figure 2c the CO (2 → 1) line (Heske et al., 1990).

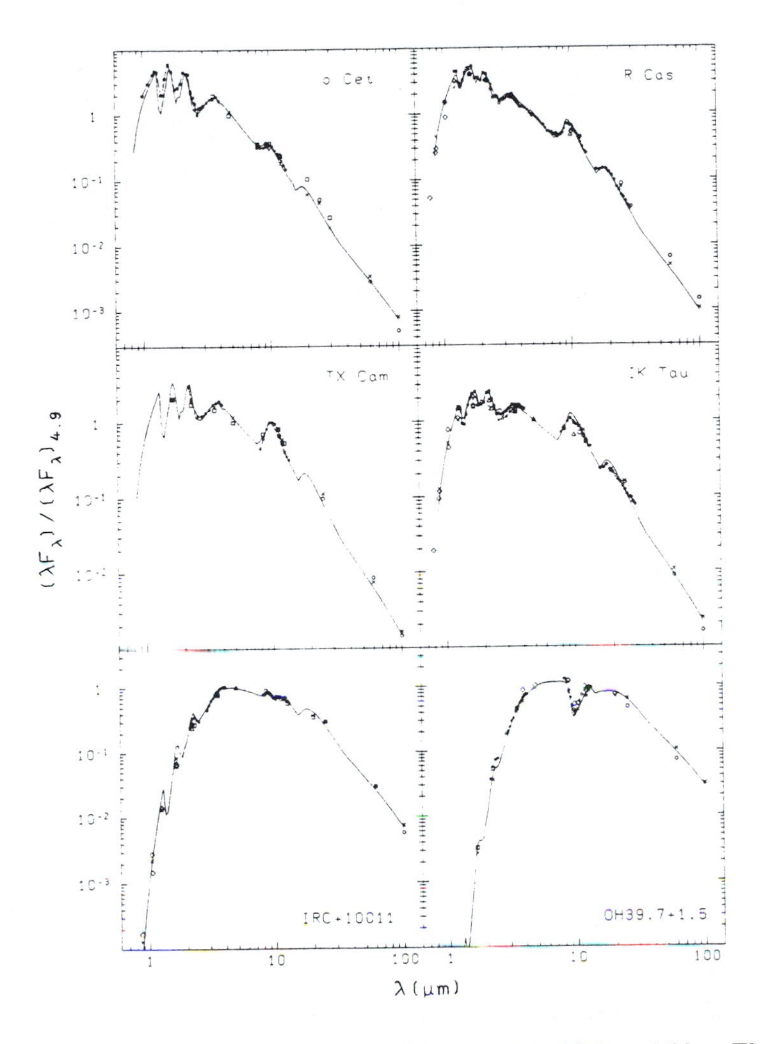

Fig. 3. Model fits to the spectra of M-type Miras and OH/IR variables. The models have optical depth in the silicate band ranging from 0.04 for the upper left to 10.0 for the lower right (Bedijn, 1987). Crosses denote calculated broad band photometric fluxes; other symbols observed photometric fluxes.

depth (Dickinson, 1991; Chengalur et al., 1992; Le Squeren et al., 1992)— and thus of the mass–loss rate. The ratio between the OH 1612 MHz maser flux and the mid–infrared flux is thus a measure of the mass–loss rate. Also the variation of the maser flux with time is interesting: this variation has a very long period and it is *con sono* with the infrared radiation. While it varies in intensity the line profile retains its shape, almost perfectly, but not completely: the blue shifted peak reaches maximum a few weeks (i.e. a few percent of the total period) before the red shifted peak. This phase difference between the light curves of the two peaks is explained by a light travel time effect and the value of the difference measures the *linear* diameter of the

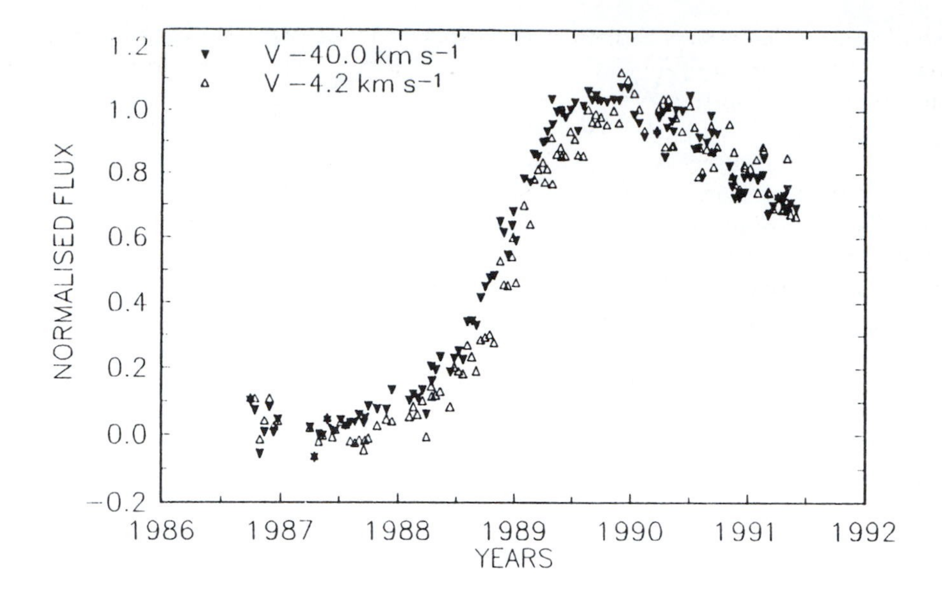

Fig. 4. Normalised light curves of blue-shifted (V=-40.0 km/s) and red-shifted (V=-4.2 km/s) points in the 1612-MHz maser spectrum of OH357.3-1.3= RAFGL 5379 (West et al., 1992).

masering shell (typically a few lightweeks). By also measuring its *angular* diameter via interferometric techniques, one finds directly the stellar distance by a geometrical method. To apply the method much patience and a high accuracy is required; nevertheless distances have been measured to several OH/IR stars. In figure 4 one recognizes easily the phase differences between the red and blue lightcurves of a very strong, southern OH/IR star at 1.4 ± 0.1 kpc (West et al., 1992). An attempt to measure the distance to OH/IR stars at the galactic center and thus to the center itself has failed, because unexpectedly strong inhomogeneities in the free electron distribution scatter the 18 cm radio waves and smear out the images of the OH/IR masers (van Langevelde et al., 1992). Nevertheless, one should continue to praise the very precise techniques of radio astronomy; there are still promises for the near future: Almost all OH/IR stars produce also SiO masers at mm wavelengths; within a few years VLBI will become operational at mm wavelengths and then one can measure the SiO maser positions with an accuracy of a few tens of μarcsec. This will allow the determination of proper motions of the masers even at the distance of the galactic center (50 km/s at 8,000 pc corresponds to 1.25 marcsec /year). Not only will such VLBI measurements give us three dimensional velocity information, but also will the statistical comparison of radial velocities with the proper motions give us a new, independent distance determination to this group of stars, that is: to the galactic center.

Figure 2c shows an observed emission line from thermally excited CO-molecules.

Several other lines of this molecule have been detected too; the same is true for several other molecules (e.g. SiO). The lines of CO are formed at large distances from the star (10^{17} cm) whereas the SiO lines are formed much deeper inward (between 10^{14} and 10^{16} cm). The parabolic shape of the line profile is typical for dense circumstellar, expanding envelopes. From the line profile one measures the stellar velocity, the expansion velocity and, from the integrated line flux and only if the stellar distance is known, the total number of molecules in the envelope. Model calculations are essential for the interpretation. Recently the models for the CO-lines have been greatly improved (at the cost of increased complexity) by calculating the excitation temperature of the molecules in a self consistent manner (Sahai, 1990; Kastner, 1992). Large systematic samples of observations of circumstellar CO have been published by Margulis et al. (1990) and by Nyman et al. (1992).
Finally: spectrographs in the ISO satellite, to be launched in late 1995, may detect various absorption bands (e.g. vibrational SiO lines) that originate in the stellar atmosphere and not in the envelope; this will give a first glimpse of the star in the center of the circumstellar envelopes.

2.2. Results from measurements

2.2.1. Source lists

OH/IR stars are strong masers; the first 500 OH/IR stars have been detected by searching in an unbiased way in the 1612 MHz line; for a compilation of these early detections see te Lintel Hekkert et al. (1989). This search technique was slow. But OH/IR stars are also strong mid–infrared sources: they are powered by giant stars and all radiative energy is emitted in the mid-infrared. A succesful search technique is to select IRAS point sources and to observe these in the 1612 MHz line. Over 1000 stars have thus been found through surveys with the Parkes and Effelsberg telescopes (te Lintel Hekkert et al., 1991); with the Arecibo telescope (Eder et al. 1988, Lewis et al., 1990, Chengular et al., 1993) and with the Nançay telescope (Le Squeren et al., 1992). In all these surveys the detection probablity never rose about 1/2; why it does not equal 1 is an unexplained fact at the moment (see e.g. Gaylard et al., 1989, and Lewis, 1992). IRAS was confused in the galactic plane, especially in the inner Galaxy, and there one expects the OH/IR source density to be highest. Thus the one thousand stars detected in the IRAS directed surveys form an essentially incomplete sample; even worse, the sample is incomplete by an unknown factor. Recently two new unbiased and ambitious 1612 MHz line surveys have been started in the galactic plane using the VLA to the north and the AT to the south of the galactic center. In total there are now over 1500 OH/IR stars known.

2.2.2. Mass loss rates, $\dot{M}$

From the observed outflow velocities and from the dust and gas density in the circumstellar envelope one obtains an estimate of the mass outflow rate— see e.g. some recent determinations by Schutte and Tielens (1989), Sahai (1990), Justtannont and Tielens (1992), and Kastner (1992). Values are found between 10^{-7} and

10^{-4} $M_\odot$/yr. In a small but significant fraction of the stars the circumstellar shell is detached and at some distance from the star— suggesting strongly that the mass loss process has stopped abruptly less than 10^5 yr ago. The first examples of detached shells were recognized in carbon–rich stars (Willems and de Jong, 1988; Olofsson et al., 1990), but recently it has been shown that also oxygen–rich stars show detached shells (Zijlstra et al. 1992).

2.2.3. Outflow velocity, $v_{\rm out}$

Analysis of the 1612 MHz line profile in some 800 OH/IR stars shows a range of velocities between 10 and 20 km/s (te Lintel Hekkert, 1990). CO- line observations give the same values; a few, possibly interesting discrepancies have been reported. The range of outflow values is thus small, and yet it appears to be significant whether the velocity is at the low end of the range or at the high end. Evidence derived from OH/IR stars in the Magellanic Clouds (Wood et al., 1992), in the galactic anticenter (Blommaert et al., 1993), close to the galactic center (Lindqvist et al., 1992), all indicate that $v_{\rm out}$ is higher when the luminosity is higher and/or when the metallicity is higher. These systematic trends can be understood if the circumstellar shell is accelerated by radiation pressure (see the discussion in section 2.1). In less reddened, related objects with 1612 MHz maser emission (Mira variables) the outflow velocities range between a few and 10 km/s and are well correlated with the stellar period (Sivagnanam et al. 1989). The correlation disappears for periods over 500 days, that is: for OH/IR stars.

2.2.4. Luminosities, L_*

Luminosities are derived from the observed flux ($L_*/4\pi d^2$), after measuring the distance. In principle the best (geometrical) distance determination is by the phase–lag method: see above. A statistical approach was taken by Habing (1988), who studied the distribution of the OH/IR stars in the Galaxy; the only scale length that enters the analysis is the distance from the Sun to the galactic center. A third way to overcome the distance problem is by using the OH/IR stars in the galactic bulge— they are all at the same distance within 25%. The net result is that luminosities have been found betweeen 3,000 and 30,000 $L_\odot$, with an average between 5,500 and 8,000 $L_\odot$ (van der Veen and Habing, 1990; Whitelock et al. 1991). A recent analysis of the new complete set of OH/IR stars discovered in Arecibo (Chengular et al., 1993) confirms these results.

2.2.5. Variability

For a significant number of OH/IR stars flux density variations have been monitored (and periods have been determined) in the OH– line or at some IR wavelength— radio and infrared fluxes vary in phase (Harvey et al., 1974). The shortest periods are around 500 days; the longest now known is over 3,000 days (Herman and Habing, 1985; West et al., 1992). In the galactic bulge the longest period is around 1000 days (Whitelock et al., 1991); longer periods are found only in stars belonging to the galactic disk. Herman and Habing (1985) studied the variation in a sample of 35

OH/IR stars in the disk of the Galaxy; about 1/4 showed small or no variation at all. These non-variables proved to have IRAS colours much redder and quite different from those of the variable OH/IR stars (Olnon et al., 1984); this led Bedijn to the suggestion that the non-variable stars had stopped —quite abruptly— to lose matter (Habing et al. 1987). The best known example of these non-variable stars is OH 17.7-2.0, now generally considered to be an object in transition from the AGB to the a planetary nebula phase. This however is a phase of very short duration and if the ratio of one non-variable out of every four OH/IR stars will stand further scrutiny, the total OH/IR phase can last only a very short time. Recent results by van Langevelde (1992) from monitoring OH/IR stars at the galactic center also show up a significant fraction of non- variables.

2.2.6. Related Objects

In the IRAS atlas of low resolution spectra (IRAS–LRS) there are over 2000 objects in the classes 2, 3 and 6, and a large fraction of these are stars with circumstellar envelopes containing silicate dust (in a recent paper Omont et al., 1993, have argued that LRS spectra 21 and 22 correspond to very cool carbon–rich stars rather than to oxygen–rich stars). Not all objects have been identified and studied in detail; yet it seems safe to assume that also in other properties they are related; Mira variables are at one end of the sequence, where the circumstellar envelope is thin and the OH/IR stars are at the other end, where the envelope has become very thick. Mira variables thus appear to be related to the OH/IR stars. It is tempting to speculate that Miras ultimately develop into OH/IR stars and that all OH/IR stars have been Miras in the beginning, but this is not at all certain.
Another group of relatives are the carbon stars: from the carbon-mira's to dust enshrouded stars like IRC+10216. The relation between "oxygen–rich" stars (such as Mira's and OH/IR stars) and these carbon–rich stars is a question that has evoked much discussion over the last few years; in this (ongoing) discussion the common ground is the prediction by the theory of stellar evolution that an oxygen–rich star can turn into a carbon–rich star when during a thermal pulse freshly-produced carbon is injected into the hydrogen–rich envelope and when this injection upsets the C/O-ratio.

2.3. INTERPRETATION

What *are* OH/IR stars? A standard answer begins by noting that OH/IR stars share important properties with the Mira variables, and that the differences are quantitative and not qualitative. Miras have late M-spectra; they are long period variables (periods between 200 and 500 days) with large amplitudes (1 to 2 bolometric magnitudes); they lose significant amounts of mass (10^{-7} to 10^{-8} $M_\odot$/yr) at a low outflow velocity (3 to 10 km/s); masers (OH, H_2O and SiO) are formed at some distance from the photospheric surface; luminosities are between 3000 and 10,000 $L_\odot$. OH/IR stars share all these properties but have more extreme values (luminosities from 3,000 to 30,000 $L_\odot$; periods from 500 to 3,000 days; mass loss rate from 10^{-6} to 10^{-4} $M_\odot$/yr; outflow velocity from 10 to 20 km/s). We miss one essential piece of information about the OH/IR stars: the photospheric temperature; we cannot observe the star directly, only its circumstellar envelope. In view of the similarities with Miras one assumes that a star of very late spectral type is also present at the center of an OH/IR star.

Late-type stars of such high luminosities must be AGB stars or supergiants. Model calculations (e.g. Iben, 1991) indicate that AGB stars consist of a core of significant mass, but of very small radius and consisting of carbon, oxygen and a degenerate electron gas. This core is surrounded by a huge envelope of hydrogen gas. Hydrogen is burned into helium in a very thin layer surrounding the core; when the mass of helium becomes larger than a critical value the helium burns into carbon in a very short time ("thermal pulse"). The core mass thus grows all the time and because the luminosity is proportional to the core mass, the luminosity increases as well; this growth is exponential with a time scale that is the same for stars of all masses: $\tau_{\mathrm{nucl}} \approx 10^6$ yr. The evolution of the star is determined by this growing core (well studied in hydrostatic models) and by mass loss on the outside. Mass loss is a hydrodynamic process that unfortunately can not (yet?) be calculated from first principles, but that we know to exist from observations. Mass-loss has a time scale $\tau_{\mathrm{ml}} = M/\dot{M}$; if $\dot{M}$ exceeds values like 10^{-6} $M_{\odot}$/yr then for a star of only a few solar masses one might have $\tau_{\mathrm{nucl}} < \tau_{\mathrm{ml}}$. In that case the star will lose its outer envelope faster than its core can grow and consequently it bleeds to death while its luminosity stays constant. This is the evolutionary stage to be associated with OH/IR stars and their relatives. It is clear that the precise chain of events in these last phases of the stellar existence depends critically on how the mass loss process evolves in time and of this history we are not yet certain. A new trend is to assume that mass loss is an interrupted process- the suggestion comes from the observations of detached shells (see above). It is tempting to associate such an interruption with the occurrence of a thermal pulse. This suggestion leads to a further speculation: At what luminosity the first thermal pulse will occur is determined by the main sequence mass of what is now the OH/IR star. Suppose that this thermal pulse inititiates the phase of rapid mass loss; then the luminosity of the star is arrested and the luminosity of an OH/IR star thus equals its luminosity at the moment of its first thermal pulse; we can thus estimate the main-sequence mass from the luminosity. Ideas similar to this have been worked out quantitatively by Vassiliadis and Wood (1993) who combine calculations of the stellar interior with an empirical (and still somewhat uncertain) mass loss rate.

In the circumstellar envelope radiation pressure is bound to be a dominating dynamic process once the dust grains have been formed; however it is clear that radiation pressure is not the *cause* of the loss of mass: Condensation of the dust particles takes place at large distances from the star and if the star had a hydrostatic atmosphere the density of gas at the condensation point would be much too low to lead to a significant rate of loss of mass. The pulsating stellar atmosphere is, however, strongly dynamic and not at all static and Bowen and Willson (1991) argue rather convincingly that stellar pulsations and the effects these have on the atmosphere lead to a sufficient enlargement of the atmospheric scale height (the same conclusions had been reached earlier by Bedijn, 1988).

In summary, what stars do become OH/IR stars? In addition to the arguments just presented there are more clues: (1) in the anticenter only young and massive stars are OH/IR stars whereas (2) in the galactic bulge relatively low luminosity objects show the strong OH 1612 MHz maser, and these objects have galactic orbits that can in no way be associated with Population I. Thus I suggest that all those stars become OH/IR star that reach the AGB, that pulsate sufficiently strongly, and that have a metallicity high enough to produce the required amount of dust. In the next section I will argue that OH/IR stars are found not only among population I objects, but also among older populations.

3. OH/IR stars as tracers of galactic populations

In the next review Dejonghe will discuss the kinematic properties of the total sample of OH/IR stars. As an introduction a few remarks: The concept of "stellar populations" is difficult to use. Yet it appears to me that in the set of 1500 known OH/IR stars we can distinguish three groups, each with its own spatial distribution and its own kinematic signature, and hence each can be called "a population":

1. *Stars belonging to the galactic disk moving on almost circular orbits.*
   Except for the very near ones they are all at small latitudes. There are very few outside the solar circle, that is at $| l | > 90°$. The radial velocities indicate large angular momenta, that is, they follow galactic rotation closely, although an "older" subgroup (smaller angular momentum, less concentrated to the galactic plane) and a "younger" subgroup can be recognised using the value of $v_{\rm out}$ as criterion (Baud et al., 1981; Chengalur et al., 1992). The first (1612 MHz) surveys for OH/IR stars in the late seventies revealed especially these disk stars and therefore the (now inadequate) notion rose that all OH/IR stars belong to Population I. Disk OH stars can be very luminous: probably up to the AGB luminosity limit of about 40,000 $L_\odot$ and beyond that as supergiants. Some AGB members have very long periods, between 1000 and 3000 days. I tentatively estimate that about 2/3 to 3/4 of the 1500 known belong to the disk.
2. *Stars in inner parts of the galaxy on orbits of low angular momentum, possibly outlining the bulge.*
   Tentatively their number is estimated at about 1/4 to 1/3 of the total number of 1500. There is no sharp criterion to distinguish the bulge stars on an individual basis from the disk stars- the distinction is made on statistical grounds (te Lintel Hekkert et al., 1991). The analysis of these stars (determination of luminosity and of pulsational period) is incomplete; the paper by van der Veen and Habing (1990) is only a first step. The stars studied by Whitelock et al. (1991; see also her review in these proceedings) are closely related to OH/IR stars, but have lower mass loss rates. One conclusion can be drawn: the total luminosity of an individual bulge star does not exceed 10,000 $L_\odot$ and its pulsational period is below 1000 days; more luminous OH/IR stars are found only in the disk population.
3. *Stars very close to the galactic center, that is within 150 pc.*
   There are some 140 known (Lindqvist et al., 1992). Kinematically they are very different from the bulge as they rotate rapidly around the center

(see the next review by Dejonghe). Infrared observations, essential to the determination of their luminosities, have been hindered by serious confusion problems and became succcesful only recently (Blommaert, 1992). The luminosities are modest (below 10,000 $L_\odot$), and the pulsational periods below 1000 days (van Langevelde et al., 1993). A surprising property is shown in figure 5. A relatively large fraction of the galactic centre population consists of stars with $v_{out} > 20$km/s, in the range of supergiants. Yet, the luminosities show that they are not supergiants and we take this as an indicator that the dust-to-gas ratio, δ, is higher in this rapidly rotating galactic population. One then speculates that this holds also for the silicium abundance (and for the overall metallicity?). On the basis of the facts just presented Blommaert (1992) speculates that these OH/IR stars might be 10^{10} yr old. An additional interesting set are the three stars discovered and discussed by van Langevelde et al. (1992) that have a radial velocity differing by more than 3 σ from the mean: true high-vlocity stars. To explain this high velocity the authors suggest that these stars are on very elongated, almost radial orbits, seen near their pericenter where the velocity is highest. Because OH/IR stars represent a large population of less evolved stars, actually a fraction significantly larger than 3/140 of the central population might be on such radial orbits.

After having discussed the places in the Galaxy where one finds the OH/IR stars it is illustrative to discuss the places where one does not find them. Surprisingly few OH/IR stars have been found outside the solar circle. The ones known appear to be very young and luminous *and* to have low outflow velocities: indicative of a young age and yet a low metallicity. This, and the fact that in the direction of the outer Galaxy carbon stars are as abundantly present as in the direction of the galactic center, suggests that low metallicity (1) forces a star of medium to low mass to become a carbon star before it can become an OH/IR star and/or (2) does not support an OH maser. Finally: what about the halo? I do not know of any OH/IR star that might be a halo object; the detection of IRAS source SSC 08546+1732, a cool envelope around a *carbon*–rich star in what is probably a halo object (Cutri et al. 1989) is most intriguing.

4. Conclusions

OH/IR stars are oxygen–rich AGB stars with heavy mass loss and with a sufficient large oxygen and silicium abundance (= exceeding the solar value?). More than 1500 of them are known, and many more remain to be detected. The total sample can be divided into three groups each with its own location in the Galaxy and with its own kinematic behaviour; each group its own population:

- OH/IR stars belonging to the galactic disk
- OH/IR stars belonging to the bulge
- OH/IR stars belonging to the nucleus of the Galaxy.

The bulge stars show little rotation, but the stars of the nucleus show fast rotation. In addition, the stars of the nucleus are probably supermetal rich.

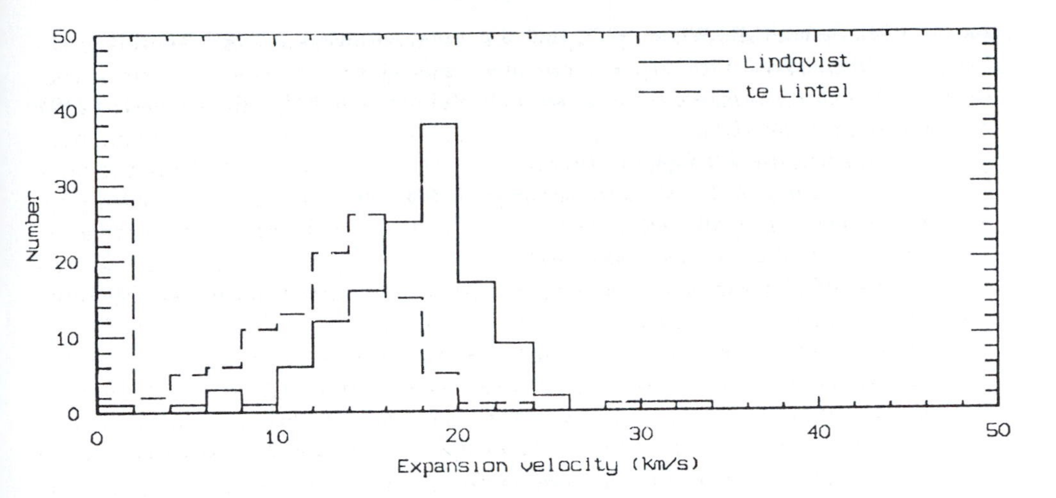

Fig. 5. A histogram of the expansion velocity for two different samples of both 134 OH/IR stars; the Lindqvist et al. (1992) sample within 1° of the centre and part of the te Lintel et al. (1991) sample within 4°.

References

Andersson, C., Johansson,L.E.B., Goss, W.M., Winnberg, A., N-Q Rieu, 1974, Astron. Astrophys. **30**, 475.

Baud, B., Habing, H.J., Matthewson, H.E., Winnberg, A. 1981, Astron. Astrophys. **95**, 156.

Becker, R.H., White, R.L., Proctor, D.D. 1992, Astron. J. **102**, 538.

Bedijn, P.J. 1987 Astron. Astrophys. **186**, 136.

Bedijn, P.J. 1988 Astron. Astrophys. **205**, 105.

Blommaert, J.A.D.L. 1992, thesis, Leiden University.

Blommaert, J.A.D.L., van der Veen, W.E.C.J., Habing, H.J. 1993, Astron. Astrophys. **267**, 39.

Bowen, G.H., Willson, L.A. 1991, Astrophys. J. Lett. **375**, L53.

Catchpole, R.M., Whitelock, P.A., Glass, I.S. 1990, Mon. Not. Roy. Astr. Soc. **247**, 479.

Chengalur, J.N., Lewis, B.M., Eder, J., Terzian, Y. 1992, preprint.

Cutri, R.M., Low, F.J., Kleinmann, S.G., Olszewski, E.W., Willner, S.P., Campbell, B., Gillett, F.C. 1989, Astron. J. **97**, 866.

David, P., Papoular, R. 1992, Astron. Astrophys. **256**, 183. Dickinson, D.F. 1991, Astrophys. J. Lett.**379**, L29.

Eder, J., Lewis, B.M., Terzian, Y. 1988, Ap.J. Suppl. **66**, 183.

Elitzur, M. 1992, "*Astronomical Masers*" (Kluwer Academic Publishers).

Forrest, W.J., Gillett, F.C., Houck, J.R., McCarthy, J.F., Merrill, K.M., Pipher, J.L., Puetter, R.C., Russell, R.W., Soifer, B.T., Willner, S.P. 1978, Ap. J. **219**, 114.

Gail, H.-P., Sedlmayr, E. 1986 Astron. Astrophys. **256**, 201.

Gaylard, M.J., West, M.E., Whitelock, P.A., Cohen, R.J. 1989, Monthly Not. Roy. Astr. Soc. **236**, 247.

Habing, H.J. 1988, Astron. Astrophys. **200**, 40.

Habing, H.J., van der Veen, W.E.C.J., Geballe, T. 1987, in "*Late Stages of Stellar Evolution*", eds. S. Kwok and S.R. Pottasch, Reidel, Dordrecht, p. 91.

Harvey, P.M., Bechis, K.B., Wilson, W.J., Ball, J.A. 1974, Astrophys. J. Suppl. **27**, 331.
Herman, J., Habing, H.J. 1985, Astron. Astrophys. Suppl. **59**, 523.
Heske, A., Forveille, T., Omont, A., Sivagnanam, P., van der Veen, W.E.C.J., Habing, H.J., 1990 Astron. Astrophys. **239**, 173.
Iben, I. 1991, Astrophys. J. Suppl. Ser. **76**, 55.
Justtanont, K., Tielens, A.G.G.M. 1992, Astrophys. J. **389** , 400.
Kastner, J.H. 1992, Ap. J. **401** ,337.
Lefèvre, J., 1989 Astron. Astrophys. **219**, 265.
Le Squeren, A.M., Sivagnanam, P., Dennefeld, M., David, P. 1992, Astron. Astrophys. **254**, 133.
Lewis, B.M. 1992, Astrophys. J. **396**, 251.
Lewis, B.M., Eder, J., Terzian, Y. 1990, Astrophys. J. **362**, 634.
Lindqvist, M., Habing, H.J., Winnberg, A. 1992, Astron. Astrophys. **259**, 118.
Margulis, M., Van Blerkom, D.J., Snell, R.L., Kleinmann, S.G. 1990, Astrophys. J. **361**, 663.
Mennesier, M.O., Omont, A. (editors) 1990, *"From Mira's to Planetary Nebulae: Which path for stellar evolution?"* Editions Frontières (Gif-sur-Yvette).
Nyman, L.-Å., Booth, R.S., Carlsström, U., Habing, H.J., Heske, A., Sahai, R., Stark, R., van der Veen, W.E.C.J., Winnberg, A. 1992, Astron. Astrophys. Suppl. Ser. **92**, 43.
Olofsson, H., Carlström, U., Eriksson, K., Gustafsson, B., Willson, L.A. 1990, Astron. Astrophys. **230**, L13.
Omont, A., Loup, C., Forveille, T., te Lintel Hekkert, P., Habing, H., Sivagnanam, P. 1993, Astron. Astrophys. **267**, 515.
Olnon, F.M., Baud, B., Habing, H.J., de Jong, T., Harris, S., Pottasch S.R. 1984, Astrophys. J. Lett. **278**, L41
Sahai, R. 1990, Astrophys. J. **362** , 652.
Schutte, W.A., Tielens, A.G.G.M. 1989, Astrophys. J. **343**, 369.
Sivagnanam, P., Le Squeren, A.M., Foy, F., Tran Minh, F. 1989, Astron. Astrophys. **211**, 341.
Spaans, M., van Langevelde, H.J. 1992, Mon. Not. Roy. Astr. Soc. **258** , 159.
te Lintel Hekkert, P., Caswell, J.L., Habing, H.J., Haynes, R.F., Norris, R.P. 1991, Astron. Astrophys. Suppl. Ser. **90**, 327.
te Lintel Hekkert, P., Dejonghe, H., Habing, H.J. 1991, Proc. Astron. Soc. Australia, **9**, 20.
van Langevelde, H.J. 1992, thesis, Leiden University.
van Langevelde, H.J., Frail, D.A., Cordes, J.M., Diamond, P.J. 1992, Astrophys. J. **396**, 686.
van Langevelde, H.J., Janssens, A.M., Goss, W.M., Habing, H.J. Winnberg, A. 1993, Astron. Astrophys. Suppl. (in press).
Vassiliadis, E., Wood, P.R. 1993, Astrophys. J. (in press).
van der Veen, W.E.C.J., Habing, H.J. 1990, Astron. Astrophys. **231**, 404.
West, M.E., Gaylard, M.J., Combrinck, W.L., Cohen, R.J., Sheperd, M.C. 1992, in *"Variable Stars and Galaxies"*, ed. B. Warner (Astronomical Society of the Pacific Conference Series no 30) page 277.
Whitelock, P.A., Feast, M.F., Catchpole, R. 1991, Mon. Not. Roy. Astr. Soc. **248**, 276.
Willems, F.J., de Jong, T. 1986, Astrophys. J. Lett. **309**, L39.
Wood, P.R., Whiteoak, J.B., Hughes, S.M.G., Bessell, M.S., Gardner, F.F., Hyland, A.R. 1992, Astrophys. J. (Letters) **397**, 552.
Zijlstra, A.A., Loup, C., Waters, L.B.F.M., de Jong, T. Astron. Astrophys. **265**, L5.

Harm HABING

HERWIG DEJONGHE
'92

KINEMATICS AND DYNAMICS OF OH/IR STARS

H. DEJONGHE
Sterrenkundig Observatorium, Universiteit Gent, Belgium

Abstract. The available kinematical data on OH/IR stars is discussed, and dynamical models for the OH/IR stars are presented.

1. Introduction

In this contribution I would like to present you one particular way of modelling the kinematics of a tracer population of stars in our Galaxy. This contribution is a continuation of the previous one. One of the things that I hope to convince you of is that the definition of such a thing as "a dynamical model for the OH/IR stars", involves quite a number of elements. Therefore I will highlight in boldface every item that is essential in defining the models presented here.

Circular orbit models will not work as representations of the kinematics of OH/IR stars, and this is immediately obvious from Figure 1. There you see a familiar (l, v) plot for a sample of OH/IR stars, compiled by Peter te Lintel Hekkert, who, very unfortunately and unexpectedly, was unable to attend this symposium in order to present these results himself. I will call it the tLH sample. It consists of almost 900 stars.

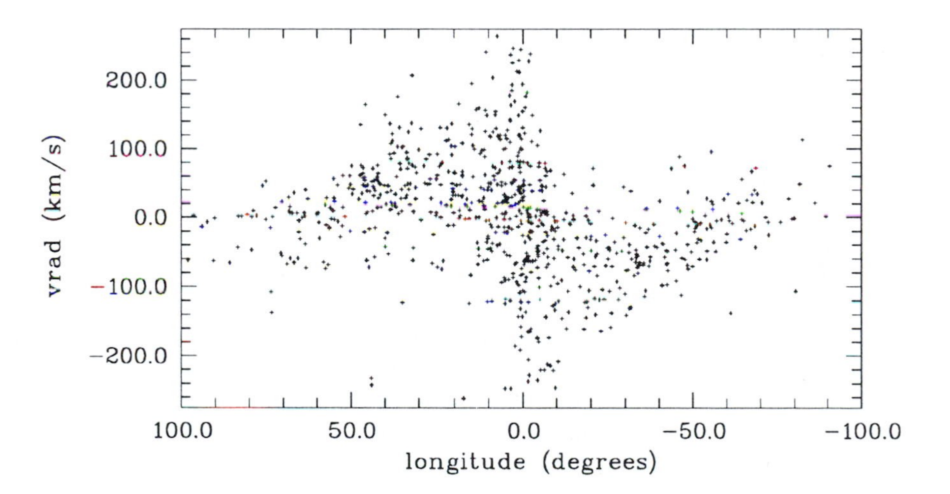

Fig. 1. The (l, v) diagram of the L sample. Note the counterrotators.

The smoking gun in Figure 1 is the presence of counterrotating stars (especially in the Bulge region), which, in any case, point to orbits that are not simply circular. It is nice to see this come out straight from the data, though nobody will argue with the statement that we cannot expect the stars in the Bulge to form only a thin disk on these small scales. By the way, I'll tentatively define the Bulge here as

H. Dejonghe and H. J. Habing (eds.), Galactic Bulges, 73–86.

a "not further specified volume inside a 3 kpc radius"... In any case, the greater variety in the orbits there does not mean that a thin disk with stars in circular orbits isn't there (after all, our solar system is a great example of a thin disk on even smaller scales), but I'll return to this in a while.

I will not expand on the way the data in Figure 1 were acquired, nor on the physics of the OH/IR's that we can learn from it, since these topics are covered by Habing (this symposium) and can also be found in te Lintel Hekkert (1990,1991a). Suffice it to say that this particular data set is a **compilation from catalogues** by Eder *et al.* (1988), te Lintel Hekkert *et al.* (1991a) and Sivagnanam *et al.* (1989), all based on the IRAS PSC from which positions were selected according to infrared colours, fluxes (f) and IRAS flux qualities. Essentially, you see here all sources with colours R21=$f_{25\mu m}/f_{12\mu m}$ between 0. and .9. This was done in an attempt to obtain a homogeneous sample of OH/IR stars with thick dust shells and a reasonably constant bolometric correction $f_{tot}/\nu/f_{12\mu m}$. Additional **selection criteria** were the presence of the two regular 1612 MHz peaks which had to be separated by more than 10 km/s, and the requirement that the $f_{12\mu m}$ had to be larger than 3 Jy. Finally, one must keep in mind that samples based on IRAS data are incomplete in the areas where IRAS was confusion limited. The models therefore will not include data from a region with approximate boundaries $|l| < 45°$ and $|b| < 2°$.

Simple models are great to answer simple questions, and such questions are likely to be of particular interest to the physicist inside the astronomer: determination of the rotation curve, estimates of the total mass, the total power emitted, etc... And yet, even the progress on these questions is slow. This is due partly because any adequate theory tends to become complex nevertheless, partly because some parameters are hard or impossible to measure (such as stellar ages, distances), partly because of the nagging problem of the extinction, partly because the parameters of interest are often statistical properties of samples (e.g. determination of the local standard of rest), and finally because there are so many interrelated parameters. It would seem that one really needs a "model for everything". Such may be a great proposal for ambitious nationals of wealthy countries who aspire large grants from well-grown funding agencies, but it's simply not very realistic.

All the problems mentioned above are present in this contribution. I will not address them any further, since this would be restating the obvious. But allow me to expand on something else: why, apart from the above question, could one be interested in dynamical models in the first place? I think that, for the cartographer inside the astronomer, it's a natural thing to do. In order to motivate this opinion meaningfully, we must agree on what we mean by dynamical modelling.

2. Elements of Dynamical Modelling

The equations that govern the motion of 10^{11} starlike objects in mutual interactions and the interpretation of their solutions are hopeless. Both theory and assumptions will be needed to make them suitable for human grasp. It is a well-known textbook topic (e.g. Binney & Tremaine 1987) that in most cases of astrophysical interest it suffices to simplify the motion of each of these objects to the motion of one (hypothetical) object, moving in the gravitational field which is generated collectively by all the others. This reduces phase space from 6×10^{11} coordinates to 6 coordinates, together with a potential function that is loosely coupled by an integral operator to the contents of that space, which is characterized by a **distribution function**, which is a probability density in phase space.

The potential generates structure in phase space, because it creates orbits. The quintessential orbit is the linear harmonic oscillator, e.g. a spring. At every moment it has a length z which changes at a rate v_z. Phase space is the 2-dimensional space (z, v_z). If we do not know the dynamical state of the spring completely, it is natural to ask what information we can single out as particularly important. The maximum length z_m must be such a quantity, because then, at least, we can confine the length of the spring, though we've lost the ability to predict its actual length at any particular moment. This z_m is an example of an integral of the motion: it is a function of phase space coordinates that remains a constant along the orbit, and therefore it can be used as a label for that orbit. Hence, we can now describe the linear harmonic oscillator with z_m (a constant) and only one rapidly changing coordinate (z, or v_z, or something else).

In 3 dimensions, one would expect 3 constants of the motion and 3 rapidly changing variables for every orbit. This is true for integrable potentials, by definition. Most potentials however are not integrable, but it is likely that for most astrophysical purposes there exist good integrable fits (Goodman and Schwarzschild 1981, Dejonghe & de Zeeuw 1988), safe possibly for tumbling triaxial figures. A very elegant class of integrable potentials are the Stäckel potentials, which have the nice property that the integrals of the motion are quadratic functions of the velocities.

In order to better understand the significance of these integrals, let's consider the following experiment. We affix many springs to a flat surface. The springs only vibrate in the z–direction (perpendicular to the surface), and hence their x and y coordinates which are markers on the surface are constants of the motion. Now we disturb the springs, for example by pushing them down simultaneously by hand. The imprint of the hand will be lost very quickly, and in the analysis of the resulting dynamical state, it will certainly not matter very much to focus on the description of the rapidly changing coordinates. If we only knew $z_m(x, y)$, then we would know the profile of the perturber, which is everything that there is to know in this experiment.

Consider next the somewhat different situation that at every location (x, y) there are a lot of springs (for example molecules), which may or may not start to vibrate due to infalling light, then the number of excitations N will be proportional with the intensity, while the degree of excitation (the z_m) will tell us something about the wavelength of the infalling light. The function $N(x, y, z_m)$ we call a distribution function. It is written here as a function in integral space. It cannot exist without a medium for which it is a probability density, though it may provide us with important information on something else (the infalling light). This function is very analogous to the concept with the same name in stellar dynamics; the medium there is called a tracer population. On a photographic plate, the distribution function is a faithful representation of the perturbing radiation, and the resulting picture is its own justification. In stellar dynamics, and now comes a more personal view, the distribution function could serve the same purpose: it is "simply" a picture of the stellar component. By analogy with ordinary pictures, this picture may very well be its own justification. One is not necessarily unimaginative if no more questions are asked, just as one does not ordinarily inquire about the "origin and evolution" of some feature in a portrait when looking at it...

Of course, it is hoped that the distribution function will teach us a lot about the origin and evolution of galaxies and bulges.

3. The input: data and assumptions

My first assumption is far reaching: I'll assume that the data can be modelled by an **equilibrium model**. This is to a lesser extend a statement about physical intuition than it is a poor man's choice: within the framework of a dynamical model with a distribution function, there is simply no other choice, for lack of adequate theories. One can show however that equilibrium models are fine for slowly evolving systems, if one translates the results to so called action space, but this is not my intention here.

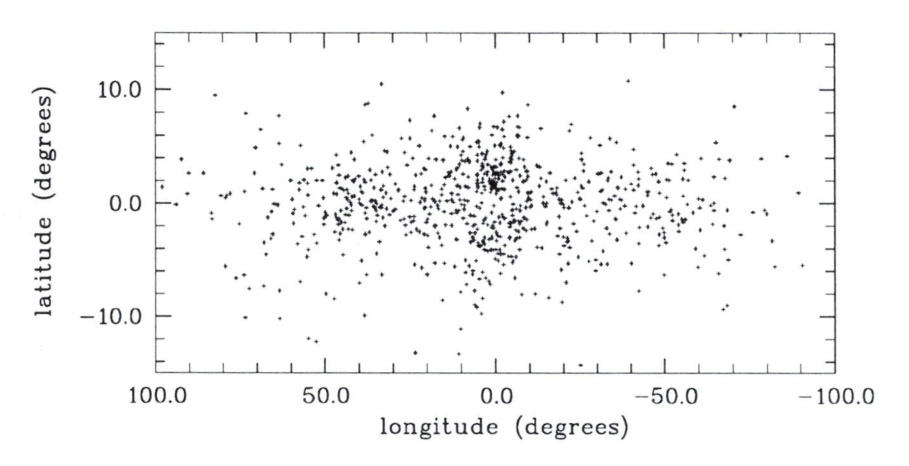

Fig. 2. The (l, b) diagram of the tLH sample.

When studying the dynamics of a tracer population, it is obvious that self-consistency is not required, i.e. the gravitational potential is not generated by the tracer population. This means that the gravitational potential can be decoupled from the original set of equations, and this potential therefore must be a given. The specification of the **gravitational potential** is the first important decision one has to make when building a dynamical equilibrium model. In particular, one has to decide on the prevailing **geometry**, i.e. whether the potential is spherical, axisymmetric or triaxial.

Figure 2 presents the (l, b) distribution of the tLH sample. There is no obvious triaxiality in the data, which is also the case for the (l, v) diagram in Figure 1. It's a very sad thing to note, especially since everybody now seems to detect more or less the same barlike structure; I suffice with referring to the index of the proceedings to illustrate this point. On the other hand, the absence of triaxiality in a tracer population does not mean that the potential isn't triaxial! Conversely, triaxiality in a tracer population does not imply triaxiality of the potential, but that is something you're not supposed to say... Well, anyway, let me say the unspeakable: I'll settle for **axisymmetric potentials**, and I'll forgive the very few who still have not quit reading for doing so now.

The IRAS satellite was severely confused in the GP, because of the high density of sources. In order to find OH/IR stars there, one must resort to mapping type surveys. Lindqvist *et.al.* (1992a), hereafter L, searched for OH/IR in a limited region close to the GC, harvesting 134 stars. None of the surveys have been completely satisfactory in their velocity coverage, for technical and feasibility reasons. This means that (a few) high velocity stars can be expected to turn up when searching for them (van Langevelde 1992).

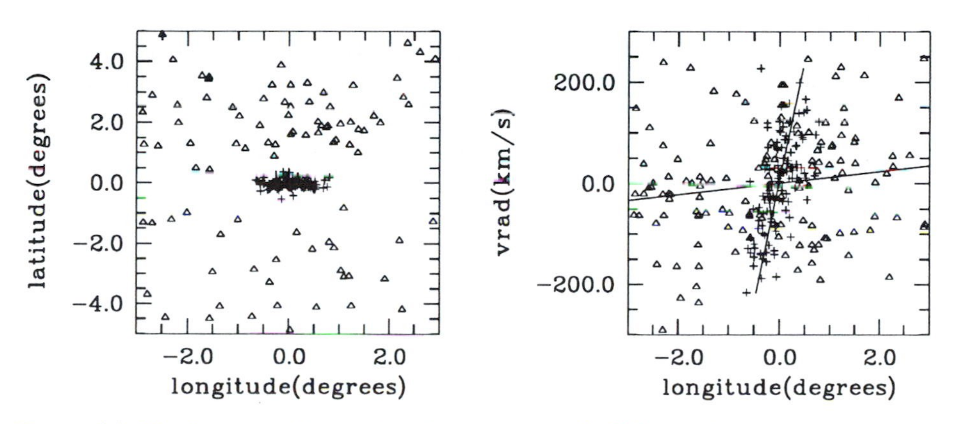

Fig. 3. (a) The L sample (crosses) and the truncated tLH sample (triangles) in (ℓ, b) on the sky. (b) The L sample (crosses) and the truncated tLH sample (triangles) in (ℓ, v_r) space.

I will not compare in any great detail the tLH and L samples. Some of that can be found in te Lintel Hekkert *et.al.* (1991b), and Lindqvist *et.al.* (1992b). Posters by Whitford and Winnberg, Lindqvist & Habing also address different aspects of this.

There is one observation I would like to make however, which is, I hope, strikingly clear in Figure 3. Figure 3a shows the (ℓ, b) diagrams for the L and the tLH samples, the latter being truncated to $\pm 3°$ in longitude and $\pm 5°$ in latitude. It is obvious that both samples are complementary, but certainly not enough so: the IRAS confusion zone is not covered. If the 2 samples are drawn from different galactic OH/IR populations, we miss the data in the important transition region. Currently a consortium headed by Habing is working on surveys at the VLA and the AT to fill in a few gaps. Figure 3b shows the (ℓ, v_r) plot for both samples (heliocentric velocities). The regression lines are the linear approximations to the "rotation curves". The L rotation curve (crosses) in this plot has a slope of about 500 km/s/degree or about 3.7 km/s/pc, using $R_\odot = 7.5$kpc. On the other hand, the slope of the tLH sample is 11 km/s/degree or about 82 km/s/kpc. Such a rotation curve reaches its presumed peak value of about 220 km/s at about 2.5 kpc, which is very reasonable.

There are two quite different opinions here. The first one answers positively on the question whether the very different regression lines have anything to do with the sampling of different galactic populations. It is convincingly argued by Habing (this symposium) that the nature of the OH/IR's in both samples are different. If this also translates into the kinematics in such a way as it does here, than this connection would constitute one of the most dramatic links between abundance and kinematics that we know of. The other opinion would merely see the different regression lines as the result of a different dynamical environment. In particular, the tLH regression line would be representative for the rotation curve as we know it, while the L regression line would indicate a (fairly modest) mass concentration in the very center. Values are given in Lindqvist *et.al.* (1992b). This second opinion implies that we would do well not to combine both samples, unless there is a mass concentration provided in the potential.

There is a lot to say about the nature of the OH/IR stars, and whether we can differentiate the class into different galactic populations (Habing, this symposium). For example, the conjecture by Baud *et al.* (1981) and Olnon *et al.* (1981) that the velocity of the expanding circumstellar shell may indicate an age, in the (statistical) sense that larger expansion velocities are associated with younger stars, has been a working hypothesis ever since it was first proposed. It looks like this simple picture may need amendment, and therefore I will in the sequel indicate this by quoting "old" and "young".

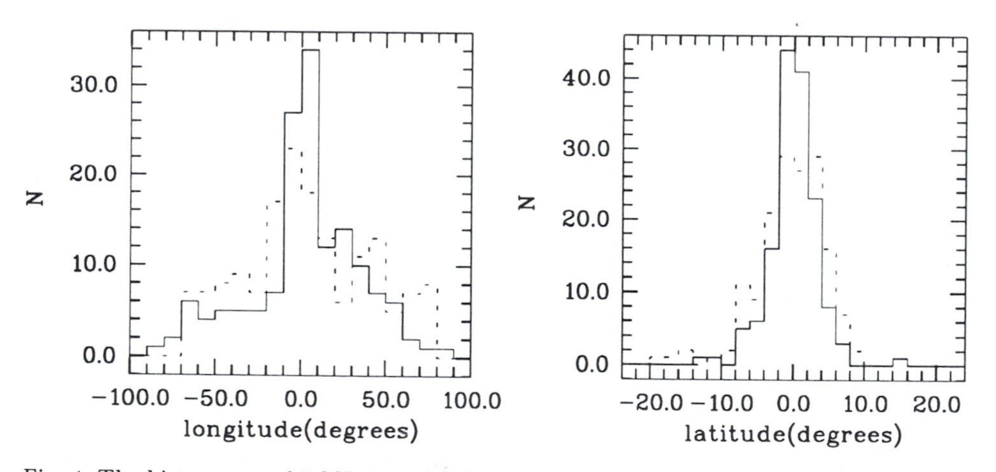

Fig. 4. The histograms of "old" stars (shells expanding at a rate between 10 and 13 km/s) in the tLH sample (dashes) and the "young" stars (solid lines, shells with expansion velocities between 16 and 20 km/s), as a function of galactic longitude and latitude

In any case, trom the analysis by Lindqvist *et.al.* (1992b) it would seem that indeed the projected velocity dispersion of the "younger" stars is smaller than the dispersion for the "older" stars, as one could expect if stars are born on primarily circular orbits. In Figure 4 we see histograms as a function of longitude and latitude for a selection of the tLH sample, divided into **two groups** defined by the "oldest" and "youngest" stars, each containing about 150 stars. If the aforementioned effect is real, it points in any case into the right direction: the "younger" population is more confined to the disk, and somewhat more bulgy, i.e. centrally concentrated. Whether you believe it or not may depend on whether you are an astronomer or a statistician.

Finally, since the available **positions and radial velocities** do not give much information on the z-component of the velocity for stars in the GP, it is natural to try **two integral models** first (based on the specific binding energy E and the z–component of the angular momentum, which are both integrals of the motion in an axisymmetric potential). Such models have the property that $\sigma_r = \sigma_z$. Only when the projected velocity dispersion turns out to be much too small (since for two–integral models σ_r is determined by the thinness of the disk), will we be able to rule out two–integral models. In view of Kent's (1992) success in modelling the Bulge with 2-integral hydrodynamical models, it is instructive to see whether this result holds when 2-integral models are considered that go all the way to the distribution function.

4. The model: method and results

So far we have made modelling choices based on arguments that could be deduced from the data alone. Now it is time to consider the options that are rather a matter of preference.

We need an explicit potential function. For this we used either a **Stäckel fit** to the **Bahcall-Soneira potential** (Dejonghe & de Zeeuw 1988) or a simple **Stäckel halo-disk potential** of the type discussed by Batsleer & Dejonghe (this symposium). The Stäckel fit is not needed at this point, but will come in handy when 3 integral models are made, which must be done eventually. It is important to note however that the adopted potentials all more or less reproduce the Burton & Gordon (1978) peak at .5 kpc. Recent analysis (Burton & Liszt 1993), allowing for non-circular motions, puts into question this peak, reminding us how little we know for sure about the potential of the Bulge.

We need a **distance to the GC**. We adopt 7.5 kpc, for no particular reason.

From the sample we can construct a body of data that can be very inhomogeneous, including star counts, mean velocities, velocity dispersions and line profiles..., on a few selected patches in the sky. The left panels in Figure 5 are smoothed renditions of the data, based on running averages (counts, mean velocities and velocity dispersions) over the 15 closest data points, which for this figure were selected among the "youngest" stars. The white (or light gray) areas correspond to the regions where there are no data because of completeness problems (IRAS confusion) or simply because of paucity of data points. The left panels were used to create the input data for the modelling.

No general theorem exists that would enable us to decide on the uniqueness of the distribution function for such data, and few analytic procedures are known to construct it. These procedures are applicable only in special cases (e.g. homogeneous data on the mass density, see a contribution by Hunter (this symposium)). Therefore, it might be useful to consider a method that is more or less independent

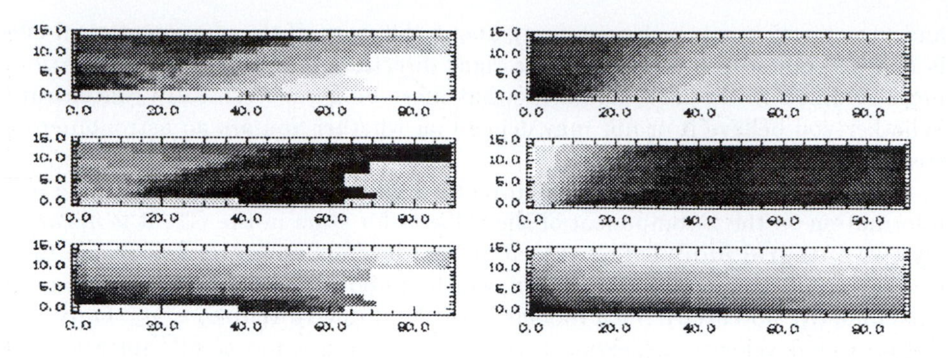

Fig. 5. A qualitative comparison of data (left panels) and model (right panels) on the sky, produced for the sample with highest circumstellar expansion velocities. The bottom panels represent the logarithm of the counts per square degree; the gray scales range from -1.5/sq.deg. to 1./sq.deg. The middle panels are the (projected) mean velocity fields, with values from 0 km/s to 220 km/s. The top panels represent the (projected) velocity dispersions, gray scales range from 10 km/s to 115 km/s.

of the diversity in the data.

In almost all cases the relation of this distribution function with the observable quantities takes on the form of an average of the distribution function:

$$\mu_{ob}(\mathbf{x}_\ell, \mathbf{v}_\ell) = \int_{\tau_\ell} \mu(\mathbf{x}, \mathbf{v}) F(E, I_2, I_3)\, d\mathbf{x}\, d\mathbf{v}, \tag{1}$$

with $(\mathbf{x}_\ell, \mathbf{v}_\ell)$ a label for a region in the space of observables (e.g. a region on the sky, a region in (ℓ, v_r) space...), τ_ℓ the region in phase space that contributes to the mean, and μ_{ob} the observable (counts/sqare degree, mean velocity,...) The operator is linear, which means that we can construct models by superposition.

One way to proceed is with **Quadratic Programming** (Dejonghe 1989). In this method it is assumed that the distribution function can be written as a linear combination of (preferably analytically simple) components, with coefficients c_i. A χ^2-type function (quadratic in the c_i) is then mimimized, subject to the constraint that the distribution function must be positive everywhere (linear constraints in the c_i). We also need to pick a series of **basis functions** to construct the linear combination, and these were **Fricke components**, which are basically powers in energy and angular momentum, and **disk components**, of the type discussed by Batsleer & Dejonghe (this symposium). For additional details, see also te Lintel Hekkert *et.al.* (1991b).

The right panels in Figure 5 show the model fits to the data (in the left panels). The gray scales give at least a qualitative idea properties of the fit. The areas where there are no data are now filled in (let's call them predictions, or extrapolations if you insist). In figure 6 a more quantitative comparison is presented. Clearly, one should not expect the fit to reproduce all details in the data, which may not be all real anyway. Problem areas may be the projected velocity dispersion for the "old" stars, which is somewhat too low, and the mean velocity which rises too fast for

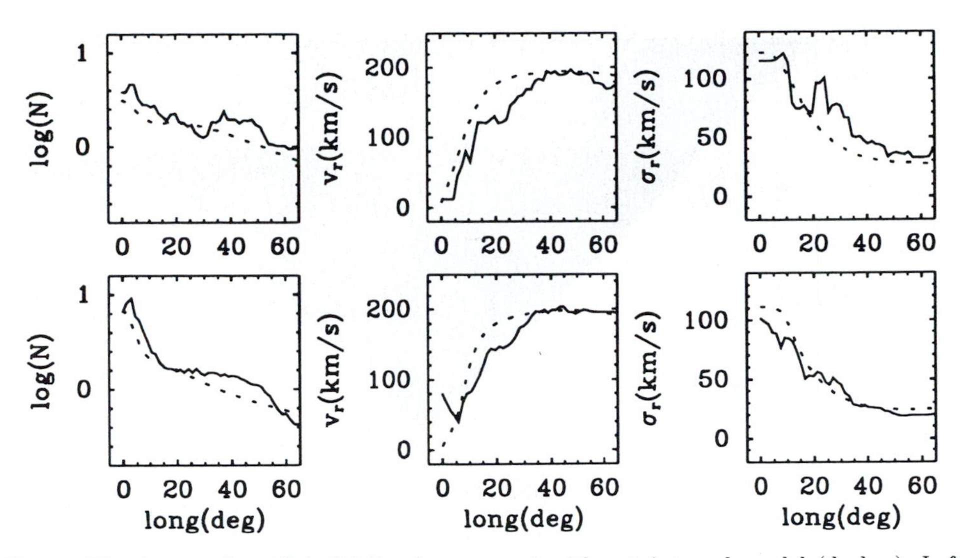

Fig. 6. The data at $b = 2°$ (solid lines) compared with a 2–integral model (dashes). Left panels: logarithm of projected star counts per square degree, middle panels: projected mean velocity, right panels: projected velocity dispersions. Top panels: the "old" stars, bottom panels: the "young" stars.

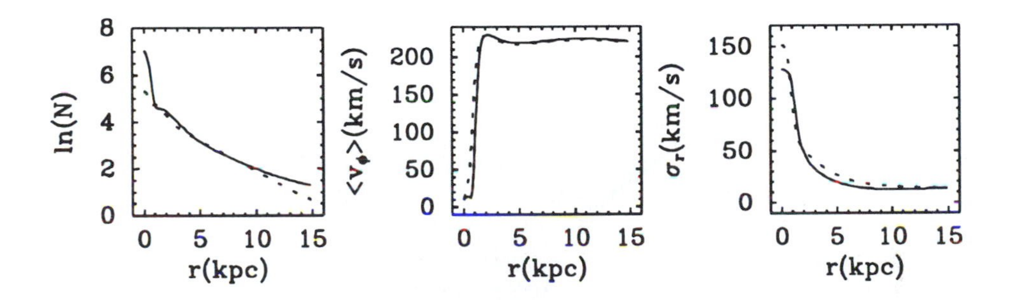

Fig. 7. The logarithm of the spatial density in the plane, the mean rotation and the radial velocity dispersion for the "young" (solid) and "old" (dashed) stars

both samples. This may indicate the need for a third integral.

In Figure 7, the spatial number density, mean rotation and radial velocity dispersion are plotted for both groups of stars. The "old" stars show a nice exponential disk with scale factor 3.5 kpc. This is no artifact of the components used in QP, since none of them showed exponential behaviour but were rather polynomial. The mean rotation follows very closely the rotation curve. This is a consequence of the 2-integral models, which, in order to produce fairly flat disks, must have small velocity dispersion. The "old" stars have overall a somewhat higher velocity dispersion, which clearly shows a bulge component.

Finally, Figures 8 and 9 show the distribution function in turning point space and in integral space (the color versions are much nicer!). All well–known compo-

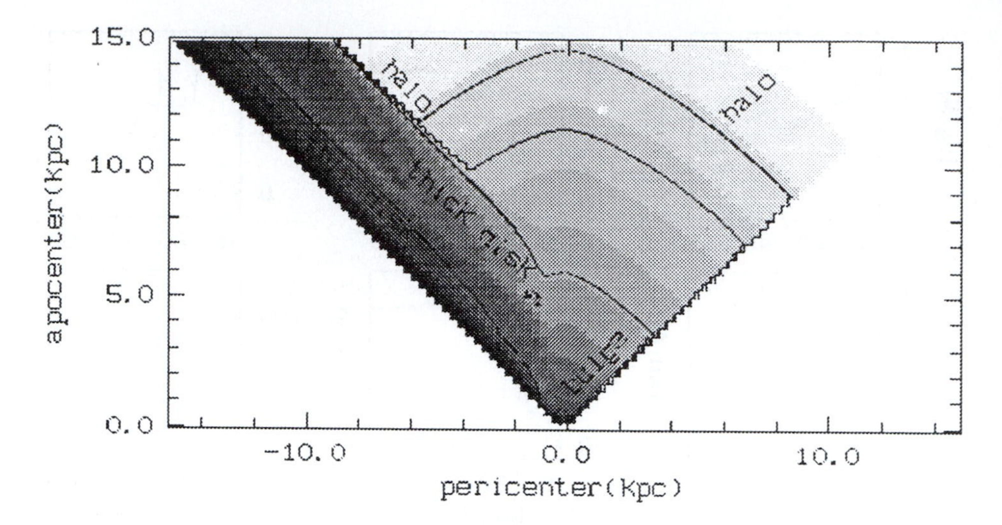

Fig. 8. The distribution function in turning point space. Every point inside the wedge-like region corresponds to an orbit with apocenter and pericenter as implied by the axes. Negative pericenter simply means negative angular momentum. Contours are chosen to delineate the different components. The dynamic range is of the order 10^{12}.

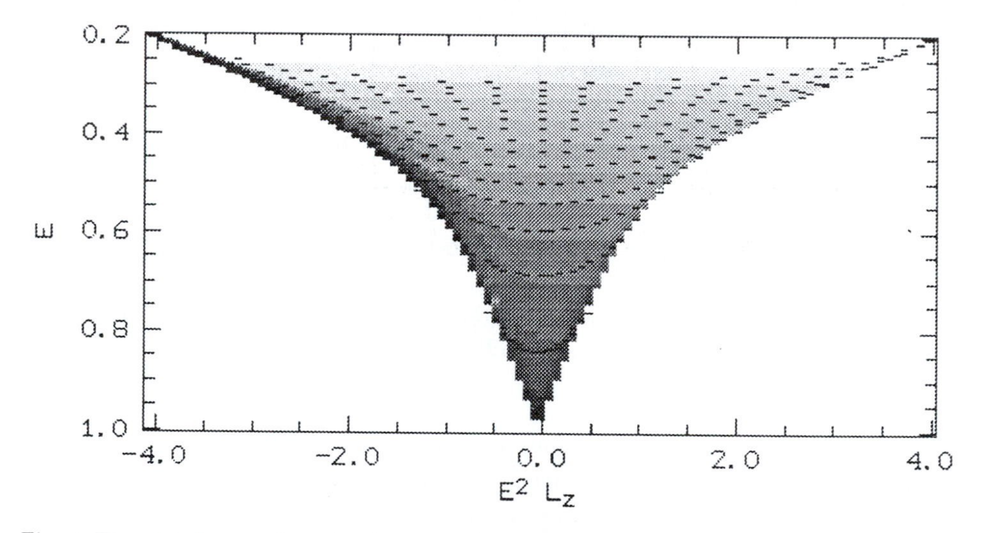

Fig. 9. The distribution function in integral space. Every point inside the wedge-like region corresponds to an orbit with integrals as implied by the axes. The boundaries of the region are the loci of the circular orbits. The distribution function is the same as in Figure 8. The dotted contours are loci of constant apocenter, given from 1 kpc (bottom) to 15 kpc (top).

nents are present, and are indicated in the turning point plot. The dynamic range is very large: the highest value is about 10^7 stars/kpc^3/(km/s)^3 in the thin disk, but, obviously, such values are very uncertain.

It is possible to make predictions on the basis of these models. For example, since we know the distribution function, all observable kinematical quantities can be calculated. As an example, Figure 10 depicts the line profiles (i.e. the (l, v) diagram) in the GP, drawn from the above distribution function for the "younger" OH/IR's, superposed on the sample. This calculation is, in a sense, a prediction, since there has been no fitting on the (l, v) diagrams directly. As is obvious, the fit is very reasonable.

5. Conclusions

The OH/IR stars are excellent probes of our dynamical Galaxy. They are fairly old, strong infrared emittors that are reasonably representative of a relaxed population. As such, they can be modelled with equilibrium dynamical models. Their strong infrared emission makes them shine right through the dusty GP, a property which is needed for a sufficient spatial coverage.

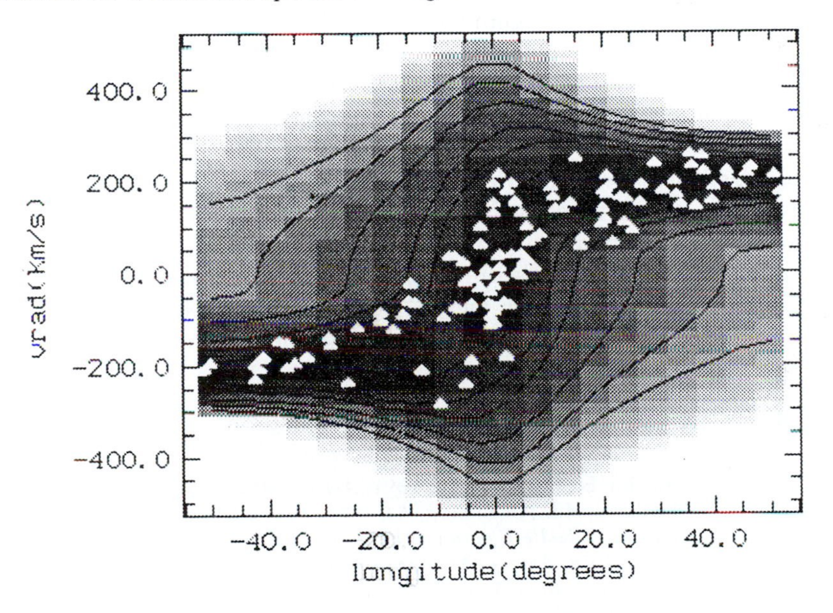

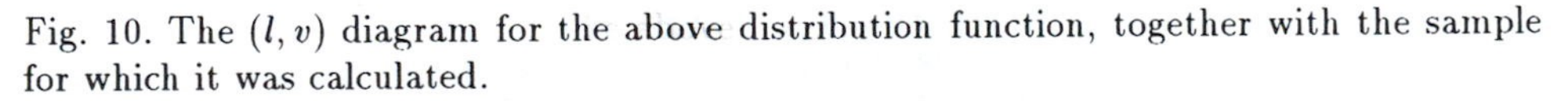
Fig. 10. The (l, v) diagram for the above distribution function, together with the sample for which it was calculated.

Differentiation of the sample towards a definition of galactic populations is possible, but no firm conclusions can be drawn, as yet.

Only numerical experience at this point can give us an idea to what degree we can have confidence in the computed distribution function. It is obvious that the more the data cover phase space, the more the distribution function will be constrained. Also, the more restrictive we are in the functional form of F (function of one, two, or three integrals), the less realistic our results may be, but the less indeterminacy we will encounter when trying to determine a distribution function. And this is, by the way, is an important reason to start with 2-integral models!

Two integral models are reasonable fits, but it is already clear that eventually three integral models will be needed. This is indicated by the fit of the model in the transition region between Bulge and Disk, where the flatness of the Disk sets fairly strong constraints on the kinematics in a 2-integral approach. This is somewhat in contrast with Kent's (1992) results, possibly because the positivity of the distribution function is taken into account here. As soon as proper motions become available (see e.g. Spaenhauer, Jones & Whitford 1992, Minniti, this symposium) on large areas in the sky, global models such as these will almost certainly need 3-integral dynamics.

The adopted potential is still very unsure. Especially for the Bulge regions, the potentials used in this analysis are probably inadequate. It would be worthwhile to reconsider the models with Bulge potentials of the kind used by Kent (1992) and Burton & Liszt (1993). In a more remote future, simple theoretical models for orbits in rotating barlike potentials may be needed, if the bar in the Bulge turns out to be dynamically significant for the stellar populations.

This contribution is only the beginning of what I consider to be long and laborious but potentially extremely rewarding stellar dynamical modelling of Galactic populations. Moreover, not only OH/IR stars are amenable to this kind of analysis. Whenever a sample is available with sufficient spatial coverage, it is presumably worthwhile to try analyses of this kind. In particular, the IRAS PSC has also been used to search for Planetary Nebulae, since these, too, are strong infrared emittors, and occupy a fairly well defined place in the $f_\nu(12\mu m)/f_\nu(25\mu m)$ versus $f_\nu(25\mu m)/f_\nu(60\mu m)$ color diagram (Pottash *et.al.* 1988, Ratag *et.al.* 1990). Radio interferometry can be used to decide on the true nature of the candidates (Zijlstra *et.al.* 1989). This method up to now yielded about 50 new PNs within 15° from the galactic center, on a total of about 400 in roughly the same region (Acker *et.al.* 1991).

References

Acker, A., Köppen, J., Stenholm, B., Raytchev, B., 1991, *Astron. Astrophys. Suppl.*, **89**, 237

Baud, B., Habing, H.J., Matthews, H.E., Winnberg, A., 1981, *Astron. Astrophys.*, **95**, 171.

Binney, J. & Tremaine, S., 1987, *Galactic Dynamics*, Princeton University Press

Burton, W.B. & Gordon, M.A., 1978, *Astron. Astrophys.*, **63**, 7

Burton, W.B. & Liszt, H.S., 1993, *Astron. Astrophys.*, submitted

Dejonghe, H., 1989, *Astrophys. J.*, **343**, 113

Dejonghe, H. & de Zeeuw, P.T., 1988, *Astrophys. J.*, **329**, 720

Eder, J., Lewis, B.M., Terzian, Y., 1988, *Astrophys. J. Suppl.*, **66**, 183

Goodman, J. & Schwarzschild, M., 1981, *Astrophys. J.*, **245**, 1087

Kent, S, 1992, *Astrophys. J.*, **387**, 181

Lindqvist, M., Winnberg, A., Habing, H.J., Matthews, H.E., 1992a, *Astron. Astrophys. Suppl*, **92**, 43

Lindqvist, M., Habing, H.J., Winnberg, A., 1992b, *Astron. Astrophys.*, **259**, 118

Olnon, F.M., Walterbos, R.A.M., Habing, H.J., Matthews, H.R., Winnberg, A., Brzezinska, H., Baud, B., 1981, *Astrophys. J.*, **245**, L103.

Pottash, S.R., Bignell, C., Olling, R., Zijlstra, A.A., 1988, *Astron. Astrophys.*, **205**, 248

Ratag, M.A., Pottash, S.R., Zijlstra, A.A., Menzies, J., 1990, *Astron. Astrophys.*, **233**, 181

Sivagnanam, P., Braz, M.A., Le Squeren, A.M., Tran Minh, F., 1989, *Astron. Astrophys.*, **211**, 341.
Spaenhauer, A., Jones, B.F. & Whitford, A.E., 1992, *Astron. J.*, **103**, 297
te Lintel Hekkert, 1990, *Ph. D. thesis*, Leiden
te Lintel Hekkert, P., Caswell, J.L., Habing, H.J., Norris, R.P., Haynes, R.F., 1991a, *Astron. Astrophys. Suppl.*, **90**, 327
te Lintel Hekkert, P., Dejonghe, H., Habing, H.J., 1991b, *Proc. of Astron. Soc. of Austr.*, **9**, 20
van Langevelde, H.J., Brown, A.G.A., Lindqvist, M., Habing, H.J., de Zeeuw, P.T., 1992, *Astron. Astrophys. Lett.*, submitted
Zijlstra, A.A., Pottash, S.R., Bignell, C., 1989, *Astron. Astrophys. Suppl.*, **79**, 329

JOINT DISCUSSION (HABING AND DEJONGHE)

de Zeeuw: How well does your distribution function for the Bulge, the two integral one, compare to simple forms that have been proposed in the literature, for example by Rowley, for other bulges?

Dejonghe: The model that comes out for the Bulge is fairly isotropic. To that degree, since Rowley's models can also be isotropic, there is likely to be agreement. I haven't at this stage checked this though, but we could probably find at least one set of parameters for his models that would produce a qualitative fit. But producing a quantitative fit, that's another matter!

Rich: If you separate your sample out just by expansion velocity, that is including the Winnberg sample, do you see the Winnberg kinematic disk falling out naturally or is there some other means of your distinction between young and old OH/IR stars?

Dejonghe: The Winnberg sample is much less extended. So I'm not sure that this kind of comparison could be easily made. But answering your question: the Winnberg disk would not merely fall out naturally, simply on the basis of selection on expansion velocity.

Rich: te Lintel had a group I and group II. I believe that group II is thought to be younger. How does that group II compare to the Winnberg stars, is it a lower expansion?

Habing: They are overlapping, so there is no clear distinction on that basis.

Sellwood: You claimed there was no evidence for triaxiality in your sample. But you threw out the one piece of evidence that is there, and that is the rapid rotation of the Winnberg sample. Surely you are looking down a bar and that is exactly what the Winnberg rapid rotation is telling us.

Dejonghe: This model does not use the Winnberg sample. Judging from the te Lintel sample, there is no obvious need for a triaxial bulge and it very well may be that the rapid rotation of the Winnberg sample could be due to high mass, but, then again, it might be a bar. For the big picture, out to 15kpc (in the te Lintel sample), I see no need for triaxiality.

A. Whitford, together with K. Freeman, H. Dejonghe and H. Habing

BULGE K AND M GIANTS

DONALD M. TERNDRUP
Department of Astronomy, The Ohio State University,
174 W. 18th. Ave., Columbus, OH 43210 USA

Abstract. Several independent studies of the abundances and kinematics of K and M giants in the inner Galaxy ($R < 2$ kpc) are assembled to trace out this region's global properties. The mean metal abundance is [M/H] $\approx +0.3$ at $R = 0.5$ kpc, and declines by about 1.2 dex out to $R = 2$ kpc. The line-of-sight velocity dispersion at $R = 0.5$ kpc is $\sigma_r \approx 115$ km sec^{-1} for all population tracers, and declines by $d\log\sigma_r/d\log R = -0.4$. It now seems fairly clear that only the most metal-poor K giants become RR Lyrae variables, while the more metal-rich ones become late M stars. There is some evidence that the most metal-rich stars are in a flatter, more rapidly rotating system. Metal abundance ratios for a few K giants suggest that the inner Galaxy may have formed rapidly.

Key words: Galactic Bulge – Late-type Stars – Metallicity Gradients – Stellar Abundances

1. Introduction

One of the recurrent themes of this meeting is that we do not have it clear in our heads what we *mean* when we use the word "bulge." Some of us use the word to describe the stellar population along lines of sight toward $|\ell|$, $|b| < 10°$ (approximately), and especially the very metal-rich K giants or highly evolved late-type stars found there. Others here refer to a component of the Galaxy that has a particular spatial distribution, boxy or roundish like the bulges of other galaxies, and perhaps containing bars or other non-axisymmetric features. The situation is further confused as we are try to distinguish the bulge from other components of the Galaxy, such as the disk or halo. Several of us here have repeatedly raised – without satisfactory answers – many questions about the whether disk, halo, and bulge stars are all mixed up in the inner Galaxy, or whether there is even any way of distinguishing these components with our current data.

I should like to argue, just for the duration of this talk, that we would be better off if we realized that the terms "bulge," "disk," and "halo" may have limited meaning in our studies of the inner Galaxy:

- Bulges and ill defined: We should remember that a given stellar population is defined either by its spatial distribution or kinematics or chemical abundance. The various structural or population meanings of bulge, disk, and halo which we are using here are not equivalent to one another, especially since the defining populations for these groups are located in very different locations in the Galaxy.

- The bulge is complex: Any stellar population can be thought of as a distribution of points in a parameter space with axes of age, chemical abundances, kinematics, etc. The case can be made (e.g., Carney et al. 1990) that *in the solar neighborhood*, the disk and halo have nearly disjoint distributions in (say) [Fe/H] and kinematics, though intermediate populations, such as the thick disk, have properties that overlap those of the disk and halo. This leads us to view the Galaxy as a superposition of well-defined populations, as Baade

H. Dejonghe and H. J. Habing (eds.), Galactic Bulges, 87–100.

(1944) did when he defined his Populations I and II. Even if we have a complete map of the inner Galaxy in parameter space, however, we may not find that the Galaxy comes apart in that region into separable groups, some of which correspond to the disk or halo that we see locally. Indeed we know (e.g., Lewis and Freeman 1989) that the disk and bulge are kinematically inseparable within $R = 1 - 2$ kpc.

- Our data are incomplete: We still need to clarify the evolutionary relationship between the various tracers of the Galaxy's population and gravitational potential within $R = 1$ kpc. We are busy studying the spatial distribution, metallicities and kinematics of K and M giants, RR Lyrae stars, LPVs, Masers, etc., which may have quite different ages or metallicities. Certain of these objects correspond to local halo stars (e.g., the metal poor ones) or to disk objects (e.g., those orbiting near the circular speed). We may, however, be examining parts of the inner Galaxy that are in different evolutionary stages, calling them all "bulge" objects, when in fact we are looking at portions of the inner Galaxy which are not related to one another in an evolutionary sense, though mixed spatially.

If we were to study the inner regions of the Galaxy or of an external system such as M 31 with a minimum of prejudices, we would probably go to the telescope and obtain long-slit spectra and surface photometry to derive the stellar spatial distribution, and radial dependence of metallicity, velocity dispersion, and projected rotation speed. We would then *model* the integrated light as the sum of a bulge and disk (or whatever other components we want); those who are in this business know that one can decompose a galaxy into bulge and disk in a variety of ways, usually reflecting somewhat the (pre)judgement of the observer. What I intend with this review is to assemble the kinds of data for our own Galaxy that we would get for an external system, then see what these data tell us about the inner Galaxy *as a whole.*

2. K Giants

2.1. Abundance indicators and Abundance patterns

The metallicities of K giants in Baade's Window have been extensively studied because stars of any abundance become K giants in old populations; a well conducted survey should therefore yield the true abundance determination. Spectroscopic surveys by Whitford and Rich (1983) and Rich (1988) have shown that the mean metallicity in Baade's Window is very high ([Fe/H] $\sim +0.3$), and that there is a wide range of abundances present, from [Fe/H] $= -1.0$ to $+0.7$ or even higher. (The precise determination of the metallicity scale for such high-abundance stars is uncertain, but independently of the calibration about one third of the stars in Rich's sample are more strong-lined at a fixed color than anything in the solar neighborhood.) Tyson (1991 and this meeting) and Geisler and Friel (1992) have derived similar metallicity distributions using Washington photometry of the K giants in Baade's Window.

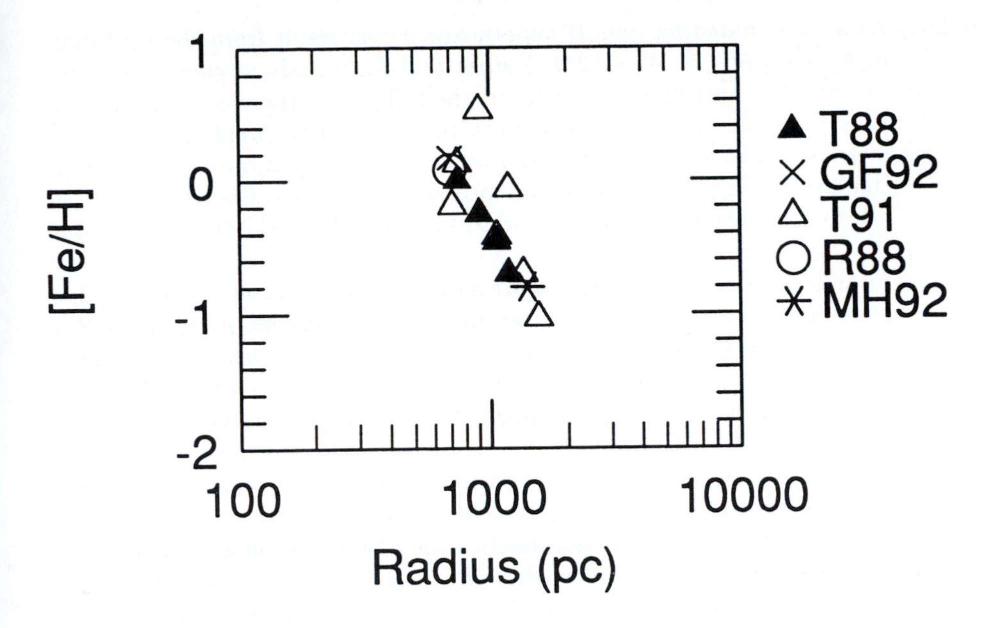

Fig. 1. Mean [Fe/H] of K giants in the inner Galaxy. References are: T88: Terndrup (1988); GF92: Geisler and Friel (1992), T91: Tyson (1991); R88: Rich (1988); MH92: Contributions by Harding and Morrison (this meeting).

In Figure 1 are plotted the mean [Fe/H] of inner-Galaxy K giants as a function of galactocentric distance. The falloff in [Fe/H] is about 1.2 dex from = 0.3 kpc to R = 2 kpc. Over this range, the surface brightness of the bulge falls off by about a factor of 100 (Terndrup 1988, Blanco 1988; see the discussion below). In each of these fields, the dispersion in metallicity is quite large, on the order of 0.75 dex.

Now all of the points on Figure 1 are from photometric studies or from low-resolution spectra, so should be considered as abundance *indicators* and not the true abundances one would derive from model atmosphere/line synthesis studies high-resolution spectra. Progress in obtaining true abundances in the inner Galaxy has been slow, since the relative faintness of bulge K giants ($V > 16.5$) has made high-resolution ($\lambda/\Delta\lambda > 20 \times 10^3$) studies of large samples very difficult. McWilliam and Rich (1992) have spectra of 12 Baade's Window K giants whose magnitudes and radial velocities give a high probability of membership in the inner Galaxy (rather than being nearby stars along the line of sight). They confirm the abundance scale found from the low-resolution spectra and find via analysis of equivalent widths that several of their stars exhibit a mild enrichment of the α-elements, on the order of $[\alpha/\mathrm{Fe}] = +0.2$ to $+0.3$. Perhaps most interesting is that they derive very high abundances of the r-process element Eu. As has been reviewed extensively (Tinsley 1980; Spite and Spite 1985; Gilmore et al. 1989; Wheeler et al. 1989), an enchancement of Eu or of the α-capture elements indicates that the chemical enrichment was dominated by type II supernovae; these result from the explosion of single, massive ($30M_\odot > M > 10M_\odot$) stars that have main sequence lifetimes of 10^9 yr or less. The enrichment process in the bulge was therefore rapid, as in the halo. It is probably reasonable to expect that the bulge could have gone to very high metallicities in a short time: if the inner Galaxy began making stars with a high initial gas density, the evolutionary products would have been trapped in the bulge's deep gravitational potential, and so be available quickly for further star formation.

Peterson and Terndrup (1992) have obtained a high-resolution ($\lambda/\Delta\lambda = 30 \times 10^3$) spectrum of one faint ($V = 17.1$) star in Baade's Window. Analysis of this spectrum was performed using model atmospheres and line synthesis as described by Peterson et al. (1992). The analysis is currently not complete, but preliminary results indicate that for this star at least, strong CN bands suggest an enchancement of nitrogen: [N/Fe] ~ 0.4 at [Fe/H] $= +0.45$. The star does not seem to have an observable oxygen enhancement, suggesting that the bulge may not be completely composed of stars formed in a rapid-enrichment scenario.

Now *rapid* formation does not necessarily imply that the bulge is extremely old, with the same age, for example, as the oldest globulars. In the discussions of the local halo, the statement is often made that the halo has to be old because halo stars have [α/Fe] > 0 and the disk is younger because [α/Fe] ~ 0 (resulting from chemical evolution over time scales longer than 10^9 yr). But strictly speaking the formation of the halo and disk could have been independent, as suggested by the nearly complete separation between the two components in metallicity and kinematics (e.g., Carney et al. 1990). The halo is old because its main-sequence stars are all of low mass; it formed quickly because the stars have [α/Fe] > 0. Even if the inner-Galaxy stars all have [α/Fe] > 0, we still have not measured the age.

The direct measures of the bulge turnoff in fields at $|b| \geq 8°$ (Terndrup 1988) and in Baade's Window (Baum et al., this meeting), give a mean age of $\approx 10 \times 10^9$ yr. Though this value is rather uncertain, depending sensitively on adopted values for the bulge metallicity scale and line-of-sight extinction, it seems that the bulge is younger than the halo even though it, like the halo, may have formed rapidly[1]. This seems consistent with the views of Carney et al. (1990), who argued that the bulge and halo should be a natural evolutionary sequence, with the bulge formed from lost halo gas.

3. M giants

The bulge M giants have been the subject of many studies, since they are relatively bright in the red and infrared, and are easily detected in grism surveys (e.g., Blanco et al. 1984, Blanco and Terndrup 1989, and references in these papers). The metallicity scale for M giants is less certain than for K giants, since the atmospheres of such cool stars contain strong absorption from molecules and it is therefore hard to generate theoretical models for comparison to observed spectra. Bulge M giants

[1] The mean age quoted by Terndrup (1988) – 11 to 14×10^9 yr – was for an adopted distance of $R_0 = 7$ kpc. For $R_0 = 8$ kpc, the distance adopted by Baum et al., the mean age would be 9 to 12×10^9 yr; the two studies are therefore in agreement.

have significantly stronger atomic and molecular absorption than do stars of the same temperature in the solar neighborhood or in globular clusters (Frogel and Whitford 1987; Frogel et al. 1990; Sharples et al. 1990; Terndrup et al. 1990, 1991); semi-quantitative analysis of the photometric and spectroscopic abundances of the Baade's Window M giants in all these papers yield a mean metallicity of [Fe/H] $\sim +0.2$ to $+0.4$ in Baade's Window, in agreement with mean metallicity of the K giants.

The M giants show a decline in mean metallicity by a few tenths dex from $R = 500$ pc to $R = 1500$ pc (Terndrup 1988; Frogel et al. 1990; Terndrup et al. 1990; Tyson 1991). To illustrate this, I reproduce as Figure 2 the correlation of the strength of TiO absorption in bulge M giants as a function of infrared $J - K$ color for three groupings of stars along the minor axis of the bulge ($\ell = 0°$); this plot is from Terndrup et al. (1990).

There is considerable evidence that the M giants, unlike the K giants, are a biased tracer of the bulge population, representing instead the metal-rich end of the metallicity distribution. The first piece of evidence, shown in Figure 3, is that the gradient in the area density of coolest M giants is significantly steeper than that of the K giants or early M giants (Blanco 1988; Terndrup 1988). The bulge M2+ giants fall off with radius according to $\Sigma(R) \propto R^{-\nu}$, with $\nu = 2.5$, like that of the integrated light (Kent 1992) or of the halo, but the falloff of M giants of spectral type M7 or later is much steeper: more like $\nu = 3.3$. The usual interpretation is that the giant branch shifts to warmer temperatures as the metallicity declines, so that fewer stars will have sufficiently cool temperatures to be late M giants.

The second piece of evidence is that there seems to be a lack of M giants that have abundances like the most metal-poor K giants wherever both types of stars have been studied. In Figure 2, for example, the are practically no M giants in the $-3°$ and $-4°$ fields that have TiO strengths like solar-neighborhood stars, and none at all with strengths like 47 Tuc stars (i.e., weaker than those of the solar-neighborhood stars at a given $J - K$), even though there are plenty of K giants in Baade's Window with this abundance. In fact several photometric and spectroscopic studies of bulge M giants (Frogel et al. 1990; Sharples et al. 1990; Terndrup et al. 1990, 1991) all show that these late-type stars show a negligible dispersion in metal abundance; specifically they show that the dispersion in abundance indicators such as CO or TiO strength *vs.* $J - K$ is not larger than is produced by observational error alone. This would be true if only the most metal-rich third (say) of the K giants (which have a dispersion in metallicity of ~ 0.7 dex) in the inner Galaxy were producing all the M giants. The precision in the determination of metallicity in the current studies is about ± 0.3 dex per star, which is about half the intrinsic dispersion of the K giants' metallicities, so all the M giants would look like they have the same abundance. Figure 4 shows a comparison of the abundance distribution for Baade's Window M giants with that of the K giants; the M giant distribution is qualitatively shown as a gaussian of width 0.4 dex centered at [Fe/H] $= +0.3$.

Finally, Harding and Morrison (this meeting) show evidence that in a field at $R \sim 2$ kpc, the kinematics of the M giants resembles that of the metal-*rich* K giants.

Given the situation in globular clusters, in which only the metal-rich ones produce M giants, it is perhaps not surprising that only those stars on the metal-rich

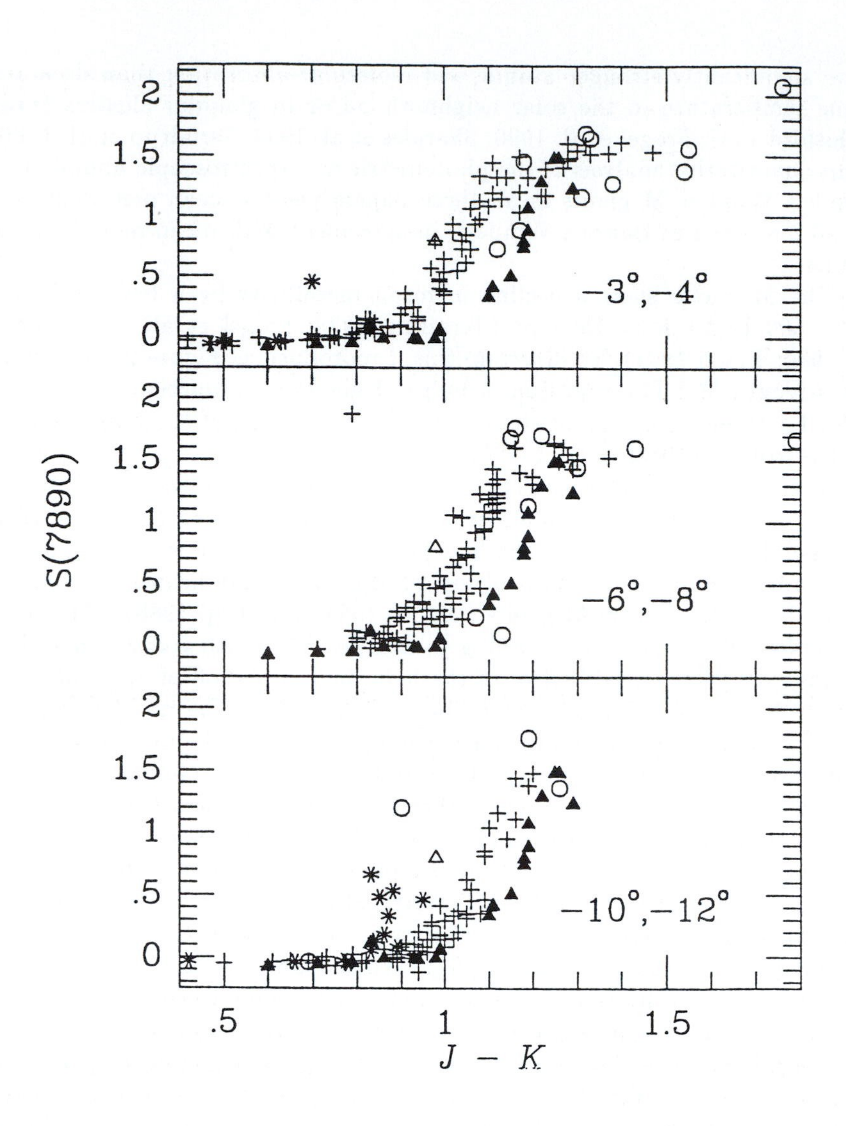

Fig. 2. Correlation between S(7890), a measure of the strength of TiO absorption near 7890Å, and dereddened $J - K$ color. The data are for bulge fields along $\ell \approx 0°$, grouped into three pairs of fields: for $b = -3°$ and $-4°$ (top panel), $b = -6°$ and $-8°$ (center), and $b = -10°$ and $-12°$ (lower panel). Symbols are: *filled triangles*, solar-neighborhood M stars; *open triangle*, the nearby M dwarf Wolf 359; *plusses*, bulge M giants that are not long-period variables; *open circles*, bulge long-period variables; and *asterisks*, foreground M dwarfs seen in front of the bulge. Data are from Terndrup et al. 1990.

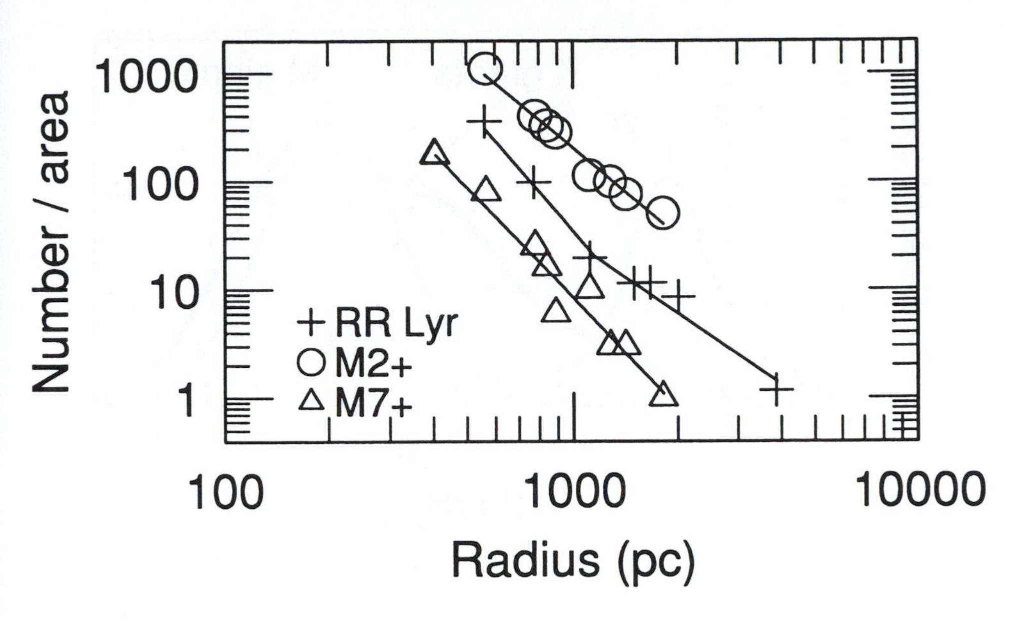

Fig. 3. Surface density of various tracers of the bulge population. The plusses are for RR Lyrae variables, the open circles are for M giants of type M2 or later, and the open triangles are for giants of type M7 or later.

tail of the bulge's [Fe/H] distribution should produce M giants. But I find it odd nevertheless that there are few or no 47 Tuc-like M giants in Baade's Window. If the bulge is somewhat younger than 47 Tuc, which is suggested by the color-magnitude diagrams of the turnoff, then the stars now going up the giant branch should be somewhat more massive, and even more likely to become M giants than those in 47 Tuc. The only way they cannot become M giants is if somehow these turnoff stars are *less* massive than those in 47 Tuc. Guided by the mass-age-metallicity relationship in VandenBerg and Laskarides (1987), the turnoff stars in that cluster should have $M = 0.85 M_{\odot}$. Suppose this is the minimum mass necessary for an M giant to be created, and we want to explain why the minimum metallicity of Baade's Window M giants is [Fe/H] ~ 0 (say); in other words we would like turnoff stars in Baade's Window to have $M = 0.85 M_{\odot}$ at [Fe/H] $= 0.0$. If the helium abundance Y is 0.25, the bulge stars at this metallicity would have an age of approximately 20×10^9 yr. On the other hand if the helium abundance were even a bit higher than 47 Tuc, say $Y = 0.28$, then stars with [Fe/H] $= 0$ and age 10×10^9 yr would have a mass of 0.9 $M_{\odot}$. For Y increasing by a few hundredths while Z goes from 47 Tuc-like to solar implies $\Delta Y / \Delta Z = 1 - 3$, not an unreasonable value.

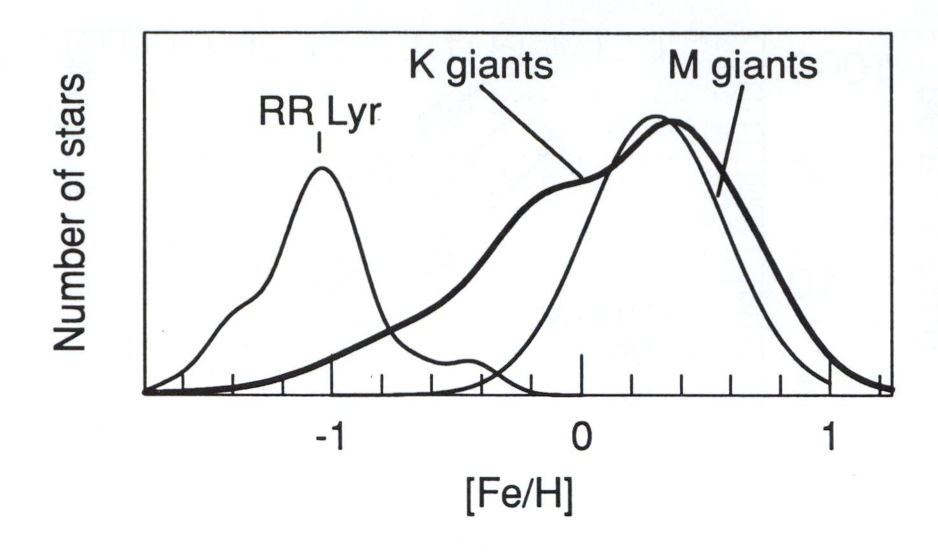

Fig. 4. Comparison of metallicity distributions in Baade's Window. The vertical scale and the relative numbers of stars under each distribution are arbitrary.

4. The RR Lyraes and the K giants

The RR Lyraes in the bulge have a very different metallicity distribution than do the K giants. In Figure 4, I have plotted the RR Lyrae distribution from Walker and Terndrup (1991) along with the K giant distribution from Rich (1988); both distributions are shown as generalized histograms computed from the tabulated metallicities of individual stars. The RR Lyraes are evidently being produced from the most metal-poor tail of the K giants.

In Figure 3 are plotted the surface density of RR Lyrae variables from Blanco (1984) and from Oort and Plaut (1975). The kink in the solid line through the points for the RR Lyrae stars suggests that the surface density gradient may be steeper in the inner bulge than at larger galactocentric distances. (A single line fit through all the points has $d \log \Sigma / d \log R = -2.5$, like that for the M2+ giants.) Now if the bulge were behaving like a "first parameter" globular cluster, then we should expect that the surface density would be less steep than for the general population; as the metallicity increases toward the galactic center, the distribution of stars on the horizontal branch would shift to the cool side of the instability strip, and there should be proportionally fewer RR Lyrae variables. That the surface density steepens within $R = 1$ kpc suggests that the inner bulge exhibits the "second parameter" effect.

Lee (1992) has recently analyzed the distribution of stars on the horizontal branches of globular clusters, in an attempt to see whether there is any variation with location in the Galaxy of the age spread among the globulars, assuming that age is the second parameter. Lee finds that globular clusters within $R = 8$ kpc are all coeval, and on average about 2 Gyr older than those with $8 \leq R < 40$ kpc; clusters in this latter range of radius have an age spread of $3 - 4$ Gyr. Such an interpretation is also suggested by Suntzeff et al. (1991) in their analysis of RR Lyraes in the field halo.

Lee also finds that the difference is mean metallicity between the RR Lyraes and the K giants in Baade's Window (see Figure 4) can only be explained if the bulge is $1{-}2 \times 10^9$ yr *older* than the halo globular clusters. The combination of old bulge and the age gradient with galactocentric radius leads Lee to propose that the bulge formed from the inside out. This is hard to reconcile with the observations from color-magnitude diagrams (Terndrup 1988; Baum et al., this meeting) that the bulge has a mean age near 10×10^9 yr, which suggests to me that there are many details of the evolution of bulge stars which remain to be understood.

5. Kinematics

5.1. Bulge kinematics

To close, I would like to summarize the current information on bulge kinematics, and draw a few conclusions about the structure of the bulge and how stars of different metallicities may differ in spatial distribution. Rich (1990) and Sharples et al. (1990) have presented similar discussions in their studies of the K and M giants in Baade's Window.

The line-of-sight velocity dispersion in the inner Galaxy is a slow function of galactocentric radius. In Figure 5 are plotted values of σ_r from several recent studies. These data are for K and M giants both along the minor axis of the bulge and for some fields several degrees away. The correlation between σ_r and galactocentric distance R is quite tight, with slope $d \log \sigma_r / d \log R = -0.4$. This behavior of σ_r is also seen in Kent's (1992) kinematical model of the bulge.

In Figure 5 are shown two points of σ_r from Rich (1990) and from Harding and Morrison (this meeting), who divide their sample into metal-poor and metal-rich stars; in each case, the metal-rich giants have a lower velocity dispersion, suggestive that they are confined to a flatter system with higher rotation speed Harding and Morrison show this explicitly in their off-axis field.

Possible correlations between metal-abundance and kinematics been explored extensively in Baade's Window. Data from this region are summarized in Table I. Rich (1990) presented velocities of 53 K giants in Baade's Window, and derived a line-of-sight radial velocity dispersion $\sigma_r = 105 \pm 11$ km sec^{-1} for his full sample. He then divided his sample into bins of metallicity, and showed (though with small number statistics) that the radial velocity dispersion for the most metal rich stars ([Fe/H $\geq +0.3$) was 92 ± 14 km sec^{-1}, much lower than 126 ± 22 km sec^{-1} for stars with [Fe/H] < -0.3. This effect was also seen in the proper motion data of Spaenhauer et al. (1992), who measured relative proper motions in the ℓ and b directions for over 400 stars in Baade's Window. (The proper motions are relative

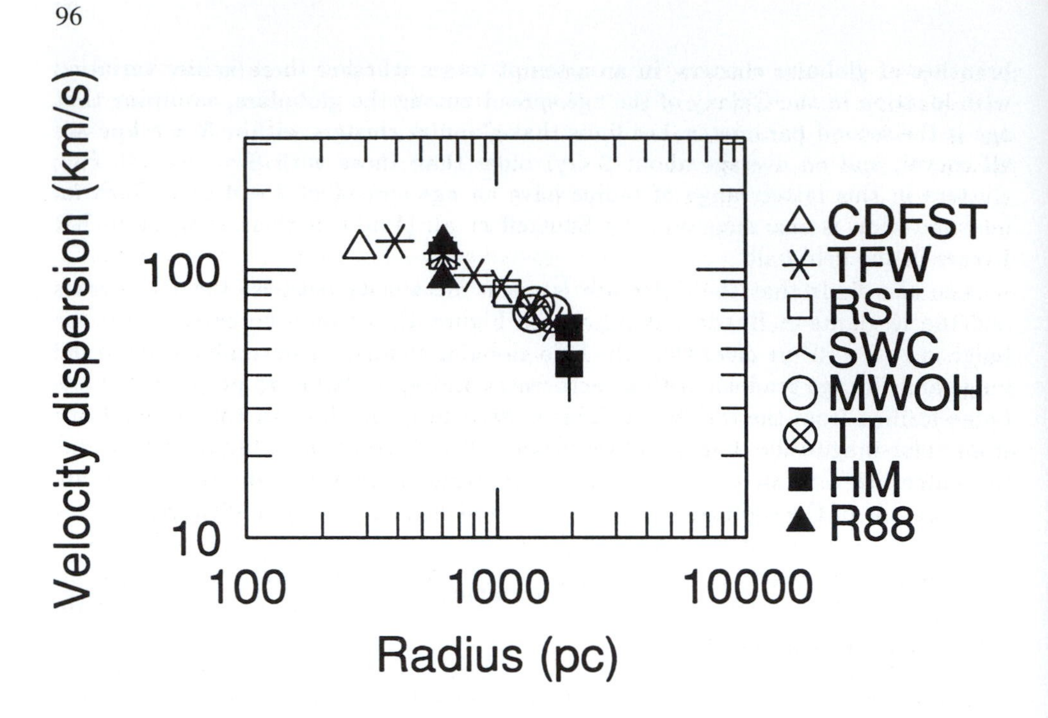

Fig. 5. Line-of-sight velocity dispersion in the bulge. The sources for the points are: CDFST: Carr et al. 1992; TFW: Terndrup et al. 1992; RST, Rich et al. 1992; SWC: Sharples et al. 1990, MWOH: Minniti et al. 1992; TT: Tiede and Terndrup 1992; HM: Harding and Morrison, this meeting; R88: Rich (1988).

TABLE I
Line-of-sight velocity dispersions in Baade's Window.

Source	σ_r	σ_ℓ	σ_b	N
Rich (1990) K giants	105 ± 11	...	...	53
Rich, [Fe/H] < -0.3	126 ± 22	...	...	16
Rich, [Fe/H] $\geq +0.3$	92 ± 14	...	...	21
Sharples et al. (1990) M5+ giants	113 ± 11	...	...	225
Spaenhauer et al. (1992) all stars	...	115 ± 4	100 ± 4	429
Spaenhauer et al. with [Fe/H] $\geq +0.0$	101 ± 15	118 ± 14	58 ± 9	34

to the average of the sample because there are no background galaxies or proper motion standards visible in Baade's Window; their study therefore is a measure of the *dispersion* in the line-of-sight proper motion.) When they add radial velocities from Rich, Spaenhauer et al. find that the velocity dispersions in Baade's Window are nearly isotropic in projection, meaning the line-of-sight velocities are nearly equal: $\sigma_r \approx \sigma_\ell \approx \sigma_b$. When they use the small number of Rich's most metal-rich stars that are also in their sample, Spaenhauer et al. show that the vertical velocity dispersion $\sigma_b \sim 60$ km sec^{-1} is smaller than the tangential velocity dispersion σ_ℓ.

That the projected velocity dispersion in Baade's Window is nearly isotropic may be a problem for some of the bar models for the bulge. A bar aimed somewhat towards us, as most of the current models indicate, ought to show $\sigma_\ell > \sigma_r$ and $\sigma_\ell > \sigma_b$. This does not seem to be the case, which suggests to me that the *majority* of Baade's Window stars cannot be in the bar. It may be the case that the expected bar signature is seen in the most metal-rich K giants in Baade's Window, but current statistics are not good enough to confirm this[2]. The only evidence that suggests that the metal-rich stars are in a bar is presented by Sharples et al. (1990), who have shown that the late-type M giants in Baade's Window, which perhaps are the most metal-rich of the stars in that region, show evidence for a tangentially anisotropic velocity distribution.

The values of projected rotation speed have measured in several places in the inner Galaxy, as summarized in Figure 6.

Bulges of galaxies like our own (between Hubble type Sb and Sc) are *not* self-gravitating bodies; the gravitational potential in the bulge has a very significant contribution from the disk. Consequently, the only proper way to explore kinematics in the bulge is to generate a mass model, as Kent (1992) has done from the Galaxy's K-band light distribution, and solve the equations relating kinematics and the density distribution. On the other hand, we can estimate how stars of different kinematics will be arranged in the bulge by making use of simple dynamical arguments, like those in Frenk and White's (1980) analysis of the globular cluster system. Suppose that stars near the mean of the bulge's metallicity distribution are arranged like $\Sigma \propto R^{-2.5}$, and are in a flattened axisymmetric bulge with axial ratio $c/a = 0.7$ (Kent 1992). At the distance of the Baade's Window, these stars would mean rotation speed of $60 - 80$ km sec^{-1}. If the metal-rich stars have $\sigma_r = 90$ km sec^{-1}, as suggested by the data in Table 1, then they would be in a significantly flatter system, one with $c/a \sim 0.4 - 0.5$ and have a projected surface density like $\Sigma \propto R^{-3.2}$, which is what the late M giants do.

So a first-order self-consistent picture the M giants have the metallicities and kinematics that would arise if they are produced by the metal-rich tail of the K giant distribution. To confirm this, we would have to get much better abundances for both the K and M giants in several locations; a confirmation would also come with have well-understood star counts from which the evolutionary paths and rates can be derived. The big picture seems increasingly clear, but many details need to

[2] I am part of a program which has obtained radial velocities and metallicities for the full Lick proper motion survey in Baade's Window; our results are now in preparation (Rich et al. 1992), and should indicate whether there really is a significant variation of kinematics with metallicity or whether there are kinematical substructures in the bulge

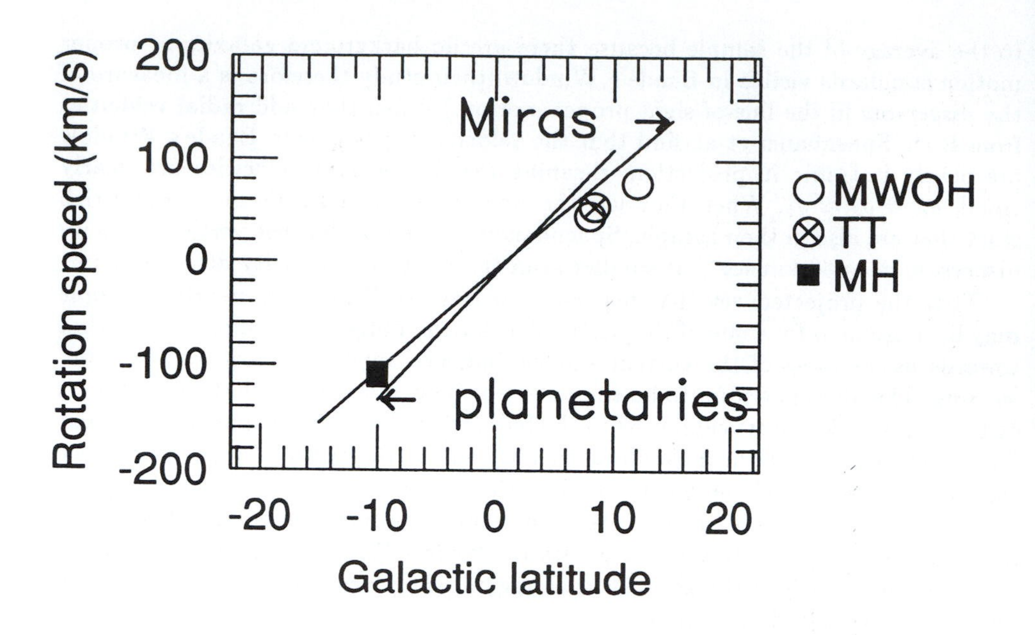

Fig. 6. Rotation velocity for various tracers in the inner Galaxy. The sources for the points are: MWOH: Minniti et al. 1992; TT: Tiede and Terndrup 1992; HM: Harding and Morrison, this meeting. The solid lines show the velocities found for Miras (Menzies 1990) and for bulge planetary nebulae (Kinman et al. 1988).

be filled in.

References

Baade, W. 1944, ApJ, 100, 137
Blanco, B. M. 1984, AJ, 89, 1836
Blanco, V. M. 1988, AJ, 95, 1400
Blanco, V. M., McCarthy, M. F., & Blanco, B. M. 1984, AJ, 89, 636
Blanco, V. M., & Terndrup, D. M. 1989, AJ, 98, 843
Carr, J., DePoy, D., Frogel, J. A., Sellgren, K., & Terndrup, D. M. 1992, in preparation
Carney, B. W., Latham, D. W., & Laird, J. B. 1990, AJ, 99, 572
Frenk, C. S., and White, S. D. M. 1980, MNRAS, 193, 295
Frogel, J. A., & Whitford, A. E. 1987, ApJ, 320, 199
Frogel, J. A., Terndrup, D. M., Blanco, V. M., & Whitford, A. E. 1990, ApJ, 353, 494
Geisler, D., & Friel, E. D. 1992, AJ, 104, 128
Gilmore, G., Wyse, R. F. G., & Kuijken, K., 1989, ARA&A, 27, 555
Kent, S. M. 1992, ApJ, 387, 181
Kinman, T. D., Feast, M. W., & Lasker, B. M. 1988, AJ, 95, 804
Lee, Y.-W. 1992, AJ, 104, 1780
Lewis, J. R., & Freeman, K. C., 1989, AJ, 97, 139
McWilliam, A., & Rich, R. M., 1992, in preparation.

Menzies, J. W, 1990, in *Bulges of Galaxies*, eds. B. Jarvis & D. Terndrup, ESO Conf. Ser., 35, 115
Minniti, D., White, S. D. M., Olszewski, E. W., & Hill, J. M. 1992, ApJ, 393, L47
Oort, J., & Plaut, L. 1975, A&Ap, 41, 71
Peterson, R. C., & Terndrup, D. M. 1992, in preparation
Peterson, R. C., Dalle Ore, C., & Kurucz, R. 1992, ApJ, in press
Rich, R. M. 1988, AJ, 95, 828
Rich, R. M. 1990, ApJ, 362, 604
Rich, R. M., Sadler, E. M., & Terndrup, D. M. 1992, in preparation
Sharples, R., Walker, A., & Cropper, M. 1990, MNRAS, 246, 54
Spaenhauer, A., Jones, B. F., & Whitford, A. E. 1992, AJ, 103, 297
Spite, M., & Spite, F. 1985, ARA&A, 23, 225
Suntzeff, N. B., Kinman, T. D., & Kraft, R. P. 1991, ApJ, 367, 528
Terndrup, D.M 1988, ApJ, 96, 884
Terndrup, D. M., Frogel, J. A., & Whitford, A. E. 1990, ApJ, 357, 453
Terndrup, D. M., Frogel, J. A., & Whitford, A. E. 1991, ApJ, 378, 742
Terndrup, D. M., Frogel, J. A., & Wells, L. A. 1992, in preparation
Tinsley, B M. 1980, Fund. Cosmic Phys., 5, 287
Tyson, N. D. 1991, Thesis, Princeton University
Walker, A. R., & Terndrup, D. M. 1991, ApJ, 378, 119
Wheeler, J. C., Sneden, C., & Truran, J. W. 1989, ARA&A, 27, 279
Whitford, A. E. & Rich, R. M. 1983, ApJ, 274, 723
VandenBerg, D. A., & Laskarides, P. G. 1987, ApJS, 64, 103

DISCUSSION

Sellwood: In The Spaenhauer *et.al.* proper motion samples, for the l and b components, did you look to see if they were gaussian?

Tendrup: Sadler has done this for the whole sample. If you look at the (l, b) plot, it forms a sort of round web. It isn't strongly tilted one way or the other.

Rich: I find it interesting that the M-giants seem to change in abundance as one goes from the inner to the outer fields. Yet, all the populations seem to share a single fall of velocity dispersion along the minor axis. This includes RR-Lyraes that were found by Rogers (1977), that seemed at a very low velocity dispersion, about 60 km/s at -10 degrees.

Tendrup: I agree. The differences in velocity dispersion between stars of 2 dex in metallicity are subtle. They were 1σ in your work and 1.5σ–2σ now, even in our new survey of over 400 stars. You must have very large samples in individual lines of sight to be able to make good statistical claims. I think that simply what is going on is that the gravitational potential has a certain shape and within that potential there were small changes in the kinematics as the chemical enrichment proceeded.

King: If they have the same velocity dispersion in the centre and if they have anything like a gaussian velocity distribution, then they must have the same run of velocity dispersion with distance from the centre. Gravitation doesn't know about abundances, they're just test particles.

Norman: Is your metal rich population, of flattened stars, old or young?

Tendrup: It hasn't been looked into and I think it would be extraordinarily difficult to do, because you would need to know the details of determining the ages for individual stars. This means building really trustworthy stellar atmospheres models that would let you point to the spectrum and say "I know the gravity of this star" and that's tough! You could do statistical analysis, for example, on the main sequence and ask why the turnoff is so fat. Is it consistent with a rapid formation or with a big spread of abundances etc...

Morrison: I think that RR-Lyrae kinematics may give more information about age, this should be looked into.

Rich: We also need more complete samples of Miras, especially in identifying the longest period Miras. This may give us more direct information about the period and kinematics and any possible correlation between them.

Working at dinner time: W. Zeilinger and F. Bertola

THE GLOBULAR CLUSTER AND HORIZONTAL BRANCH CONTENT OF THE BULGE AND HALO

George W. Preston
OCIW
Carnegie Institution of Washington
Pasadena, California, USA

1. The Transition From Halo to Bulge

The radial variations ("gradients") of three properties of the horizontal branch (hereafter HB) across the transition region ("edge") between Halo and Bulge place modest constraints on the early history of these regions and possible interactions between them. The extant data are fragmentary, and the conclusions correspondingly uncertain. The observables are HB morphology, heavy element abundance, and space density. The data base is constrained by sampling techniques. According to current dogma HB stars are descendents of low-mass stars that grow degenerate He cores during post-main sequence evolution, so they are old. Unfortunately, we cannot sample the whole HB. In most parts of the Galaxy our knowledge of the field HB is limited to those species that are amenable to discovery by simple survey techniques, namely, the BHB and RR Lyr components. This is an annoyance, because it is the *total* HB that traces the density of its parent population (Preston, Shectman, and Beers 1991a) and the RHB component must be inferred by indirect methods. Furthermore, well above the Galactic plane, where nearly all searches have been conducted (see Preston *et al* 1991a for references), these two species occur only in that subset of the population for which [Fe/H]<-0.8 [a generalization, incidentally, that does not apply to RR Lyr stars in the solar neighborhood (Preston 1959) or in the Bulge (Walker & Terndrup 1991)]. So, bear in mind that our deductions flow from an incomplete sampling of the metal-poor component of the HB. Finally, by "edge" of the Bulge I refer to the intuitive boundary seen in maps of IRAS 12-25 micron point sources (Habing *et al* 1985, Harmon and Gilmore 1988) or, equivalently, to the FWHM of the areal distribution of M-giants (Blanco & Terndrup 1989), which define an elliptical region of enhanced density with axis ratio $c/a \sim 0.7$ and characteristic dimension ~ 1 kpc. I leave the assessment of Weinberg's (1992) recent counter view to others at this symposium.

1.1. THE HB MORPHOLOGY GRADIENT

Three independent lines of evidence indicate that the color distribution of the HB in the Galactic Halo reddens with increasing Galactocentric distance (R):

(1) The HB color distributions, described by the parameters B/(B+R) (Mironov 1972) or (B-R)/(B+V+R) (Lee 1989), of families of globular clusters defined by Galactic location, shift redward at fixed composition as R increases (Searle & Zinn

H. Dejonghe and H. J. Habing (eds.), Galactic Bulges, 101–115.

1978, hereafter SZ; Zinn 1980, 1986; Lee 1992a). Recall that B, V, and R denote the blue, RR Lyr, and red components of the HB. The redward shift is accompanied by an increase in dispersion. This seminal discovery by SZ introduced HB morphology, however defined, as a parameter of Galactic structure.

(2) The mean *B-V* color of field BHB stars, appropriately defined, increases with R on $4 < R < 12$ kpc in spite of a small negative abundance gradient that, by itself, would produce an effect of opposite sign, were mean HB mass constant throughout the halo (Preston, Shectman, & Beers 1991a).

(3) The space density of field BHB stars appears to decline outward more rapidly than that of the RR Lyr stars in the vicinity of the solar radius, which implies a reddening of the field HB with increasing R (Preston *et al* 1991a).

SZ cautiously suggested the simplest explanation proffered by HB theory in the context of Galactic structure, namely, that mean HB mass increases (age decreases) outward in the Halo. The observational basis for this conclusion was subsequently refined by Zinn (1980, 1986), and strengthened more recently by the construction of semi-empirical correlations between age and HB morphology among the globular clusters (Sarajedini and King 1989, VandenBerg, Bolte & Stetson 1990, Sarajedini and Demarque 1990, Preston *et al* 1991a). Particular attention (King, Demarque & Green 1988, Bolte 1989, Demarque *et al* 1989, Green & Norris 1990) has been paid to the globular clusters NGC 288 and NGC 362, which possess similar abundances and Galactocentric distances, but different HB morphologies ("blue" and "red", respectively) and ages (older and younger, respectively).

Although the FHB is very "blue" at the solar radius, n(BHB)/n(RR Lyr)~6 [n = number density], and appears to become even bluer as R decreases, the HB in Baade's Window (hereafter BW) is certainly "red" (Terndrup 1988). I have used the scant data at my disposal to quantify how the morphology gradient changes sign abruptly at the "edge" of the Bulge.

We have identified some 700 BHB candidates in the Curtis Schmidt objective prism field CS30321, which overlaps the Oort-Plaut field at l = 0, b = -10 reinvestigated by Wesselink (1987). *UBV* photometry of 150 of these indicate that virtually all of the fainter candidates are BHB stars (Preston, Shectman, & Beers 1991a, 1991b), so I used apparent magnitudes derived from a photoelectric calibration of candidate brightness classes (Beers, Preston & Shectman 1988) to derive an observed ratio n(BHB)/n(RR(Lyr) ~ 4 near R = 2 kpc from a plot of RR star and BHB candidate counts *versus* distance modulus in the region of overlap. Wesselink's RR Lyr counts are virtually complete in the apparent magnitude interval of interest, but the objective prism data are incomplete by a factor of 2 or more at B = 15, so the true ratio is probably 8 or more: the field HB continues to become "bluer" to R ~ 2 kpc.

To quantify morphology of the metal-poor HB in BW, I counted BHB, RR Lyr, and RHB stars in a strip 0.2 mag. high [~FWHM for an R^{-3} density law at $l=0$, $b=-4$] superposed on Terndrup's CMD at V(RR Lyr)=16.8 [calculated from B(RR Lyr)=17.55 (Blanco 1984), $(B-V)_0$ =0.30, and $E(B-V)$=0.45 (Terndrup 1988)]. I took the shape of the HB blueward of the RR Lyr gap from the globular clusters (Preston *et al* 1991a) and terminated the RHB at $(B-V)_0$=0.85, the approximate red end of the HB in 47 Tuc (Lee 1976, Hesser *et al* 1987). To count RR Lyr stars observed at random phases, I supposed that RR Lyr stars lie in an inclined strip appropriate for a V light amplitude of 1 mag. and a linearly correlated $B-V$ variation of amplitude 0.35 mag. Such stars will be found with greatest probability near minimum light, i.e., in the lower right corner of the strip. Finally, I counted stars in boxes with dimensions 0.05 in $B-V$ and 0.5 in V to derive areal densities of contaminators above and below the HB and interpolated to derive corrections along the HB. The corrected numbers of BHB, RR Lyr, and RHB stars are 3.5, 2.1, and 18, respectively -- much work for small numbers. The RR Lyrae value, 2.1, is to be compared with 1.5 estimated from Blanco's (1984) areal density of 0.10/square arc minute and Terndrups's field of 15 square arc minutes. These meager results verify that the metal-poor HB in BW is "red" and that n(BHB)/n(RR Lyr) ~ 1 at R = 0.5 kpc, i.e., lower by a factor of 8 than it is at R=2 kpc. These estimates can be improved by additional photometry in BW, but it will be difficult to make accurate counts of the HB stars of higher abundance. The HB of the relatively metal-rich globular cluster NGC 6553 studied by Ortolani, Barbuy, & Bica (1991) appears to overlap the giant branch, and one can see vague evidence of this phenomenon in Terndrup's CMD of BW. Subtraction of a giant branch contribution from total counts in the region of the HB by interpolation of giant branch counts into the HB region from above and below would produce a first approximation. As of now we have neither theoretical nor empirical locations for HB's of solar metallicity or greater.

1.2. THE ABUNDANCE GRADIENT

Exterior to the solar radius the Halo abundance gradient is undetectably small by all techniques that have been employed to measure it (Zinn 1985; Suntzeff, Kinman & Kraft 1990, hereafter SKK; Carney *et al* 1990). However, on $2 < R < 8$ the gradient derived from field RR Lyr stars, -0.10 dex/kpc, is twice that of the Halo globular clusters (SKK). The gradient derived from RR Lyr stars in clusters binned in intervals of R differs from the cluster gradient in the same manner, as shown in Fig. 1, a phenomenon that appears to be produced entirely by the gradient in HB morphology (Zinn 1986, SKK): clusters of lower-than-average abundance near R_0 produce copious numbers of RR Lyr stars, while those of comparable abundance in

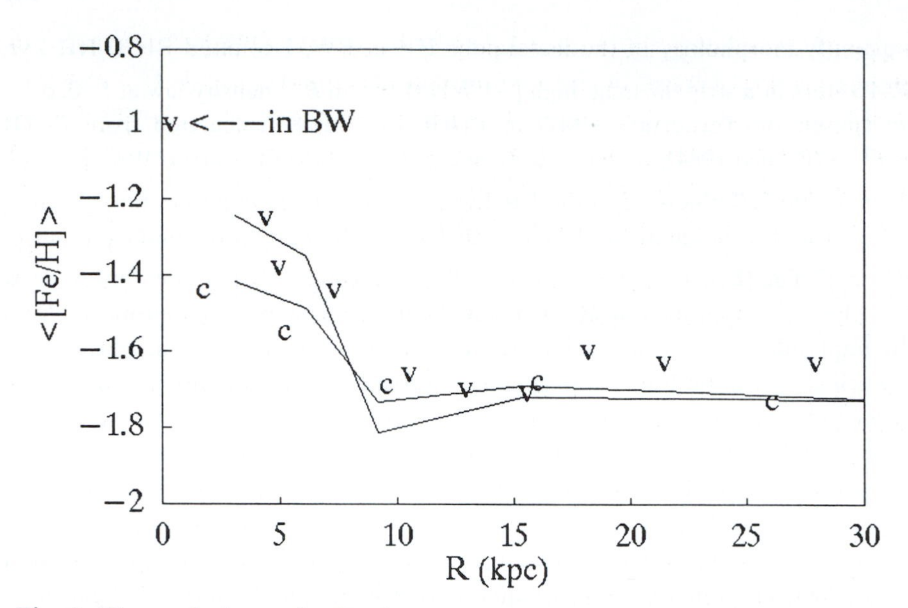

Fig. 1.-The variations of <[Fe/H]> for the globular cluster data ("c") of Zinn (1985) and the RR Lyr data ("v") of SKK, binned in intervals of Galactocentric *R*. The segmented curves represent the mean abundances of the SKK cluster sample and of the RR Lyr stars contained in that sample, also binned in intervals of *R*. The gradient of HB morphology biases the RR Lyr sample against low [Fe/H] at small *R*.

the inner Halo produce few or none at all because of their "blue" morphologies. Thus, the cluster gradient is more nearly representative of the Halo field gradient exterior to $R = 2$ kpc. Direct measurement of the Halo gradient by measurements of *bona fide* tracers of the field, e.g., red giants or turnoff stars, would provide welcome, though laborious, confirmation of this surmise. As noted below, the distinction between properties of clusters and co-spatial field populations may be important in the central region of the Galaxy.

The mean abundance of RR Lyr stars in BW (Walker & Terndrup 1991) lies along a linear extension of the SKK data for the inner Halo (see Fig. 1). If the foregoing interpretation of observed Halo gradients is correct, this behavior must be a purely accidental consequence of the interplay between partially correlated changes in HB morphology and abundance distributions that accompany the transition from Halo to Bulge. I refer to "partial correlations", because HB morphology depends on three composition parameters, core mass, and mean HB mass, in addition to assumptions about the sensitvity of progenitor mass-loss to abundance. Critics of variable age as the cause of the morphology gradient can try to invent defensible alternatives with other combinations of these parameters.

The particular assumptions made recently by Lee (1992a, 1992b) lead him to the conclusion that the decrease of mean HB mass (increase of age) with decreasing R in the Halo continues into the Bulge, which, accordingly, contains the oldest stellar population in the Galaxy. Such an old population could only be reconciled with Harmon and Gilmore's (1988) relatively young (age < 10 Gy) Miras by an extended period of star formation in the Bulge, but such reconciliation may not be necessary in view of the greater ages for Miras reported by Blommaert *et al* (1992) at this symposium. Spectroscopic determination of the extent of the [O/Fe] plateau in the Bulge may constrain the allowable age spread (Matteucci & Brocato 1990).

Globular clusters cannot be used to extend the Halo gradient into the Bulge, because so few clusters have been found there. The abundance distribution of those few clusters that can be located within a kiloparsec of the Galactic Center does not resemble that of Rich's (1990) field giants. These remarks follow from the presentations in Figs. 2 and 3. To construct the Figures I used the Galactocentric distances of Webbink (1985) and abundances from, in order of preference, Armandroff (1989), Zinn & West (1984), and Webbink (1985).

Aguilar, Hut, & Ostriker (1988) calculate that globular clusters formed at $R < 2$ kpc are destroyed efficiently in a Hubble time, and those that survive may sink inward and out of sight due to dynamical friction. Their calculations warn us that some globular clusters that we now see in the inner Halo and Bulge may have spiraled into these regions from the outside, so the abundance gradient across the edge of the Bulge is better estimated by use of Rich's K giants at $R = 0.5$ kpc and the Halo globular clusters just exterior to $R = 2$ kpc, in which case the gradient out of the plane, near the edge of the Bulge, is ~ -0.8 dex/kpc, some 20 times steeper than that of the inner Halo.

1.3. THE DENSITY GRADIENT OF THE METAL-POOR HORIZONTAL BRANCH

Survey techniques define the "metal-poor HB" as one that contains RR Lyr and/or BHB stars. Exterior to $R = 2$ kpc the upper abundance bound from globular cluster data (SKK 1991) is ~ -0.8. In BW the upper bound appears to be larger by about 0.3 dex (Walker & Terndrup 1991). From BHB and RR Lyr counts near the solar radius and a small RHB correction inferred from local globular clusters we estimated the total HB density at R_0 to be 42 kpc^{-3} (Preston *et al* 1991a). At $R = 2$ kpc the HB is so blue, n(BHB)/n(RR Lyr)~8, that the RHB correction is probably miniscule, i.e., the sum of Wesselink's (1987) RR Lyraes and our BHB estimate is very nearly the total HB. In BW we estimate the total density from Blanco's (1984) RR Lyr data and the n(BHB)/n(RR Lyr) and n(RHB)/n(RR Lyr) number ratios given in Section 1.1 above. From these data we can sketch the run of density of the

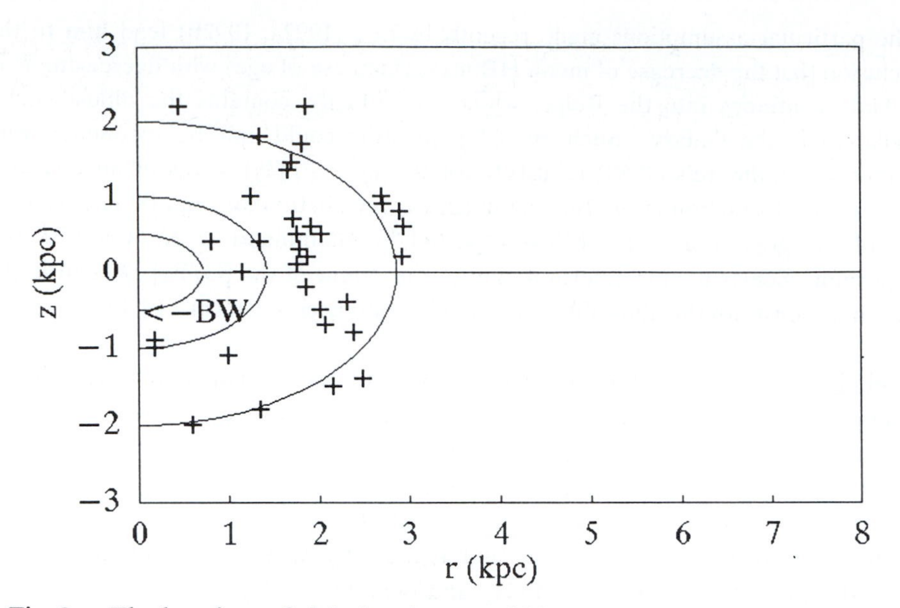

Fig. 2. - The locations of globular clusters within R = 3 kpc, according to Webbink (1985), are plotted in cylindrical coordinates. Isodensity contours inferred from cool giants (c/a = 0.7) are shown for c = 0.5, 1.0, and 2.0 kpc. All of the clusters are exterior to the tangent point in BW.

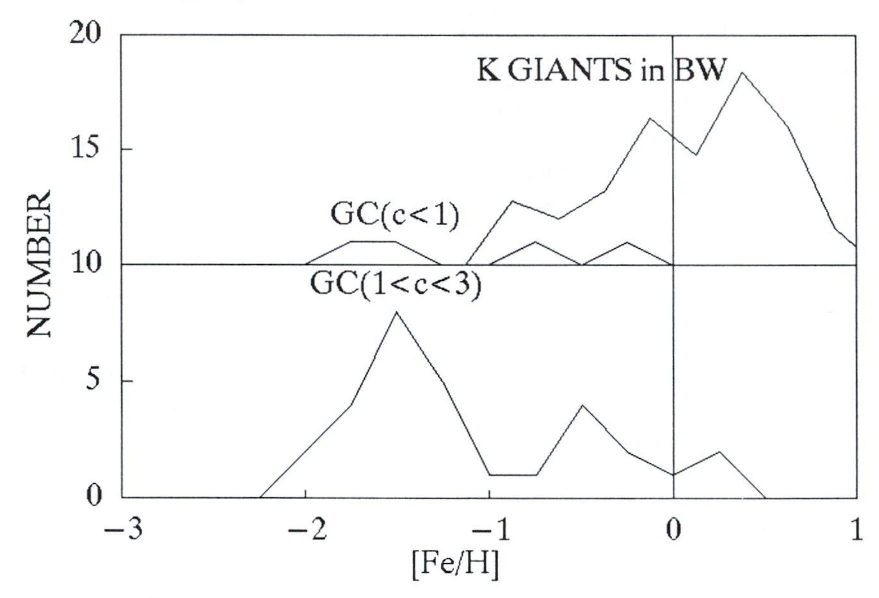

Fig. 3. - The abundance distribution of the four globular clusters interior to the ellipsoid c = 1.0 kpc does not match that of Rich's (1990) K giants in BW. It resembles a subset of the cluster abundance distribution in the adjacent ellipsoidal shell.

total HB in Fig. 4. After the RR Lyr counts are corrected for the gross changes in HB morphology that occur near the edge of the bulge, the total HB density in BW appears to lie, approximately, on an inward extension of the Halo density distribution.

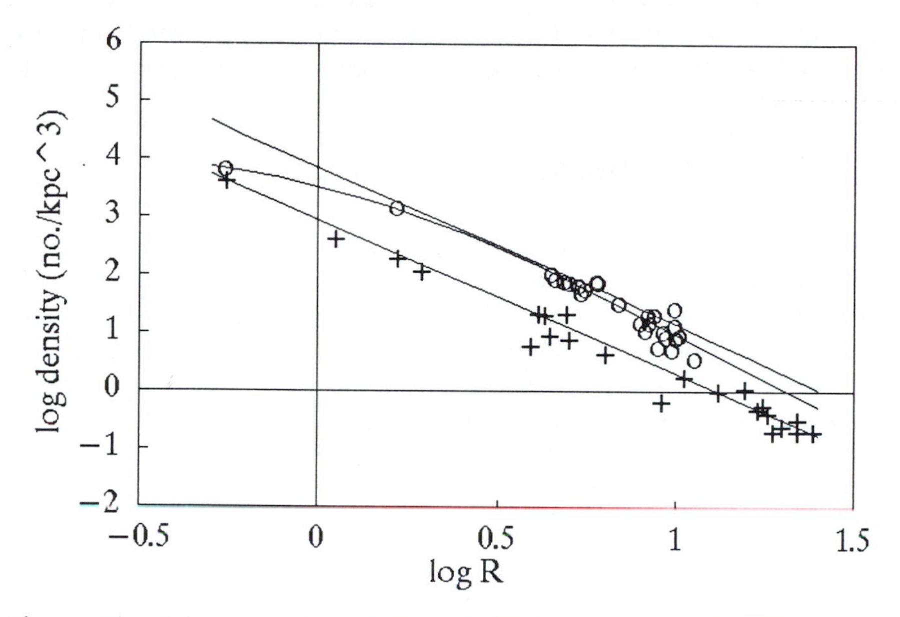

Fig. 4. - The Galactic density variations of RR Lyr stars (+) and BHB stars (o). The uppermost of the three curves is a schematic representation of the density of the total metal-poor HB derived from fragments of information, as described in the text.

I used the density law of the Blanco & Terndrup (1989) spheroid component, scaled to the HB densities at R = 0.5 kpc and at R = 2.0 kpc, to estimate the mass of the stellar population associated with the metal-poor HB within R = 1 kpc. Converting HB numbers to mass by use of L_V/N_{HB} = 500 (Preston *et al* 1991a) and M/L_V = 2.5, I obtain metal-poor masses in the range 2.6E+8 to 3.4E+8. In these calculations ~75% of the contribution to the integral is located inside of 0.5 kpc, where there are no HB data, so the result is largely an artifact of the adopted density law. These masses, ~ 2 to 3 percent of the total mass derived by Blanco & Terndrup, are comparable (to within a factor of ~ 2) to those expected for the metal poor component in a simple model of yield twice solar. The HB data, poor as it admittedly is, corroborates Rich's (1990) conclusion that there is no gross G-dwarf problem in the Bulge. I offer this crude calculation not as a polished result but rather as an illustration of how improvements in the HB data can be used to explore the chemical evolution of the Bulge.

In summary, the behavior of the HB seems to be in reasonable accord with a suggestion made by Searle (1979) in Liege thirteen years ago. The basic features of abundance distributions throughout the Halo-Bulge system can be attributed to variations of Hartwick's (1976) gas-depletion rate during star formation, large and somewhat variable among Searle's primordial fragments of the outer Halo, and declining steeply, perhaps to zero, in the densest portions of the central attracting mass, where star formation first took place, if Lee's (1992a, 1992b) arguments should prove to be correct.

2. Variable Stars in the Galactic Bulge

The RR Lyr and Mira variable stars have been used extensively to explore properties of the Bulge. These are but two of the several classes of variable stars long known to inhabit globular clusters (Hogg 1973). More recently a whole new literature has arisen about the occurrence of binary stars and blue stragglers, their putative progeny, and theoretical treatment of the capture/ejection processes that alter binary populations and accelerate the dynamical evolution of clusters (see Leonard 1989 and Mateo *et al* 1990). The Bulge offers a rather different environment in which to study physical properties of the variable stars for their own sake, and from these there will be feedback in the form of information about the structure, kinematics, age, and chemical properties of the Bulge. These remarks are preamble to a description of work in progress at the Las Campanas Observatory.

A major photometric search for variable objects in the Galactic Bulge began in April of this year as a Carnegie-Princeton-Warsaw collaboration, led by Bohdan Paczynski, at the LCO 1-m telescope equipped with a Ford 2048^2 chip. The ultimate goal is detection of gravitational microlensing of Bulge stars by all compact objects (stars, brown dwarfs, planets) that contribute to the Galactic Disk. Success in this endeavor is precursor to the search for dark matter in the Galactic Halo.

Eleven fields are measured on each suitable night in the *V* and *I* passbands. Sixty nights have been assigned to the project in the current calendar year and we anticipate this level of support for an additional three years. Because the quality of the photometry for faint stars in crowded fields is very sensitive to "seeing", observations are restricted to nights when the FWHM of the PSF is less than 1.5". Under these conditions each of the nine CCD images in BW produces photometry of 100,000 to 150,000 stars, of which ~ 40% have formal errors < 0.10 magnitudes. The errors increase with apparent magnitude as indicated below:

V	sigma	N
20.0	0.07	33000
20.5	0.10	55000
21.0	0.15	80000

Further details of the project (observations, reductions, preliminary results) are presented in a poster paper at this symposium prepared by Mario Mateo on behalf of the project participants.

In a lensing project intrinsic variable stars are undesirable distractions that spawn false hope. Therefore, all of the variables must be found and their properties established, if only to ignore them on a regular basis. These will be collected in a catalogue of nuisances and stored in a NASA-sponsored data base for use by the community for sundry purposes. For example:

Bar structure: Whitelock & Catchpole (1992) have presented evidence for longitudinal asymmetry in the density distribution of Bulge Miras that can be understood in terms of bar structure. RR Lyr stars in fields that straddle the Galactic Center offer an opportunity to search for such structure in a population of different abundance and, presumably, age.

RR Lyrae Stars: Blanco (1992) finds that the <P> of RRa type variables in WI ($l = 0.6$, $b = -5.5$) is larger (0.537 days) than that in BW (0.497 days), a difference she suggests may be due to the abundance gradient. Mean periods elsewhere in the Bulge will test the generality of this apparent variation of period with location.

Cataclysmic variables: Some 85% of all M31 novae occur in its bulge. Cappacioli *et al* (1989) estimate 29 $\pm$ 4 novae outbursts per year (see also Ciardullo *et al* 1987). According to the scenario developed by Shara and his collaborators (1986), a population rich in novae also should contain a rich population of dwarf novae which should appear on $19 < V < 20$ during outbursts in the Bulge fields observed by Terndrup (1988). Detection of a population of mass-accreting white dwarfs in the Bulge would be important, as they are promising candidates for precursors of SNI.

Dwarf cepheids & anomolous cepheids: Dwarf cepheids have been discovered in 2 globular clusters: Omega Cen (Jorgensen & Hansen 1984) and NGC 5466 (Mateo *et al* 1990). Their masses, estimated from pulsation theory, are ~ 1.3 solar masses. It has been proposed that dwarf cepheids in globular clusters are consequences of stellar mergers, produced either by star-star collisions or by coalescence of close binaries. The former process should be ineffective in the Bulge. Comparison of the number ratio of dwarf cepheids to close binaries in the Bulge and in globular clusters should permit evaluation of the relative importance of these mechanisms for the production of dwarf cepheids and of a related kind of variable, the anomolous cepheids found in dwarf galaxies and globular clusters (Nemec *et al* 1988, Wallerstein & Cox 1984).

Eclipsing binaries: With the advent of very large telescopes it will be possible to contemplate mass determinations at the main sequence turnoff by spectroscopic observations of detached eclipsing binaries. These will be found in the photometric survey. Meanwhile, more luminous semi-detached Algol systems (hereafter EASD) can provide immediate gratification, because they can be observed with existing telescopes to set a lower limit to the mass of stars at the main sequence turnoff in the Bulge, as outlined below.

A long-neglected paper by S. Gaposchkin (1954) assures us that such stars exist in BW. Of 45 eclipsers in BW identified by Baade (1946, 1951) and investigated by Gaposchkin, 30 exhibit the shallow secondary minima and lie in the period range 1 to 4 days characteristic of EASDs in the 4th edition of the *General Catalogue of Variable Stars* (Kholopov 1985). The apparent magnitude distribution of these stars is reasonably represented by an error-broadened power-law density fitted to the RR Lyr stars in the same field and shifted by 0.6 mag. in B (Fig. 5). If accepted at face value, this means that the dispersions in luminosity of the two species are similar.

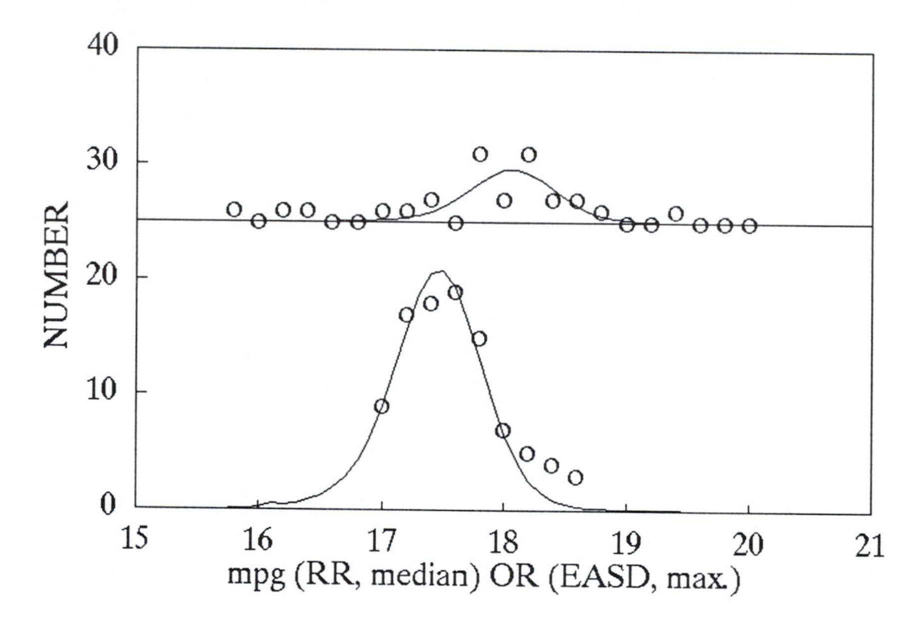

Fig. 5. The apparent magnitude distributions of EASDs (top panel) and RR Lyr stars (bottom panel) in BW. A power law density distribution, broadened by a Gaussian error function, was fitted to the RR Lyr data. The same curve, scaled and shifted by 0.6 mag., is used to represent the EASD apparent magnitude distribution.

The notion of essentially mono-luminous EASDs is a surprise in itself, for it implies, at the least, a preferred mass-ratio among the initial primaries. In time we shall learn whether this feature of Gaposchkin's data is real or is merely an indication of magnitude-dependent incompleteness. Meanwhile, consider the consequences of mono-luminous EASDs at $B = 18.1$.

(1) Absolute magnitudes of EASD primaries in BW. Estimate with $M_B(\mathrm{h}) = M_B(\mathrm{RR}) + 0.6 + DB$, in which h refers to the hot component and $DB \sim 0.2$ is a typical increment to B(EASD) that results from subtraction of secondary light outside of eclipse (Cester et al 1978, Amman & Walter 1973). Using $M_V = 0.85$ for the relatively metal-rich RR Lyr stars in BW (Walker & Terndrup 1991) with $(B\text{-}V)_0 = 0.3$ for the intrinsic colors of the RR Lyr stars, obtain $M_B(\mathrm{h}) = 1.95$, which is 0.17 mag. smaller than the mean for the three least luminous(least massive) EASDs tabulated by Popper (1980a) and similar to the bright limit ($M_V \sim 1.9$) of the composite luminosity function of blue stragglers in the globular clusters Omega Cen, M3, NGC 5466, NGC 5053, and NGC 6101 (Sarajedini & Da Costa 1991).

(2) Lower limit to stellar mass at MS Turnoff in BW. Using no more than conservation of mass we may write

observed EASD	=	initial binary	+	hypothetical mass loss
$m_\mathrm{h} + m_\mathrm{c}$	=	$m_\mathrm{p} + m_\mathrm{s}$	+	m_l

in which h and c denote the hot(primary) and cool(secondary) components of the EASD, p and s denote the primary and secondary components of the initial, unevolved binary, and m_l is mass lost from the system during or prior to mass transfer. From the theory of mass-transfer binaries (Kippenhahn & Weigert 1967, Paczynski 1971) we expect that EASDs arise as a consequence of the binary-perturbed evolution of initial primaries from the main sequence, so $m_\mathrm{p} = m_\mathrm{TO}$, and

$$m_\mathrm{TO} = m_\mathrm{h}(1 + q)/(1 + Q - L)$$

in which Q, q, and L are the secondary masses and mass loss expressed as fractions of their respective primaries. The quantities Q and L are unknown, so we set $Q = 1$ and $L = 0$ to summarize what can be learned in the form of a lower limit to m_TO (upper limit to age) of the parent population,

$$m_\mathrm{TO} > m_\mathrm{h}(1 + q)/2,$$

i.e., the turnoff mass must be greater than half of the total mass of the initial binary. The true limit must be a bit larger, because $Q = 1$ cannot produce an EASD. Measurement of m_h and q for EASDs in BW is barely within the grasp of 4-m class telescopes. To entice observers I applied this formalism to three of the least massive EASDs in the solar neighborhood (Popper 1980a, 1980b). Interpolation of masses in the lifetime *versus* [Fe/H] relations for [Fe/H] = 0.0 and Y = 0.30 in Fig. 5 of VandenBerg & Laskerides (1987) leads to upper limits on age of <7, <10, and <25: Gy for AS Eri, TW And, and RY Aqr, respectively. Limits smaller than 10 Gy in the Bulge would be interesting. In view of the considerable uncertainties that attend other methods of age estimation in the Bulge, I reckon that this one is worthy of pursuit.

Finally, the co-spatial counts of RR Lyr stars and Algols in the Bulge may provide a norm for judging the effects of stellar encounters on binary populations in the dense environments of globular clusters. The number ratios of these two species in the general field at $|b| > 30^o$ (Kholopov 1985), in BW (Gaposchkin 1954), and in all globular clusters (Hogg 1973) are tabulated below:

Sample	n(RR)	n(EASD)	n(RR/n(EASD)
GCVS4$\|b\|>30^o$	526	51	4.5*
BW	113	35	3.2
Glob. Clusters	1900	2**	950

* The number of EASDs in GCVS4 was arbitrarily increased by the factor $10^{0.6DB}$.

** One in Omega Cen (Jensen & Jorgensen 1985) and one in NGC 5466 (Mateo *et al* 1990)

I removed the RR Lyr stars of Kinman, Wirtanen & Janes (1966) in Coma from the statistics because the search technique (Kinman 1965) used to find them was biased against discovery of variable stars with $P > 1$ day. The ratio is uninterpretable at low latitudes because of the large numbers of massive Algols of the Young Disk that are found there. After application of a modest volume correction factor to the EASDs in the *General Catalogue of Variable Stars* based on the absolute magnitude difference $DM_B = M_B(\text{EASD}) - M_B(\text{RR}) \sim 0.6$ in BW, the ratios at high latitudes and in the Bulge are comparable. Both are smaller than the ratio in globular clusters by more than two orders of magnitude, which means that the cluster antecedents of EASDs (1) never existed in significant numbers, (2) have been destroyed efficiently, or (3) have sunk to the inadequately explored cluster centers.

REFERENCES

Aguilar, L., Hut, P. & Ostriker, J. P. 1988, ApJ, 335, 720
Amman, M. & Walter, K. 1973, A&A, 24, 131
Armandroff, T. E. 1989, AJ, 97, 375
Baade, W. 1946, PASP, 58, 249

Baade, W. 1951, Pub. Obs. U. of Michigan, 10, 7

Beers, T. C., Preston, G. W., & Shectman, S. A. 1988, ApJS, 67, 461

Blanco, B. M. 1984, AJ, 89, 1836

Blanco, B. M. 1992, AJ, 103, 1872

Blanco, V. M. & Terndrup D. M. 1989, AJ, 98, 843

Blommaert, J., Brown, A., Habing, H., van der Veen, W., & Ng, Y. K. 1992, in Galactic Bulges (IAU Symp. No. 153) ed. H. Habing (Dordrecht: Kluwer Academic Publishers)

Cappacioli, M., Della Valle, M., D'Onofrio, M. & Rosino, L. 1989, AJ, 97, 1622

Carney, B. W., Aguilar, L., Latham, D. W.& Laird, J. B. 1990, AJ, 99, 201

Cester, B., Fedel, B., Giuricin, F., Mardirossian, F. & Mezzetti, M. 1978, A&A, 62, 291

Ciardullo, M., Ford, H. C., Neill, J. D., Jacoby, G. H. & Shafter, A. W. 1987, ApJ, 318, 520

Demarque, P., Lee, Y.-W., Zinn, R., & Green, E. M. 1989, in The Abundance Spread in Globular Clusters: Spectroscopy of Individual Determinations, eds. G. Cayrel de Strobel, M. Spite & T. L. Evans (Paris: Paris Observatory), 97

Gaposchkin, S. 1954, Variable Stars Bulletin, 10, 337

Habing, H. J., Olnon, F. , Chester, T., Gillett, F., Rowan-Robinson, M. & Neugebauer, G. 1985, A&A, 152, L1

Harmon, R. & Gilmore G. 1988, MNRAS, 235, 1025

Hartwick, F. D. A. 1976, ApJ, 209, 418

Hesser, J. E., Harris, W. E., VandenBerg, D. A., Allwright, J. W. B., Shott, P. & Stetson, P. B. 1987, PASP, 99, 739

Hogg, H. S. 1973, Pub. DDO, U. of Toronto, Vol. 3, No. 6

Jensen, K. S. & Jorgensen, H. E. 1985, A&AS, 60, 229

Jorgensen, H. F. & Hansen, L. 1984, A&A, 133, 165

Khopolov, P. N. 1985, General Catalogue of Variable Stars, 4th edition (Moscow: Nauka)

King, C. R., Demarque, P. & Green, E. M. in Calibration of Stellar Ages, ed. A. G. D. Philip (Schenectady: Davis), 211

Kinman, T. D. 1965, ApJS, 11, 199

Kinman, T. D., Wirtanen, C. & Janes, K. 1966, ApJS, 13, 379

Kippenhahn, R. & Weigert, A. 1967, Z. fur Ap, 65, 251

Lee, S.-W. 1977, A&AS, 27, 381

Lee, Y.-W. 1989, PhD thesis, Yale University

Lee, Y.-W. 1992a, in Stellar Populations of Galaxies (IAU Symposium No.149) eds. B. Barbuy & A. Renzini (Dordrecht: Kluwer Academic Publishers) 446

Lee, Y-W 1992b, First Hubble Symposium (STScI), PASP (in press)
Leonard, P. T. J. 1989, AJ, 98, 217
Mateo, M., Harris, H. C., Nemec, J. & Olszewski, E. W. 1990, AJ, 100, 469
Matteucci, F. & Brocato E. 1990, ApJ, 365, 539
Mironov, A. V. 1972, Soviet Astr.--AJ, 16, 105
Nemec, J. M., Wehlau, A. & de Oliviera, C. M. 1988, AJ, 96, 528
Ortolani, S., Barbuy, B. & Bica, E. 1990, A&A, 236, 362
Paczynski, B. 1971, ARA&A, 9, 183
Popper, D. 1980a, in IAU Symposium No. 88, eds. M. J. Plavec, D. M. Popper, & R. K. Ulrich (Dordrecht: D. Reidel), 203
Popper, D. 1980b, ARA&A, 18, 115
Preston, G. W 1959, ApJ, 130, 507
Preston, G. W., Shectman, S. A., & Beers, T. C. 1991a, ApJ, 375, 121
Preston, G. W., Shectman, S. A., & Beers, T. C. 1991b, ApJS, 76, 1001
Rich, R. M. 1990, ApJ, 362, 604
Sarajedini, A. & Da Costa, G. S. 1991, AJ, 102, 628
Sarajedini, A. & Demarque, P. 1990, ApJ, 365, 219
Sarajedini, A. & King, C. R. 1989, AJ, 98, 1624
Searle, L. 1979, in Les Elements et Leur Isotopes Dans L'Univers, (Liege: Universite de Liege), 437
Searle, L. & Zinn, R. 1978, ApJ, 225, 357
Shara, M. M., Livio, M., Moffat, F. J., & Orio, M. 1986, ApJ, 311, 163
Suntzeff, N. B., Kinman, T. D., & Kraft, R. P. 1991, ApJ, 367, 528
Terndrup, D. M. 1988, AJ, 96, 884
VandenBerg, D. A., Bolte, M.& Stetson, P. B. 1990, AJ, 100, 445
VandenBerg, D. A. & Laskerides, P. G. 1987, ApJS, 64, 103
Walker, A. R. & Terndrup, D. M. 1991, ApJ, 378, 119
Wallerstein, G & Cox, A. N. 1984, PASP, 96, 583
Webbink, R. F. 1985, in Dynamics of Star Clusters (IAU Symposium No. 113), eds. J. Goodman & P. Hut (Dordrecht: D. Reidel), 541
Weinberg, M. D. 1992, ApJL, 392, L67
Wesselink, T. 1987, A Photometric Study of Variable Stars near the Galactic Center (Nijmegen: Brakkenstein)
Whitelock, P. & Catchpole, R. 1992, in The Center, Bulge and Disk of the Milky Way, ed. L. Blitz (Dordrecht: Kluwer Academic Publishers) in press
Zinn, R. 1980, ApJ, 241, 602
Zinn, R. 1985, ApJ, 293, 424
Zinn, R. 1986, in Stellar Populations, eds. C. A. Norman, A. Renzini, & M. Tosi (Cambridge: Cambridge Univ. Press), 73
Zinn, R. & West, M. J. 1984, ApJS, 55, 45

DISCUSSION

King: You use the word spheroid exactly as I recommended, you used it geometrically and you did not talk about a spheroid population, which was the thing that I abominated.

Preston: Thank you Ivan.

Tyson: Rich and I have a poster, where there is an abundance distribution given for a field interior to Baade's window, one of the clear windows in Sagittarius at -2.5 degrees (Baade's window is at -4 degrees) and in that field we do not find the mean abundance to go up to infinity as your graph had shown. So abundances scaling up to infinity may be premature at this point, interior to Baade's window!

Preston: I stand corrected, the abundances do not go up to infinity!

Ng: In my poster I show the C-M diagram for a quarter of a million stars in the same field, and I don't think that the blue horizontal branch stars are as blue as you proposed in that field.

Preston: I don't know what to say, I measured 150 of them myself and I believe my own colors. I measured them with an aperture photometer one by one and I'm proud of those colors, and they're just as blue as I said they were...

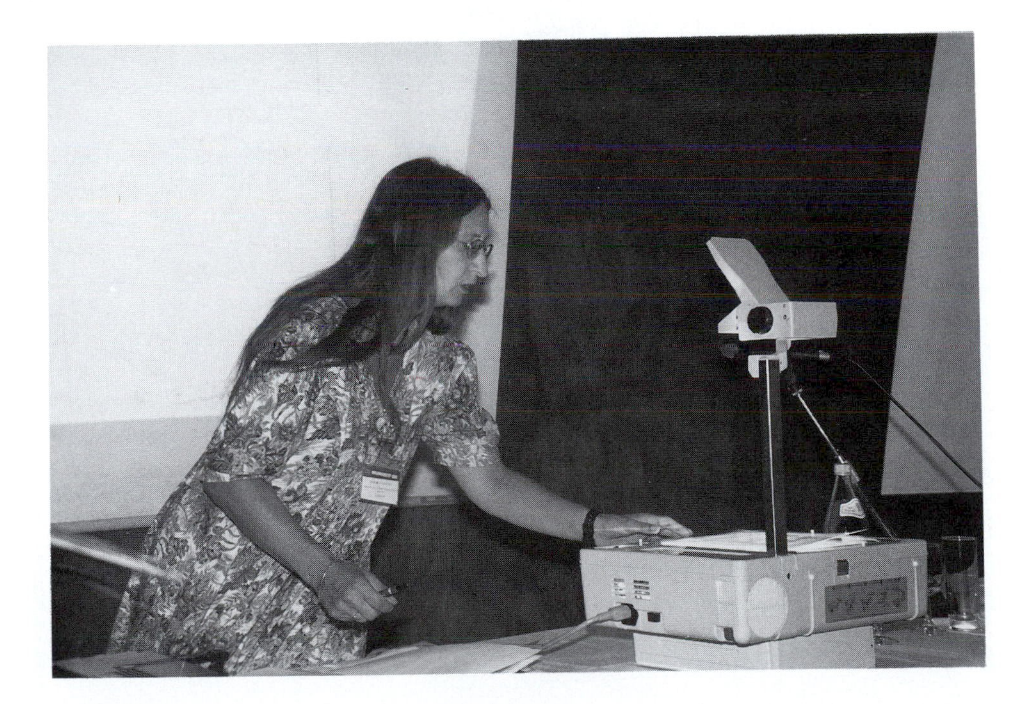

G. Stasinska

George PRESTON

THE GLOBAL PROPERTIES OF PLANETARY NEBULAE IN THE GALACTIC BULGE

G. STASIŃSKA
DAEC, Observatoire de Paris-Meudon, France

ABSTRACT. After a discussion of the methods used to derive the physical parameters of planetary nebulae and their central stars, the global properties of planetary nebulae in the Galactic bulge are reviewed, and compared to those of planetary nebulae in different contexts.

1. Introduction

Planctary nebulae (PN) are thought to be the evolution product of intermediate mass stars - 0.8 to $8M_{\odot}$ - before they become white dwarfs (see recent reviews by Habing 1990 or Pottasch 1992). They are formed of matter ejected by these stars after the asymptotic giant branch stage and are gradually ionized as the core temperature grows.

It is natural to expect that the study of a population of PN should reveal some of the characteristics of the parent stars: ages, stellar masses and chemical composition, kinematics. Such was indeed the motivation of some of the former studies of Galactic bulge planetary nebulae (GBPN) like those of Webster (1975, 1976) or Isaacman (1983).

Further, the study of GBPN is a necessary step towards the understanding of the final evolutionary stages of intermediate mass stars. Indeed, most of the observational tests of existing theories require the knowledge of PN distances. Unfortunately, because PN are essentially transient objects, there is no quantity which remains constant during their evolution and could be used as a distance indicator. Together with the Magellanic Clouds, the Galactic bulge offers the possibility to study a population of PN at known distance, and this has been the motivation of many recent studies of GBPN (Gathier et al 1983, Pottasch and Acker 1989, Pottasch et al 1988, 1990, Pottasch 1990, Dopita et al 1990, Ratag 1991, Stasińska et al 1991a, b, Tylenda et al 1991, Pottasch and Zijlstra 1992). Unexpectedly, these studies have perpetuated some of the controversies regarding the interpretation of PN.

H. Dejonghe and H. J. Habing (eds.), Galactic Bulges, 117–131.

2. The GBPN sample

2.1. METHOD OF DISCOVERY

The first systematic PN searches were made in the optical, either with objective prisms to identify characteristic emission lines (e.g. Minkowski 1948-1951, see references in Minkowski 1965), or on Palomar plates where are seen PN of large angular diameters and low surface brightness (Abell 1955). Deeper surveys of selected areas have been performed since then (see references in the Strasbourg-ESO catalogue of Galactic planetary nebulae, Acker et al 1992).

Interstellar extinction towards the bulge is large, hampering the detection of faint PN, except in some regions like the Baade Window which has been extensively studied by Kinman et al (1988).

Planetary nebulae can also be discovered by their radio continuum emission, but the risk of confusion with compact HII regions and especially extragalactic sources is important.

A more promising way is to select potential PN from IRAS sources on the basis of their infrared colours (Pottasch et al 1990) and then confirm their nature by further observations. The first results of a vast program of this kind are given in Pottasch et al (1988), Ratag et al (1990), Ratag and Pottasch (1991) for radio continuum measurements, and Manchado et al (1989) and Garcia-Lario et al (1990) for near-infrared photometry .

An optical spectrophotometric survey of all candidate GBPN published by 1988 has been conducted by Acker et al (1991), resulting in the confirmation of 335 PN in the direction of the bulge. The criteria adopted to classify an object as a PN are presented in Acker et al (1987). While it is easy to distinguish an emission line galaxy from a PN, thanks to the redshift, it is more difficult to decide between a very young PN and a symbiotic star or a compact HII region. Any criteria presuppose a certain representation of the PN phenomenon. It is instructive to compare the lists of misclassified PN of Acker and Stenholm (1990) and Kohoutek (1992).

2.1. SORTING OUT TRUE GBPN

As noted by Minkowski (1965), the strong concentration of known Galactic PN towards the bulge indicates that most of the PN seen in that part of the sky actually belong to the bulge. The construction of separate histograms for PN with large (>20") and small (<20") angular diameters led Gathier et al (1983) to estimate that about 80% of all the small PN within 10° of the Galactic center are physically close to it. After rejecting those PN with radio flux F(6cm) above 100 mJy, among which a large proportion of foreground object is expected, the resulting sample should contain 90 - 95% of genuine GBPN (Pottasch and Acker 1989, Stasińska et al 1991). Further arguments, like for example a small distance as measured from the reddening(Gutiérrez-Moreno et al 1991, Ratag et al 1992) can be used to eliminate a few additional objects.

3. A first glance at the GBPN sample

3.1. RADIAL VELOCITIES

Using radial velocity measurements of about 300 galactic PN, Minkowski (1965) showed that in the Galactic bulge, many PN are on non-circular orbits, contrary to the majority of PN situated in the Galactic disk. Data on radial velocities are now available for 577 PN (Acker et al 1992), and confirm that most GBPN are of extreme Population II. According to Pottasch (1984), their central stars should thus have the same age as globular clusters, i.e. about 10-15 Gyrs. Since the stars had to take this long to evolve into PN, their initial masses must have been about 0.8 $M_\odot$. In comparison, the PN velocity dispersion in the solar neighbourhood leads to an age of 3-5 Gyrs, corresponding to a progenitor masses of 1.4 - 2 $M_\odot$.

The average velocity of GBPN relative to their ambient medium is about 140 km/sec(Pottasch,1984) compared to 40 km/sec for disk PN.

3.2. EXCITATION CLASSES

The "excitation class" of a nebula is an easily observed spectral property, as it relies only on the brightest lines. Therefore, it has become customary to use excitation classes to compare samples of PN in different environments. Several classification systems exist. Aller's (1956) favours line ratios independant of elemental abundances while Feast's (1968) minimizes reddening effects by considering lines in the same wavelength range.

Compared to PN located in the disk, GBPN have, on average, lower excitation classes (Acker et al 1991, Ratag 1991). They have also smaller excitation classes than PN in the LMC (Webster 1988). This is interpreted as due to a smaller mean effective temperature of the exciting stars in GBPN. However, this interpretation may not be correct: Schönberner and Tylenda (1990) have shown that the excitation class is not necessarily a good indicator of the effective temperature, because of the coupled evolution the star and the surrounding nebula.

4. Determination of the physical properties of PN and their nuclei

The methods mentionned below are classical and widely used in nebular studies (see e.g. Pottasch 1984). To enlighten the further discussion, we briefly recall the basics and limitations of the most important ones.

4.1. NEBULAR DENSITIES AND TEMPERATURES

Nebular densities can be obtained from the ratios of collisionnally excited forbidden lines ([O II], [S II], [Cl III], [A IV]), which are sensitive to this parameter in the density range 10^2 - 10^5 cm^{-3} approximately.

For PN at known distances, densities can also be derived from the angular diameter ϕ and the total reddening-free Hβ (or radio) flux since $n_e \propto F(H\beta)^{0.5} \phi^{-1.5} d^{-0.5}$. This method is believed to give a value more representative of the whole nebula than the line ratio method, which is very sensitive to density fluctuations .

The temperature of the ionized gas is given by the ratio of [O III] 4363/5007 and sometimes [N II] 5755/6583 lines. In the case of densities above 10^5 cm^{-3}, these ratios can be misinterpreted, overestimating the temperatures. The electron temperatures of PN in the bulge range from 8000K to 15000K. It is not excluded that some PN have lower temperatures which cannot be measured because the [O III]4363 line is too faint. This could be the case of GBPN with abundances much higher than solar.

4.2. MASSES OF THE NEBULAE

In general, only the masses of the ionized portions of the PN can be readily obtained. Their derivation requires the knowledge of the distance and of the total Hβ (or radio) flux. The first method uses the density n_e derived from forbidden line and the relation $M_i \propto F(H\beta)\, n_e\, d^2$. The other one uses the angular diameter ϕ and the relation: $M_i \propto F(H\beta)^{0.5} \varepsilon^{0.5} \phi^{1.5} d^{2.5}$, where ε is the filling factor, assumed of the order of 0.5. Both methods have their drawbacks. The first one because of the inhomogeneous structure of PN, the second one because of the large dependance of M_i on the angular diameter which is uneasy to measure.

This last decade, observations of neutral and molecular gas in PN (see review by Huggins 1992) have started giving some information on the mass of the neutral gas may surround the optically shining PN (see Pottasch 1992), but there are yet no measurements for PN in the bulge.

4.3. TEMPERATURES OF THE CENTRAL STARS

Temperatures of PN nuclei can be estimated by different means.

One way is to use the nebular recombination lines to count the stellar ionizing photons, as proposed by Zanstra in 1931. Provided that the nebula is in balance between ionization and recombination (which is generally true) and absorbs all the stellar photons with energies above 13.6eV (which is not always true), the star effective temperature can be derived from the ratio of the nebular Hβ flux and the stellar flux in the V band. The nebular HeII 4686 line can be used instead Hβ to derive another estimate of the temperature. The chances for the nebula of being optically thick to He^+

ionizing photons is much larger, but the star must be hotter than 5 10^4K to give a detectable nebular He II 4686 flux. A variant of this method, which does not require the measurement of the star magnitude, is to use the HeII 4686/Hβ ratio.

Another way to derive the star temperature is from the mean energy of the photons absorbed by the nebula, like in the energy balance method. This energy is given by the sum of all the collisionnally excited lines emitted by the nebula, normalized to Hβ. This method has the advantage of being only weakly dependant on the optical thickness of the nebula in the Lyman continuum. It has been recently worked out by Preite-Martinez and Pottasch (1983), Preite-Martinez et al (1989) and Köppen and Preite-Martinez (1991). If the available spectra are restricted in wavelength (and this is generally the case, especially for GBPN) some corrections have to be introduced for the unobserved lines (mainly the UV carbon lines and the strongest infrared lines). An extreme version of the energy balance method - used recently by Méndez (1992) for GBPN - considers only the [O III]/Hβ ratio, after calibration by an independent method.

Both these methods break down in the situation where dust competes with the gas in absorbing ionizing photons. This, however, appears to be thc case only in very young PN (Lenzuni et al 1989).

The temperatures of PN nuclei can also be obtained as a result of photoionization modelling of the surrounding nebulae, such as was done by Ratag (1991) for about 100 GBPN. This method is not completely independent of the previous ones, since the first requirement of a model is to reproduce the HeII/Hβ line (when observed), as well the intensities of the strongest lines, taking into account all the physical processes occuring in the nebula - which are schematized in the former two approaches.

The derived effective temperatures depend on the model atmospheres used for the exciting stars. Considerable progress has been made recently in the modelling of such atmospheres, with inclusion of non LTE effects in hydrogen and helium as well as in the metals, consideration of extended atmospheres and winds (see Kudritzki 1989 and 1992 for a review). It seems that, grossly, the effect of a stellar wind superimposed on a NLTE atmosphere is to restore the energy distribution towards a blackbody. Therefore, it is still sufficient for many purposes to assume that PN nuclei radiate as blackbodies (Dopita et al 1990, Méndez et al 1992).

Finally, these last years, it has become possible to derive the stellar parameters of the PN cores by direct spectroscopic non-LTE methods for about 20 objects (see Méndez et al 1992). These methods, however, are not devoid of problems (see Kudritzki 1992, Pottasch 1992). They are difficult to apply to GBPN, because they require good signal-to-noise spectra allowing the measurements of stellar line profiles.

The discrepancies between the different derivations of star temperatures are often large (see e.g. Ratag 1991, or Pottasch 1992). They contain potential information on the properties of the PN and their exciting stars. Meanwhile, one should remain cautious about the significance of star temperature estimates.

4.4. LUMINOSITIES OF THE CENTRAL STARS

For PN at known distance, once the core temperatures are known, their luminosities are easily derived from their observed magnitudes by applying a bolometric correction. If the magnitudes are not available - which is the case of about one third of known GBPN - core luminosities are sometimes estimated from the total nebular Hβ or radio flux (Pottasch and Acker 1989, Zijlstra and Pottasch 1989, Ratag 1991). If the nebula is density bounded or if dust absorbs part of the ionizing radiation, this procedure underestimates the stellar luminosity. Ratag (1991) has used the infra-red fluxes measured by IRAS to correct for dust absorption, but the problem of leakage of ionizing photons remains.

4.5. MASSES OF THE CENTRAL STARS

During the evolution of a PN, the luminosity of its nucleus varies by several orders of magnitude, and its temperature by a factor 10. A PN is intrinsically characterized by its core mass and its nebular mass and age.

The core mass can be derived by comparing measured stellar parameters with a grid of evolutionary models for post-AGB stars. By now, only two such grids are available: one for PN nuclei with a hydrogen burning shell (Schönberner 1979, 1983, Blöcker and Schönberner 1989), the other for both hydrogen and helium burning nuclei (Wood and Faulkner 1986). These grids cannot be combined, however, because the calculations were made under different assumptions (see a discussion by Schönberner 1989). A new grid is in preparation (Vassiliadis an Wood 1992), for different chemical compositions.

Several types of diagrams can be used to derive the PN core masses. All have their drawbacks (see a review by Tylenda 1992).

Errors in the luminosities or temperatures of PN nuclei affect the derived stellar masses. The application to density bounded PN of methods valid for ionization bounded ones leads to systematic effects. Tylenda et al (1991) have attempted to minimize them by comparing the position of the data points to "apparent theoretical evolutionary tracks" corresponding to models of nuclei surrounded by a standard expanding nebula and treated exactly like true objects in the derivation of the stellar parameters.

Finally, when the stars are still on their horizontal track in the H-R plane, a core-mass luminosity relation is expected. When considering a population of PN, the maximum luminosity corresponds to a maximum mass of the exciting stars. This is difficult to use in practice since, as noted by Pottasch (1992), different authors obtain different core-mass relations.

4.6. CHEMICAL COMPOSITION OF THE NEBULAE

Planetary nebulae are excellent sites for determining elemental abundances. Indeed, the process of formation of the emission lines used as abundance indicators is relatively well understood.

The simplest methods are empirical ones. The intensities of emission lines lead directly to ionic abundances (once the electron temperature is known). Elemental abundances are then obtained after correcting for unseen ionization stages. These corrections are adopted from simple consideration of ionization potentials (e.g. Köppen et al 1991), from grids of photoionization models (Walton et al 1992), or from modelling the ionization structure of individual PN (Aller and Keyes 1987, Ratag 1991).

Another method is to produce for each PN a photoionization model fitting *all* the lines. This painstaking method has seldom been applied. A simplified version was used by Dopita et al (1990) for 6 GBPN with isobaric nebular models and by Ratag (1991) for 120 GBPN with uniform density models. A perfect fit to all the observed lines was not looked for.

In GBPN, the abundances of He are considered to be accurate within 10-20%, those of O within about 20-30%, those of N and Ne within about 30-40%. The C abundance has been determined only recently in a dozen of GBPN (Walton 1992), from IUE measurements. Abundances of S and A can also be also determined, the main problem lying in the ionization correction factors. Typical quoted errors are of 50-60%. It is interesting to compare the abundances derived recently by 4 different groups of observers for M2-29, a halo PN located in the bulge.

He/H	C/H (x10^4)	N/H (x10^4)	O/H (x10^4)	Ne/H (x10^4)	S/H (x10^6)	Ar/H (x16^4)	
0.15		0.13	0.6	0.2	2.5	2.0	Dopita et al 1990
0.13		0.10	0.295	0.064	0.68	0.44	Ratag 1991
0.093	0.01	0.172	0.277	0.039			Walton et al 1992
0.091		0.01:	0.38	0.57	0.26		Köppen et al 1992
0.116		*0.115**	*3.88*	*1.01*	*1.25*	*.90*	*mean*
25%		*17%**	*38%*	*86%*	*87%*	*107%*	σ

* *: without Köppen's estimate*

The scatter between the different abundance sets is larger than the uncertainty claimed by each group.

As shown by Peimbert (1967), temperature fluctuations in the nebula can lead one to underestimate the abundances of the heavy elements (C, N, O, Ne...) relative to hydrogen. Up to now, there is no compelling observational or theoretical evidence for the existence of large temperature fluctuations but the problem is still open (see e.g. Ratag et al 1992).

5. Some controversies in GBPN studies

Even though the severest problem in PN studies - distances - is considerably reduced in the case of GBPN, their general properties are hard to establish because of difficulties in interpreting the observed data in terms of accurate physical parameters, and because objects of different evolutionary status may coexist. We describe below a few hot problems.

5.1. MEASUREMENT UNCERTAINTIES

A good example is that of the angular diameters, for which optical and VLA determinations may differ by a factor 2 or more (Stasińska et al 1991, Pottasch and Zijlstra 1992).

Another difficulty is the extinction correction, which must be applied to optical data. As shown by Stasińska and Tylenda (1992), the Balmer decrement method with the standard extinction law may overestimate the true fluxes by up to a factor 2-3 if the reddening is large.

5.2. INTERPRETATION OF OBSERVATIONAL DIAGRAMS

A typical case is that of the mass-radius relation. Several authors (Pottasch and Acker 1989, Dopita et al 1990) have noted a clear correlation between the nebular ionized masses and diameters. The slope of 1.5 is exactly the one expected if the nebulae were gradually ionized by stars of constant ionizing flux, seeming to indicate that most GBPN are ionization bounded.

Zijlstra (1990) argued that observational selection must affect this diagram. Numerical simulations (Stasińska and Tylenda 1992) demonstrate that the observed diagram can be reproduced for a population of PN with only 0.3 dex dispersion in total masses, taking into account a 10% dispersion in the GBPN distances, combined with observational uncertainties and selection effects. In the simulation, more than 50% of PN turn out to be density bounded. For them, Zanstra temperatures and luminosities would be undersetimated.

5.3 EVOLUTION OF THE CENTRAL STARS

Recently, a number of number of authors (Pottasch and Acker 1989, Zijlstra and Pottasch 1989, Ratag et al 1992, Pottasch 1992) have reported the existence of young PN of low stellar luminosities, corresponding to core masses below $0.546M_\odot$. According to current evolutionary models of post AGB stars, such low mass nuclei would require more than 10^5 years to become hot enough to ionize the ejected nebula. This is larger by about a factor five than the estimated expansion age of the PN. In fact, in the above papers, the luminosities were obtained under the assumption that the nebulae were ionization bounded. Tylenda et al (1991), as well as Méndez (1992), reanalyzing the data of Pottasch and Acker (1989), did not obtain any significant discrepancy with the theoretical evolution of PN nuclei.

However, some objects may effectively deviate from standard evolutionary tracks, as reviewed by Mazzitelli (1992). For example, some PN may have been ejected during helium-shell burning. Others may have nuclei which experienced a helium-shell flash, describing a loop in the HR diagram. Differences in post AGB mass loss efficiencies affect the evolution of PN nuclei in that high mass loss speeds up the evolution: Trams et al (1989) have shown that a mass loss rate of $10^{-7} M_\odot$/yr shortens the evolution of a $0.546 M_\odot$ post AGB star by a factor 5.

6. GBPN versus other populations of PN

In the following, we briefly comment on some possible differences and similarities between GBPN and PN in other environments.

6.1 CHEMICAL COMPOSITION

The abundances of elements such as O, Ne, S, Ar indicate the composition of the matter out of which the progenitor stars were formed, thus providing constraints to galactic chemical evolution studies. He, C, and N have been partly manufactured in the interiors of the stars and then brought to the surface by dredge-up mechanisms occuring during the red giant and AGB phases. They probe the physical processes occuring during the evolution of intermediate mass stars, and can be used as indicators of the progenitor masses (see Clegg 1989, Ratag et al 1992). For example, the classical subdivision of PN into type I and non type I according to whether the N/O ratio is larger or smaller than 0.5 (Peimbert and Torres-Peimbert 1983) is interpreted as a subdivision of the progenitor stars above and below $2M_\odot$ respectively. This, in turn, leads to an indication of the ages (above and below 3Gyrs on the Maeder and Meynet 1988 scale).

The following table is a compilation of the mean abundances found in the Galactic bulge, in the Galactic disk and in the Magellanic Clouds. The values given for the GBPN are taken from Ratag (1991) and are based on about 120 objects.The ones for the Magellanic Clouds come from Walton et al (1992) and concern over 80 PN. The values quoted for the Galactic disk come mainly from Aller and Czyzak (1983) and Aller and Keyes (1987), and concern about 100 objects.

	log He/H+12	log N/H+12	log O/H+12	log Ne/H+12	log S/H+12	log Ar/H+12
bulge [1]	11.08±0.08	8.50±0.54	8.71±0.43	7.99±0.40	7.00±0.41	6.54±0.36
[2]			(9.17±0.06)	(8.33±0.11)	(7.18±0.04)	(6.80±0.06)
disk [3]	11.06±0.11	8.35±0.45	8.64±0.23	8.02±0.29	7.02±0.35	6.52±0.33
LMC[4]	11.02±0.06	8.05±0.71	8.40±0.15	7.66±0.20	6.55±0.18	5.91±0.18
SMC[4]	11.02±0.08	7.58±0.24	8.18±0.20	7.37±0.23	6.42±0.29	5.61+0.22

1 Ratag et al 1992
2 extrapolation of Galactic radial gradients towards the bulge, Ratag et al 1992
3 compilation of disk sample, as quoted in Ratag et al 1992
4 Walton et al 1992a

The values in parenthesis are an extrapolation towards the bulge of Galactic gradients derived for PN in the disk. They are on average a factor 3 larger than the abundances effectively measured in the bulge, a fact already noted by Webster (1988). This is not surprising, since kinematically, GBPN are from a different population than disk PN.

The next table gives the average abundances relative to oxygen, in the Galactic bulge and disk and in the Magellanic Clouds.

	log C/O	log N/O	log Ne/O	log S/O
bulge [1,2]	-0.39±0.41	-0.16±0.41	-0.72±0.26	-1.63±0.31
disk [3]	0.06±0.21	-0.30±0.38	-0.60±0.21	-1.60±0.32
LMC[4]	0.31±0.33	-0.27±0.33	-0.73±0.15	-1.79±0.34
SMC[4]	0.53±0.00	-0.43±0.13	-0.78±0.20	-1.72±0.14

1 Ratag et al 1992
2 Walton et al 1992 b
3 Aller and Keyes 1987 as quoted in Ratag et al 1992
4 Walton et al 1992a

Clearly, type I PN are more frequent in the bulge than in the disk, contrary to what would have been expected for an old population (Ratag et al 1992). Note that, in contrast with Galactic disk PN (Perinotto 1991), large N/O ratios are not preferentially found for objects with a high O/H. Ratag's conclusion is that, in the bulg, N/O is not an indicator of the progenitors masses, but rather of an earlier N enrichment of the gas out of which the progenitor stars were formed.

The C/O ratios obtained by Walton et al (1992b) for a dozen of GBPN are significantly smaller than in disk PN, showing that the third dredge up did not operate. This is consistent with the idea that GBPN do not have massive progenitors.

Comparison of abundances in PN and other stars is a risky task, because the determinations use methods which are not calibrated against each other (see a discussion in Ratag et al 1992 for the bulge and Pagel 1992 for the Magellanic Clouds). But it is meaningful to compare abundances derived for the same types of objects in different environments. For example, Rich (1988) finds strong evidence that K-giants in the Baade window have higher iron abundances than solar neighbourhood giants.

How is it possible to reconcile the following facts: i) O/H ratios similar in the bulge and in the solar neighbourhood ii) ages of the PN populations (as derived from velocity dispersion) of 10-15 Gyrs for the bulge (Pottasch 1984, van der Veen and Habing 1990) and 3-5 Gyrs in the solar neighborhood (Pottasch 1984) iii) N/O and Fe/H ratios larger in the bulge than in the solar neighbourhood? According to Ratag et al (1992), the first two conditions imply that element enhancement must have occured earlier in the bulge than in the disk, the third one indicates a steeper stellar initial mass function in the bulge than in the solar neighbourhood and - possibly - a higher frequency of binary systems.

6.2. NEBULAR MASSES

The values of M_i defined above may be different from the true PN masses, but their largest estimates give an idea of the characteristic total nebular masses. Ratag (1991) finds an upper limit of about $0.6M_\odot$ for GBPN. In the LMC, the upper limit is also $0.6M_\odot$ (Barlow 1987, Meatheringham et al 1988). However, it is premature to conclude that PN envelopes have similar masses in the bulge and in the LMC, since Ratag's determinations use angular diameters while those of Barlow and Meatheringham use the densities derived from [O II]. Besides, it is hasardeous to consider upper values in samples of limited size where some natural scatter is expected. Barlow has divided Magellanic Clouds PN into thick or thin, and found that optically thin PN have a mean M_i of $0.27 \pm 0.06M_\odot$. A similar study should be made for GBPN. Note that, simulating the sample of GBPN observed with the VLA, Stasińska and Tylenda (1992) obtain satisfactory agreement with observational diagrams for a distribution of total masses of $0.1M_\odot \pm 0.3$dex (but other solutions may be acceptable).

6.3 CORE MASSES

Here again, it is essential to compare values that have been obtained through similar methods based on the same stellar evolutionary tracks.

Present determinations suggest that PN core masses are lower for objects located in the bulge than for PN in the disk: Tylenda et al (1991) find a distribution of central star masses of 0.593 ± 0.025 $M_\odot$ in their bulge sample, and 0.615± 0.036 $M_\odot$ in their disk sample. However, any conclusion regarding the true distribution of central star masses must fully take into account selection effects.

It seems, though (Ratag 1991), that there is a real lack of GBPN with core masses above 0.6$M_\odot$ in agreement with the upper end of the luminosity distribution of evolved AGB stars in the bulge (Van der Veen and Habing 1990).

6.4 MASSES OF PROGENITORS

An insight into the masses of PN progenitors is traditionally provided by the consideration of N/O ratios, as mentionned above. However, we have seen that in the bulge, N/O ratios are rather indicative of the early chemical evolution.

Weideman and Koester (1983) have established the existence of a monotonic (though rather flat) relation between the masses of PN nuclei and of their progenitors. In the case of a unique initial-final mass relation, the O/H ratio in GBPN should be correlated with the central star mass. Indeed, a PN with a more massive core would have a more massive - thus younger - progenitor, therefore made of chemically enriched material. Ratag (1991) finds no correlation between O/H and the core mass, and my own analysis based on the core masses derived in Tylenda et al (1990) supports this conclusion. Probably, variations in mass loss efficiencies on the AGB, significantly alter the initial-final mass relation (Van der Veen 1989, but see also Weideman n1990).

Perhaps a detailed study of the kinematics of GBPN could bring some clues to the progenitor masses.

6.5 PROPERTIES OF THE GBPN DISCOVERED BY IRAS

The results discussed sofar concerned mostly PN which were known before the follow up of IRAS sources. An analysis of radio measurements of about 120 GBPN discovered by IRAS is presented in Ratag (1991).

Compared to previously known GBPN, these new GBPN present systematic differences: they have lower radio fluxes and the distribution of infra red exesses is skewed towards much higher values. A first order interpretation would be that IRAS discovered earlier stages of PN, characterized by lower temperature central stars. But Ratag noted that, for the same star temperatures, known GBPN have higher infra red excesses than known disk PN, so that the conditions in the bulge must be different. He retained two possible explanations. One involving the effect of the much stronger (by a factor 50) interstellar radiation in the bulge. The other invoking enhanced 60μ emission by small iron particles which are likely to be present in the bulge in view of the high Fe abundance.

6.6 THE TOTAL NUMBER AND THE FORMATION RATE OF GBPN

From the 26% radio detection rate in a subsample of the 1683 IRAS sources with PN colours (Ratag 1991), one expects a total of 450 IRAS detected GBPN. Pottasch (1992) estimates that the total number of GBPN with ages below 8000 years is about 700.

Usually, the PN birthrate in a given site is obtained by dividing the total number of PN by their mean lifetime. Assuming an expansion velocity of 15km/sec, Pottasch (1992) finds a GBPN birthrate of 0.1 PN per year and a specific birthrate of 1.5 - 3 10^{-11} PN/$M_{\odot}$/yr assuming a total mass of 1.5 - 3 10^{9} $M_{\odot}$ within 3° of the Galactic center.

However, this number should be considered with care because, for density bounded PN, the time to reach the detection limit depends on the expansion velocity (yet not measured in most GBPN), and for ionization bounded PN, this time is a strong function of the mass of the core and its evolution. Finally, as shown by Isaacman (1979), because of their high velocities, GBPN may be disrupted by ram pressure when crossing a molecular cloud.

Simulations of the GBPN such as initiated by Stasińska et al (1991b and 1992) should provide a better understanding of the PN birthrate and its relation with the evolution of the stellar population in the bulge.

7. Final remarks

Many results on the global properties GBPN are still tentative. The firmest ones concern chemical abundances and kinematics.

GBPN studies are developing rapidly. Progress is expected from i) increasing the number of well observed GBPN, ii) including the newly detected GBPN in statistical studies, iii) observing additional quantities like expansion velocitis, molecules... and performing deeper radio surveys iv) interpreting observational diagrams by taking full account of selection effects.

However, it may be that our quest for understanding the properties of GBPN will never be fulfilled. Indeed, post AGB star evolution depends on many parameters (mass loss efficiencies, chemical composition, binarity...). Will the total number of GBPN be sufficient to perform a statistical analysis taking all this into account?

REFERENCES

Abell G. O. 1955, PASP 67, 258
Acker A. , Chopinet M., Pottasch S.R., Stenholm B. 1987, A&AS 71, 163
Acker A., Köppen J., Stenholm B., Raytchev B. 1991, A&AS 89, 237

Acker A., Ochsenbein F., Stenholm B., Tylenda R., Marcout J., Schon C., 1992, "Strasbourg-ESO catalogue of galactic planetary nebulae, ESO
Acker A., Stenholm B. 1990, A&AS 86, 219
Aller L.H. 1956, "gaseous nebulae", Chapman & Hall, London
Barlow, M.J., 1987, MNRAS 227, 161
Blöcker T., Schönberner D. 1990 A&A 240, L11
Clegg R.E.S. 1989, IAU Symp 131, "planetary nebulae", ed. Torrès-Peimbert, Kluwer
Dopita M. A., Henry J. P., Tuohy I.R., Webster B.L., Roberts, E.H., Byun Y.I., Cowie L.L., Songaila A. 1990, APJ 365, 640
Feast M.W. 1968, MNRAS 140, 345
Garcia -Lario P., Manchado A., Pottasch S.R., Suso J., Olling R. 1990, A&AS 82, 497
Gathier R., Pottasch S.R., Goss W.M., van Gorkom J. 1983 A&A 128, 325
Gutiérrez-Moreno A., Moreno H., Cortés G. (1991) 383, 174
Habing H.J. 1990, "From Miras to planetary nebulae", ed. M.O.Ménessier & A.Omont, éditions Frontières
Huggins P.J. 1992, IAU Symp 155, "planetary nebulae", ed Acker, Kluwer
Isaacman R. 1981, A&A 95, 46
Isaacman R. 1983, IAU Symp 103, "planetary nebulae", ed Flower, Reidel
Kinman T.D., Feast M.W., Lasker B.M. 1988, AJ 95, 804
Kohoutek, L., 1992, IAU Symp 155, "planetary nebulae", ed Acker, Kluwer
Köppen J., Acker A., Stenholm B. 1991, A&A 248, 197
Kudritzki R.P. 1992, IAU Symp 155, "planetary nebulae", ed Acker, Kluwer
Kudritzki R.P., 1989, IAU Symp 131, "planetary nebulae", ed. Torrès-Peimbert, Kluwer
Manchado A., Pottasch S.R., Garcia-Laro P., Esteban C., Mampaso A. 1989, A&A 214, 139
Mazzitelli I. 1993, preprint
Meatheringham S.J., Dopita M., Morgan D.H. 1988, APJ 329, 166
Méndez R.H., 1993 A&A in press
Méndez R.H., Kudritzki R.P., Herrero A. 1993 A&A 260, 329
Minkowski R., "Galactic structure", ed Blaauw & Schmidt, unic. of Chicago press
Pagel B.E.J. 1992, "New aspects of Magellanic Cloud research", ed Klare, Springer-Verlag
Peimbert M. 1977, APJ 150, 825
Peimbert M., Torrès-Peimbert S. 1983, IAU Symp 103, "planetary nebulae", ed Flower, Reidel
Perinotto M., 1991, APJS 76, 687
Pottasch S.R. 1984, "Planetary Nebulae", Reidel,

Pottasch S.R. 1990, A&A 236, 231
Pottasch S.R. 1992, preprint
Pottasch S.R., Acker A. 1989, A&A 221, 123
Pottasch S.R., Bignell C., Olling R., Zijlstra A.A. 1988, A&A 205, 248
Pottasch S.R., Ratag M.A., Olling R.,1990, "From Miras to planetary nebulae", ed. M.O.Ménessier & A.Omont, éditions Frontières
Pottasch S.R., Zijlstra A.A. 1992, A&A 256, 251
Ratag M.A. 1991, thesis, University of Groningen
Ratag M.A., Pottasch S. R. 1991, A&AS 91, 481
Ratag M.A., Pottasch S.R., Dennefeld M., Menzies J. 1992, A&A 255, 255
Ratag M.A., Pottasch S.R., Zijlstra A.A., Menzies J. 1990, A&A 233, 181
Rich R.M. 1988, AJ 95, 828
Schönberner D. 1979 A&A 79, 108,
Schönberner D. 1983, APJ 272, 708
Schönberner D., 1989, IAU Symp 131, "planetary nebulae", ed. Torrès-Peimbert, Kluwer
Schönberner D., Tylenda R., 1990, A&A 234, 439
Stasińska G., Fresneau A., Gameiro J..F., Acker A. 1991b, A&A 252, 762
Stasińska G., Tylenda R., 1992, IAU Symp 155, "planetary nebulae", ed Acker, Kluwer
Stasińska G., Tylenda, R., Acker A., Stenholm B. 1991a, A&A 247, 173
Trams N.R., Waters L.B.F.M., Waelkens H.J.G.L.M., van der Veen W.E.C.J., 1989, A&A 218, L1
Tylenda R., 1992, IAU Symp 155, "planetary nebulae", ed Acker, Kluwer
Tylenda, R., Stasińska G., Acker A., Stenholm B. 1991, A&A 246, 221
van der Veen W.E.C.J., 1989, A&A 210, 127
van der Veen W.E.C.J., Habing H.J. 1990, A&A
Vassiliadis E., Wood P.R., 1992, in preparation
Walton, N.A., Barlow M.J., Clegg R.E.S. 1992b, IAU Symp 155, "planetary nebulae", ed Acker, Kluwer
Walton, N.A., Barlow M.J., Monk D.J., Clegg R.E.S. 1992a, "New aspects of Magellanic Cloud research", ed Klare, Springer-Verlag
Webster B.L. 1975, MNRAS 173, 437
Webster B.L. 1976, 174, 513
Webster B.L. 1988, 230, 377
Weidemann V., 1990, A R A&A 28, 103
Weidemann V., Koester D. 1983, A&A121, 77
Zijlstra A.A. 1990 A&A 234, 387
Zijlstra A.A., Pottasch S.R. 1989, A&A 216, 245

Nobuo ARIMOTO
'92

STELLAR POPULATION SYNTHESIS

Application To Galactic Bulges

N. ARIMOTO
Institut für Theoretische Astrophysik der Universität Heidelberg, Im Neuenheimer Feld 561, W6900, Heidelberg 1, Germany

and

Institute of Astronomy, The University of Tokyo, Mitaka, Tokyo 181, Japan

Abstract.
The stellar populations give traces of the formation history of the bulges. The metallicity distribution of K-giants in the Galactic bulge resembles to that of the giant ellipticals. There seems to be no conspicuous colour-magnitude relation intrinsic to the bulges. This can be explained if the bulges formed by the dissipative collapse of central regions of proto-galaxies followed by the supernova-driven bulge wind which was induced later than the dwarf ellipticals of the similar mass (the biased wind). Unfortunately, the observational data available at present of stellar populations of the bulges are not yet sufficient to get a firm conclusion on the origin of the bulges.

Key words: population synthesis – colour-magnitude relation – metallicity distribution – bulge formation – disc galaxies

1. Introduction

To study star formation history of a galaxy, it is of vital importance to know what kind of stars are of it major constituents. Two different approaches have so far been applied to analyse the population structure of galaxies: One is called as an empirical population synthesis and the other is called as an evolutionary population synthesis. Both are referred to as the stellar population synthesis.

The empirical method is to synthesize spectra of galaxies by using stellar and/or cluster spectral libraries and tries to get the best fits to the observed spectra. The spectral library should include the stellar populations with different spectral types, luminosity classes, and chemical compositions (cf., Spinrad & Taylor 1971; Faber 1972). If the cluster library is used, the synthesis can be performed by using the star clusters of various ages and metallicities (Bica & Alloin 1986; Bica 1988). The best fit is usually seeked by a trial-and-error method (eg., Faber 1972) or by the sophisticated minimization method (eg., Schmidt et al. 1991). Quite often the solution with negative contribution of certain stellar groups results. To avoid this happen, some of astrophysical constraints have been employed. These conditions include 1) non-negative contribution, 2) smooth distribution along the main sequence on the HR diagram, and 3) simple metal enrichement history (eg., Pickles 1985; Bica 1988). The resulting best fitted solution could reveal the stellar population structure of galaxies *at the present epoch* and would make it possible to trace back the star formation history in galaxies, unless galaxies contain a

H. Dejonghe and H. J. Habing (eds.), Galactic Bulges, 133–149.

significant amount of *possible* intermediate-age stars, in which case the empirical method suffers a serious problem of non-uniqueness (cf.; Schmidt et al. 1991). Unfortunately under the scheme of empirical method, it is impossible to predict the time variation of galaxy spectra as a function of lookback time. Major achievments have been accomplished by Spinrad & Taylor (1971), Faber (1972), O'Connell (1976), Pickles (1985), Rose (1985), and Bica (1988).

The evolutionary population synthesis assumes a certain scenario of galaxy formation and evolution, and predicts chemical and photometric (spectroscopic) properties of galaxies under the context of the presumed star formation history (eg., Tinsley 1972; Bruzual 1983; Arimoto & Yoshii 1986). The resulting theoretical predictions are then confronted with observational properties, and the validity of the adopted formation scenario is studied in detail. The fundamental parameters that describe particular star formation history include star formation rate (SFR), initial stellar mass function (IMF), and accretion rate (ACR). The best set of these parameters is seeked until a reasonable agreement is attained between the predictions and the observed properties.

The evolutionary method is further split into somewhat different two approaches. One is chemcial evolution study that predicts chemical compositions of gas and stars in galaxies by using theoretical stellar nucleosynthesis theory (eg., Schmidt 1959; Talbot & Arnett 1971; Tinsley 1974; Hartwick 1980; Lacey & Fall 1985; Matteucci & François 1989). The other is photospectroscopic evolution study that synthesizes spectra of galaxies by using stellar evolutionary tracks and stellar (cluster) spectral-photometric libraries (Tinsley 1972; Bruzual 1983). Both approaches assume the same context of galaxy formation and evolution. By combining two approaches, it is possible to get the most plausible scenario of galaxy formation and history of star formation that explains the observed chemical and photometric properties entirely consistently (Arimoto & Yoshii 1986). The method predicts the time variation of galaxy spectra as a function of lookback time. This makes it possible to confront the predictions with the observational data of the colours, the redshift distribution, and the faint galaxy counts of galaxies at cosmological distance, as well as the back ground radiation in the universe. Mile stones and recent extensive studies include Tinsley (1972), Bruzual (1983), Arimoto & Yoshii (1986), Guiderdoni & Rocca-Volmerange (1987), and Buzzoni (1989).

In this review article, we present a brief description of the results of the stellar population synthesis applied to galactic bulges.

2. Origin of bulges

The origin of bulges is controversial. They either could form by dissipative collapse of proto-cloud with low angular momentum or could form at the center of disc galaxies which captured smaller satelites like dwarf irregular galaxies.

2.1. Dissipative collapse

Eggen et al. (1962), based on the kinematic and metallicity correlations, suggested that the formation of the Galaxy occurred by collapse from a larger volume, with progressive metal enrichment as the collapse proceeded, leading to the formation of central bulge with somewhat less dissipatively and to the subsequent formation of disc by cloud-cloud dissipative collisions. The timescale of the collapse has been a matter of debate. Eggen et al. (1962) concluded that the collapse must have taken place within a few free-fall timescales, while Yoshii & Saio (1979) derived the much longer timescale, of order 1 Gyr, from the same observational data used by Eggen et al. (1962). The 2D-hydrodynamical collapse model by Hensler et al. (1992) shows that already after a few free-fall timescales a clearly distinct central spheroid forms. A progressive shrinkage, spinning up, and enrichment of a gas cloud would end up in forming the rapidly rotating metal-rich bulge.

2.2. Mergers

Schweizer & Seitzer (1988) have presented evidence that ripples occur not only in ellipticals but also in disc galaxies of type S0, S0/a, and Sa, and probablly even in Sbc galaxies. The geometry of ripples suggests that the matter consisting of ripples is likely to be external in its origin. Since the lifetime of ripples is rather short, of order 1 – 2 Gyr, the capture or the merger event producing these ripples must happen recently. The numerical simulations indicate that the resulting merger remnants would look like ellipticals if merging disc galaxies are of nearly equal mass (Toomre & Toomre 1972). This suggests that the ripples in disc galaxies must have been produced by a capture of a small galaxy by a massive disc galaxy. Stellar discs seems to be less fragile than is often thought. Since the stars of infalling galaxy will be trapped not by the disc of the main galaxy but by its bulge and halo, the presence of ripples in disc galaxies suggest that their bulges must have experienced episodic mass growth (Schweizer & Seitzer 1988). Therefore, it is possible that bulges of some disc galaxies may be accumulated from the merger of smaller systems such as globular clusters or dwarf irregular galaxies. If the bulge mass grows gradually with a timescale comparable to the Hubble time (slow merging), a bulk of bulge stars must be composed of *intermediate-age* stars. On the other hand, if most of accumulation occurred very early (rapid merging), it would be difficult to distinguish the subsequent evolution from the dissipative collapse models (Rich 1990).

3. Ellipticals vs bulges

Are bulges small ellipticals? In many respects bulges are very similar to ellipticals. Table I shows a list of characteristics that have been used in arguing on the origin of ellipticals. Some characteristics are common in both ellipticals and bulges (E&B), while others are known either in ellipticals (E) alone or only in bulges (B). Among the properties shown in Table I, those concerning the stellar populations generally suggest the dissipative collapse as the origin of ellipticals, while the kinematics and structures of ellipticals could be regarded as supporting evidence for merger hypothesis. In Table I, (E) does not necessarily mean that bulges do not share that property with ellipticals, instead it means that the characteristic is not yet confirmed observationally in bulges. Therefore, the table is not complete, and it is possible that many of characteristics listed here are indeed shared by both ellipticals and bulges, although some of them, like the cluster systems, may not relevant for studying the origin of bulges.

TABLE I
Similarities between ellipticals and bulges

	observations	similarities
Stellar population	Colour-magnitude relation	E
	Colour gradients	E
	Intermediate-age stars	E&B
	UV flux upturn	E&B
	Metallicity distribution	B
Global structures	Isophotal deviations	E&B
	Scaling laws	E&B
	Anisotoropy	E
	Triaxiality	E&B
Substructures	Shells & ripples	E&B
	Counter rotating cores	E
	Dust	E
Emissions	X-ray emission	E&B
	Radio emission	E
Activity	Star bursts	E
Cluster systems	Globular cluster systems	E
Interactions with	Morphology segregation	E
environments	Colour evolution	E
	Butcher-Oemler effect	E
	Heavy elements in ICM	E&B

4. Traces of bulge formation

We confine ourselves to studying the stellar populations of bulges. In spite of many observational studies, it is not yet clear if the stellar populations of bulges are identical to those of ellipticals. Here we discuss several important features relevent to the population synthesis study.

4.1. Stellar metallicity distribution

The metallicity of bulge stars is not uniform. Rich (1988) analysed spectra of 88 K-giants in Baade's window, most of them are members of the Galactic bulge, and showed that the strong-lined bulge giants are metal-rich with respect to the solar neighbourhood stars. The metallicity distribution shows the range $-1 \leq [Fe/H] \leq +1$. The mean stellar metallicity is about twice solar, and in contrast to the local solar neighbourhood a sufficient number of relatively low metal stars are detected, indicating that there is no G-dwarf problem in the Galactic bulge. Geisler & Friel (1990) enlarge the number of K-giant to a few hundreds and confirm Rich's result.

However, Frogel et al. (1990) have conducted the infrared photometry of the Galactic bulge M-giants and have found that the infrared colour-magnitude diagrams of the bulge giants show rather small dispersion in $J-K$ colour, and the range in metallicities that would be deduced from $J-K$ colour is $\Delta[Fe/H] \simeq 0.5$, thus considerablly smaller than that of Rich (1988). The infrared CO index at $2.29\mu m$ also show significantly smaller scatter, suggesting smaller dispersion in metallicity (Frogel at al. 1990). This may be due to the unidentified blanketing agents or peculiarities in the atmosphere of the bulge stars (Frogel & Whitford 1987). Alternatively, since Rich's spectroscopic estimates for most metal-rich stars have to rely on the extrapolation of calibration scale, the spread in metallicity found by Rich (1988) may be overestimated.

The relative abundance ratio of the Galactic bulge stars has not yet been fully studied. Preliminary analysis by Barbuy (1992) has shown that *bulge-like* nearby local metal-rich stars have $[C/Fe] \simeq 0.0$, $[N/Fe] \simeq 0.0$, and $[O/Fe] \simeq 0.0$ to 0.2. Rather normal $[O/Fe]$ ratio seems to suggest that these stars were born from the gas which had already been contaminated effectively by the ejecta of type Ia supernovae (SNIa). It would be difficult to explain this if the bulge had formed before SNIa produced iron significantly, ie $\sim$ 1 Gyr. However, it is not certian if these stars belong to the same bulge population as Rich's K-giants. Detailed analyses of oxygen abundance of bulge K-giants are required.

A confusing result is coming from the abundance distribution of the Galactic bulge planetary nebulae (PNe). PNe originating from stars with main sequence mass less than $2M_{\odot}$ are classified as Type II, and as believed not to experience the second dredge-up during the AGB evolution. Their ages, on average, are thought to be older than 3 Gyr. Ratag et al. (1992) have analysed the abudance distribution of Type II PNe near the Galactic centre. The distribution of bulge PN oxygen abundance is very similar to that of non-bulge PNe in the local solar neighbourhood, the mean value of bulge PNe is only 20% higher, although the excess of objects in the higher abudance range is remakable. This apparently is in conflict to the metallicity distribution of K-giants. Other elements such as neon, sulfer, and argon also show the very similar distribution to that of non-bulge PNe. Only helium and nitrogen show systematically higher abundances than non-bulge PNe, but this is very likely due to the dredge-up and may not reflect the abundance distributions of progenitor stars.

How can we reconcile the PNe abundance distribution with that of K-giants measured by Rich (1988)? Do both objects reflect the real abundance distributions? Two points are to be emphasized. 1) The spreads in abundance distribution larger than the observational errors are observed both in PNe and K-giants; however, the dispersion of PNe is significantly smaller than the bulge giants. 2) The oxygen distribution of disc PNe has a considerablly lower mean than the solar value. The mean of PNe is 50% lower, while the bulge PNe is about 40% lower, implying unidentified systematic calibration errors in PN abundance measurement. Possible causes are discussed by Ratag et al. (1992). lternatively, it is possible that K-giants reflect the metallicity distribution of the nuclear bulge, while PNe indicate that of outer bulge, thus the difference between two distributions may come from the possible metallicity gradient in the Galactic bulge. Until we obtain more data of the metallicity gradients, it would be rather dangerous to use both PNe and K-giants metallicity distributions simultaneously as the constraints to chemical evolution study of the Galactic bulge.

4.2. Colour-magnitude (metallicity-luminosity) relation

It is highly uncertain if bulges of disc galaxies show the identical colour-magnitude (CM) relation to that of elliptical galaxies. The problem comes from the fact that it is rather difficult to estimate how much the measured bulge light is contaminated by disc light.

Turnrose (1976) performed the population synthesis of nuclear regions of Sc galaxies based on the narrow-band spectrophotometry covering wavelength range $3300-10400\AA$. With the stellar birth function of the form $B(m,t) = cm^{-x}e^{-t/\tau}$, he found that the Salpeter-like IMF and the SFR gradually decreasing in time with a characteristic timescale $\tau = 3-6$ Gyr can reproduce the observed spectral energy distribution. No clear indication of the bulge metallicity - luminosity relation was given.

Pritchet (1977) analysed the spectra of nucler bulges of 7 galaxies, M31 (Sb), M32 (E2), M51 (Sc), M81 (Sb), M86 (E3), M87 (E0p), M94 (Sb), NGC3115 (E7/S0), and NGC5195 (Irr), of the wavelength range $4200-6800\AA$ with a resolution $\Delta\lambda = 15-50\AA$. His population model suggested that the nuclear bulges of intrinsically luminous galaxies have quite similar late-type stellar populations. However, the metallicity – luminosity relation of bulges was not explicitly discussed.

It is well known that ellipticals and S0's share the idenitcal CM relation (Visvanathan & Sandage 1977; Persson et al. 1979; Bower et al. 1992). Bower et al. (1992) have shown that the scatter of early-type galaxies about the CM relation is typically as small as 0.05 mag, of which 0.03 mag can be accounted for by observational measuring errors alone, suggesting no intrinsic scatter. S0 galaxies show slightly larger dispersion than ellipticals. It is not clear whether the CM relation of S0 galaxies is due to (1) an intrinsic CM relation of their bulges or (2) different degree of disc light contribution to the galaxy light within the aperture size adopted in the observations.

Some evidence suggest that Mg_2 indices of luminous S0 discs are greater than those of fainter ones (R.Bender, private communication). Since $J - K$ colours of bulges and discs of S0 galaxies are nearly identical, bulges and discs of S0's of identical luminosities are likely to have the similar metallicity (Bothun & Gregg 1990), provided that $J - K$ is a good metallicity indicator for gas-poor galaxies (Arimoto & Yoshii 1987). This suggests that bulges of more luminous S0's have higher metallicities, and thus redder colours, than those of less luminous ones.

It is Bica (1988) who first studied systematically the metallicity – luminosity relation of bulges of spiral galaxies. Based on the cluster spectra library of Bica & Alloin (1986), Bica analysed spectra of nuclei of 169 galaxies, of which 85 galaxies are spirals, in the wavelength range $3700 - 10000\mathring{A}$ at $12\mathring{A}$ resolution. According to characteristic features in spectra, spiral galaxies were binned into seven different groups S1 - S7. The groups S1 - S4 migrate simultaneously towards lower luminosity and later morphological type. These are reddest spiral groups in his sample and their spectra are very similar to the red elliptical groups E1-E4 (ie, without any significant evidence for star forming activity). The best fitted solutions were obtained by assuming simply that the metallicity of galaxy nucleus increases monotonically as a function of time. The solutions indicate that the maximum metallicities reach values a factor 4 solar (S1), 2 solar (S2, S3), and solar (S4). Bica thus suggested that *the maximum metallicity of stars in the nucleus is related to the bulge luminosity.*

If the bulge's metallicity – luminosity relation exists, as Bica suggested, it is interesting to see whether the relation is identical to that of ellipticals. In Fig.1 we plot a sum of equivalent widths of CN lines at $4200\mathring{A}$ and of Mg+MgH lines at $5175\mathring{A}$, $W(CN, Mg + MgH)$, as a function of *total* absolute magnitude by using Bica's S1-S4 spirals. Approximate locations of ellipticals, E1-E4, are also indicated for comparison. Surface decomposition parameters are not available for most of spirals shown in Fig.1; thus we are obliged to plot the total absolute magnitudes instead of bulge's intrinsic ones. Therefore, spirals should locate left to ellipticals even if the bulge's metallicity – luminosity relation is identical to that of ellipticals. In Fig.1, the scale given at the lower-left corner indicates the amount of luminosity decrease as a function of L_B/L_D – the bulge-to-disc light ratio. L_B/L_D amounting to 0.1 at $W(CN, Mg + MgH) \sim 10\mathring{A}$ and to 0.5 at $W(CN, Mg + MgH) \sim 20\mathring{A}$ could explain the apparent shift of S1-S4 spirals with respect to E1-E4 ellipticals. Although the precise comparison is impossible due to a lack of bulge's intrinsic luminosity, one can point out the followings: 1) there are very few spirals showing the line strengths as strong as giant ellipticals of metal-rich E1 group, 2) few spirals show such weak equivalent widths as dwarf ellipticals belonging to E4 group. If the spread in these line widths is due to the metallicity of bulge stars, the bulge metallicity is restricted to the range much smaller than that of ellipticals. *This means that the bulge's metallicity – luminosity relation, if it exits, is not exactly*

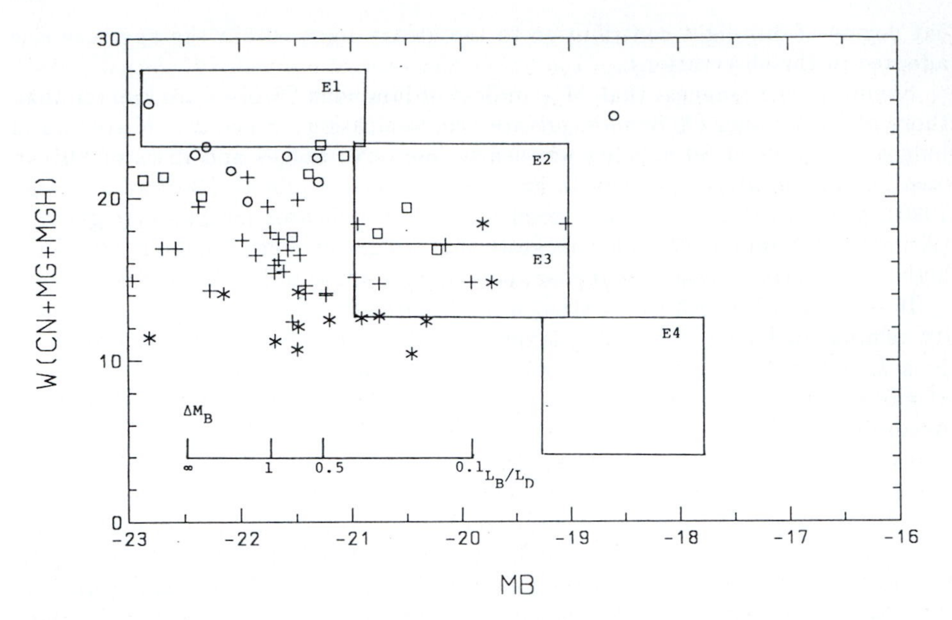

Fig. 1. Equivalent widths (Å units) of CN, Mg, and MgH lines verse total absolute blue magnitudes for Bica's (1988) red spirals: S1 (open circles), S2 (open squares), S3 (crosses), S4 (asterisks). Four boxes indicate the regions where Bica's red ellipticals E1-E4 locate. The scale shown in the lower-left corner gives the amount of luminosity decrease for various values of the bulge-to-disc light ratio.

the same as that of ellipticals.

Although the aperture size used by Bica (1988) is relatively small, 0.8 – 1.4 Kpc in average, it is not clear if there is no contamination of light from disc stars. His solutions show rather strange metallicity distribution of *old stars* (Table 4 in Bica). The solutions for S1-S4 all show the sharp peak and sudden drop at the higher metallicity end. This does not change when the luminosity contribution is converted into the mass contribution (Bica, Arimoto, Alloin 1988). Bica's metallicity distribution is completely different from that of bulge giants measured by Rich (1988), and such metallicity distribution is rather unexpected from the view point of chemical evolution. An exception is a galactic wind model, which predicts a sharp peak at the higher metallicity end (Yoshii & Arimoto, 1987), if the wind expells gas before the maximum stellar metallicity exceeds the yield of nucleosynthesis. However, it is more likely that Bica's metallicity distribution of old stars is *an artifact*, mainly due to the very simplified astrophysical constraint adopted for the metal enhancement history. The monotonous metallicity increase is assumed in the bulge, but since the observed spectra may be contaminated by disc light, at least two different enhancement histories must be incorpolated in seeking the solution (cf., Arimoto & Jablonka 1992).

It is essential to decompose the bulge and disc components of Bica's (1988) S1-S4 spirals and to evaluate the absolute magnitudes of bulges. This will make it possible to argue precisely the bulge metallicity – luminosity relation. Once such relation, if any, is established, we would be able to discuss conclusively whether bulges are small ellipticals of the same origin.

5. Theoretical considerations

5.1. Metallicity distribution

The metallicity distribution of stars is sensitive to three fundamental parameters of star formation history; ie., the SFR, the IMF, and the ACR (Köppen & Arimoto 1990). If the accretion is much faster than the star formation, the distribution is rather broad and peaks at $Z_{Fe} = y_{Fe}$, as can be shown easily for the simple closed-box model too. If the ACR is the same order as the SFR, the number of low metallicity stars is diminished, which is why infall models solve the G-dwarf probelm, but still show a peak very close to $Z_{Fe} = y_{Fe}$. However, if the ACR is slow enough, the distribution becomes markedly different: It rises towards higher metallicity monotonously, but is cut at $Z_{Fe} = y_{Fe}$, where it also has its maximum value (see Fig.2a of Köppen & Arimoto 1990). Therefore, from the metallicity distribution of stars in the Galactic bulge, one can learn directly the yield of stellar nucleosynthesis there. This yield shows the efficiency of iron production by SNII if the timescale of bulge formation is shorter than 1 Gyr, and shows the combined efficiency by both SNII and SNIa if the bulge formation lasted longer than the effective lifetime of SNIa, ie., ~ 1 Gyr (cf., Matteucci & Brocato 1990).

Unfortunately, the stellar metallicity distribution alone cannot provide a crucial criterion distinguishing two controversial hypotheses. The metallicity distribution of the dissipative collapse model is essentially the same as that of the so-called simple model (closed one-zone, time and space constant IMF, initially metal-free). Tinsley & Larson (1979) have shown that the metallicity distribution of merger models becomes identical to that of the simple model if two pre-merging subunit galaxies are identical. The metallicity distribution does not differ too much from the simple model even if Tinsley & Larson's condition is relaxed (Arimoto & Jablonla 1993, in preparation).

5.2. CM relation

So far there has been no theoretical consideration on the CM relation of the bulges. This is mainly because the bulge's CM relation has not been clearly identified. Since the CM relation of elliptcials has played a fundamental role in arguing their origin, we briefly describe theoretical models for the CM relation of ellipticals and introduce a theoretical attempt to explain the CM relation of the bulges.

5.2.1. Galactic wind models

The CM relation of ellipticals can be understood in terms of increasing metallicity with luminosity (or equivalently mass). One possible explanation for this metallicity – mass relation is a galactic wind induced by supernova explosions during the dissipative collapse phase of galaxy formation (cf., Larson 1974; Arimoto & Yoshii 1987; Matteucci & Tornambè 1987). Following Carlberg's (1984) dissipative collapse model, Arimoto & Yoshii (1987) constructed models which assume that an elliptical was initially a homogeneous gas sphere in which star formation was triggered by cloud-cloud collisions. A galactic wind is assumed to happen when the thermal energy provided by SNe exceeded the binding energy of the remaining gas. With a wind, the residual gas was swept away and the star formation was to stop. Because of the larger binding energy, the epoch of wind was delayed in an elliptical of larger initial mass. As a result, the mean stellar metallicity increased with the initial galaxy mass. The latest wind blew at 0.85 Gyr in the model for giant elliptical galaxy of initial mass 2 $10^{12} M_{\odot}$. Therefore, Arimoto & Yoshii's wind models imply that almost all iron in ellipticals had been produced by SNII explosion. Needless to say, a bulk of subsequent iron formation should follow in ellipticals after the wind because of the delayed SNIa explosions (Matteucci & Tornambé 1987). The lack of gas in ellipticals and the exsistence of hot X-ray halo around ellipticals and hot gas in clusters of galaxies suggest that iron from SNIa is now in the outside of ellipticlas.

5.2.2. Merger models

The other possibility to explain the metallicity – mass relation is the mergers (Tinsley & Larson 1979; Arimoto & Jablonka 1993). Elliptical galaxies are assumed to form by the hierarchical mergers of smaller gas-rich systems. It is assumed that stars to form only when star bursts are induced during violent mergings. If (only if) the SFR per unit mass is increasing in proportion to the total galaxy mass with the power-law index $p = 1/3$, the observed metallicity – mass relation is reproduced. Tinsley & Larson (1979) assumed the mergers of the identical gas-rich galaxies; later Arimoto & Jablonka (1993) have studied chemical evolution of more generalized mergers by taking into account various combinations of pre-merger units whose star formation histories differ considerablly. The resulting Mg_2 verse M_V relation is shown in Fig.2. Although the merger process itself is quite stochastic, the resulting metallicity – mass relation shows surprisingly small dispersion at a fixed luminosity. Thus it is of no surprise that the CM relation of ellipticals does not show any significant dispersion (Bower et al. 1992).

However, there is one difficulty to be overcome in the merger hypothesis. The gas fraction in the merger remnant is too large to be an elliptcial (Fig.3). The chemical evolution of the merger is essentially given by the simple model (Tinsley & Larson 1979). Therefore the merger remnant of small mass, having low metallicity, still keeps significant amount of residual gas. This amounts to 70% in the remnant of $10^{10} M_{\odot}$, and to 10 – 20% even in the most massive remnant of $10^{12} M_{\odot}$. If

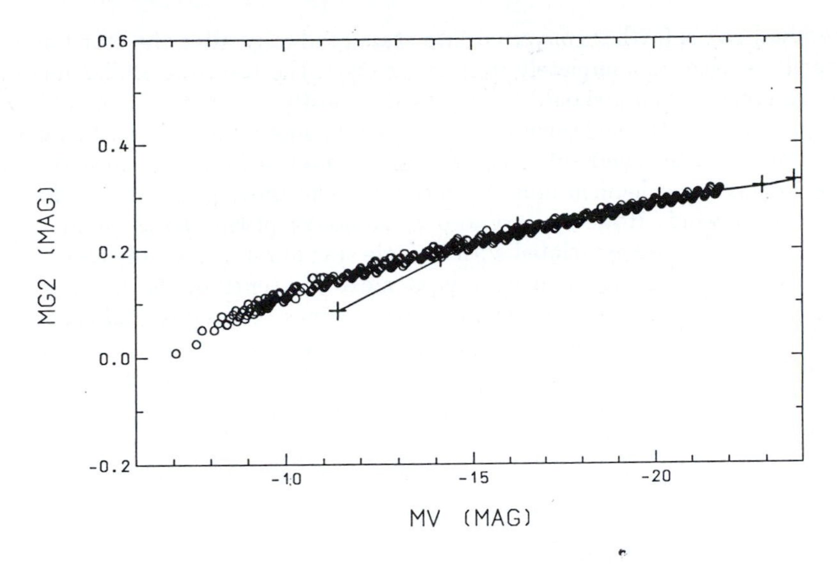

Fig. 2. The Mg_2 verse M_v relation of the stochastic merger models (open circles) by Arimoto & Jablonka (1993). Crosses linked by a solid line indicate the galactic wind models by Arimoto & Yoshii (1987) for comparison.

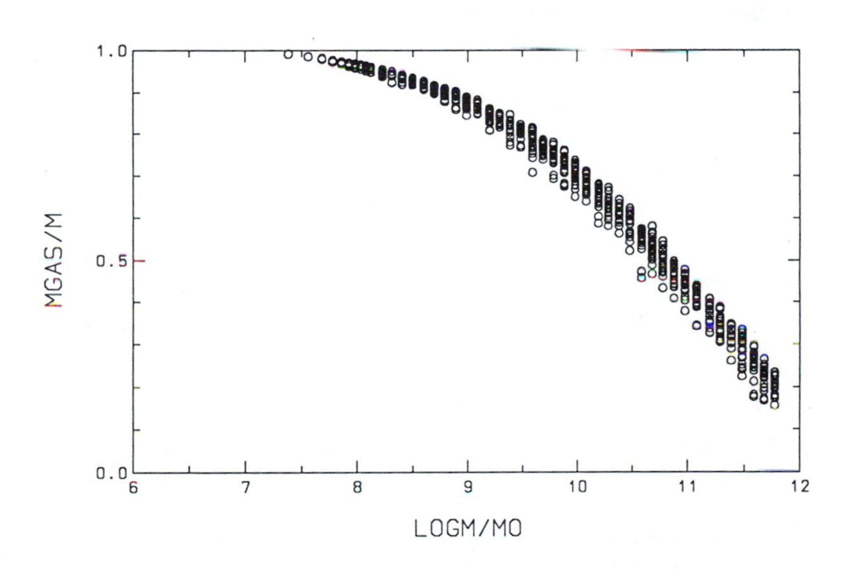

Fig. 3. The residual gas mass fraction of the merger models by Arimoto & Jablonka (1993). The upper envelope corresponds to the twin mergers (mergers of identical galaxies) originally developed by Tinsley & Larson (1979).

the residual gas is further converted into stars gradually after the star bursts and eventually is used up completely in forming stars, the resulting stellar metallicity in the merger remnants should be indetical regardless of their mass. This would destroy completely the metallicity – mass relation. Therefore, the residual gas must be removed from the remnant. Since ellipticals do exist in both field and clusters, the gas removal mechanism must be intrinsic. The most plausible mechanism is an SN-driven wind. However, Arimoto & Jablonka (1993) have found that the thermal energy release associated with a single star burst is not sufficient to induce a wind. The thermal energy at peak is at most one-tenth of the binding energy, and cools quickly once the star burst activity declines. Therefore, to induce a wind the successive events of mergers are required to happen within a short timescale before SN-ejecta of the previous generations cool down. Rough estimate suggests that the whole process must be completed within the same timescale of galaxy formation as that of Arimoto & Yoshii's (1987) wind model, ie., ~ 1 Gyr. In this case it would be rather difficult to distinguish between the wind models and the merger ones.

5.2.3. Biased galactic wind

Although not conclusive, there are several lines of evidence suggesting rather *weak or no* CM relation for the bulges of spirals and S0's: 1) Nuclei of spirals (S1-S4; Bica 1988) show the narrow range in the equivalent width of metallic lines, $W(CN, Mg + MgH) \simeq 10 - 24$, while normal red ellipticals (E1-E4) show much wider spread, $W(CN, Mg{+}MgH) \simeq 5-30$ (Fig.1). This implies that the bulges of spirals have more or less similar metallicity regardless of their mass. 2) The optical – infrared colours of the brightest S0's are identical to those of giant ellipticals (Persson et al. 1979). Since discs would hardly contribute to the lights of these luminous S0's, this indicates that the bulges of luminous S0's show the same colours as giant ellipticals, in spite of the fact that masses of S0's bulges are smaller than giant elliptcials. 3) The metallicity distribution of K-giants in our Galactic bulge (Rich 1988) is very similar to that of the galactic wind model of a giant elliptical with the initial mass $10^{12}M_{\odot}$ (Arimoto & Yoshii 1987). This metallicity distribution can be reproduced only if chemical evolution within the bulge is nearly complete (ie., at least more than 90% of gas must be converted into stars), although the mass of the Galactic bulge is far much smaller than giant ellipticals. The facts (1) - (3) seem to suggest that the stellar metallicities of the bulges of spirals and S0's are, irrespective of their mass, nearly identical to those of metal-rich giant ellipticals.

The mass of bulges is in the similar range to that of dwarf ellipticals. Nevertheless, metallicities of bulge stars are much higher than stars in dwarf ellipticals. The mean stellar metallicity of K-giants in our Galactic bulge is at least as high as twice solar, while that of dwarf ellipticals is at most solar or much less (cf., Yoshii & Arimoto 1987). This could be explained by *the biased bulge wind.* At an epoch of SN-driven wind, the bulge must have been surrounded by the halo in which the primordial gas had been still abundant. In such a case the condition for the wind must be written as:

$$E_{th} = E_b + \int P_{ex} dV,$$

where E_{th} is the thermal energy of gas heated up by SN explosions, E_b is the binding energy of gas within the bulge, and P_{ex} is the external pressure of halo gas around the bulge. The second term $\int P_{ex} dV$ gives the amount of work that the bulge gas is required to do. This term is virtually negligible for ellipticals. Therefore, to induce a wind, the bulge must provide more thermal energy compared to ellipticals of the same mass. A larger number of SNe must have been exploded before the delayed wind was eventually triggered. The amount of gas that was converted into stars must be larger in the bulge than in the dwarf elliptical. This could explain why the bulges have higher metallicities, redder colours, and brighter surface brightness (because they lose smaller amount of gas) compared with the dwarf ellipticals (cf., Burkert & Arimoto 1993, in preparation).

6. Bulge model

Under the context of dissipative collpase hypothesis, Arimoto & Jablonka (1991) have constructed a bulge model that reproduces the metallicity distribution of K-giants obtained by Rich (1988). Within an uncertainty of measurements, the overall shape of the metallicity distribution of the Galactic bulge stars is best reproduced by the infall model that is characterized by a slope of the IMF $x_B = 1.05$ (corresponding to the yield of three times solar), the SFR per unit mass $k_B = 10$ Gyr^{-1}, and the ACR $a_B = 8$ Gyr^{-1}. A bulge wind is assumed to occur at $t = 0.7$ Gyr, according to Arimoto & Yoshii's wind model for a giant elliptical of an initial mass $10^{12} M_\odot$.

TABLE II
Bulge model properties

f_g	$[Fe/H]_\ell$	$U-B$	$B-V$	$V-R$	$V-I$	$V-J$	$V-H$	$V-K$	$V-L$
0.054	+0.210	0.682	1.039	0.939	1.788	2.330	3.042	3.259	3.356

Table II gives the resulting properties of the bulge model. In the $U-B$ verse $B-V$ diagram, the bulge model locates at the reddest (lower-right) end of a sequence defined by normal galaxies, showing almost identical colours to giant ellipticals. By integrating over the observed luminosity function of stars in the Galactic bulge, Terndrup et al. (1990) have found that M-giants contribute significantly to the total light of the Galactic bulge and have obtained the infrared colours of the galactic bulge as $V-K = 3.32$, $J-K = 0.86$, and $H-K = 0.21$, in excellent agreement with theoretical values $V - K = 3.26$, $J - K = 0.93$, and $H - K = 0.22$ given in Table II. Bothun & Gregg (1990) have measured optical-infrared colours of the bulges of S0's. After correcting the contamination of disc lights, they obtain $B - H = 4.05 \pm 0.25$ and $J - K = 0.92 \pm 0.08$ as mean values of their sample galaxies. Table II gives $B - H = 4.08$, thus showing nearly identical colours to Bothun & Gregg's value.

We note that once the bulge model is constructed by using chemical properties of the bulge as constraints, its photometric properties show an excellent agreement with the observed ones. As a conclusion, we stress that the dissipative collapse hypothesis is quite successful in reproduing both chemical and photometric properties of the bulges of spirals and S0's.

7. Application to spiral galaxies

A lack of a clear indication of the bulge's metallicity – luminosity relation suggests that the star formation history has been nearly identical in bulges of all spirals (and probablly of S0's) regardless of their luminosities and morphological types. The available data of bulge colours, as we have seen in section 6, seem to support this *unique bulge* hypothesis. By using the bulge model discussed above, Arimoto & Jablonka (1991) have built two component – bulge and disc – models of spirals to check if the unique bulge hypothesis brings any serious discrepancies between the theoretical predictions and the observed properties of spirals to which more various data are availabe than to the bulges. It is assumed that central regions of a proto-galaxy had collapsed rapidly and had formed the bulge, followed by the infall of loosely bound materials in the halo onto the equatorial plane and formed the disc. The star formation in the bulge stopped when the bulge wind ejected the residual gas containing the processed heavy elements into the halo. The ejecta were mixed with the primordial gas in the halo; therefore the gas infalling later onto the disc was pre-enriched.

TABLE III
Disc model properties

Type	$[Fe/H]_\ell$	$U-B$	$B-V$	$V-R$	$V-I$	$V-J$	$V-H$	$V-K$	$V-L$
Sa	0.061	0.037	0.665	0.770	1.538	2.054	2.784	2.961	3.061
Sb	-0.095	-0.055	0.575	0.691	1.353	1.817	2.464	2.659	2.761
Sc	-0.209	-0.125	0.505	0.629	1.207	1.619	2.196	2.393	2.496
Sd	-0.333	-0.154	0.474	0.607	1.163	1.566	2.116	2.328	2.430

In a similar way to the bulge model, the parameters are fixed for each morphological type by using the observed characteristics of chemical properties, such as the stellar metallicity distribution and the hydrogen mass to stellar luminosity ratio, of the discs of spiral galaxies. The following values are chosen: the IMF slope $x_D = 1.45$ (the yield corresponds to a half solar), the ACR $a_D = 0.17$ Gyr^{-1}, and the SFR $k_D = 0.69, 0.32, 0.25, 0.17$ Gyr^{-1} for the type Sa, Sb, Sc, and Sd, respectively. The resulting properties of the disc models are given in Table III.

Assuming de Vaucouleurs' $r^{1/4}$-law for the surface brightness distribution of the bulge and an exponential-law for the disc, the local bulge-to-disc light ratios are calculated in B-band as a function of the aperture size for 79 spirals to which decomposition parameters of the surface brightness profile are empirically known (Simien & de Vaucouleurs, 1986; Kent, 1985). The local bulge-to-disc ratios are then used as a weight for synthsizing the photometric properties of individual 79 spirals. Regarding these spirals as a sample of normal spirals of different morphological types, we have compared the model properties with the observed ones. It has been shown that the disc mass of spirals is confined in rather a small range around $10^{11} M_\odot$ (a Hubble constant $H_0 = 50$ km s^{-1} Mpc^{-1} is assumed), while the bulge mass shows clear decrease from $\sim 3\ 10^{11} M_\odot$ to $\sim 3\ 10^{9} M_\odot$ towards later types along the morphological sequence, although the dispersion at a fixed type amounts to one order of magnitude.

The bulge-to-disc ratio is the single dominant parameter that leads to narrow distributions of spirals on the colour-colour diagrams (such as $U - B$ verse $B - V$) and that leads to general trends of optical-infrared colours along the morphological sequence. The scatter in colours of spiral at a fixed type arises from the intrinsic scatter in the bulge-to-disc ratio. This clearly demonstrates that if the bulge-to-disc ratio is chosen as a major factor of galaxy classification the photometric properties of spirals would correspond uniquely to the classified type. As is clearly shown in Tables II and III, the bulges and the discs have almost the same JHK colours; as a result, the near-infrared colours do not depend on the bulge-to-disc ratio and therefore show no clear trend along the morphological sequence as was already pointed out observationally by Gavazzi & Trincheri (1989). In early type spirals, S0/a-Sab, the CM relation in the $U - V$ verse M_v diagram has been known to exit (Griersmith 1980). This has been regarded as an evidence for the bulge's metallicity - luminosity relation. However, it is shown that this CM relation arises from a variation of the bulge-to-disc ratio and not from the bulge's metallicity - luminosity relation. Therefore we do not find any evidence that requires the metallicity - luminosity relation that is intrinsic to the bulges of spirals.

8. Conclusions

It is not clear whether the stellar populations of the bulges are identical to the ellipticals exists. Bica (1988) suggested an intrinsic CM relation of the bulges of normal spirals. However, it cannot be considered as real until one reanalyse his data by carefully removing the contaminated disc light and by introducing more realistic metal enhancement history in the bulges. Needless to say, accurate measurements of the bulge's intrinsic luminosities are required to be done by decomposing the surface brightness distribution into the bulge and disc components.

Apart from Bica's population synthesis study, there are no other works that suggest the CM relation of the bulges. Instead, the metallicity distribution of K-giants in the Galactic bulge (Rich 1988) indicates that the star formation history must have been very similar to the giant ellipticals, although the mass of the former is much close to the dwarf ellipticals than to the latter. This suggests that chemical evolution in the bulges of any mass is nearly identical to the giant ellipticals. We therefore conclude that there is no conspicuous CM relation intrinsic to the bulges and that the stellar populations of the bulges are more or less identical to the giant ellipticals.

The bulges could form either by the dissipative collapse of the central regions of proto-galaxies or by the capture of smaller satelite galaxies by massive disc galaxies. Our present knowledge of the stellar populations of the bulges is confined to the stellar metallicity distribution and the CM relation, which is not sufficient to reveal the origin of the bulges. Observational data of the colour gradients would provide important constraint to the bulge formation. Search for the intermediate-age stars and a systematic study of the UV-flux from the bulges are also fundamental. The relative abundance ratios, such as $[O/Fe]$ and $[Mg/Fe]$, would eventually tell the timescale of star formation in the Galactic bulge. Of course more careful analyses of the stellar metallicity distribution as well as the CM relation would be fruitful.

Acknowledgements

I am grateful to P.Jablonka for her energetic contribution to our works on the population synthesis. R.Bender and R.Guzman kindly provided me their recent photometric data of elliptcials before publication. Special thanks go to F.Leeuwin for many discussions on the kinematics of the ellipticals and the bulges, without her contribution I could not make better approach to the problem. The University of Tokyo kindly allowed me to stay in Germany longer so that I could attend this fruitful meeting. Finally financial support from the Deutsche Forschunsgemeinschaft (SFB 328) is gratefully acknowledged.

References

Arimoto, N., Jablonka,J.: 1991, *A&A* **249**, 374
Arimoto, N., Yoshii, Y.: 1987, *A&A* **173**, 23
Arimoto, N., Yoshii, Y.: 1986, *A&A* **164**, 260
Barbuy,B.: 1992, **IAU Symp. No.149, The Stellar Populations of Galaxies**, B.Barbuy, A.Renzini
Kluwer, Dordrecht 143
Bica, E.: 1988, *A&A* **195**, 76
Bica, E., Arimoto, N., Alloin, D.: 1988, *A&A* **202**, 8
Bica, E., Alloin,D.: 1986, *A&A* **162**, 21
Bothun, G.D., Gregg, M.D.: 1990, *ApJ* **350**, 73
Bower, R.G., Lucey, J.R., Ellis, R.S.: 1992, *MNRAS* **254**, 601
Bruzual,G.A.: 1983, *ApJ* **273**, 105

Buzzoni,A.: 1989, *ApJS* **71**, 817
Eggen,O.J., Lynden-Bell,D., Sandage,A.: 1962, *ApJ* **136**, 748
Faber,S.M.: 1972, *A&A* **20**, 361
Frogel,J.A., Terndrup,D.M., Blanco,V.M., Whitford,A.E.: 1990, *ApJ* **353**, 494
Frogel,J.A., Whitford,A.E.: 1987, *ApJ* **320**, 199
Gavazzi,G., Trincheri,G.: 1989, *ApJ* **342**, 718
Geisler,D., Friel,E.: 1990, **Bulges of Galaxies**, B.Jarvis, D.Trendrup
ESO 77
Griersmith,D.: 1980, *AJ* **85**, 1295
Guiderdoni,B., Rocca-Volmerange,B.: 1987, *A&A* **186**, 1
Hartwick,F.D.A.: 1980, *ApJ* **236**, 754
Hensler,G., Burkert,A., Truran,J.W., Dünhuber,H., Theis,C.: 1992, **The Stellar Population of Galaxies**, B.Barbuy, A.Renzini
Kluwer, Dordrecht 119
Kent,S.M.: 1985, *ApJS* **59**, 115
Köppen,J., Arimoto,N.: 1990, *A&A* **240**, 22
Lacey,C.G., Fall,S.M.: 1985, *ApJ* **290**, 154
Larson,R.B.: 1974, *MNRAS* **166**, 585
Matteucci,F., Brocato,E.: 1990, *ApJ* **365**, 539
Matteucci,F., François,P.: 1989, *MNRAS* **239**, 885
Matteucci,F., Tornambé,A.: 1987, *A&A* **185**, 51
O'Connell,R.W.: 1976, *ApJ* **206**, 370
Persson,S.E., Frogel,J.A., Aaronson,M.: 1979, *ApJS* **39**, 61
Pickles,A.J.: 1985, *ApJ* **296**, 340
Pritchet,C.: 1977, *ApJS* **35**, 397
Ratag,M.A., Pottasch,S.R., Dennfeld,M., Menzies,J.W.: 1992, *A&A* **255**, 255
Rich,R.M.: 1990, *ApJ* **362**, 604
Rich,R.M.: 1988, *AJ* **95**, 828
Rose,J.A.: 1985, *AJ* **90**, 1927
Schmidt,A.A., Copetti,M.V.F., Alloin,D., Jablonka,P.: 1991, *MNRAS* **249**, 766
Schmidt,M: 1959, *ApJ* **129**, 243
Schweizer,F., Seitzer,P.: 1988, *ApJ* **328**, 88
Simien,F., de Vaucouleurs,G.: 1986, *ApJ* **302**, 564
Spinrad,H., Tayler,B.J.: 1971, *ApJS* **2**, 445
Talbot,R.J., Arnett,W.D.: 1971, *ApJ* **170**, 409
Tinsley,B.M., Larson,R.B.: 1979, *MNRAS* **186**, 503
Tinsley,B.M.: 1974, *ApJ* **192**, 629
Tinsley,B.M.: 1972, *A&A* **20**, 383
Terndrup,D.M., Frogel,J.A., Whitford,A.E.: 1990, *ApJ* **357**, 453
Toomre,A., Toomre,J.: 1972, *ApJ* **178**, 623
Turnrose,B.E.: 1976, *ApJ* **210**, 33
Visvanathan,N., Sandage,A.: 1977, *ApJ* **216**, 214
Yoshii,Y., Arimoto,N.: 1987, *A&A* **188**, 13
Yoshii,Y., Saio,H.: 1979, *PASJ* **31**, 339

ALVIO RENZINI

FORMATION AND EVOLUTION OF STARS IN GALACTIC BULGES

ALVIO RENZINI
Dipartimento di Astronomia, CP 596
I-40100 Bologna, Italy

ABSTRACT. A fair fraction of stars in the Galactic Bulge (a possibly in bulges in general) appears to be more metal rich than the sun. Some of the current limitations in quantitatively modelling such super metal rich (SMR) stars are briefly recalled, including the question of the helium enrichment, of the metallicity dependence of mass loss, and of the metal opacity. Recent color-magnitude diagrams for stars in the Galactic Bulge are show that the bulk of Bulge stars must be very old, although current data do not allow to determine the age with sufficient accuracy to establish the relative age of the Halo and of the Bulge. The question of the nature of the most luminous (AGB) stars in bulges and in M32 is then addressed in some detail, discussing a series of methodological aspects which would need careful consideration before using bright AGB stars as age indicators. It is concluded that – for the time being – none of the claims for the presence of an intermediate age component in the Galactic Bulge, in M32, and in the bulge of M31 is completely exempt from ambiguities, and ways for elimitating such ambiguities are suggested. Finally, from the evidence that bulges are dominated by a very old stellar population it is concluded that star formation in bulges probably started and was essentially completed *before* the completion of star formation in the halo: bulges are likely to *on average* older than halos.

1. Introduction

There is still no general consensus on how the Galaxy formed, and how long it took to built up its various components. We all agree that the Halo is old, and the disk contains young stars. But what about the Bulge? In which genetic relation is it with the rest of the Galaxy? Is it younger or older than the Halo? We cannot really say we understand the Galaxy until we find the answers to these questions, since the Bulge is really the *core* of the Galaxy, and then its formation is the core of the Galaxy formation problem. Moreover, the Bulge of our own Galaxy together with that of our Local Group companion M31 are reasonable prototypical of bulges in general (Frogel 1990), and the bulges of spirals are generally regarded as stellar systems sharing a number of properties with early-type galaxies. It follows that answering the questions above will help a great deal in understanding galaxy formation in general. The study of stellar populations in resolved bulges can therefore help making progress in one of the central issues of modern astrophysics and cosmology. In this brief review I will address just three points: (1) the quest for the evolution of super metal rich (SMR) stars in the Bulge, (2) the color-magnitude diagrams of field and globular cluster stars in the Bulge, and (3) the interpretation of the bright end of the luminosity funcion of the Gralactic Bulge, of M32, and of the bulge of M31. From these considerations about the stellar content of bulges I will finally draw some speculative inferences about their formation.

H. Dejonghe and H. J. Habing (eds.), Galactic Bulges, 151–168.

2. Problems with the Evolution of SMR Stars

It is now well established that the Galactic Bulge contains stars exceeding the solar metallicity (Whitford & Rich 1983; Rich 1988). These are very rare objects in the solar neighborhood, and their evolution still presents several aspects which are not (quantitatively) fully dominated.

2.1 THE COMPOSITION OF SMR STARS

A first problem that we encounter when dealing with SMR stars concerns the actual chemical composition to adopt for them, i.e. the detailed proportions of the various elements for given overall metal abundance Z. For example, to construct models we need to specify the relative oxygen abundance [O/Fe], and the helium abundance Y. Here I concentrate on the helium problem.

There is little doubt that along with metals mass losing stars and supernovae contribute also some helium to the enrichment of the interstellar medium (Peimbert 1983; Pagel 1989). However, the actual size of this enrichment remains rather uncertain, i.e. value of the helium enrichment parameter $\Delta Y/\Delta Z$ is still poorly constrained by either theory or observations, with most popular values ranging from 1 to 3. The actual helium abundance for given metal abundance Z is given by the standard relation:

$$Y(Z) = Y_{\rm p} + \frac{\Delta Y}{\Delta Z} Z, \tag{1}$$

where $Y_{\rm p}$ is the primordial helium abundance, for which I assume $Y_{\rm p} = 0.24$. For illustration purposes I will explore values of $\Delta Y/\Delta Z$ between 0 and 3, although larger values have been occasionally proposed. It is important to realize that SMR stars may be rather unusual objects indeed. For example, assuming $\Delta Y/\Delta Z = 2$ (a conservative value!) and solar proportions ([M/Fe]=0) hypothetical SMR stars with [Fe/H]=1 (such as the most extreme SMR stars in the Bulge, according to Rich 1988) would consist of

20% METALS
65% HELIUM
15% HYDROGEN

with hydrogen having been reduced to a minority constituent. Notice that the increase in metals and helium tend to have opposite, partially compensating effects on several evolutionary properties. More metals means more opacity, and fainter ZAMS stars. More helium means larger mean molecular weight, and therefore more compact and brighter ZAMS stars for given mass. It is easy to realize that the choice of $\Delta Y/\Delta Z$ will decide which of the two effects will dominate over the other, and therefore will determine for example whether the stellar lifetime (for given mass) is an increasing or a decreasing function of Z, or similarly for the mass of evolving stars (for given age) as a function of Z.

The case is illustrated in Fig. 1, which shows the (initial) mass of evolving stars (i.e., the mass $M_{\rm RG}$ of stars having just reached the base of the red giant branch, RGB) as a function of metallicity Z, for a 15 Gyr old stellar population (adapted from Renzini and Greggio 1990). We can easily notice that – not surprisingly – for $Z \lesssim Z_\odot$ the effect of the helium enrichment is virtually negligible, and the

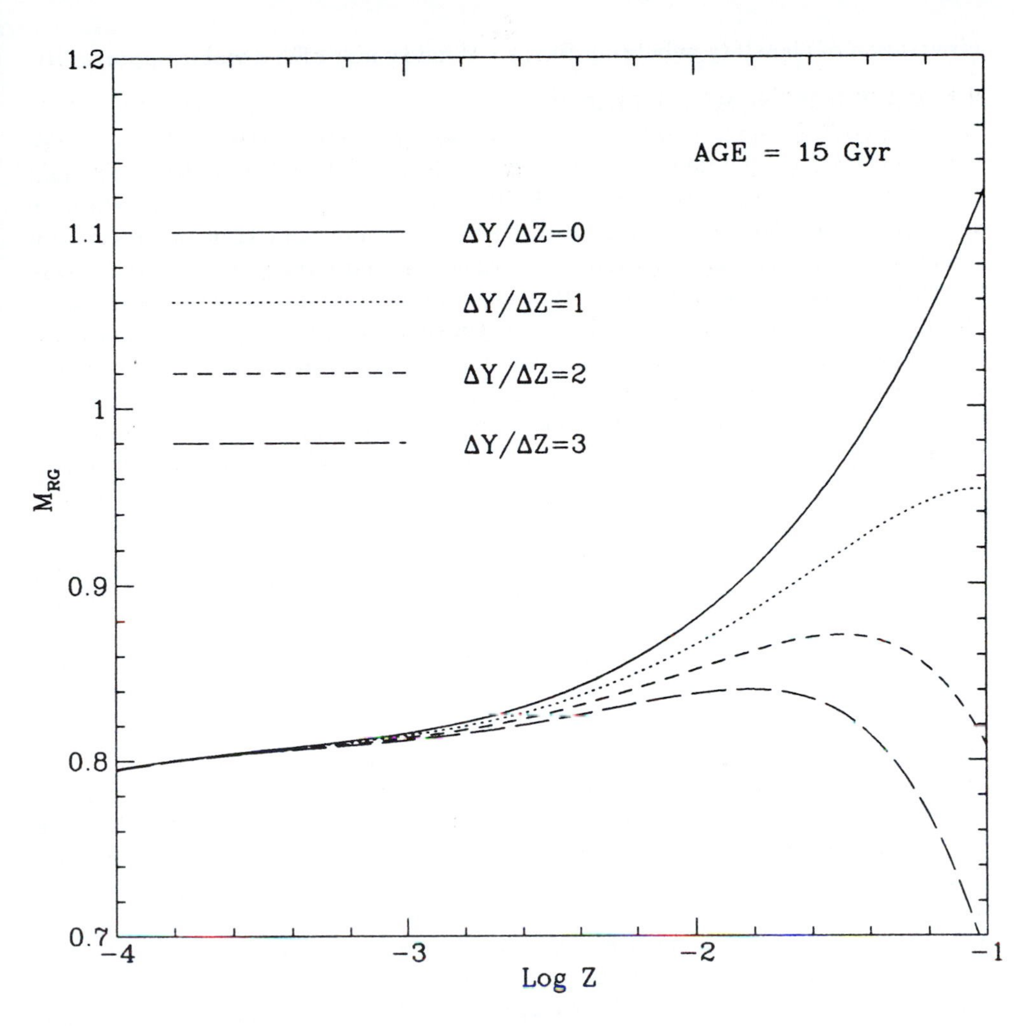

Fig. 1.– The mass of 15 Gyr old stars that evolve off the main sequence (or the mass at the base of the RGB) as a function of the metal abundance Z, and for various values of the helium enrichment parameter $\Delta Y/\Delta Z$.

various options run close to each other. But as Z grows beyond solar $M_{\rm RG}$ becomes increasingly sensitive to the actual value of the $\Delta Y/\Delta Z$ parameter, and for $Z = 5Z_\odot$ the various relations dramatically diverge.

In turn, the smaller $M_{\rm RG}(Z)$ the bluer (hotter) the subsequent horizontal branch (HB) phase, while the larger $M_{\rm RG}(Z)$, the more fuel will be available during the asymptotic giant branch (AGB) phase, and thus the brighter the AGB temination. We very clearly see how the choice of the parameter $\Delta Y/\Delta Z$ is going to dramatically affect at once the UV output and the luminosity of the brightest and coolest stars in old stellar populations. The direct determination of $\Delta Y/\Delta Z$ in a SMR environment would then be of great value for our understanding of SM-R populations, such as those dominating in giant elliptical galaxies. Perhaps the observation of planetary nebulae in Baade's Window may offer a viable opportunity.

2.2. MASS LOSS vs METALLICITY

Unfortunately $\Delta Y/\Delta Z$ is not the only ill known parameter able to affect the evolution of SMR stars. Red giants lose mass via a low velocity stellar wind, and again how much mass is lost during the RGB phase determines first the effective temperature at which stars later spend their HB phase, and then the maximum luminosity that they reach on the AGB. Empirical mass loss rates for RGB stars are rather uncertain, and we don't have any empirical indication whatsoever as to whether the mass loss rate has a direct dependence on metallicity. I will now illustrate how sensitive is the post-RGB evolution to small variations in the adopted mass loss rates, especially in the case of SMR stars. Following Greggio and Renzini (1990), to describe mass loss along the RGB I adopt a slightly modified version of the standard parameterization (Fusi Pecci and Renzini 1976) of the empirical rate (Reimers 1975):

$$\dot{M} = -4 \times 10^{-13}\eta\left(1 + \frac{Z}{Z_{\rm crit}}\right)\frac{L}{gR} \qquad (M_\odot\ {\rm yr}^{-1}), \tag{2}$$

where the factor $(1 + Z/Z_{\rm crit})$ is introduced so as to mimic a direct metallicity dependence, with $|\dot{M}|$ increasing with Z by an amount which – by construction – reaches a factor of 2 at $Z = Z_{\rm crit}$. I do not pretend to have a specific physical or astrophysical justification for this choice*, but just notice that for $Z \ll Z_{\rm crit}$ the standard rate is recovered, and as we are interested in the SMR range fairly high values of $Z_{\rm crit}$ will be explored. With such assumption ($Z_{\rm crit} \gtrsim Z_\odot$) there is no influence whatsoever on the HB morphology at the metallicities spanned by galactic globulars. Even for $Z \gtrsim Z_\odot$ the implied increase of $\dot{M}$ over standard values is very modest, i.e., less than a factor of 2, well within the present observational uncertainties. Yet, such a small increase can dramatically affect the HB and post-HB evolution of low mass stars, as we are going to see. The case is conveniently illustrated in Fig. 2, where – for an age of 15 Gyr – the mass of stars on the HB ($M_{\rm HB} = M_{\rm RG}$ minus the mass lost along the RGB) is displayed for several combinations of $\Delta Y/\Delta Z$ and $Z_{\rm crit}$. The parameter η has been fixed to 0.35 by demanding to $Z = 0.001$ HB models to lie within the RR Lyrae strip, so as to mimic the even HB morphology of intermediate metallicity globular clusters.

From Fig. 2 we can fully appreciate how a modest increase of mass loss with metallicity – such as that described by Eq. (2) – can lead to diverging predictions for the post-RGB evolutionary phases of SMR stars. Suffice to mention that at high metallicity a mass difference of $\lesssim 0.03 M_\odot$ is sufficient to move a star from the red to the far blue side of the HB. As for Fig. 1, we see again that assumptions that have no effect at low to intermediate metallicities, can have dramatic consequences in the SMR regime, with variations as small as 10-20% in the adopted mass loss rate being able to turn a red HB into a very blue one. Seemingly, a difference of a few

---

* A trend of this kind could in principle be produced by e.g. a mass loss enhancement due to dust grain formation.

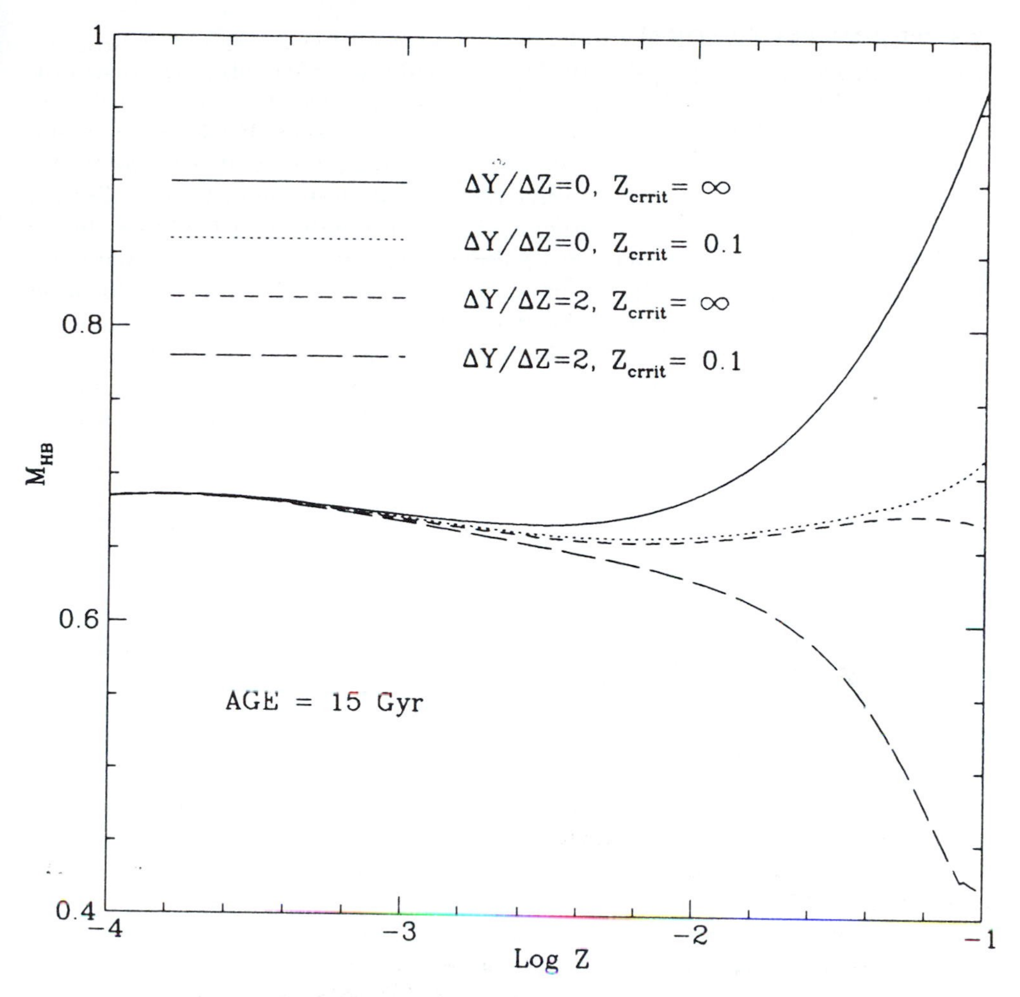

Fig. 2.– The mass of HB stars as a function of metallicity, for a population 15 Gyr old and for four different combinations of the $\Delta Y/\Delta Z$ *and* $Z_{\rm crit}$ *parameters. Clearly,* $Z_{\rm crit} = \infty$ *corresponds to a mass loss rate with no direct dependence on metallicity. The mass loss rate parameter* $\eta = 0.35$ *has been adopted.*

$0.01 M_\odot$ in the core mass at the end of the AGB would correspond to a difference up to one magnitude in the maximum AGB luminosity.

We don't know what particular combination of the $\Delta Y/\Delta Z$ and $Z_{\rm crit}$ parameters nature has chosen for SMR stars. Thus we are stuck when we try to predict the color of the HB and the maximum AGB luminosity of such stars, but the theory of stellar evolution cannot be blamed for this unconfortable situation. The two parameters in question are in fact *external* to the theory, and should be independently determined. Once the parameters are specified, the stellar evolution theory has no difficulty in accurately predicting the evolution of SMR stars during the RGB phase and beyond, with just one reservation that I discuss in the next section.

2.3. THE METAL OPACITY

At very low metallicities (say $Z \simeq 10^{-4}$) the contribution of metals to the opacity is small compared to that of electron scattering over most of stellar interior, including the important temperature range between ~few 10^5 to ~few 10^6 K, where metal opacity comes from the last ionization stages of abundant elements such as O, Ne, Mg, Si, and Fe. However, as metallicity increases so does the metal contribution to opacity while the electron scattering contribution clearly stays the same. In the SMR regime one then encounters the opposite situation, with the electron scattering opacity becoming a (small) fraction of the metal opacity. It has been recognized for years (cf. Iben & Renzini 1984) that the metal opacity in this particular temperature range represents perhaps the most uncertain ingredient in the construction of stellar models (apart from convection), an obvious consequence of the huge number of ionization stages, energy levels and electronic transitions which must be taken into account, and of the complexities introduced by the perturbation of these levels due to the high particle density. It is now well known that the so called *Livermore* opacities recently computed by Iglesias & Rogers (1991) can be a few times larger than the old *Los Alamos* opacities. Of course, the difference comes from the contribution of the metal in the critical temperature range. Fig.s 1 and 2 have been constructed using stellar models buit up with the old opacities, and no doubt the new opacities will have a strong impact on the run of quantities such as $M_{\rm RG}$ and $M_{\rm HB}$ with metallicity. For the reasons above, such an impact will be small at low metallicity, and may be very large in the SMR regime. Again, opacity adds further uncertainty especially where helium enrichment and mass loss already make very difficult to risk detailed predictions. To my knowledge Livermore opacities are not yet available for SMR compositions, and we should be aware that existing SMR isochrones may give the wrong age, as they all are based on the old opacities.

For the three reasons detailed in this section it appears that we are in serious trouble in trying to predict the HB morphology and AGB termination of SMR stars in galactic bulges. Given this situation an empirical approach appears to be wise, and I will recall some of the main results in the next section.

3. Color-Magnitude Diagrams of the Galactic Bulge

Terndrup (1988) was first in obtaining fairly deep and extensive CCD photometry of stars in Baade's Window (BW). From the main sequence turnoff – and isochrones based on the old opacities – Terndrup concluded that the mean stellar age is 11-14 Gyr, while the fraction of stars younger than 5 Gyr must be negligible, as anticipated by Rich (1985). Yet, photometric errors do not allow to delineate the turnoff region very accurately, and Terndrup result can be taken as evidence that the dominant population in the Bulge is very old, while open the question of the precise age is left open. Terndrup's CMD also shows that the bulk of HB stars are red, which would exclude those combination of the $\Delta Y/\Delta Z$ and $Z_{\rm crit}$ parameters which give a blue HB (supposing we know the age) for the metallicity of the surveyed stars (1 to 2 times solar, according to Terndrup).

One limitation of studying field stars in BW is that there exist a range of metallicities, and it is difficult to establish whether the scatter in the CMD is entirely accounted by the scatter in metallicity, or if also a distribution of ages is

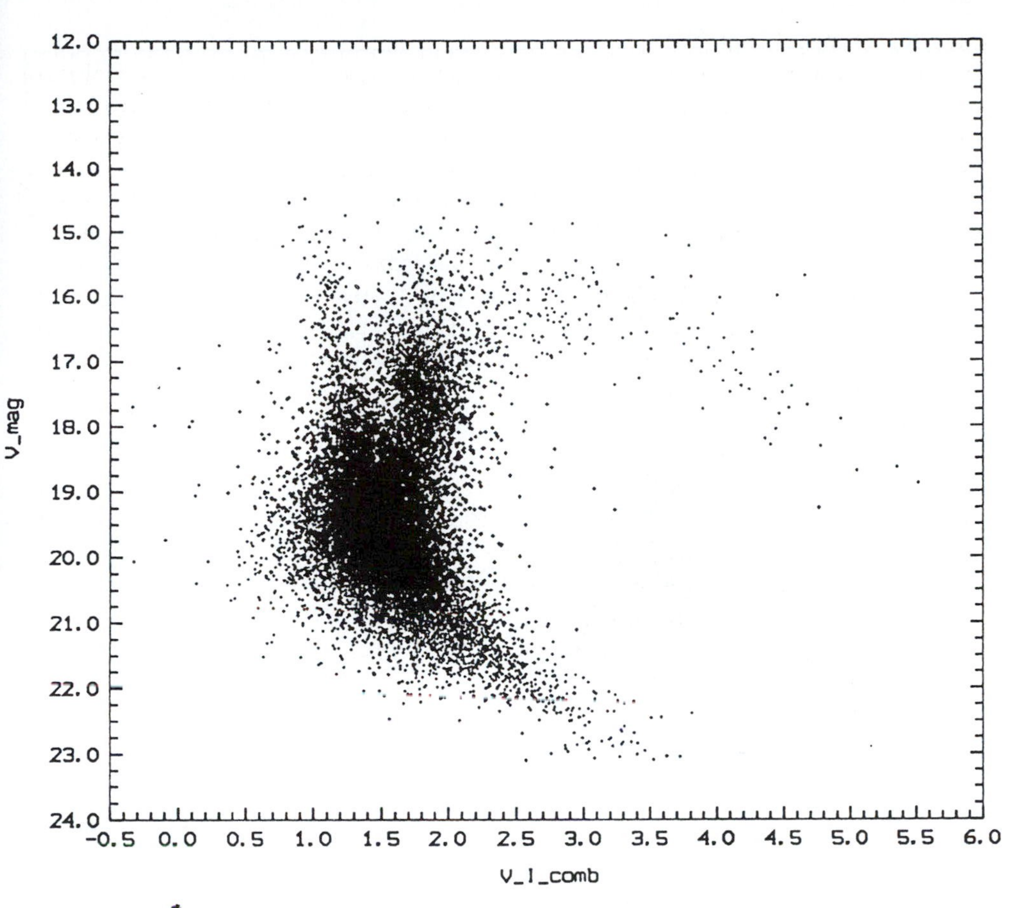

Fig. 3.– The $V-(V-I)$ CMD for stars in Baade's Window. Courtesy of Ortolani & Rich (1992).

necessary (even supposing to have taken photometric errors out). Being chemically homogeneous, the metal rich globular clusters of the Bulge offer an attractive perspective for an accurate dating of bulge stars, but unfortunately the clusters are projected against a very densely populated field, and to avoid field contamination one is forced to push the survey towards the cluster center, where crowding degrades the photometric accuracy. Ortolani *et al.* (1990, 1992a,b) have obtained CMD's of the metal rich clusters NGC6553, NGC6528, and Terzan 1. Undoubtely these are old clusters, but how old it is still difficult to say given the mentioned theoretical and observational uncertainties. All the three clusters have very red HB's, but what is most stryking in their CMD is the morphology of the red giant branches. For example, in the V vs $V-I$ plot the upper RGB and the AGB become fainter in V for increasing $V-I$, to the extent that the tip of the RGB becomes fainter (in V) than the HB itself. This behavior is due to the strong blanketing of the TiO molecules, and demonstrates that the upper RGB and AGB are both

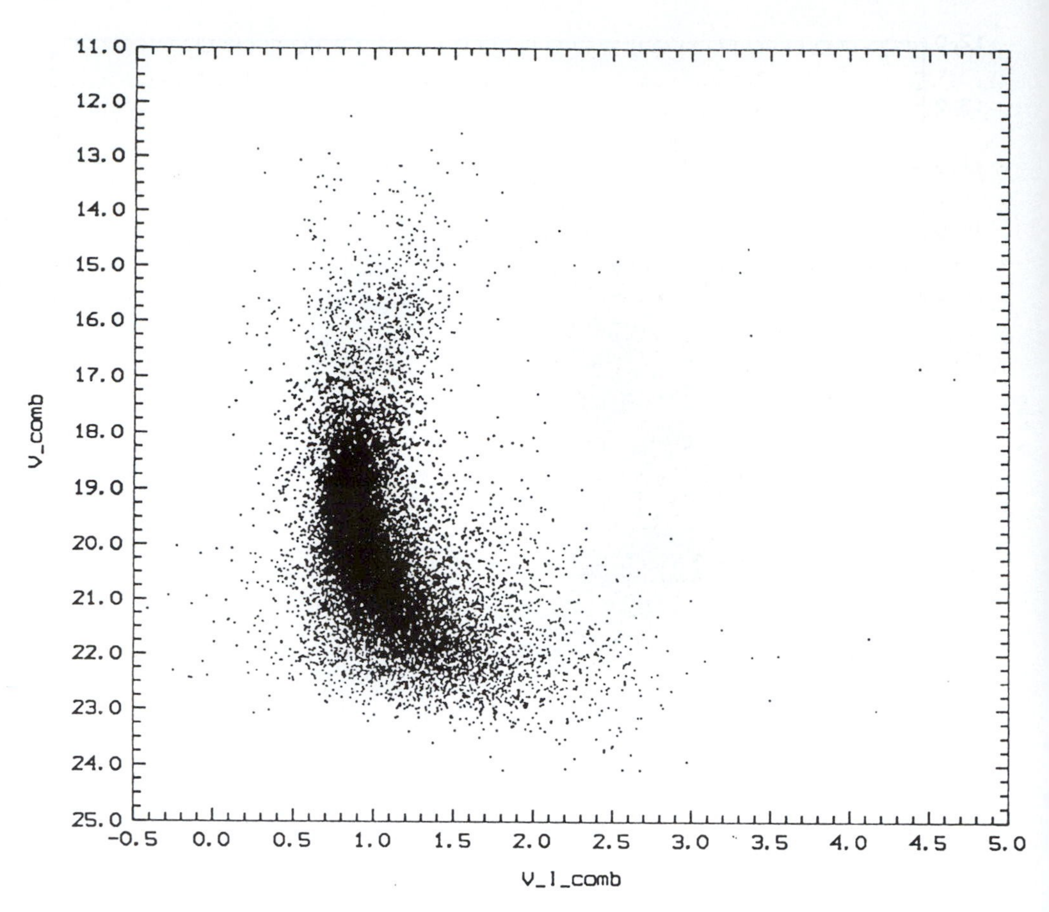

Fig. 4.– The $V-(V-I)$ CMD for stars in a Bulge field $\sim 8°$ from the Galactic center. Courtesy of Ortolani & Rich (1992).

composed of M-type stars, contrary to the case of more metal poor clusters where both branches are made up of K-type giants. This is a crucial aspect, especially for population synthesis studies, as all evolutionary population synthesis models have so far assumed the RGB stars to be K giants.

Ortolani & Rich (1992) have recently obtained high resolution, deep NTT images of several Bulge fields at various galactocentric distances. Fig. 3 shows their $V-(V-I)$ CMD for a field in BW, $-4°$ from the Galactic center. The main sequence turnoff at $V-I \simeq 1.5$ is fainter than $V \simeq 20$, unfortunately too close to the frame limit to put stringent constraints on age. Very evident is the foreground contamination by disk main sequence stars, while very prominent are the red HB clump around $V \simeq 17$, and the peculiar RGB+AGB which extend to $V-I \simeq 5.5$ and $V \simeq 19$. Clearly, contrary to the case of galactic globular clusters (see Renzini & Fusi Pecci 1988), the upper RGB must contribute very little light to the integrated V-band luminosity of the Bulge. Fig. 4 shows a somewhat deeper

CMD that Ortolani and Rich have obtained for a less crowded field $\sim 8^\circ$ from the Galactic center. Crowding, reddening and absorption are significantly lower than in the BW field ($A_V \simeq 0.7$ mag against ~ 1.8 mag). The red HB clump can be recognized at $V \simeq 16$, while the turnoff is somewhat fainter than $V \simeq 19$. The HB to turnoff luminosity difference is therefore 3.0–3.5 magnitudes, consistent with the typical value of halo globular clusters, and therefore there is no *appreciable* difference in age (cf. Renzini 1991). I should caution, however, that a difference of a few 0.1 mag (corresponding to a few Gyr age difference) could hardly be noticed on the basis of these data. I conclude that judging from the existing CMD's the Bulge is definitely old, but "how old is old" cannot yet be said with sufficient accuracy to tell whether there is an appreciable age difference with respect to halo globulars. Unaberrated HST observations will certainly help.

4. The AGB Termination and the Age of Youngest Stars in Bulges

Given the difficulties encountered in accurately dating Bulge stars using the turnoff clock, the attempt to use other clocks is certainly a commendable effort. The luminosity of the AGB termination has been widely used for dating purposes in a variety of astrophysical objects, including Magellanic Cloud globular clusters, M32, the bulge of M31, and our own Galactic Bulge. In this latter case *bona fide* bright AGB stars are represented by long period variables (LPV) and OH/IR sources, objects discussed in greater detail by Whitelock and Habing at this meeting. In this section I will address a few questions concerning their use as age indicators.

4.1. THE LUMINOSITY OF SMR AGB STAR

In a previous review (Renzini & Greggio 1990) it was pointed out that the core mass-luminosity relation for TP-AGB stars still remains to be explored for SMR compositions, to the extent that the question " *Are bright AGB luminosities in bulges a result of young ages or of large $Z + Y$?*" is still unsettled (Renzini 1992). According to Boothroyd & Sackmann (1988) the luminosity of AGB stars would scale as the cube of the mean molecular wieght, i.e.:

$$L(M_{\rm H}) = \mu^3 f(M_{\rm H}), \tag{3}$$

where $M_{\rm H}$ is the core mass. On the basis of this relation we can estimate how brighter a SMR AGB star would be compared to stars of the same core mass and the composition of e.g. the cluster 47 Tuc. For the latter I adopt $(Y, Z) = (0.24, 0.004)$ and correspondingly $\mu = 0.59$. For illustration purposes I further adopt $\Delta Y/\Delta Z = 3$, which for $Z = 5Z_\odot = 0.1$ implies $Y = 0.54$ and $\mu = 0.85$. Therefore, for given core mass, such SMR thermally pulsing AGB stars would be $(0.85/0.54)^3 = 3$ times brighter than TP-AGB stars in 47 Tuc, which corresponds to 1.2 mag. Since the AGB in 47 Tuc and similar clusters extends to $M_{\rm bol} \simeq -4.5$ (Frogel & Elias 1988), Eq. (3) would easily predict an AGB extension up to $M_{\rm bol} \simeq -5.5$ for a SMR composition.

I should caution, however, that relation (3) comes from fitting a rather restricted number of models, and that the SMR regime was not explored by Boothroyd & Sackmann. My suspicion is also that the effect of metals and helium cannot be simply cumulated through the mean molecular weight. Indeed, the effect of the CNO abundance on the strength of hydrogen burning shell is not described by μ, and therefore the actual relation must be more complicated than Eq. (3). Such a relation can be established only after a more extensive esploration of the composition parameter space.

4.2. THE LUMINOSITY-TIME-NUMBER RELATION

Occasionally bright stars are found in old stellar populations, and we would like to know their significance for the age distribution of the constituent stars, and therefore for the formation process of galaxies. Crucial for understanding the nature of such bright stars is considering their frequency with respect to the bulk population in which they appear. In a coeval population the number of stars in a generic post-MS evolutionary stage is given by:

$$N_{\rm j} = B(t)L_{\rm T}t_{\rm j}, \tag{4}$$

where $L_{\rm T}$ is the total bolometric luminosity of the population and $t_{\rm j}$ is the duration of the generic phase "j" (Renzini & Buzzoni 1986). The *specific evolutionary flux* $B(t)$ is a slow function of age, and for ages in excess of ~ 1 Gyr we have $B(t) \simeq 2 \times 10^{-11}$ stars per year per $L_{\odot}$ of the parent population. To my understanding Eq. (4) is one of the most robust predictions of stellar evolutionary theory, almost completely exempt from the unceratinties plaguing other specific aspects of the theory, such as e.g. those mentioned in §2. Thus, from number counts Eq. (4) allows to estimate the duration of a given evolutionary phase, provided all evolving stars experience such a phase. If a specific category of stars is produced by only a fraction of the population, but we know the duration of the phase, then Eq. (4) allows to estimate the fraction of the population which possess the ability to generate such stars. I will now exemplify with a few specific – albeit miscellaneous – applications how this relation can be used to study the stellar populations in bulges.

4.2.1. Estimating Lifetimes (e.g. of LPV and OH/IR Stars). Knowing the total luminosity of a population, than star counts allow to derive the duration of specific evolutionary phases. For example, the galactic globular 47 Tuc contains 4 LPV stars and its total luminosity is $8 \times 10^5 L_{\odot}$. From Eq. (4) one immediately derives that in this cluster the average duration of the LPV phase is $(2.5 \pm 1.2) \times 10^5$ yr.

Let me make another example. At this meeting Habing has reported that there are (at least) 250 OH/IR stars in the Bulge field with $|\ell| < 3^{\circ}$ and $|b| < 5^{\circ}$. What is the average lifetime of Bulge OH/IR stars? If all evolving stars in the Bulge go through the OH/IR phase, then:

$$t_{\rm OH/IR} = \frac{250 \times 10^{11}}{2\, L_{\rm T}(6^{\circ} \times 10^{\circ})},$$

where $L_{\rm T}(6^{\circ} \times 10^{\circ})$ is the total luminosity sampled by the $6^{\circ} \times 10^{\circ}$ field of coverage. Alternatively, suppose that from other means we know the average duration of the OH/IR stage, then the ratio of actual number of OH/IR stars to that predicted by Eq. (4) gives the fraction of the Bulge population which actually produces OH/IR stars. Crucial to both estimates is the preliminary determination of the intrinsic sampled luminosity L_T, a quantity that it should not be difficult to obtain, e.g. from a model of the light distribution in the Bulge*.

4.2.2. Normalizing Luminosity Functions. Luminosity functions (LF) of globular clusters, bulges, and other resolved galaxies are often compared to each other in order to derive astrophysical inferences from their similarities and differences. A comparison of this kind needs the various LF's to be normalized in some way. A way which has been frequently adopted consists in matching the faint end of two or more LF's, so as to emphasize differences in the bright portion. This procedure should be avoided, because it can lead to erroneous conclusions. The faint end of an empirical LF is in fact seriously affected by incompleteness, and in a way which differs from one studied stellar system to another (e.g. the faint end of the LF of M giants in the Galactic Bulge is certainly affected by a different degree of incompleteness compared to IR bright giants in the bulge of M31). There is instead only one correct way of normalizing the LF's of different stellar populations, and this is to refer star numbers to sampled luminosities: the number to luminosity ratio (N_j/L_T) is in fact proportional to durations, i.e. it is equal to $B(t)t_j$, an intrinsic property of the stars in the population.

As an example, I will consider the case of the bright end in the LF of M32, which extends to $M_{bol} \simeq -5.5$ (Freedman 1992). This is one mag brighter than in galactic globular clusters of comparable metallicity, and Freedman discusses various possibilities for extending by this amount the AGB LF. Eventually she inclines in favor of an intermediate age component, thus adding new fuel to a still burning debate (cf. Greggio & Renzini 1990, and references therein). Freedman's LF of M32 refers to a $40'' \times 100''$ field of view, with an alleged typical surface brightness of $V = 21$ mag/$\square''$, or $B = 21.9$, given the color of M32 ($B - V = 0.9$). With a true modulus 24.4 to M32, the absolute B magnitude of the sampled area is $M_B = 21.9 - 24.4 - 2.5\mathrm{Log}(4000) = -11.5$, or $L_B = 5.3 \times 10^6 L_\odot$. This corresponds to a total bolometric luminosity ~ 3 times larger, or $L_T \simeq 1.6 \times 10^7 L_\odot$. Freedman lists 10 stars with $-5.5 < M_{bol} < -4.5$, and their lifetime from Eq. (4) is $\simeq 3 \times 10^4$ yr if all the stars in M32 were to climb the AGB up to $M_{bol} = -5.5$. This is an exceedingly short time for a one magnitude interval on the AGB, which instead is covered in $\sim 1.5 \times 10^6$ yr by an individual AGB star (e.g. Renzini & Voli 1981). This means that in M32 only one evolving star every ~ 50 actually succeds in climing above $M_{bol} = -4.5$, and therefore the bright end of the LF is produced by just a trace component in the M32 population. This trace component can either be an intemediate age contaminant to the dominant, old population, or the progeny of blue stragglers – coeval to the old population – resulting from the merging of binary components as suggested by Renzini & Greggio (1990), or some combination thereof (also a contamination from the disk of M31 cannot be excluded in the outer parts, Davidge & Jones 1992). Actually, Renzini & Greggio estimated the number of bright AGB stars progeny of blue stragglers to be $\sim 6 \times 10^{-13} L_T t_{AGB}$, that for the estimated sampled luminosity ($1.6 \times 10^7 L_\odot$) and $t_{AGB} = 1.5 \times 10^6$ gives a total

* For the bolometric luminosity of an old population one can use $L_T \simeq 3\,L_B$, where L_B is the blue luminosity (see Fig. 1.4 in Renzini 1993). This should be accurate within 10-20%.

of 14 AGB stars in the $-5.5 < M_{\rm bol} < -4.5$ range. This beutifully compares to 10 stars in this range observed by Freedman (1992)*, and I would be tempted to conclude that there is no need invoking an intermediate age component to explain the LF of M32. Crucial for this conclusion is the estimated luminosity $L_{\rm T}$ sampled by the covered area in Freedman's study. If the actual average surface brightness of the covered area is much fainter than $V = 21$ mag/□″, then there would be room for an intermediate age component. Accurate surface photometry for M32 has been recently obtained by Peletier (1992), who gives $R = 23.2$ (corresponding to $V = 23.7$) mag/□″ at $r = 109''.4$ from the center. Since Freedman's field of view was centered 120″ from the center, it is indeed quite possible that the actual sampled luminosity is significantly lower than $1.6 \times 10^7 L_\odot$ as estimated above. On the other hand, it is worth noting that – scaling from 47 Tuc – with $t_{\rm PLV} = 2.5 \times 10^5$ yr and $L_{\rm T} = 1.6 \times 10^7 L_\odot$ Eq. (4) gives $N_{\rm LPV} \simeq 80$ LPV stars. Freedman's counts give about 100 stars in the magnitude range $-4.5 < M_{\rm bol} < -3.6$, and again there appears to be consistency†.

An intermediate age component is also favored by Elston & Silva (1992). Theirs is a $4' \times 4'$ field of view centered $3'$ from the center of M32. They count 135 stars in the luminosity range $-5.5 < M_{\rm bol} < -4.5$, and argue that 25% of main sequence stars should be blue stragglers in order to explain them in terms of blue straggler progeny. I do not understand which kind of calculation is behind this figure, but I would agree with their conclusion if the total luminosity sampled by their field of view is significantly lower than $135/(6 \times 10^{-13} \times 1.5 \times 10^6) = 1.5 \times 10^8 L_\odot$. Life would be much simpler if observers would preliminarily estimate the *sampled luminosity* $L_{\rm T}$ ($\simeq 3\,L_{\rm B}$) of the covered area of their targets. Additional motivations for this wishful expectation are presented next.

4.2.3. Estimating the Stellar Population of a Pixel. CCD and IR-array stellar photometry in nearby galaxies such as M32 and the M31 bulge are becoming common practice. Like in the photometry of galactic globular clusters, crowding is certainly a problem when measuring stellar magnitudes in such distant objects. However, the observational conditions are much different compared to those prevailing for galactic globulars, the targets for which existing photometric packages have been first conceived and then optimized. Table 1 illustrates the case.

Column 1 gives the distance from the center of the galaxy and column 2 gives the corresponding blue surface brightness. The third column gives the absolute blue magnitude that is sampled by one □″ having assumed a distance modulus 24.4 mag, while columns 4 and 5 give the corresponding blue and bolometric luminosity (i.e. again sampled by one □″). For this latter conversion I have adopted $L_{\rm T} = 3\,L_{\rm B}$.

* This number comes from Freedman's Table 2, which may not list all stars brighter than $M_{\rm bol} = -4.5$. Her Table 1 lists 25 stars which in K are brighter than the faintest star in Table 2, and therefore the actual number of stars in the quoted magnitude range must be between 10 and 25: still consistent with my estimate for blue straggler progeny AGB stars.

† Freedman has successively applied incompleteness corrections to her actual counts, but such corrections are not explicitly given. Were they such to significantly increase over ~ 100 the number of stars, then this conclusion would be invalid.

TABLE 1
STELLAR POPULATION SAMPLING IN M32 AND THE BULGE OF M31

M32						
r	$SB_{\rm B}$ mag/□″	$M_{\rm B}$ mag/□″	$L_{\rm B}$ $L_\odot$/□″	$L_{\rm T}$ $L_\odot$/□″	$N_{\rm LPV}$ stars/□″	$N_{\rm RGT}$ stars/□″
5″	17	−7.4	1.2×10^5	3.6×10^5	2	7
27″	18	−6.4	4.8×10^4	1.4×10^5	0.7	3
2′	21.9	−2.5	1.3×10^3	4.0×10^3	0.02	0.08
M31 Bulge						
2′	19.8	−4.6	9×10^3	2.7×10^4	0.14	0.5
4′	21	−3.4	3×10^3	9×10^3	0.05	0.18

LPV's and of RGB stars in the last quarter of magnitude below the RGB tip. To get these numbers I have used Eq. (4), with $t_{\rm LPV} = 2.5 \times 10^5$ yr (see §4.2.1), and $t_{\rm RGT} = 10^6$ yr, the time RGB models spend in the corresponding luminosity interval (Sweigart & Gross 1978; see also Fig. 1 in Renzini 1992), thus implicitly assuming that the stellar population of M32 is similar to that of 47 Tuc. I recall that LPV's such as those in 47 Tuc reach a peak $M_{\rm bol} \simeq -4.5$, or $\sim 5000L_\odot$, while the RGB tip is at $M_{\rm bol} \simeq -3.8$, or $\sim 2600L_\odot$. For the M32 surface brightness at $r = 2'$ Table 1 gives the value taken by Freedman (1992) to be representative of a $100'' \times 40''$ field, but see the discussion in §4.2.2.

Numbers in Table 1 are self-explanatory. In the sequel I will call for short "pixel" the area of one resolution element, and therefore sampled luminosities and numbers of stars per pixel are obtained by multiplying values in Table 1 by the actual area of the pixel in □″ units. Of course, the area of the so-defined pixel depends on the specific observation, and should not be confused with the physical pixel of the detector (when observations are seeing-limited the pixel area is roughly the seeing squared). Table 1 shows that 5″ from the center of M32 each □″ pixel samples a luminosity of $\sim 3.6 \times 10^5 L_\odot$, like that of a fairly populous globular cluster. Eq. (4) correspondingly predicts that in each pixel one finds on average 2 LPV's and 7 RGB tip stars, a very crowded pixel indeed. Running a photometric package on such a frame may produce a list of magnitudes, but I doubt that they will actually refer to individual stars. Rather, the package may call stars what actually are the 2, 3, or 4σ fluctuations in the number of bright stars per pixel, thus producing a bright extension of the LF which in fact is just an artifact of the observational conditions. The situation is only marginally better at $r = 27''$, with $\sim$ 3 RGB tip stars per pixel and a 70% chance to find an LPV in a pixel. At $r = 2'$ the situation is now far better, with only a 2% chance to find an LPV in a given 1□″ pixel. Contrary to the previous two cases, the corresponding stellar photometry should be reasonably accurate, and along with it the resulting LF.

Field locations in Table 1 are not randomly chosen. In fact, Davidge & Nieto

(1992) have obtained near-IR CCD images with their frame extending from $r = 5''$ to $r = 27''$. They find an extended LF that they attribute to an intermediate age component. The FWHM size of their stellar images was only $0''.4$, and in today's terminology with such an outstanding resolution the area of their "pixel" is $\sim 0.16 \square''$. Thus, luminosities and numbers in Table 1 should be devided by ~ 6. Even so, in the less crowded part of their $65'' \times 110''$ field each $\sim 0.16 \square''$ "pixel" has a $\sim 10\%$ chance to contain one LPV, and thus $\sim 1\%$ of the "pixels" should contain 2 LPV's. Since their field encompasses $\sim 44,700$ pixels, this means that ~ 447 of them contain 2 LPV's, ~ 45 of them 3 LPV's, etc., which I suspect may account for a fair fraction of the bright portion of the resulting LF without necessarily appealing to an intermediate age component.

I have already discussed the $2'$ field studies by Freedman (1992). She also had very good seeing, with FWHM resolution of $0''.6$, roughly corresponding to pixels of $\sim 0.3 \square''$. Numbers in Table 1 should then be divided by ~ 3, and no doubt observing conditions were much better in this field. Crowding could have been a little worse for the study of Elston & Silva (1992), whose field of view extends from $1'$ to $5'$ from the center. They had worse resolution, with $\sim 2 \square''$ "pixels", but their bright star photometry should still be reasonably accurate, a conclusion which should also apply to the observations of Davidge & Jones (1992), pointing $100''$ from the center and with $\sim 1 \square''$ pixels.

Table 1 also gives data for two locations in the bulge of M31. The outer one, at $4'$ from the center, was covered with near-IR imaging by Rich & Mould (1991), who found a significant extension of the AGB luminosity function up to $M_{\rm bol} \simeq -5.5$. They conclude that an intermediate age component is most likely necessary to explain the LF, although Davies *et al.* (1992) argue for a contamination from the disk of M31 being responsible for most if not all the effect, a possibility not completely excluded by Rich & Mould. With $1 \square''$ pixels – such as in the R & M study – the $4'$ field looks very crowded, and I wonder if overlapping LPV's and RGB tip stars may have concurred in arificially extending the bright portion of the LF. For example, from Table 1 data we see that $\sim 0.05 \times 0.18 = 1\%$ of the pixels should contain a LPV+RGB tip star blend, and seemingly $\sim 3\%$ of them should contain 2 RGB tip stars, etc. Since there is a total of ~ 3000 pixels in the field of view, we see that a non trivial fraction of very bright stars may actually result from blends. The situation looks three times worse in a field $2'$ from the center, and I suspect that only with really outstanding seeing one can cope with stellar photometry in such very crowded field.

I am not in the position to definitely state that the bright part of the M31 bulge LF obtained by Rich and Mould is entirely due to blended images, but I would be very reassured if the LF were to remain the same even using data obtained from observations with far better resolution, e.g. such as those of Davidge & Nieto (1992) of M32. I would also be very interested in the results of simulated observations, for example cloning from the $4'$ field frame a mock $2'$ field doubling or so the surface brightness of the $4'$ field, something not too difficult to do starting from existing data.

Given the great variety of observational conditions that are encountered in stellar population studies (surface brightness of the target galaxy, distance, resolution of the telescope + camera + atmosphere), it would be certainly very useful for the producers of CMD's and LF's – as well as for their consumers – if we could dispose of a simple *rule of thumb* saying which photometry we should safely believe, and which we should look at with some more concern of how crowding may have affected the results. Without demonstration, I propose the following algorithm, with the proviso that it should be tested and calibrated by means of adequate simulations. Safe photometry of stars with luminosity L_* requires that

$$L_* \gg \Sigma_{L<L_*} \cdot \text{pixel area}, \tag{5}$$

where $\Sigma_{L<L_*}$ is the average surface brightness of the target in $L_\odot/\square''$ having subtracted the stars brighter than L_*, and the pixel area in $\square''$ has been defined above. Inequality (5) should apply to distant galaxies in which we aim at resolving the *brightest* stars in the population, as well as to nearby globular clusters where on the contrary the point is to get accurate magnitudes for the *faintest* stars in the population.

If criterion (5) is correct, then from Table 1 we see that there should be no problem with the mentioned M32 observations at $z \simeq 2'$, but I suspect that those at $5''$ and $27''$ may be in trouble, while those in the $4'$ field in M31 appear to be on the borderline. Certainly, at $5''$ from the center of M32 to reach the photometric quality reached by Freedman in her $120''$ field the pixel area should be ~ 100 times smaller, or the resolution $\sim 0''.06$, something that even an unaberrated HST can barely reach.

I wish these considerations have shown that one of the first questions to ask about a stellar field is *"what is the total luminosity sampled by the whole frame and by each resolution element?"*

5. Did Bulges Form First?

From a CMD of the Galactic Bulge such as that shown in Fig. 4 one can reasonably conclude that the bulk of stars are very old, quite possibly as old as galactic globulars, or thereabout. Unfortunately, existing CMD's can hardly set a tighter limit. From the LF of the brightest stars in the Galactic Bulge (Frogel & Withford 1987) and in the bulge of M31 one can also reach a similar conclusion, with still the reservation that a trace population of intermediate mass (age) stars cannot be entirely excluded. What can we conclude from this about the formation of bulges, i.e. about the star formation history in bulges?

I do not understand scenarios in which stars first start froming in a halo, and then successively the leftover, chemically enriched gas flows quietly in and goes to form the bulge. Seemingly, I do not understand scenarios in which disks form first, and then some dynamical instability depletes and heats up the inner disk to form the bulge. I do not understand why star formation should have started first in the low density halo, rather than in the central, high density regions of a protogalaxy. Seemingly, I do not understand why stars should have formed first in a disk, and then brought in some way to the center about which the disk itself rotates. It makes more sense to me if star formation was much more violent at the bottom of the galactic potential well, where high gas and cold cloud densities were certainly first achieved. What else was the *center* otherwise? Here chemical enrichment by supernovae, and heating of the residual gas – eventually discontinuing further stars formation – were probably much more rapid than in the outer parts of a galaxy like our own, and very high metallicities were soon achieved. The duration of the star formation epoch in a bulge was probably of the order of the local free fall time ($\sim$few 10^8 yr), since cold, star forming gas can hardly survive longer in a pressure supported system in which the *dynamical* temperature is of the order of several million degrees. Thus, stars in the metal poor halo came later, as the local free fall time is significantly longer, up to a few billion years in the outer parts. Thus, in this scenario the *spheroid* component of a spiral was buit up starting from the center, in such a way that the super metal rich bulge is – on average – older than the metal poor halo (Renzini & Greggio 1990), just the contrary of what one may naively expect. Rather than a latecomer, in this view a bulge is seen instead as the first seed about which the rest of a galaxy has then grown.

There is still no conclusive evidence for a significant fraction of stars in the Galactic Bulge and in the bulge of M31 to belong to an intermediate age population. A *contamination* of this kind cannot be excluded either, and one can imagine a number of ways in which it might have been produced. But in any event one should attempt to distinguish between the main episode of stars formation and any possible subsequent addition. Considerations such as those developed in the previous sections may help quantifying these arguments, and set more precise limits on the age distribution of stars in bulges.

Before concluding this review, I would like to mention a few problems by which I am still intrigued, but which may soon be satisfactorily solved by means of dedicated observations:

• Is the difference in the upper LF between the Galactic Bulge (Frogel & Withford 1987), the bulge of M31 (Rich & Mould 1991), and M32 (Freedman 1992) real?

• The Bulge CMD's such as that displayed in Fig. 3 show a well populated, fairly bright MS which is due to foreground contamination by disk stars. How many bright red giants, Miras, and OH/IR stars in the same fields do not belong to the Bulge, but are the RGB/AGB progeny of the foreground population?

• If the extension of the LF in M32 is due to a blue straggler progeny, why such an extension is not seen in the Galactic Bulge LF?

• Why the Bulge LF of Frogel & Whitford does not show the expected drop at $M_{\rm bol} \simeq -3.8$ associated to the RGB tip (see Renzini 1992)?

• Are all Miras really AGB stars? After all, pulsation is an evelope phenomenon, and the deep structure does not matter. When metallicity increases the RGB moves to lower temperatures and larger radii, thus favoring pulsational instability. In the SMR regime could this suffice to cause the brightest RGB stars to become Miras?

• Are the two previous problems related to each other? i.e. is the drop in the Bulge LF washed out by the SMR stars near the RGB tip being variable?

I wish to thank Sergio Ortolani and Mike Rich for their permission to reproduce their CMD's of the Bulge, and for our frank, entertaining debates on this subject.

REFERENCES

Boothroyd, A.I., Sackmann, I.-J. 1988, ApJ, 328, 641

Davidge, T.J., Jones, J.H. 1992, AJ, 104, 1365

Davidge, T.J., Nieto, J.-L. 1992, ApJ, 391, L13

Davies, R.L., Frogel, J.A., Terndrup, D.M. 1992, AJ, 102, 1729

Elston, R., Silva, D.R. 1992, AJ, 104, 1360

Freedman, W.L. 1992, AJ, 104, 1349

Frogel, J.A. 1990, in Bulges of Galaxies, ed. B.J. Jarvis & D.M. Terndrup (Garching: ESO), p. 177

Frogel, J.A., Elias, J.H. 1988, ApJ, 324, 823

Frogel, J.A., Whitford, A.E. 1987, ApJ, 320, 199

Fusi Pecci, F., Renzini, A. 1976, A&A, 46, 447

Greggio, L., Renzini, A. 1990, ApJ, 364, 35

Iben, I.Jr., Renzini, A. 1984, Phys. Rep. 105, 329

Iglesias, C.A., Rogers, F.J. 1991, ApJ, 371, 408

Ortolani, S., Barbuy, B., Bica, E. 1990, A&A, 236, 362

Ortolani, S., Bica, E., Barbuy, B. 1992a, A&AS, 92, 441

Ortolani, S., Bica, E., Barbuy, B. 1992b, A&A, in press

Ortolani, S., Rich, R.M. 1992, in preparation

Pagel, B.E.J. 1989, in Evolutionary Phenomena in Galaxies, ed. J.E. Beckman & B.E.J. Pagel (Cambridge Univ.), p. 368

Peimbert, M. 1983, in Primordial Helium, ed. Shaver, P.A., Kunth, D., Kjär, K. (Garching: ESO), p. 267

Peletier, R.F. 1992, ESO Preprint No. 851

Reimers, D. 1975, Mem.Soc.Roy. Liège, 6th Ser., 8, 369

Renzini, A., 1991, in Observational Tests of Cosmological Inflation, ed. T. Shanks *et al.* (Dordrecht: Kluwer), p. 131

Renzini, A., 1992, in The Stellar Populations of Galaxies, ed. B Barbuy & A. Renzini (Dordrecht: Kluwer), p. 325

Renzini, A., 1993, in Galaxy Formation, ed. J. Silk & N. Vittorio (Amsterdam: North Holland), in press

Renzini, A., Buzzoni, A. 1986, in Spectral Evolution of Galaxies, ed. C. Chiosi & A. Renzini (Dordrecht: Reidel), p. 135

Renzini, A., Greggio, L. 1990, in Bulges of Galaxies, ed. B.J. Jarvis & D.M. Terndrup (Garching: ESO), p. 47

Renzini, A., Voli, M. 1981, A&A, 94, 175

Rich, R.M. 1985, Mem. S.A.It. 56, 23

Rich, R.M. 1988, AJ, 95, 828

Rich, M.R., Mould, J.R. 1991, AJ, 101, 1286

Sweigart, A.V., Gross, P.G. 1978, ApJS, 36, 405

Terndrup, D.M. 1988, AJ, 96, 884

Whitford, A.E., Rich, R.M. 1983, ApJ, 274, 723

DISCUSSION

Baum: Let's assume that star formation got underway very very early in the centre and I think your arguments for that are good. That it should proceed very rapidly where the material is dense, is also very logical. But, by what reasoning can we conclude that it was over and done with very rapidly?

Renzini: My reasoning is that the Bulge is a hot type system that is not rotationally supported and has velocity dispersion of order 100 km/s. What I find difficult is to understand how you can keep molecular clouds at the density of the stars we see today (spatially averaged) with such high random velocities without forming stars on a short time scale. The remaining gas will go up to several million degrees and flow out due to supernovae explosions.

Baum: Would you consider that that conclusion might be kept open until we finish the observations?

Renzini: Of course! I was making a bet, indeed.

Rich: A comment on the G-dwarf problem: in fact the number of 47-Tuc like stars, from my abundance distribution, should be around 10 to 15%. So you have to fold in lifetimes on the giant branch and so on. The jury is still out, but I would say from the surveys, that when you normalize them and remember that the mean abundance is twice solar (so the normalization pushes it to higher abundances), I still think there is a bit of G-dwarf problem at this time. There are no luminous carbon stars seen anywhere in the Bulge, so if it's the progeny of the bright main sequence that does superpose that field which is there, it isn't making luminous carbon such as we see in reasonable volumes in the solar neighborhood.

Rich On the issue of Todry M31: in fields taken two years apart, we see individual stars vary by more than more than a magnitude in K, in a field that is more than 400 parsec from the nucleus.

Norman: How sure are you that the Bulge is extremely old?

Renzini: I would just reverse the question, show me the young stars.

THE STELLAR POPULATION OF THE INNER 200 PARSECS

R. MICHAEL RICH*
Department of Astronomy, Columbia University
538 West 120th Street, New York, NY 10027, USA

ABSTRACT. The central 200 pc of the Galaxy is obscured by 5-30 visual magnitudes of extinction and therefore may be studied only in the infrared. The central parsec consists of a mostly red star cluster (IRS 16) and a cluster of $\approx$ 10 HeI Wolf-Rayet or Of-like stars; this cluster is the best evidence for star formation in this region. The central 5 pc has an extended giant branch (to M_{bol}=−6) and long period variables. These stars are not extraordinarily bright; in the central 200 pc luminous asymptotic giant branch (AGB) stars ($M_{bol} < -5$) concentrate toward the nucleus in a distribution more flattened than the general light. Candidates for relatively young objects include a rapidly rotating population of high outflow velocity OH/IR stars. The AGB luminosity function in the central 200 pc resembles that recently found in M32 and M31 bulge fields. Depending on how one interprets the extended giant branch, the population may be intermediate age with some ongoing star formation, or mostly old and metal rich, with star formation confined to the nucleus.

1. Introduction

During the July, 1945 dark run Stebbins and Whitford (1947) began to search for the Galactic nucleus by drift scanning across the Sagittarius region using a CsO photocell and ballistic galvanometer attached to the 60-inch telescope on Mt. Wilson. They were first in observing the central 200 parsecs, finding a central bulge-like stellar population but not uncovering the nucleus; they noted (with foresight) that a search for the nucleus would likely be more successful at 2μm. Success arrived some 22 years later, when Becklin & Neugebauer (1968) discovered the central star cluster at 2.2μm, comparing its surface brightness profile to that of the M31 nucleus.

Since its discovery, the Galactic center has been revealed as the site of tremendous activity, possibly an active galactic nucleus in its own right. This review does not address those properties; rather, recent conference

* Alfred P. Sloan Foundation Fellow.

H. Dejonghe and H. J. Habing (eds.), Galactic Bulges, 169–188.

volumes (see below) address this wide range of complex activity. The aim of this article is to review the properties of and define what we need to learn about the relatively neglected stellar population of the inner 200 pc, or 1.4° (for $R_0 = 8$ kpc). This region of the bulge lies deep within the zone of obscuration, suffering 10-30 mags of visual extinction. Figure 1a shows a composite infrared image of the central region (Gatley *et al.* 1989). The penetrating power of infrared is great, but is still blocked by the heaviest dust clouds. The high extinction of the region is shown in Figure 1b, from Catchpole, Whitelock, & Glass (1990); at 140 pc/deg ($R_0 = 8\ kpc$) one sees that entire inner 200 pc cannot be studied in the optical.

Figure 2 shows Gatley's new emission line images of the center, revealing some of the remarkable structures arising from the activity there; while not the subject of this review, I include this image to remind the reader of the special conditions at the center.

The following section discusses the population of the central parsec, where activity and possible star formation generate conditions of such intensity that the properties of the stellar atmospheres are affected. §3 examines the stellar content and distribution of the inner 200 pc. We explore similar populations in the local group in §4: M32 and the bulge of M31. Finally, we try to apply what we have learned to understanding the formation of the bulge.

This review concerns only the stellar population, and makes no efforts to address the source of the central activity or the existence of a black hole there. These issues are addressed in IAU Symp. 136 (Morris, (ed.) 1989), AIP Conf. 155 (Backer (ed.) 1987) and by Genzel & Townes (1987). Other aspects of the population are addressed in the recent volume edited by Blitz (1992). This article only addresses dynamics as they concern the stellar populations, not the central mass distribution. Recent efforts on a dynamical models include those of Lindqvist *et al.* (1992) and Kent (1992).

2. The Central Parsec

Within this region, a source known as IRS 16 has been resolved (de Poy *et al.* 1991; Eckart *et al.* 1992) and is the best candidate for the stellar nucleus; a cluster of late-type stars. It is $1''$ distant from SgrA*, the nonthermal source that is the best black hole candidate, not an unusual circumstance given that the nucleus of M31 is not on centered on the bulge light. Sellgren's (1989) baseline review on the central star cluster discusses the basic properties, including her discovery (Sellgren *et al.*

Fig. 1 (a). *Left.* Composite J,H,K (coded as blue, green, and red) infrared image of the Galactic center, 30 ′ = 70 pc on a side (Gatley *et al.* 1989). North at top, East to the left. Notice the large and patchy obscuration, even at infrared wavelengths.

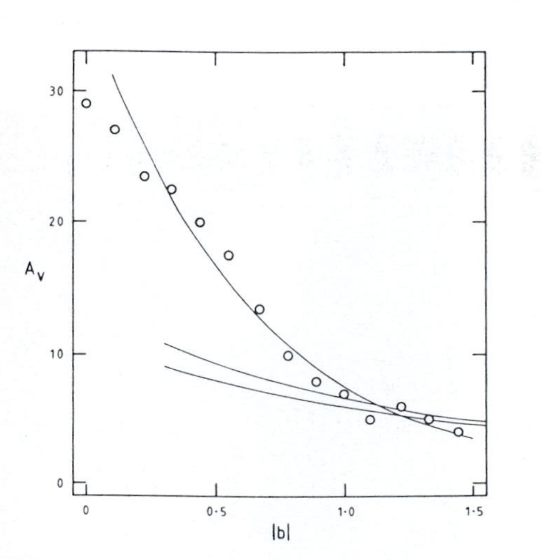

Fig. 1 (b). *Top.* Visual extinction in the central 200 pc from Catchpole, Whitelock, and Glass (1990); solid line is an exponential fit with scalelength 0.64°; shallow curves illustrate an extinction law of van Herk (1965).

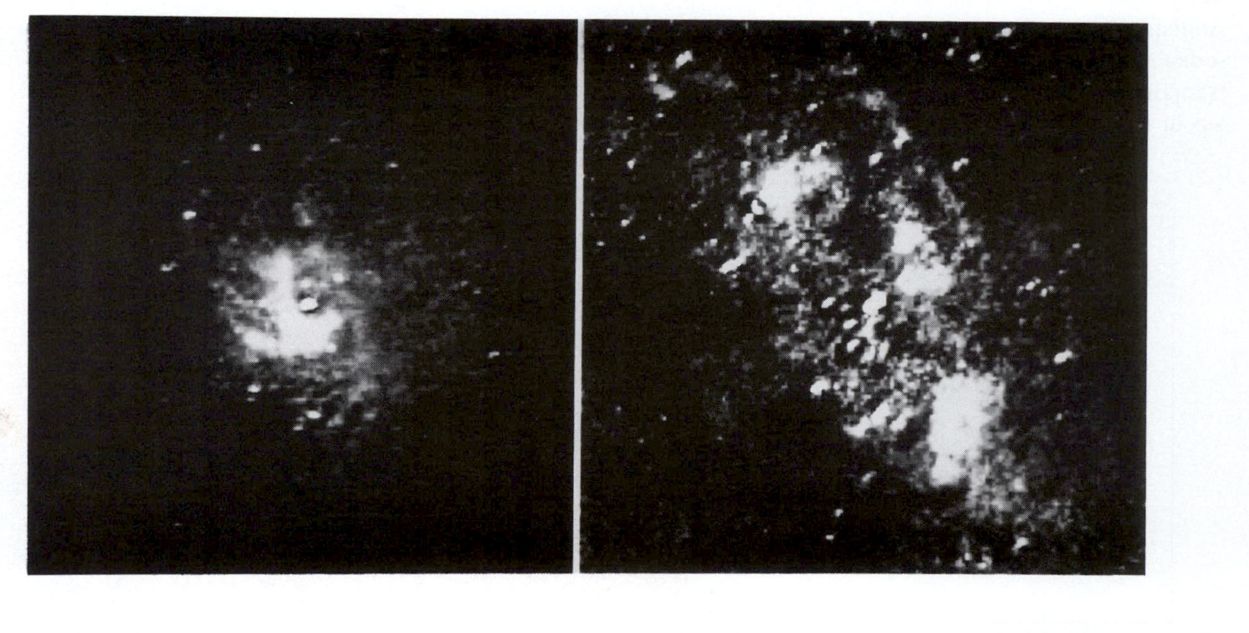

Fig. 2. The Kitt Peak Cryogenic Optical Bench (Merrill & Gatley, 1992) gives new views of gas in the galactic center. These narrow band images are obtained by subtracting off from on band images obtained with a HgCdTe detector. *Left:* Ionized gas, Brackett gamma image of the center; Sgr A* is the bright point source; notice the similarity to radio images. *Right:* molecular hydrogen (v=1-0 s(1)) clearly reveals the molecular ring (responsible for some of the large extinction SE of the nucleus in Figure 1) North at top, East to left; 2 ′ on a side.

1990) that stars within the central pc have weak 2.3μm CO bands. It is interesting to note that for $R_0 = 8$ kpc the central parsec subtends 24″; it may be studied in great detail from the ground, except for the ≈ 30 V mag of extinction toward this line of sight.

It is in the central pc that we find the best candidates for very young luminous stars; the only direct evidence in this region for star formation. Allen, Hyland, & Hillier (1990) and Krabbe *et al.* (1991) find a cluster of He I emission-line stars interpreted to be either Of or Wolf-Rayet. It is significant that star formation and the central star cluster may now account for the total output of far-infrared emission from the center ($\approx 10^7 L_\odot$), a credible alternative to the proposal (cf. Gatley, 1987) that a central engine must account for this emission. It is interesting that the He I star cluster is much more extended than the central portion of IRS 16. While line imaging and spectroscopy of the He I stars looks convincing, their certification as genuine young stars remains controversial; Morris (1992) suggests that the apparent stars may actually be compact objects with their luminosity from an accretion process.

Some combination of this star cluster and SgrA* creates a wind strong enough to blow the atmosphere of IRS 7 (the only certain supergiant) into a spectacular comet-like tail (Yusef-Zadeh & Morris, 1991). This wind may ablate the atmospheres of the giants in the central pc, or the blue star cluster might dissociate the CO, either of which could cause the decreased CO linestrength seen in the giants. The drop in CO might also be due to the high stellar density of $\approx 10^6 M_\odot$ pc^{-3}; encounters might strip the atmospheres of half the giants in the central 0.4 pc (Phinney, 1989). On top of the reddening, these effects make it risky to use observed colors and luminosities to infer the age of the red stars in the central pc.

The key question about the population of the central parsec applies to the global population as well: we cannot easily distinguish between a population of intermediate age stars of solar abundance, and possible peculiar evolution of super metal rich stars. Stars of high luminosity are present, but it is hard to tell if they are bright because they are young, or metal rich.

In summary, the central region consists of an unresolved ($< 1''$) red star cluster (IRS16) which is likely the stellar nucleus and is offset from SrgA*, the nonthermal radio source. The central parsec also contains an extended cluster of luminous HeI stars which can by themselves account for nearly all of the emergent luminosity of the central region. There is evidence for HeI emission (hence possibly star formation) centered on the

IRS16 cluster. The late-type stars alone account for $\approx 2 \times 10^6 L_\odot$ in the central pc, or 10% of the total luminosity.

2.1. THE LATE-TYPE STARS

The high reddening makes it difficult to measure accurate colors and luminosities to compare the central stars with known populations like the outer bulge. Haller (1992) attacks this problem by measuring both JHK photometry and variability of stars in the central $5' \times 5' = 6$ pc; the reddest stars are LPV (long period variable) candidates. Figure 3 shows a mosaic image constructed by Haller and Rieke in which some 1000 stars are found, including 59 variable star candidates. Assuming that $L_{TOT} \approx 10^7 L_\odot$ (Becklin & Werner, 1982), we may use the Fuel Consumption Theorem (Renzini & Buzzoni, 1986) to calculate the expected number of Miras, assuming a lifetime of 2×10^5yr. One expects to find ≈ 40 Miras in the central few pc. Adopting a lifetime of 1.3×10^6yr per magnitude of evolution for AGB stars, 10^3 AGB stars should be found in the central region; both predictions agree with Haller's numbers. The number of luminous AGB stars is consistent with the total luminosity: they are *not* merely a trace population.

There are two baseline studies of critical importance in analyzing this population: Frogel & Whitford (1987); FW87 (the Baade's Window M giants) and Whitelock, Feast & Catchpole (1991); WFC91 on LPV's selected from the *IRAS* database. Late M giants in the bulge first isolated by Blanco (1965) comprise the FW87 optically selected sample, while WFC91 selected a set of cool stars from the *IRAS Point Source Catalog* with $0.7 < f_{25}/f_{12} < 2.0$. The work of van der Veen & Habing (1990) completes the picture, addressing the properties of the rare OH/IR stellar population in the bulge.

Figure 4a shows an effort at normalizing the FW87 and Haller (1992) luminosity functions based on L_{TOT} for the stellar population using a modulus of 14.2 to the center (current estimates favor a modulus 0.3 mag larger, hence bolometric magnitudes 0.3 brighter). Figure 4b shows that the actual number of luminous stars is large and that the variables are consistent with known properties of Miras. This galactic center population does not exist in isolation, however. Lindqvist *et al.* (1992) discover numerous OH/IR stars there, and a population of luminous stars is found to exist over the entire inner 200 pc.

The population of the central 5 pc is not extraordinary; the long period variable star candidates have the colors and luminosities of Miras from the WFC91 sample, which itself lies 1000 pc from the nucleus. Haller

Fig. 3. Infrared image of the Galactic center in the K (2.2μm) band by Haller & Rieke (1993); 6 $'$ = 12 pc on a side, N at top, E to the right. The luminosity function in Figure 4 is drawn from this sample.

finds this population in fields 8$'$ perpendicular to the plane, where one has continuity with the surveys described below.

3. The Inner 200 Parsecs

The techniques of Stebbins & Whitford were fundamentally sound, and had a detector of sufficient sensitivity been available, they could have mapped the central population and discovered the nucleus. An infrared map covering a 1 deg $\times$ 2 deg field of the center was constructed from DC drift scans of a specially constructed InSb photometer (Glass, Catchpole, & Whitelock, 1987).

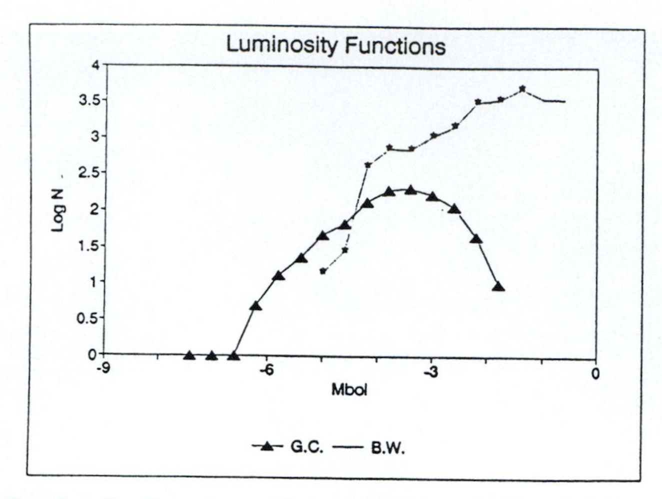

Fig. 4 (a). Luminosity function of late-type giants in the Galactic center field (Figure 3) triangles, and of optically selected M giants in Baade's Window; Frogel & Whitford, 1987, asterisks. Both are for $R_0 = 7$kpc; the Baade's Window luminosity function is scaled to the luminosity of the Galactic center. From Haller (1992) by permission. Notice the similarity of the Galactic center luminosity function to that of the M31 bulge (Figure 7).

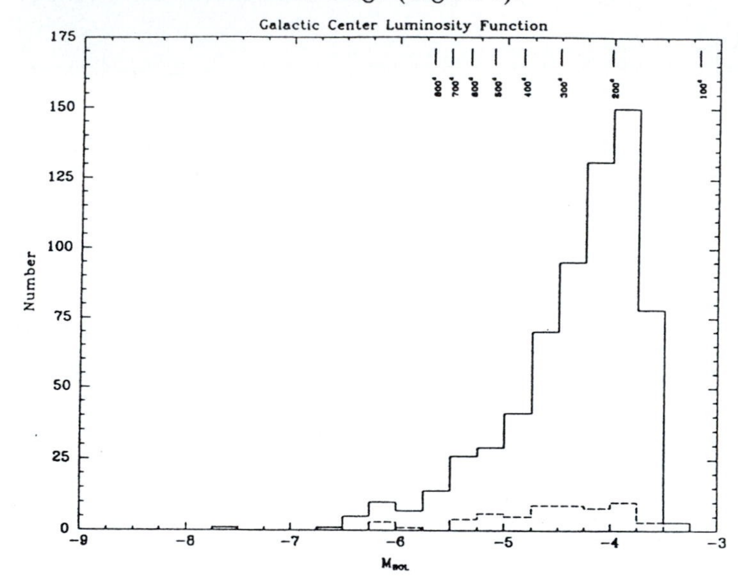

Fig. 4 (b). Bolometric luminosity function for 659 stars in the Haller (1992) Galactic center sample $m_K < 11.0, (H - K) > 1.50$. Solid histogram shows the total distribution while the dashed histogram is for the LPV candidates with $\delta m_K > 0.27$mag. Periods at top are from the period-luminosity relationship for LMC and Galactic Miras; bulge Miras may be ≈ 0.5 mag fainter at a given period (Whitelock, 1992, private communication).

Analyzing these scans, Catchpole, Whitelock & Glass (1990) find that the brightest stars ($M_{bol} < -5.5$) have a spatial distribution that is both flattened and concentrated to the center relative the bulk of the bulge. The luminosity function changes approaching the center. Figure 5a illustrates the flattened isophotes of the distribution, while 5b shows the different luminosity functions. We are dealing here with a population extending $\approx$ 100 pc out of the plane, and clearly not highly flattened. It is difficult to attribute this structure to a population that has been forming much mass within the last 2 Gyr, as some of the most luminous stars might require.

3.1. STELLAR CONTENT AND STRUCTURE

Is the central population dominated by the products of relatively recent ($<$ 2 Gyr) star formation, or is it the metal rich population formed during the final stages of dissipative collapse $\approx$ 10 Gyr ago? The luminous population favors the star formation hypothesis, but it is noteworthy that the luminous stars have a scale height much larger than 20 pc. Armed with the Catchpole *et al.* results, we can certainly argue continuity between the central 10 pc and the next 100 pc. Very luminous stars such as IRS 7 and the He I cluster remain the exception rather than the rule.

Lindqvist, Habing & Winnberg (1992); LHW, report radial velocities and shell expansion velocities for 134 OH/IR stars close to the Galactic center. LHW's group II (high velocity outflow) OH/IR stars rotate $\approx$ 10 times faster than the bulge stellar population and are no more than 30 pc from the nucleus. Divided into low and high velocity expansion groups, OH/IR stars show the classic dichotomy of the higher expansion velocity (younger? more metal rich?) group have lower velocity dispersion, higher rotation, and more concentration to the plane and nucleus. It is interesting that while LHW make their group I/II cut at 18 km/sec, van Langevelde (1990) makes a cut at 14.5 km/sec and finds the same dichotomy, except this time over scales of $\approx$ 1000 pc, with rotation for his OH/IR stars following that of the general stellar population (cf. 100 km/sec/deg).

It is well known that abundances and kinematics are correlated in the K giants at 500 pc (Rich, 1990). Sadler, Terndrup, and Rich (1993) are analyzing a sample of 400 bulge K giants with line strengths, radial, and proper motion velocity dispersions, and find that stronger lined stars have smaller vertical and radial velocity dispersions. Minniti (1993) finds greater rotation for the metal rich stars. The observed correlations between abundances and proper motion dispersions are almost certainly due to the increased rotation speed for the metal rich stars; the metal

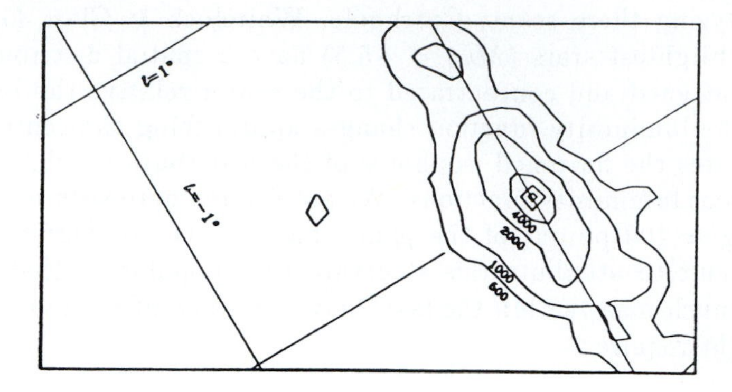

Fig. 5 (a). Contour map of the brightest stars ($5 < K < 6 = -6 < M_{bol} < -5$) in the sample of Catchpole, Whitelock, & Glass (1990). The K magnitude quoted is dereddened. Contours are marked by the number of stars per square degree. Note the flattened elliptical distribution centered on the Galactic Center (by permission).

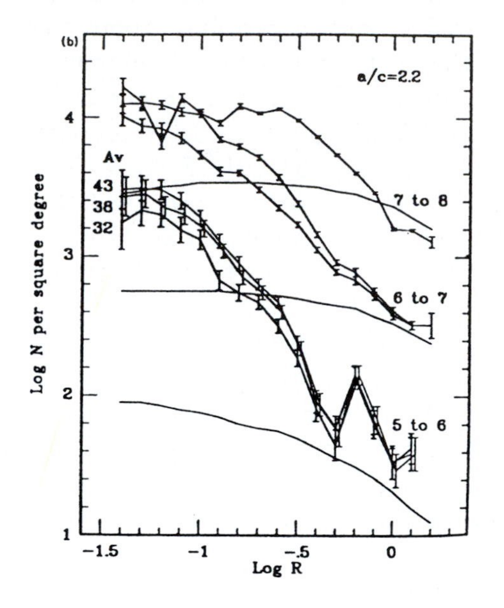

Fig. 5 (b). The concentration of the most luminous stars toward the Galactic center is evident (Catchpole, Whitelock & Feast, 1990). The concentration of stars with $M_{bol} < -5$ takes place over a scale of 100 pc. Number of stars per sq. deg shown as a function of log R, angular distance in deg. from the Galactic center. Smooth curve are disk models for each luminosity class; because K=7-8 stars are seen only to 32 mag of visual extinction, 3 sets of counts were made in order to make certain the counts are comparable.

rich stars must also have a more flattened spatial distribution (Figure 6; Zhao, Applegate & Rich, 1992).

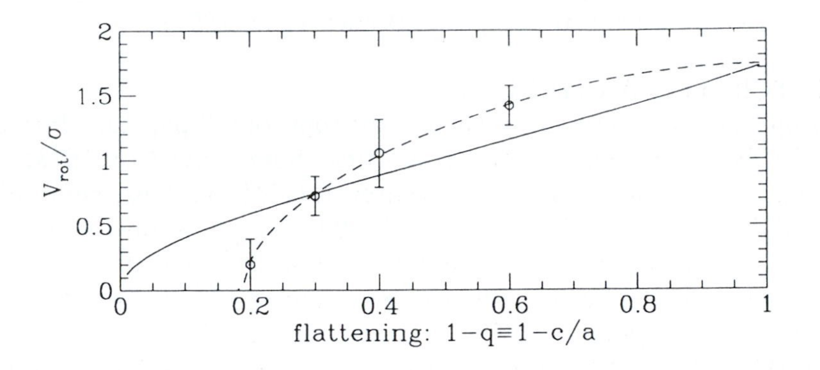

Fig. 6. Proper motion data in the bulge are consistent with more metal rich stars having a faster rotating, more flattened distribution than the general population. This possibly favors an enrichment and spin-up formation scenario which would lead to a super metal rich flattened population in the central 200 pc. Data points derived from Spaenhauer, Jones, and Whitford (1992) and Sadler, Terndrup, and Rich (1993). More metal rich stars are indicated with a larger value of flattening q. Solid line indicates a series of Maclaurin oblate rotator models, while the dashed line treats metallicity subgroups as massless tracers. Rotation velocity is derived from the assumption that observed proper motion anisotropy can be understood in the context of a flattened oblate rotator model for the bulge (Zhao, Applegate, and Rich, 1993).

There is no evidence for a highly flattened stellar system or discontinuity in stellar properties other than the obvious He I star cluster in the central few parsecs. Stars with $M_{bol} < -5$ are concentrated to the nucleus; however such luminous stars are seen in the bulges of M31, M32 and M33. It is not clear at this time whether these luminous giants are intermediate age or metal rich.

3.1.1. Effect of Triaxiality. There is growing evidence that the bulge is a triaxial, possibly barred structure (Blitz & Spergel, 1992; Binney *et al.* 1991; Whitelock & Catchpole, 1992). Whitelock & Catchpole's study is particularly compelling because of its use of the *IRAS* selected Miras to reveal a clear asymmetry as a function of Galactic longitude. As Spergel (1992) points out, triaxial potentials permit a class of orbits that intersect at the nucleus. It is possible that this orbit family is particularly effective at delivering gas to the center and fuels some of the activity. This may partially account for why the Milky Way appears to be more active than M31, and may have supported some star formation in the central 100 pc. The molecular gas occupies "forbidden" regions in the $l-V$ diagram that virtually require a triaxial potential (Binney *et al.* 1991).

3.2. INTERSTELLAR MEDIUM

Gas properties in the central 100 pc are the topic of full papers, addressing both the dynamics and physical properties. Some 10% ($5 \times 10^8 M_\odot$) of the Galaxy's molecular gas is in the center; this is in pressure equilibrium with 10^8K X-ray gas (Yamauchi *et al.* 1991). It is argued that while there is much molecular gas in the bulge region, high gas pressures (and consequent high internal velocity dispersions) are particularly unfavorable for star formation (Spergel & Blitz, 1992; Morris, 1992).

In this volume, Whitford suggests that nature knows how to make stars in this region; he proposes that the rapidly rotating group of high outflow velocity OH/IR stars were formed in this region further proposing that the disk continues within the central 100 pc. If this is so, how is it that the molecular gas in the middle of the Galactic potential lies so precisely in the Galactic plane? Is there any evidence for star formation in the plane outside the central parsec? Given the gap in HI from 1 to 3 kpc (Burton & Gordon, 1978) can one reasonably connect the Galactic center HI with the general disk?

4. Bright Stars in Local Group Spheroids: A Distant Mirror

Infrared arrays have recently been used to image the bulges of M31, M32, and M33, with the consequent discovery of luminous giants in these populations. Figure 7 shows the first such discovery (Rich & Mould, 1991), that giants in a field 500 pc from the nucleus greatly exceeded the M_{bol}=–4.2 considered to be the AGB tip in the original Frogel & Whitford (1987) luminosity function. While adoption of an 8 kpc distance to the center effectively brightens this by 0.3 mag, the FW87 luminosity function still lacks bright stars. Figure 8 shows resolution of a field 2′ from the nucleus of M31; bright AGB stars are now also seen in M32 (Freedman, 1992a,b; Elston & Silva, 1992). Using Kent's (1989) surface brightness model for M31, we can rule out the possibility that the bright stars belong to the M31 disk, as was suggested by Davies *et al.* (1991). Figure 9 shows the latest discovery, an $r^{1/4}$-law bulge population of M_{bol}=–6 stars in M33 (Minniti & Rieke *et al.* 1993).

In this volume, Renzini (his Table 1) points out that observers have been attempting photometry on stellar populations of dangerously high

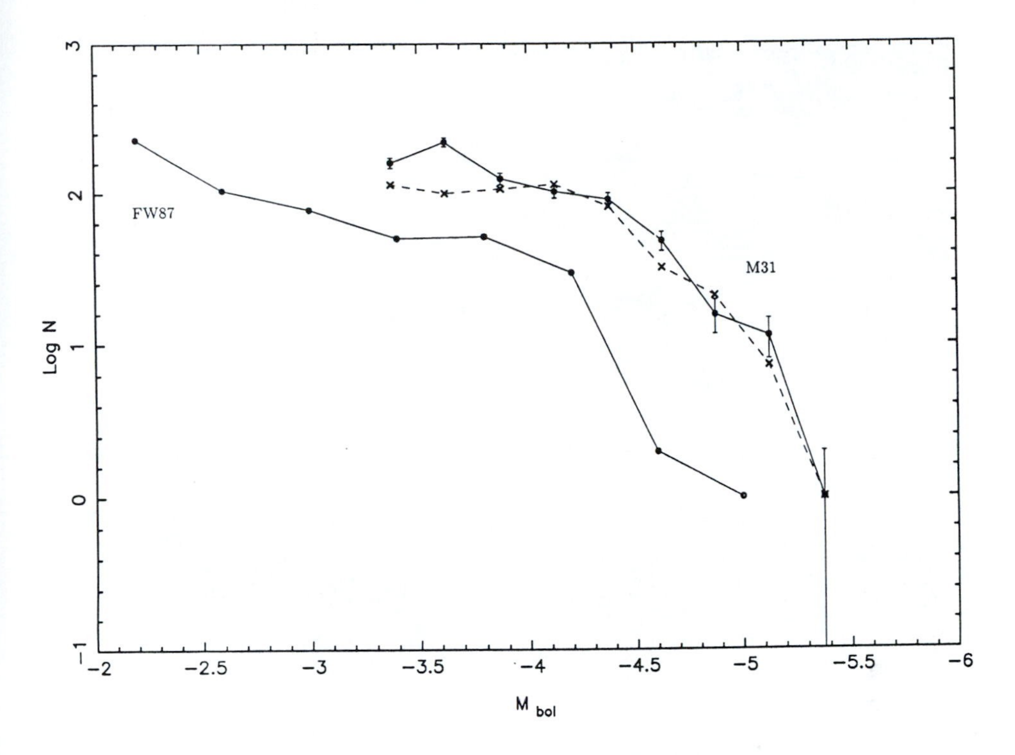

Fig. 7. Bolometric luminosity functions for a bulge field 500 pc SE of the M31 nucleus (Rich & Mould, 1991) and for Baade's Window (Frogel & Whitford, 1987). The FW87 luminosity function is for $R_0 = 7$kpc; the assumed modulus to M31 is 24.2. Currently favored distance moduli are 0.2 larger for both systems. Notice that the M31 luminosity function is extended to $M_{bol} = -5$. Rich, Mould, & Graham (1992) find that fields closer to the M31 nucleus may have even brighter stars and that these stars are not disk members (Figure 8).

surface brightness: some of my M31 fields have total luminosities of $10,000\ L_\odot$ per sq arcsec. Is it possible that many of these "bright stars" are actually blended images, measured as a single star? I offer 3 arguments in support of my published photometry. First, Figures 8 and 9 both have $\approx 5,000\ L_\odot$ per sq arcsec and are clearly resolved. Elston & Silva (1992) publish an image centered 3′ E of M32, where there is $500\ L_\odot$ per sq arcsec. The field is notably uncrowded, and one could easily increase the star density by a factor of 10 and still do photometry. Finally, one can perform sensitive artificial star tests on these fields covering a wide range

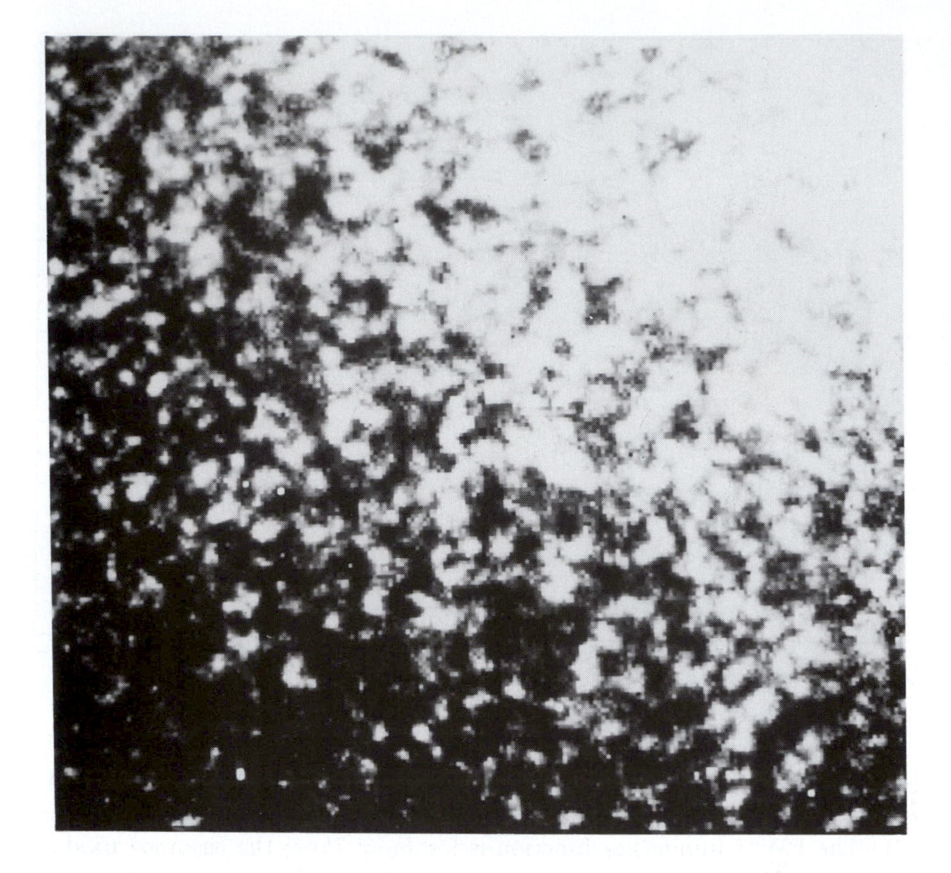

Fig. 8. 2μm image of the M31 bulge, 2 $'$ SW of the nucleus. North at top, East to right, 75 $''$ on a side. The lower left edge of the field has $L_{TOT} \approx 10,000\ L_\odot$ per sq. arcsec. Obtained with the Palomar infrared imager at the 4-m telescope (Rich, Mould & Graham, 1992).

in surface brightness. Rich, Mould & Graham (1992) observe M31 disk and bulge fields less than 1000 pc from the nucleus. The artificial star tests find the innermost field (2$'$ SE of the nucleus) to be overcrowded, but the other fields are relatively well behaved. Note that the luminosity functions of 2 different M32 fields are indistinguishable from the Rich & Mould (1991) luminosity function illustrated in Figure 7. The extension

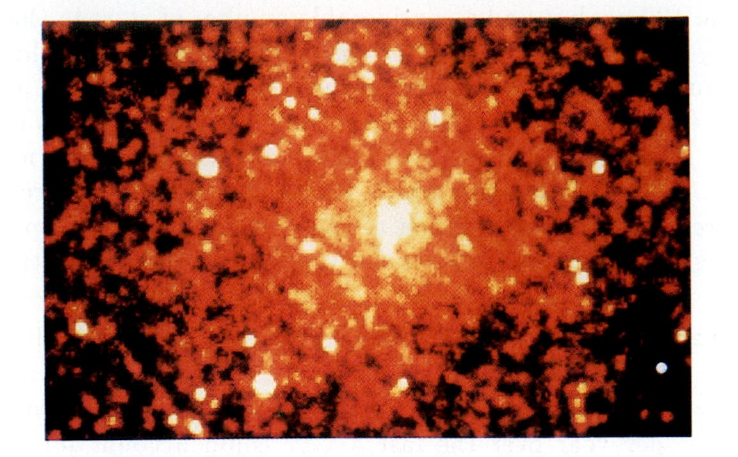

Fig. 9. In addition to M32, M31, and NGC 205, M33 joins the club of luminous bulge populations. This infrared frame by Minniti, Rieke *et al.* (1992) reveals an $r^{1/4}$-law distributed luminous stellar population of AGB stars with M_{bol} =–6. Image is 2.5 ′ in length; North at top, East to left. This is part of a 10 ′ long mosaic which also samples the disk of M33. While the surface brightness is 2000 $L_{\odot}$ per sq arcsec, notice the clear resolution into stars.

to M_{bol}=–5.5 is similar to that seen in the complete infrared surveys of Baade's Window reported by Glass *et al.* and Whitelock in this volume. Further, applying the Fuel Consumption Theorem one finds the the counts of the brightest stars per unit luminosity consistent with lifetimes of $\approx 10^5$yr, or that of Miras. Despite a factor of 10 range in surface brightness, all of the bright stars in the local group populations studied so far have this approximate lifetime.

5. What is the Central Population?

Once we compare the population of the central 200 pc with that of the M31 bulge and the surveys of Glass and Whitelock described in this volume, we no longer find the stars anomalously luminous. Even in the central 5 pc, where it might be argued that some star formation must have occurred, the luminosity function is only slightly brighter than is found in the central 500 pc of the M31 bulge. Colors and luminosities of the stars are similar Whitelock *et al.* 's (1991) IRAS-selected Miras in the outer bulge. Glass *et al.* (1990) do not publish a luminosity function for the central 100 pc, but find a considerable number of stars in the interval $-5 < M_{bol} < -6$ remain even after correction for the disk. Bloemmaert (this volume) finds OH/IR stars brighter than M_{bol}=–5 in the outer bulge. As Catchpole *et al.* (1990) beautifully illustrate, there is no abrupt change in the central 200 pc, but a gradual increase in the number of luminous AGB stars, a trend which Haller continues to the central few pc. The remarkably high rotation of the central OH/IR star population remains as the strongest evidence for young stars outside of the central parsec.

We have the problem of explaining whether the central $10^8 M_{\odot}$ has formed in the last few Gyr or is largely the remnant of a more ancient event in which the most of the bulge formed. A star formation rate of 0.02 $M_{\odot}$ per year over the last 4 Gyr could account for all of the mass. On the other hand, it is also possible that the concentration of high luminosity stars toward the center are the AGB progeny of a very metal rich ancient population. The observed flattening may relate to the collapse and spin-up inferred from the correlations between abundances and kinematics observed in the outer bulge; the central 200 pc may be mostly the result of the final stage of a dissipative collapse.

In this volume, Baum *et al.* report a bright main sequence turnoff at Baade's Window, confirmed by Ortolani & Rich (1993). If the bulge was formed in a starburst 10 Gyr ago (Rich, 1992) it is relatively easy to explain the high AGB luminosities. If the starburst formed a massive stellar disk (unstable to bar formation) then a thick bulge could have formed via the mechanisms described herein by Sellwood and Norman. The high luminosities and mass loss rates of the AGB stars still pose (in my opinion) a daunting challenge to those who would like the bulge to be the oldest population in the Galaxy. The key to learning the history of the central population is to understand the nature of the luminous AGB stars found in the Galactic bulge and other local group spheroids.

The effort to define the nature and origin of this central population precisely mirrors our struggle at this meeting to define what the bulges of galaxies are, and how they form. We are unsure whether they best represent disk-like structures, continuation of the spheroid, or a relative of the thick disk. In the spirit of this delightful confusion, I close this contribution with a humorous figure, and my apologies to the Belgian artist René Magritte.

I am grateful for Kris Sellgren's insights into this subject. I also acknowledge very lively discussions with Alvio Renzini on the topics of bulge formation and the nature of the AGB stars. I am grateful to Joe

CECI N'EST PAS UN BULBE

M. RICH
(APOLOGIES TO
RENE MAGRITTE)

Haller and Dante Minniti for their release of data prior to publication, and to Wendy Freedman and George Djorgovski for reviewing the manuscript. I am indebted to Francisco Feliciano for preparation of the manuscript.

REFERENCES

Allen, D.A., Hyland, A.R., & Hillier, D.J. 1990, *M.N.R.A.S.*, **244**, 706.

Backer, D.C., ed., *The Galactic Center*, AIP Conf. 155, (AIP:NY).

Becklin, E., Gatley, I., and Werner, M.W. 1982, *Ap.J.*, **258**, 135.

Becklin, E.E., & Neugebauer, G. 1968, *Ap.J.*, **151**, 145.

Binney, J. *et al.* 1991, *M.N.R.A.S.*, **252**, 210.

Blanco, V.M. 1965, in Vol. 5, *Stars & Stellar Systems*, A. Blanco & M. Schmidt, eds. (Univ. of Chicago Press, Chicago) p. 241.

Blitz, L., ed., 1992, *The Center, Bulge, and Disk of the Milky Way*, (Dordrecht:Kluwer).

Blitz, L., & Spergel, D.N. 1991, *Ap.J.*, **379**, 631.

Burton, W.B., & Gordon, M.A. 1978, *Astr. Ap.*, **63**, 7.

Catchpole, R.M., Whitelock, P.A., & Glass, I.S. 1990, *M.N.R.A.S.*, **247**, 479.

Davies, R.L., Frogel, J.A., & Terndrup, D.M. 1991, *A.J.*, **102**, 1729.

Depoy, D., & Sharp, N. 1991, *A.J.*, **101**, 1324.

Eckart, D. *et al.* 1992, *Nature*, **335**, 526.

Elston, R., & Silva, D.R. 1992, *A.J.*, **104**, 1360.

Feast, M. W. 1981, in *Physical Process in Red Giants*, A. Renzini & I. Iben, eds. (Reidel: Dordrecht) p. 64.

Frogel, J.A. & Whitford, A.E. 1987, *Ap.J.*, **320**, 199.

Freedman, W.L. 1992a, in *IAU Symp. 149 The Stellar Populations of Galaxies* B. Barbuy and A Renzini, eds. p. 169.

Freedman, W.L., 1992b, *A.J.*, **104**, 1349.

Gatley, I. in *The Galactic Center*, D.C. Backer, ed. AIP Conf. 155, (AIP:NY) p. 8.

Glass, I.S., Catchpole, R.M., & Whitelock, P.A. 1987, *M.N.R.A.S.*, **227**, 373.

Kent, S.M. 1992, *Ap.J.*, **387**, 181.

Kent, S.M., Dame, T.M., & Fazio, G. 1991 *Ap.J.*, **378**, 496.

Kent, S.M. 1989, *A.J.*, **97**, 1614.

Krabbe, A., Genzel, R., Dropatz, S., & Rotaciuc, V. *Ap.J. (Letters)*, **382**, L19.

Lindqvist, M., Habing, H.J., & Winnberg, A. 1992, *Astr. Ap.*, **259**, 118.

Minniti, D.M. 1993, in preparation.

Minitti, D., Rieke, M. *et al.* 1993, *Ap.J. (Letters)*, in press.

Morris, M., ed., 1989, *The Center of the Galaxy*, IAU Symp. 136 (Dordrecht:Kluwer) .

Morris, M. 1992, in preparation.

Ortolani, S., & Rich, R.M. 1993, in preparation.

Renzini, A. & Buzzoni, A. 1986 in *Spectral Evolution of Galaxies*, eds. C. Chiosi & A. Renzini (Dordrecht: Reidel), p. 195.

Renzini, A., & Fusi-Pecci, F.F. 1988 *ARAA*, **26**, 199.

Rich, R.M., Mould, J.R., & Graham, J. 1993, *A.J.*, in press.

Rich, R. M. 1992, in *The Center, Bulge, and Disk of the Milky Way* L. Blitz, ed. (Reidel: Dordrecht) p. 47.

Rich, R.M. & Mould, J.R. 1991, *A.J.*, **101**, 1286.

Rich, R.M. 1990, *Ap.J.*, **362**, 604.

Sadler, E., Terndrup, D., & Rich, R.M. 1993, in preparation.

Sellgren, K., McGinn,M.T., Becklin, E.E., & Hall, D.N.B. 1990, *Ap.J.*, **359**, 112.

Sellgren, K. 1989, in *The Center of the Galaxy*, M. Morris, ed., IAU Symp. 136 (Dordrecht:Kluwer) p.477.

Spaenhauer, A., Jones, B., & Whitford, A.E. 1991, *A.J.*, **103**, 297.

Spergel, D., & Blitz, L. 1992, *Nature*, **357**, 665.

Spergel, D. (1992) in *The Center, Bulge, and Disk of the Milky Way* L. Blitz, ed. (Reidel: Dordrecht) p. 77.

Stebbins, J., & Whitford, A.E. 1947, *Ap.J.*, **106**, 235.

van der Veen, W., & Habing, H.J. 1990, *Astr. Ap.*, **231**, 404.

van Herk, G. 1965, *B.A.N.*, **18**, 71.

van Langevelde, H. 1990, Ph.D. Thesis, Leiden University.

Whitelock, P. & Catchpole, R. 1992, in *The Center, Bulge, and Disk of the Milky Way* L. Blitz, ed. (Reidel: Dordrecht) p. 103.

Whitelock, P.A., Feast, M.W., & Catchpole, R.M. 1991, *M.N.R.A.S.*, **248**, 276.

Wood, P.R., & Bessell, M.S. 1983, *Ap.J.*, **265**, 748.

Yamauchi, S. *et al.* 1991, *Ap.J.*, **365**, 532.

Yusef-Zadeh, F., & Morris, M. 1991, *Ap.J. (Letters)*, **371**, L59.

Zhao, H.S., Applegate, J.H., & Rich, R.M. 1993, in preparation.

DISCUSSION

Tyson: Could you comment further on the highly flattened concentration of stars near the centre? In particular, are the flattened isophotes isolated, or do they blend smoothly with the general bulge?

Rich: It would be interesting to re-analyze the Catchpole work or push it with area detectors. They published this distribution of bright stars that have flattened ellipsoids at K=5 to 6, it would be interesting to look at the distributions of 6 to 7 and 7 to 8 at K, and to see if there is any difference in the spatial distribution. Mould asserted (1986) that there is a concentration of bright stars towards the nucleus of M31.

Habing: We see a flattening, but that's on a slightly larger scale in the OH/IR stars, where the high outflow velocities seem to be more concentrated to the plane. One way to find out whether the high outflow velocities are due to metallicity or to luminosity, is simply by obtaining the apparent luminosity of the stars.

Sellwood: The Winnberg and Lindqvist sample of OH/IR stars have very distinct kinematics from the outer OH/IR sample. Is that telling us that the inner 200 parsec is something entirely different?

Rich: I think it is concentrated on a scale of about 100 parsec. I don't know because coexisting in that spatial volume are hotter populations. This particular sample has cold kinematics, there is no question of that.

Franx: No one has mentioned Tonry's work (on the luminosity variance method) in the last two days. I wonder if you can relate the result in our bulge and the M31 bulge to his work.

Rich: I think that what you see in the population depends sensitively on what wavelength you are working at. Tonry I believe is working in the R band and the luminosity function works very close to the centre, where it is calibrated, in M32. In the Virgo cluster one sees some reasonably correlation with what we see in M32, because you are looking at the fluctuations of the brightest giants in the R band. In the K band you start seeing rarer objects.

Tendrup: Tonry and I are talking about how to reconcile the bulge of M31 and other galaxies, in the brightness fluctuations method. If you do a simple calculation based on a bulge luminosity function, you get the same results as you do using M31 and M32. We plan to do this in the K band over the next year or so.

MIKE RICH
'92

Tim DE ZEEUW

Dynamics of the Galactic Bulge

TIM DE ZEEUW
Sterrewacht Leiden

Abstract. Recent work on the intrinsic shape, the internal kinematics of stars and gas, and the dynamics of the Galactic Bulge is discussed. Starcounts, measurements of the integrated light and the kinematics of the atomic and molecular gas all provide strong evidence that the Bulge is triaxial, and is rotating fairly rapidly. To date, there is little evidence for triaxiality in the stellar kinematics: the available stellar velocities are consistent with Kent's (1992) oblate model. This unsatisfactory situation is expected to improve rapidly.

1. Introduction

The Galactic Bulge is one of the major components of the Galaxy. A careful study of its morphology, kinematic and dynamical properties is required to answer questions such as: Can we think of the Bulge as a small elliptical galaxy? Was it formed before or after the disk? What is its relation to the metal–poor stellar halo, and to the dark halo? An extensive discussion of these and related issues was given by Freeman (1987), Gilmore, King & van der Kruit (1990), and Spergel (1992). Here we concentrate on two specific areas of research on the Galactic Bulge, namely, i) what is its intrinsic shape, and ii) what is its internal velocity structure? We first consider measurements of the integrated light and starcounts, as well as observations of the gaseous and the stellar kinematics, then discuss the dynamics of the Bulge, and finally compare the results briefly with what we know about the dynamics of bulges of other disk galaxies. Detailed discussions of the properties of other bulges, and of the relation between bulges and elliptical galaxies, are given by Kormendy and Franx, respectively, elsewhere in this volume.

2. Morphology

2.1. INTEGRATED LIGHT

The Bulge is heavily obscured at optical wavelengths, and is best studied in the infrared. Early measurements showed that the integrated surface brightness distribution between 2 and 2.4 micron is flattened, and has a central cusp (Becklin & Neugebauer 1968; Matsumoto *et al.* 1982). A variety of functions have been proposed as fits to the observed surface brightness profile, including power laws, exponentials and even de Vaucouleurs' profiles. Sellwood & Sanders (1988) reviewed much of the early work, and showed that the volume emissivity profile between 0.5 and 500 pc from the Galactic Center is a power–law with slope −1.8, while at larger radii the profile steepens considerably, and the slope approaches −3.7. Kent (1992) analyzed the Spacelab 2.4 micron data (Kent, Dame & Fazio 1991) and showed that foreground contamination by the disk amounts to 40% of the observed light. The corrected profile (Fig. 1) is similar to the one proposed by Sellwood & Sanders, has an inner slope of −1.85, and a minor axis scale–length of ~400 pc. There is no component with a scale–length of 2.7 kpc, as is sometimes claimed.

H. Dejonghe and H. J. Habing (eds.), Galactic Bulges, 191–208.

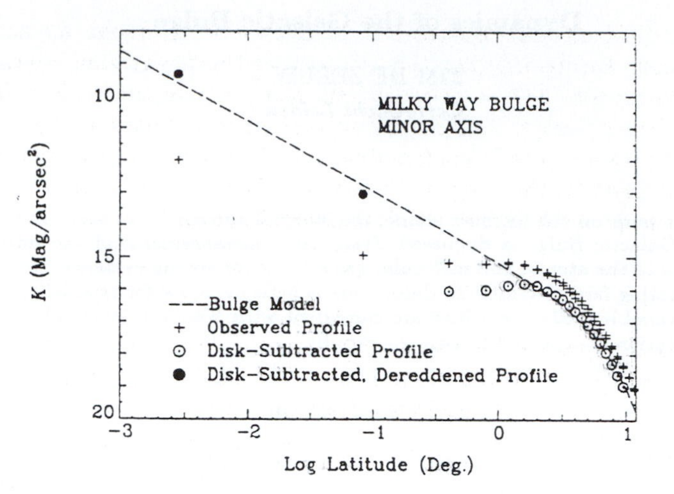

Fig. 1. **Minor axis surface brightness profile of the Bulge in the near-infrared, as given by** Kent (1992).

Most previous studies have assumed that the Bulge is oblate, with its short axis perpendicular to the Galactic plane. This means we observe the Bulge edge-on, and the surface brightness distribution can be deprojected uniquely to give the three-dimensional luminosity distribution (Rybicki 1987). Kent (1992) finds that the Bulge has an axis ratio of 0.61, and is slightly box- or peanut-shaped. Inside one kpc the Bulge may be more flattened.

Blitz & Spergel (1991b) reanalyzed the balloon measurements of Matsumoto *et al.* (1982), and found that the Bulge is in fact brighter at positive galactic longitude ℓ than at negative longitude. They argued that this is not due to extinction variations, but is caused by the Bulge being triaxial rather than oblate, with the near side at $\ell > 0°$. If this is indeed the case, then the observed surface brightness distribution no longer determines the three-dimensional shape uniquely (Stark 1977).

If we assume that the surfaces of constant volume emissivity are approximately ellipsoidal with semi-axes a, b and c, then in order to determine the intrinsic shape of the Bulge, we must derive its luminosity profile, the scale-length a, the two axis ratios b/a and c/a, and also the orientation of the ellipsoid. If the Galactic plane is one of the symmetry planes of the triaxial Bulge, then its orientation is fixed by the angle ϕ between the major axis of the Bulge and the line-of-sight from the Sun to the Galactic Center. In our convention c/a is the axis ratio perpendicular to the Galactic plane, and b/a is the axis ratio in the plane. Both axis ratios, and also ϕ, may depend on radius. The deprojection of the observed surface brightness is not unique because a change in b/a can be compensated by a change in ϕ so as to give the same projected distribution. It is therefore not surprising that the analysis by Blitz & Spergel (1991b) does not give an accurate value for ϕ, and that it does not provide a significant constraint on b/a.

While the non-uniqueness of the deprojection of the Bulge light is unavoidable, progress can be made by using measurements of higher quality. Combination of observations at different wavelengths should allow one to derive an accurate extinction correction, and to improve the estimate of the foreground contamination. The COBE measurements will be of great help here, as will the various large-scale surveys with infrared arrays that are being carried out (Glass 1993). Analysis of this data should provide useful constraints on the intrinsic shape of the Bulge, and should also shed light on the suggestion that the Bulge may be tipped relative to the Galactic plane (Blitz & Spergel 1991b; Spergel 1992).

2.2. STARCOUNTS

Much of the classical work on individual stars in the Bulge has been restricted to a small number of special windows that are not heavily obscured by interstellar extinction and hence gives only limited information on the shape and structure of the entire Bulge (Frogel 1988). IRAS improved this situation dramatically. Habing *et al.* (1985) showed that a simple criterion based on the observed flux densities at 12 and 25 micron allows one to select AGB stars from the IRAS point source catalog. The distribution of these sources on the sky clearly shows the disk of the Galaxy, and the flattened Bulge (Habing 1988). After correction for the effects of confusion near the Galactic plane, the derived luminosity profile and axis ratio of the Bulge are consistent with the integrated light measurements (Harmon & Gilmore 1988).

The recent interest in triaxiality has spurred re-analysis of the properties of various populations of Bulge stars. Nakada *et al.* (1991) investigated the luminosity function of a subsample of the IRAS AGB stars in the Bulge, and found that the stars at $\ell > 0°$ are brighter on average—and hence nearer to us—than those at $\ell < 0°$, in agreement with an earlier suggestion by Harmon & Gilmore (1988).

Weinberg (1992a) re-analyzed the IRAS point source catalog, considered only objects with galactic latitude $|b| < 3°$, and used selection criteria which differ slightly from those of Habing *et al.* (1985). He assumed that all these stars have an absolute bolometric luminosity of 8000 $L_{\odot}$, calculated individual bolometric corrections based on the observed IRAS colors, and assumed uniform extinction throughout the Galaxy. This gives distances for all the objects. The resulting galactic distribution appears to be lopsided. Weinberg argued that this is due to the apparent luminosity cutoff of the sample, which causes the exclusion of nearly all stars beyond 10 kpc. Analysis of a deeper, but incomplete, sample showed a more symmetric bar-like distortion, again with the near side at $\ell > 0°$. Instead of fitting a specific model to the distribution of stars, Weinberg calculated the coefficients of the harmonic expansion that fits the data best. He reconstructed a smooth density distribution from his expansion coefficients, ignoring the asymmetric terms. This yields a symmetric bar, out to about 5 kpc, with the near side at $\ell > 0°$, an axis ratio $b/a \sim 0.6$, and an orientation $\phi = 36{\pm}10°$. Varying the prescriptions for the bolometric correction and for the extinction, and allowing for a realistic spread in the intrinsic luminosities of the stars, did not change the main result—the existence of a bar—but influenced its properties. It remains to be seen whether the spatial extent of 5 kpc found by Weinberg can be reconciled with the smaller size inferred from studies of the gas kinematics (§3.1). It is also not clear how the derived properties are influenced by confusion of sources near the Galactic plane.

Weinberg (1992b) compared the counts of IRAS AGB stars at $|b| > 3°$ with simulated samples, and found that the central bulge seen in the distribution on the sky could be an artefact of extinction variations combined with the limited sensitivity of IRAS. Whereas in the Galactic plane IRAS barely detects the AGB stars at the center, at increasing $|b|$ the extinction decreases, and such stars can be detected along ever longer lines–of–sight, so that the counts are systematically larger. This effect is strongest towards the center of the Galaxy. The result is an apparent bulge in the distribution of stars in the IRAS map of the Galaxy.

The measurements of the integrated light (§2.1), and notably the COBE data (Hauser *et al.* 1990), show clear evidence for a Bulge in the center of the Galaxy, in agreement with observations of nearby galaxies. The result of the experiments reported by Weinberg (1992b) do not indicate that the Galaxy contains no Bulge, but demonstrate that extinction is important even at 12 and 25 micron, and, more importantly, that results based on the IRAS database can be biased strongly by its limited sensitivity. These effects should not influence the detected asymmetry between the counts at positive and negative ℓ.

A much improved analysis of the shape of the Bulge based on starcounts will be possible in the near future. The 2MASS (Kleinmann 1992) and DENIS (Epchtein, Guglielmo, & Burton 1992) sky surveys near 2 micron are expected to detect every AGB star in the Galaxy, and will provide ideal datasets for application of Weinberg's expansion method. Measurements in three bands will allow accurate correction for extinction, and the large number of sources will make a three–dimensional study feasible which takes into account a realistic luminosity function for the AGB stars. This should then also shed light on the question whether the structure detected by Weinberg (1992a) in his two–dimensional analysis is an elliptic thick disk, as suggested by Spergel (1992), or a cross–section through a triaxial Bulge.

Whitelock & Catchpole (1992) investigated a sample of 104 Mira variables for which individual distances are known from the observed periods (Whitelock, Feast & Catchpole 1991). The objects are located in two strips parallel to the Galactic plane with $-15 < \ell < 15°$ and $7 < |b| < 8°$. The Miras at $\ell > 0°$ have distance moduli that are ~0.4 magnitudes smaller on average than those at $\ell < 0°$. There is no difference between the distribution of distances within the two strips. A prolate bulge with $b/a = c/a \sim 0.25$ and $\phi \sim 45°$ fits the data, in rough agreement with the results mentioned above. This work also shows that the elongated component extends to at least 1 kpc above the Galactic plane. The Whitelock & Catchpole study is a major step forward, because it is based on accurate individual, rather than statistical, distances. It is important to extend this work to smaller $|b|$. This is difficult—not only because of the increased extinction but also because many observations per star are required to determine the period—but is worth the effort.

Finally, we note that the RR Lyrae stars with known distances also show marginal evidence for a triaxial distribution, but with its near end at $\ell < 0°$ (Wesselink 1987; Le Poole & Habing 1990). It will be interesting to see whether this result holds up when a larger sample is studied, since the RR Lyraes are old (Lee 1992), and—unlike the Miras and the AGB stars—presumably belong to the metal–poor halo. This may itself be triaxial, and indeed elongated in a direction opposite to the Bulge (Blitz & Spergel 1991a; but see Kuijken & Tremaine 1991, 1993).

3. Kinematics

3.1. GAS MOTIONS

It has been known for a long time that the inner rotation curve of the Galaxy displays a prominent hump (Combes 1991; Liszt 1992). A natural explanation is to assume that the central part of the Galaxy is not axisymmetric, but contains a bar (de Vaucouleurs 1964) or a triaxial bulge. The simple closed orbits available to the gas are then elongated, and the gas velocity varies along the orbit. When viewed from the proper direction, i.e., by proper orientation of the Bulge, one may observe gas velocities that are larger than the circular velocity, and hence see a hump in the rotation curve. Liszt & Burton (1980) gave an early description of the observations in terms of a simple kinematic model which is equivalent to motion on elliptic orbits (Kent 1992). Gerhard & Vietri (1986) showed that the inner rotation curve can be reproduced by closed orbits in a prolate bulge with a stationary figure, a realistic density profile, and seen nearly broadside on. Burton & Liszt (1993) have shown recently that the HI measurements are contaminated by absorption against the continuum radiation from the Galactic core. They correct for this effect by using observations of OH and H_2CO, and again confirm the presence of strong non–circular motions. The resulting azimuthally averaged velocity field gives a circular velocity curve that is consistent with the 2 micron light profile (§2.1) and a constant mass–to light ratio $\mathcal{M}/L$.

Binney *et al.* (1991) have constructed the most comprehensive model to date of the motions of both the atomic and the molecular gas in the inner kpc of the Galaxy. These authors assume the gas moves on non–selfintersecting closed orbits, and show that a rapidly rotating inner bar with $b/a = 0.75$, a figure rotation rate of 63 km/s/kpc—so that the corotation radius lies at 2.4 ± 0.5 kpc—and seen nearly end–on at $\phi = 16° \pm 2°$, provides a natural explanation for the observed kinematics of the gas. The derived density profile resembles closely the one adopted by Kent (1992): the logarithmic slope is –1.75 in the inner Bulge, and approaches –3.5 at large radii. Designating the elongated central component as a bar or as a triaxial bulge may be a matter of semantics: the gas motions in the Galactic plane do not constrain the density distribution of the bar outside the plane, and, as Binney *et al.* point out, it is possible that the rapidly rotating bar is identical to the triaxial box–shaped Bulge evident in the integrated light and the star counts. The bar and the Bulge may also be separate Galactic components (§4).

Proper modeling of the gas kinematics requires a careful hydrodynamical treatment, because the closed orbit approximation breaks down when the orbits become very elongated, or when the gas switches from one orbit family to another. Early hydrodynamical studies that discuss the effects of a triaxial bulge on the Galactic HI kinematics include van Albada (1985), Yuan (1984), and Mulder & Liem (1986). The former author uses a bar which is similar to the one found by Binney *et al.* (1991), but the latter authors require a bar which has its nearest side at $\ell < 0°$ rather than at $\ell > 0°$. These studies use a rather crude approximation to the Galactic potential. Recent work by Jenkins & Binney (1993) demonstrates that it is worthwhile to redo this kind of study, with a more up–to–date potential and state–of–the–art hydrodynamical schemes.

3.2. STELLAR RADIAL VELOCITIES

Much of the classical spectroscopy of Bulge stars was restricted to Baade's Window. At present, radial velocities are available for a variety of populations in a number of windows. These include K and M giants, carbon stars, planetary nebulae and RR Lyrae stars. Kent (1992) summarizes many of the measurements, including those of the integrated light. More recent studies include work on K giants in fields between 1.5 and 2 kpc from the center (Minniti *et al.* 1992; Harding & Morrison 1993).

This body of work shows that the Bulge rotates, and has a mean line-of-sight velocity $\langle v_{los} \rangle$ of 5–10 km/s/°, so that $\langle v_{los} \rangle$ may reach about 80 km/s at 1 kpc along the major axis. Different gradients of $\langle v_{los} \rangle$ found for different samples are sometimes taken as evidence for a dependence of the kinematics on other properties of the stellar population, such as abundance or age. However, such differences may also be caused by the customary but highly suspect fitting of linear regression lines to samples that often have different radial extent. There is no a priori reason to expect a linear dependence of $\langle v_{los} \rangle$ on radius (§4).

The line-of-sight radial velocity dispersion σ_{los} in the Bulge increases from ~80 km/s at 10 pc to ~115 km/s at a few hundred pc from the center. At larger distances σ_{los} decreases again, in a manner which is consistent with the inward increase of σ_{los} measured in the old disk (Lewis & Freeman 1989; Carney, Latham & Laird 1990). There is evidence that σ_{los} increases in the inner 3 pc (Sellgren *et al.* 1990), but not all authors agree (Rieke & Rieke 1988). Such an increase could be caused by a central black hole or nuclear star cluster with a mass of a few times $10^6\ M_\odot$.

In order to investigate the stellar kinematics of the entire Bulge, tracers are needed that can be detected at radio wavelengths, such as OH/IR stars and planetary nebulae. The OH/IR stars are most useful: these AGB objects are surrounded by an expanding dust shell, which can be readily identified by means of its OH emission at 18 cm (Fig. 2). The spherical geometry of the shell causes the line profiles to be double-peaked, so the measured velocity width is twice the expansion velocity v_{exp} of the shell, and the mean velocity is the radial velocity v_{los} of the embedded star. Because radio measurements are not hampered by Galactic obscuration, v_{los} can be determined for OH/IR stars throughout the Galaxy. By contrast, radial velocities of planetary nebulae are generally based on optical spectroscopy, so that foreground extinction remains a problem (Stasińska 1993).

Early radio surveys of the central few hundred pc (Habing *et al.* 1983; Winnberg *et al.* 1985) found a few dozen OH/IR stars, and showed that the inner Bulge may well rotate rapidly, with $\langle v_{los} \rangle$ ~100 km/s at less than 100 pc from the center, and with σ_{los} between 60 and 140 km/s. Lindqvist *et al.* (1992a, b) have recently completed a large VLA survey which produced 134 OH/IR stars in the central 100 pc. Their kinematic properties seem to depend on the physical properties of the stars: those with $v_{exp} < 18$ km/s, which are thought to be low mass old objects, have $\sigma_{los} = 82 \pm 7$ km/s, while stars with $v_{exp} > 18$ km/s, which are thought to be more massive, and younger, have $\sigma_{los} = 65 \pm 6$ km/s. The fast rotation found earlier is confirmed: $\langle v_{los} \rangle$ reaches over 100 km/s at ~40 pc from the center. This is comparable to the circular velocity, and suggests that at least part of this OH/IR star population may be in a rotationally supported disk.

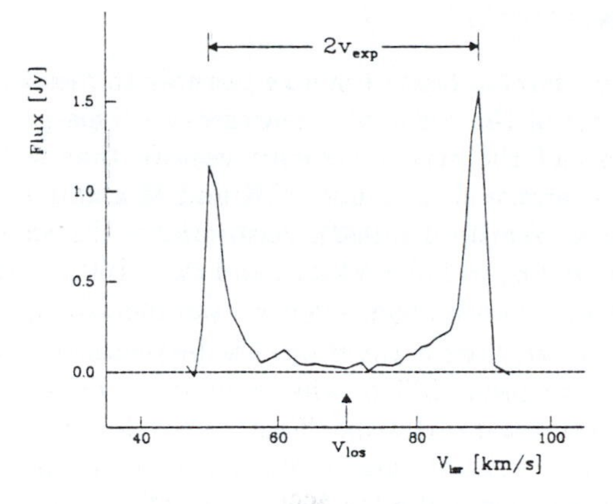

Fig. 2. Line profile of an OH/IR star at 18 cm. The velocity difference between the two peaks is twice the expansion velocity of the circumstellar dust shell, and the mean velocity of the peaks equals the radial velocity of the central star. The data are for the source OH359.954–0.041 (van Langevelde 1992).

A number of surveys of IRAS AGB stars have been carried out at 18 cm, resulting in radial velocities for several hundred OH/IR stars in the Bulge (te Lintel–Hekkert *et al.* 1989, 1991). This sample has $\sigma_{\rm los}$ ~110 km/s and $\langle v_{\rm los}\rangle$ ~10 km/s/° (Dejonghe 1993b), in good agreement with other measurements, notably those of Mira variables (Menzies 1990) and planetary nebulae (Kinman, Feast & Lasker 1988; Acker *et al.* 1991), which are in evolutionary stages bracketing the OH/IR stage. Radial velocities based on the SiO maser emission of OH/IR stars with dense dust shells also give similar values for $\langle v_{\rm los}\rangle$ and $\sigma_{\rm los}$ (Nakada *et al.* 1993).

It is unfortunate that the IRAS selected OH/IR sample is incomplete near the Galactic plane, due to the confusion limitations of the IRAS point source catalog. As a result, there is hardly any overlap with the fast-rotating Lindqvist *et al.* (1992a) sample near the Galactic Center (see Fig. 1 of Dejonghe 1993b), so that the possible presence of an extended disk component in the Bulge remains uncertain. The systematic, sensitive, and unbiased radio survey of the entire Bulge at 18 cm, which is being carried out by Habing and his collaborators, will improve this situation considerably.

The distribution of $v_{\rm los} - \langle v_{\rm los}\rangle$ for the known OH/IR stars in the inner Bulge appears to be nearly Gaussian, but it should be realized that surveys for these stars are often done in a limited velocity range, due to the finite bandwidth of receivers. Van Langevelde *et al.* (1992b) sampled a large velocity interval, and showed that in addition to Baud's star, two more OH/IR stars close to the center have $v_{\rm los}$ ~350 km/s. Since deviations from a Gaussian velocity distribution are expected (§4), especially if the Bulge is triaxial, it is important to search for OH/IR stars in a large velocity interval, even if the bulk of the objects lies in a more limited range. By the same argument, "obvious outliers" in samples of stars with optical radial velocities should be deleted with caution.

3.3. PROPER MOTIONS

In a very exciting recent development, it is now possible to measure proper motions of Bulge stars, which are of the order of 3 marcsec/yr. This provides information on the two components of the stellar velocity vectors that so far could not be measured directly. An example is the study of K and M giants in Baade's Window by Spaenhauer, Jones & Whitford (1992). Assuming a distance to the Galactic Center of 7.7 kpc, they find $\sigma_\ell = 115 \pm 4$ km/s and $\sigma_b = 100 \pm 4$ km/s in this minor axis field; these values may depend somewhat on the metallicity of the stars.

Proper motions of OH/IR stars in the Bulge can be measured by VLBI techniques in the near future. In the inner 100 pc this cannot be done at 18 cm, because interstellar scattering limits the resolution (van Langevelde *et al.* 1992a). However, since the scattering scales with λ^2, use of the H_2O (λ1.3 cm) or SiO (λ0.7 cm) maser emission of these objects will allow accuracies of ~10 km/s in 5–10 yr.

4. Dynamics

Can we reconcile the morphology of the Bulge with the kinematics? We have seen in §§2.1, 2.2 and 3.1 that the integrated light, the starcounts, and the kinematics of the gas all seem to point to an elongated Bulge, and that the various studies generally agree on its properties, even though differences in detail remain. What about the stellar motions? This is a difficult question to answer, because many different intrinsic velocity distributions may be consistent with the same mass model, so that it is not easy to constrain the shape, the density profile and the entire intrinsic velocity distribution in the Bulge all at once just from the measured stellar kinematics. For this reason the usual approach is to first specify a potential and a mass distribution, and then to calculate the observed kinematics of the stellar component for various velocity distribution functions f. The result of this *dynamical modeling* is then compared to the observations. If no model velocity distribution can be found that fits the data, the potential and/or mass model was chosen incorrectly. If there are solutions, then this does not prove that the model is correct, but only shows that it is at least consistent with the available data, and allows a certain range of possible distribution functions. Whether the inferred distribution functions are in fact plausible is then a matter for theories of bulge formation.

To date, dynamical modeling of the Bulge has been restricted to spherical or axisymmetric geometries, and has made use of the Jeans equations, i.e., considered only the first and second moments of the velocity distribution function f. The most comprehensive model that is available is due to Kent (1992). His mass model for the Bulge fits the 2.4 micron integrated light distribution (Fig. 1) for a constant $\mathcal{M}/L$. The associated rotation curve agrees with the observed gas kinematics (§3.1). Kent assumes that the distribution function f is of the special form $f = f(E, L_z)$, where E is the orbital energy, and L_z is the component of angular momentum around the short axis of the Bulge, taken as the z–axis. This guarantees that the second velocity moments $\langle v_R^2 \rangle$ and $\langle v_z^2 \rangle$ in the R and z directions, respectively, are equal everywhere. They follow from the Jeans equations, together with the remaining second moment $\langle v_\phi^2 \rangle$. By symmetry, there can be no mean streaming in the R and

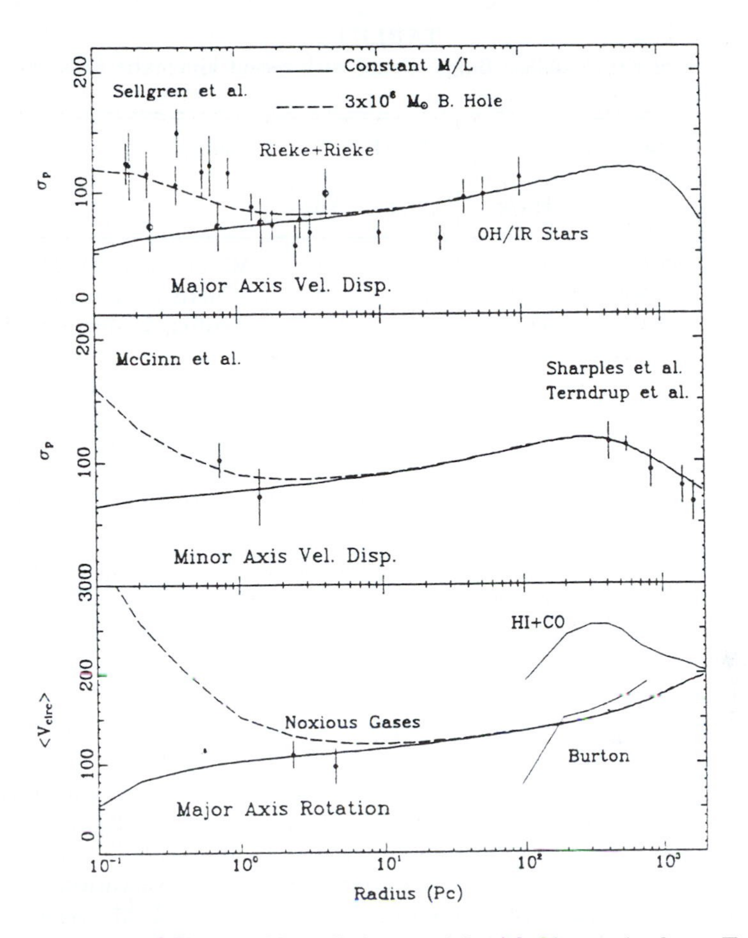

Fig. 3. Comparison of Kent's oblate Bulge model with kinematic data. Top & middle panels give $\sigma_{\rm los}$ along the major and the minor axis, respectively. Bottom panel gives v_c along the major axis. Solid lines are for constant $\mathcal{M}/L$, and dashed lines indicate a model with a central black hole of $3 \times 10^6 M_\odot$. See text and Kent (1992) for more details.

z directions, so the velocity dispersions σ_R and σ_z are equal everywhere. Kent now *chooses* the mean streaming velocity in the ϕ–direction in such a way that $\langle v_\phi^2 \rangle - \langle v_\phi \rangle^2 = \langle v_R^2 \rangle = \langle v_z^2 \rangle$. This makes all three velocity dispersions equal at every point. Such a model is often referred to as an *oblate isotropic rotator.*

Figure 3 shows $\sigma_{\rm los}$ along the major and the minor axis of Kent's model, as well as the circular velocity v_c as function of radius. The solid lines are for a model with constant $\mathcal{M}/L$. The curves rise proportional to $r^{0.075}$ in the inner Bulge, as expected for a density profile that decreases as $r^{-1.85}$. Specifically, this means an increase by a factor of 1.4 in $\sigma_{\rm los}$ between 5 and 500 pc. Beyond the knee in the density profile (Fig. 1) $\sigma_{\rm los}$ decreases, as expected, while v_c continues to increase until it reaches ~200 km/s at 2 kpc from the center. The dashed lines are the expected curves when a central black hole of $3 \times 10^6 M_\odot$ is included. The data

TABLE I
Comparison of Kent's oblate Bulge model with recent kinematic observations.

ℓ °	b °	Observed σ_{los} km/s	Observed $\langle v_{los} \rangle$ km/s	Model σ_{los} km/s	Model $\langle v_{los} \rangle$ km/s	Source
8	7	85 ± 7	45 ± 10	91	42	Minniti *et al.* (1992)
12	3	68 ± 6	77 ± 9	65	90	Minniti *et al.* (1992)
–10	–10	67 ± 6	82 ± 8	77	47	Harding & Morrison (1993)

points are from Kent's compilation of recent observations, and are well–fit by the model. The value of $\mathcal{M}/L$ has been set by requiring that the model reproduces the observed velocity dispersion in Baade's Window. This gives $\mathcal{M}/L_K = 1.0 \pm 0.15$ in solar units.

Any oblate model with $f = f(E, L_z)$ predicts $\sigma_b = \sigma_{los}$ along the minor axis, as a consequence of the fact that $\sigma_R = \sigma_z$ everywhere. This can be tested for Kent's model by comparing it with the proper motions in Baade's Window. For the distance to the Galactic Center of 8 kpc favored by Kent, the Spaenhauer, Jones & Whitford (1992) measurements are: $(\sigma_\ell, \sigma_b) = (119 \pm 4, 104 \pm 4)$ km/s, while $\sigma_{los} = 113 \pm 6$ km/s (Mould 1983; Sharples, Walker & Cropper 1990; Rich 1990; Tyson & Rich 1991). Kent (*priv. comm.*) predicts $(\sigma_\ell, \sigma_b, \sigma_{los}) = (120, 113, 113)$ km/s for his model, which agrees remarkably well with the observations. A further test of the model is provided by the radial velocities obtained by Minniti *et al.* (1992) and by Harding & Morrison (1993). Kent kindly made available his unpublished predictions for these fields. They are compared with the observations in Table 1. The agreement is excellent, except for $\langle v_{los} \rangle$ in the Harding & Morrison field. It is not clear whether this is caused by an inaccurate foreground correction, has to do with the definition of the sample, or is a first hint that the Bulge is not an oblate isotropic rotator.

Kent's model is the simplest oblate Bulge model that can be constructed, and yet it fits essentially all the available data. This is quite remarkable, not only because many other velocity distributions are possible in principle, but also in view of the strong indications that the Bulge is not oblate but triaxial (§§2 and 3).

At the same time, the observed stellar kinematics do constrain the shape, the density profile, and the velocity distribution of the Bulge: not every model fits the available data. To illustrate this, we consider a simple scale–free model for the inner Bulge with a potential $\psi \propto 1/(R^2 + z^2/q^2)^{\alpha/2}$, where q and α are constants, and we ignore the putative central black hole. The associated density profile is a power law with logarithmic slope $-(2 + \alpha)$. The even part of the self–consistent distribution function $f(E, L_z)$ of such models is given by $f(E, L_z) = AL_z^2 E^{4/\alpha - 3/2} + CE^{2/\alpha - 1/2}$, with A and C constants (Evans 1993b). The limiting case $\alpha \to 0$ has a logarithmic potential and is discussed by Toomre (1982) and Evans (1993a). We note that none

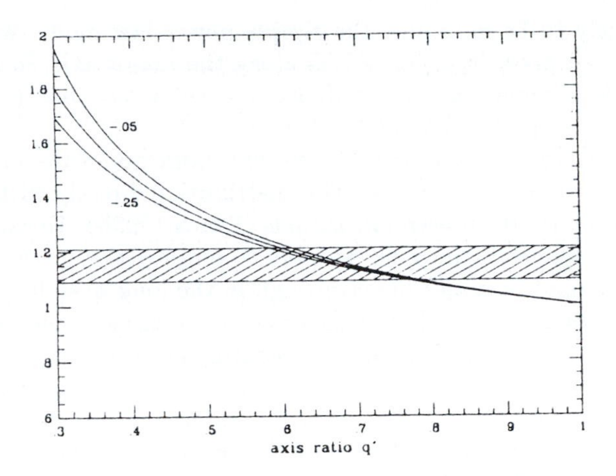

Fig. 4. Relation between σ_ℓ/σ_b on the minor axis and the observed axis ratio q' of the simple power law models for the inner Bulge, seen edge–on. The curves correspond to density profiles $\rho \propto 1/r^{2+\alpha}$, with $\alpha = -0.05$, -0.15, and -0.25. The hatched area indicates the range of σ_ℓ/σ_b compatible with the measurements in Baade's Window.

of these models have Gaussian velocity distributions, because f is not a simple exponential of the energy. The value of q determines the flattening of the models; they become increasingly peanut–shaped with decreasing q. For all these models the observed second moments of the line–of–sight velocities and the proper motions can be given explicitly at any point. The same is true for the first moments of the oblate isotropic rotator along the principal axes. The full expressions will be given elsewhere. Here we limit ourselves to one example, and show in Figure 4 the observed value of σ_ℓ/σ_b along the minor axis of these models as function of the observed flattening q', for various values of α. For edge–on observation the relation between σ_ℓ/σ_b and q' is given by:

$$\left(\frac{\sigma_\ell}{\sigma_b}\right)^2 = \frac{q^2+1-2\alpha}{3q^2-1-2\alpha}, \qquad q' = q\left(\frac{q^2-\alpha}{1-\alpha q^2}\right)^{1/(1+\alpha)}.$$

We conclude that the value $\alpha = -0.15$ used by Kent (1992), or the value $\alpha = -0.25$ advocated by Binney *et al.* (1991), can only be reconciled with $\sigma_\ell/\sigma_b = 1.15 \pm 0.06$ in Baade's Window for q' between 0.6 and 0.75, in good agreement with the observations (§2.1). This by no means proves that the Bulge is oblate with $f = f(E, L_z)$, but illustrates that meaningful modeling of the intrinsic velocity distribution requires accurate knowledge of the shape of the Bulge. The figure furthermore shows that using simple models with a $1/r^2$ density profile is adequate for some purposes, but not when they are restricted to be spherical.

Kent's model is similar in spirit to the oblate models for elliptical galaxies constructed by Binney, Davies & Illingworth (1990) and van der Marel (1991). These authors restrict themselves to $f = f(E, L_z)$ but, following Satoh (1980), they consider a variety of mean streaming motions by taking $\langle v_\phi \rangle^2 = k^2(\langle v_\phi^2 \rangle - \langle v_R^2 \rangle)$, with k a constant. The oblate isotropic rotator has $k = 1$. In the inner Bulge, the value of k is likely to be near one: the simple power law model with $\alpha = -0.15$, $q' \sim 0.65$ and $k = 1$ gives $\langle v_{\rm los} \rangle / v_c \sim 0.35$ along the major axis, in agreement with the few available observations and with Kent's statement that $\langle v_{\rm los} \rangle \sim 40$ km/s throughout the inner part of his model (Fig. 3).

Instead of working with the first and second moments of the observed velocity distribution, one would like to compare this distribution directly with the model prediction. This is easy for the power–law models (Evans 1993b). To calculate $f(E, L_z)$ for more realistic models it may be possible to employ the method by Hunter & Qian (1993), who made a major breakthrough in the long–standing problem of the practical calculation of $f(E, L_z)$ for axisymmetric systems. Their work transforms the calculation of $f(E, L_z)$ from an often frustrating chase through tables of integral transforms and special functions, to the straightforward (numerical) evaluation of f as a well–defined contour integral. It will be interesting to see whether this can be used to obtain f for a model like the one proposed by Kent.

One disadvantage of using velocities of individual stars to study the dynamics of the Bulge is the need for many measurements before a velocity distribution can be determined—or even just its first and second moments. For this reason it is valuable to redo and extend the spectroscopic measurements of the integrated light (Freeman *et al.* 1988), preferably in the calcium triplet region, or in the infrared (McGinn *et al.* 1989). Various methods now exist to extract not just $\langle v_{\rm los} \rangle$ and $\sigma_{\rm los}$ but also the entire line–of–sight velocity distribution from such data (e.g., Rix & White 1992; van der Marel & Franx 1993).

Before considering truly anisotropic models in which $\langle v_R^2 \rangle \neq \langle v_z^2 \rangle$ and $\langle v_R v_z \rangle \neq 0$, it is useful to investigate simple models with an embedded disk. This can be done following the approach of Evans & Collett (1993), who give two–integral distribution functions for exponential disks in a logarithmic halo potential, or by using the Jeans equations (Cinzano & van der Marel 1993). The resulting models will help to elucidate the properties of the population of fast–rotating OH/IR stars in the inner 100 pc (Lindqvist *et al.* 1992b). A general method to test whether an edge–on oblate model has $f = f(E, L_z)$ is described by Merrifield (1991). Its application will require further kinematic observations, however.

Anisotropic oblate models have $f = f(E, L_z, I_3)$ where I_3 is a(n approximate) third integral of motion. Selfconsistent calculation of such f's is sometimes possible semi–analytically (Binney & Petrou 1985), but generally requires numerical techniques such as linear programming (Schwarzschild 1979; Richstone 1980, 1984; Levison & Richstone 1985a, b; Fillmore & Levison 1989). More direct methods are available for the solution of the Jeans equations when $\langle v_R^2 \rangle \neq \langle v_z^2 \rangle$ and $\langle v_R v_z \rangle \neq 0$ (Bacon 1985; Fillmore 1986), and these should be applied to the Bulge. Kent's (1992) simple estimate based on spherical models already shows that the observations in Baade's Window will provide strong constraints on the allowed anisotropy of the velocity distribution in the Bulge.

Te Lintel–Hekkert, Dejonghe & Habing (1991) have taken a slightly different approach. They choose a potential, define certain smooth components with simple distribution functions f, calculate the observed properties for each of these separately, and then use a quadratic programming method (Dejonghe 1989) to determine which combination, if any, reproduces the observations. Application to the

OH/IR stars in the Bulge (excluding the Lindqvist *et al.* 1992a sample) again shows that—so far—$f(E, L_z)$ components in an oblate potential are consistent with the observations (Dejonghe 1993b). This approach has considerable promise. The effects of an anisotropic velocity distribution and of a triaxial shape can be incorporated by considering models with separable potentials (cf. Dejonghe & Laurent 1991). Unfortunately, self-consistent models of this kind of necessity have a finite density core rather than a central cusp such as observed in the Bulge, and also must have stationary figures. Separable models with cusps can be built non-consistently (Dejonghe 1993a), but so far there is little evidence that $\mathcal{M}/L$ varies in the Bulge, at least outside 3 pc. Furthermore, the triaxial Bulge may have a figure rotation rate of 63 km/s/kpc (Binney *et al.* 1991). Although stationary scale-free triaxial models of the kind constructed by Levison & Richstone (1987) might be useful for the innermost part of the Bulge (but see Schwarzschild 1993), models for the entire Bulge must include non-zero figure rotation, a density profile which is not a power law, and must allow for the possibility of a separate flat bar in a triaxial Bulge. Construction of such models is a non-trivial project, which will require substantial numerical effort.

Dynamical modeling of the Bulge should be accompanied by careful N-body simulations, in order to investigate its evolution and interaction with other Galactic components (Hernquist & Weinberg 1992). Studies of this kind will allow an exploration of various formation scenarios: they have shown already that an initially flat bar in a disk galaxy may grow fatter in time, become box-shaped, and either form a bulge, or a metal-rich component in a pre-existing bulge (Combes *et al.* 1990; Pfenniger & Norman 1990; Raha *et al.* 1991; Pfenniger & Friedli 1991). The effects of (the growth of) a central point mass can be investigated also.

The success of the oblate $f(E, L_z)$ models shows that it is not easy to detect the signature of the triaxiality of the Bulge in the available stellar kinematic data, and hence to constrain more sophisticated dynamical models and simulations. One of the first hints may be the observation of high velocity OH/IR stars near the Galactic Center. Van Langevelde *et al.* (1992b) show that the observed number is consistent with these stars being on the elongated orbits needed to support a triaxial Bulge, but only if the Bulge is seen nearly end-on, which is precisely the geometry favored by Binney *et al.* (1991). Various other tests for triaxiality have been proposed. These include searching for stars in the solar neighborhood that are on very elongated orbits which bring them close to the Galactic Center. The expected number of such stars is influenced by the shape of the Bulge (Spergel 1992). Along similar lines, the present distribution of orbital elements of the globular clusters may still contain evidence for enhanced destruction of such clusters in the past due to dynamical friction in a triaxial potential (Long, Ostriker & Aguilar 1992). Another signature of triaxiality would be the detection of a gradient in $\langle v_{\rm los} \rangle$ along $\ell = 0°$. This could be due to the motion of stars on the long axis tube orbits expected in a triaxial Bulge, but could also be caused by a tipping of the Bulge with respect to the Galactic plane (Spergel 1992). There clearly is room for a lot more work on the dynamics of the Bulge.

5. Other Bulges

Various lines of evidence suggest that spiral bulges as a class are not oblate. The position angles of the apparent major axis of the bulge and the disk of spiral galaxies often differ from each other. This is a natural consequence of triaxiality, and is caused by projection. A well-known example is the bulge of M31 (Stark 1977). Bertola, Vietri & Zeilinger (1991) studied a sample of 32 bulges, and showed that if disks are round, then bulges as a class are indeed triaxial, and have shapes similar to elliptical galaxies. The derived distribution of shapes may be incorrect, however, as photometrically the disks of spirals are not round, but instead are slightly elongated, with an axis ratio close to 0.9 (Binney & de Vaucouleurs 1981; Kuijken & Tremaine 1991, 1993; Franx & de Zeeuw 1992). Derivation of the intrinsic shapes of bulges will require inclusion of kinematic data, just as was done for elliptical galaxies (Binney 1985; Franx, Illingworth & de Zeeuw 1991).

Individual bulges also show signs of triaxiality. The regular gas velocity field of NGC 4845 is well-fit by motion on elongated closed orbits in a triaxial bulge, with axis ratios $b/a = 0.74 \pm 0.06$ and $c/a = 0.60 \pm 0.06$ (Bertola, Rubin & Zeilinger 1989; Gerhard, Vietri & Kent 1989). Stellar absorption line measurements of bulges are consistent with rotationally supported axisymmetric models, when the disk potential is taken into account (Jarvis & Freeman 1985; Rowley 1988; Kent 1989). The one known exception is the curious galaxy NGC 4550, which has two counter-rotating stellar disks, and a stationary bulge (Rix *et al.* 1992). The data are consistent also with triaxial shapes with substantial internal streaming, and/or figure rotation. The various indications that the Galaxy contains a triaxial bulge, even though the stellar kinematics so far appears to be well-described by $f(E, L_z)$ axisymmetric models, are therefore fully in line with what we know about other bulges.

6. Conclusions

Observations of the integrated light, starcounts, and measurements of the kinematics of the atomic and molecular gas in the inner region of the Galaxy all indicate that the Galactic Bulge is triaxial, with its near side at positive longitude, and its long axis close to the line-of-sight to the Galactic Center. In the inner regions the density profile is a power law with logarithmic slope -1.8 ± 0.05, which steepens to -3.7 ± 0.2 beyond ~400 pc along the minor axis. The observed axis ratio in the direction perpendicular to the Galactic plane is 0.65 ± 0.05. The COBE observations, and the starcounts to be done with the DENIS and 2MASS surveys, will further delineate the shape and orientation of the Bulge. Studies of populations of variable stars for which accurate individual distances can be determined will also be very useful for this purpose. Improved modeling of the gas kinematics will require detailed hydrodynamical simulations, and should provide better constraints on the figure rotation and the elongation of the Bulge.

The consequences of triaxiality for the dynamics of the Bulge remain largely unexplored, for two reasons. First, radial velocities are available for a modest number of Bulge stars only, and are mostly restricted to certain windows. So far, the data are consistent with the simple oblate $f = f(E, L_z)$ model of Kent (1992), which is well-approximated by a simple power law model in the inner Bulge. Second, construction of anisotropic triaxial models with realistic density profiles and non-zero figure rotation is difficult and time-consuming.

This unsatisfactory situation should improve in the near future. New radial velocity surveys, such as the unbiased radio survey of OH/IR stars in the entire Bulge, and especially the work on proper motions, will provide a superior stellar kinematic dataset for the Bulge, even though it remains non-trivial to correct for foreground contamination by the disk. These observational programs should also clarify the nature of the sample of fast rotating OH/IR stars seen in the inner 100 pc, and in particular whether the Bulge contains an extended disk component.

On the theoretical front one should construct velocity distribution functions for the triaxial mass model that best fits the gas kinematics, the starcounts and the integrated light measurements, and then investigate which of these are preferred by the different populations of stars in the Bulge. This will require considerable numerical effort, but should help to constrain different formation scenarios for the Bulge, such as formation by direct gaseous infall from the halo, or by thickening of the disk. It is crucial to complement this dynamical modeling with careful N-body simulations which incorporate a realistic disk, halo and Bulge. This will allow a study of the evolution of e.g., an initially flat bar into a triaxial Bulge. Comparison with the observed correlations between kinematics and abundances will result in a much improved understanding of the formation history of the Bulge.

Acknowledgements

It is a pleasure to acknowledge enlightening discussions with Marijn Franx and Huib-Jan van Langevelde, and especially with Wyn Evans and Steve Kent, who also generously gave permission to quote some of their unpublished results.

References

Acker, A., Köppen, J., Stenholm, B., & Raytchev, B., 1991. A&AS, **89**, 237.
Bacon, R., 1985. A&A, **143**, 84.
Becklin, E.E., & Neugebauer, G., 1968. ApJ, **151**, 145.
Bertola, F., Rubin, V.C., & Zeilinger, W.W., 1989. ApJ, **345**, L29.
Bertola, F., Vietri, M., & Zeilinger, W.W., 1991. ApJ, **374**, L13.
Binney, J.J., 1985. MNRAS, **212**, 767.
Binney, J.J., Davies, R.L., & Illingworth, G.D., 1990. ApJ, **361**, 78.
Binney, J.J., & de Vaucouleurs, G., 1981. MNRAS, **194**, 679.
Binney, J.J., Gerhard, O.E., Stark, A.A., Bally, J., & Uchida, K.I., 1991. MNRAS, **252**, 210.
Binney, J.J., & Petrou, M., 1985. MNRAS, **214**, 449.
Blitz, L., & Spergel, D.N., 1991a. ApJ, **370**, 205.
Blitz, L., & Spergel, D.N., 1991b. ApJ, **379**, 631.
Burton, W.B., & Liszt, H.S., 1993. A&A, in press.
Carney, B.W., Latham, D.W., & Laird, J.B., 1990. AJ, **99**, 572.
Cinzano, P.-A., & van der Marel, R.P., 1993. In *Structure, Dynamics, and Chemical Evolution of Early-Type Galaxies*, eds I.J. Danziger, W.W. Zeilinger, & K. Kjär (ESO Garching), p. 105.
Combes, F., 1991. ARAA, **29**, 195.
Combes, F., Debbash, F., Friedli, D., & Pfenniger, D., 1990. A&A, **233**, 82.
Dejonghe, H.B., 1989. ApJ, **343**, 113.
Dejonghe, H.B., 1993a. In *Structure, Dynamics, and Chemical Evolution of Early-Type Galaxies*, eds I.J. Danziger, W.W. Zeilinger, & K. Kjär (ESO/EIPC, Garching), p. 337.

Dejonghe, H.B., 1993b. In *IAU Symposium 155, Planetary Nebulae*, eds R. Weinberger, A. Acker (Kluwer, Dordrecht), in press.
Dejonghe, H.B., & Laurent, D., 1991. MNRAS, **252**, 606.
de Vaucouleurs, G., 1964. In *IAU Symposium 20, The Galaxy and the Magellanic Clouds*, eds F.J. Kerr & A.W. Rodgers (Sydney: Australian Academy of Science), p. 195.
Epchtein, N., Guglielmo, F., & Burton, W.B., 1992. In *IAU Symposium 149, The Stellar Populations of Galaxies*, eds. B. Barbuy & A. Renzini (Kluwer, Dordrecht), p. 414.
Evans, N.W., 1993a. MNRAS, **260**, 191.
Evans, N.W., 1993b. In preparation.
Evans, N.W., & Collett, J.L., 1993. MNRAS, in press.
Fillmore, J.A., 1986. AJ, **91**, 1096.
Fillmore, J.A., & Levison, H.F., 1989. AJ, **97**, 57.
Franx, M., Illingworth, G.D., & de Zeeuw, P.T., 1991. ApJ, **383**, 112.
Franx, M., & de Zeeuw, P.T., 1992. ApJL, **392**, L47.
Freeman, K.C., 1987. ARAA, **25**, 603.
Freeman, K.C., de Vaucouleurs, G., de Vaucouleurs, A., & Wainscoat, R.J., 1988. ApJ, **325**, 563.
Frogel, J.A., 1988. ARAA, **26**, 51.
Gerhard, O.E., & Vietri, M., 1986. MNRAS, **223**, 377.
Gerhard, O.E., Vietri, M., & Kent, S.M., 1989. ApJ, **345**, L33.
Gilmore, G., King, I.R., & van der Kruit, P.C., 1990. *The Milky Way as a Galaxy*, 19th Saas Fee Advanced Course, eds R. Buser & I.R. King (Mill Valley: Univ. Science Books).
Glass, I., 1993. This volume.
Habing, H.J., 1988. A&A, **200**, 40.
Habing, H.J., Olnon, F.M., Winnberg, A., Matthews, H.E., & Baud, B., 1983. A&A, **128**, 230.
Habing, H.J., Olnon, F.M., Chester, T., Gillett, F., Rowan-Robinson, M., & Neugebauer, G., 1985. A&A, **152**, L1.
Harding, P., & Morrison, H., 1993. This volume.
Harmon, R., & Gilmore, G., 1988. MNRAS, **235**, 1025.
Hauser, M.G., *et al.*, 1990. NASA photograph G90–03046.
Hernquist, L., & Weinberg, M.D., 1992. ApJ, **400**, 80.
Hunter, C., & Qian, E., 1993. MNRAS, in press.
Jarvis, B.J., & Freeman, K.C., 1985. ApJ, **295**, 324.
Jenkins, A., & Binney, J.J., 1993. MNRAS, in press.
Kent, S.M., 1989. AJ, **97**, 1614.
Kent, S.M., 1992. ApJ, **387**, 181.
Kent, S.M., Dame, T.M., & Fazio, G., 1991. ApJ, **378**, 131.
Kinman, T.D., Feast, M.W., & Lasker, B.M., 1988. AJ, **95**, 804.
Kleinman, S.G., 1992. In *Robotic Telescopes in the 1990s*, ASP Conference Series No. 34, ed. A.V. Filippenko, p. 203.
Kuijken, K., & Tremaine, S.D., 1991. In *Dynamics of Disk Galaxies*, ed. B. Sundelius (Göteborg, Sweden), p. 71.
Kuijken, K., & Tremaine, S.D., 1993. ApJ, in press.
Lee, Y.-W., 1992. In *IAU Symposium 149, The Stellar Populations of Galaxies*, eds. B. Barbuy & A. Renzini (Kluwer, Dordrecht), p. 446.
Le Poole, R.S., & Habing, H.J., 1990. In Proceedings of the ESO/CTIO Workshop on *Bulges of Galaxies*, eds B.J. Jarvis & D.M. Terndrup, p. 33.
Levison, H.F., & Richstone, D.O., 1985a. ApJ, **295**, 340.
Levison, H.F., & Richstone, D.O., 1985b. ApJ, **295**, 349.
Levison, H.F., & Richstone, D.O., 1987. ApJ, **314**, 476.
Lewis, J.R., & Freeman, K.C., 1989, AJ, **97**, 139.
Lindqvist, M., Habing, H.J., Winnberg, A., & Matthews, H.E., 1992a. A&AS, **92**, 43.
Lindqvist, M., Habing, H.J., & Winnberg, A, 1992b. A&A, **259**, 118.
Liszt, H.S., 1992. In *The Center, Bulge, and Disk of the Milky Way*, ed. L. Blitz, (Dordrecht: Kluwer), p. 111.
Liszt, H.S., & Burton, W.B., 1980. ApJ, **236**, 779.
Long, K., Ostriker, J.P., & Aguilar, L., 1992. ApJ, **388**, 362.
Matsumoto, T., Hayakawa, S., Koizumi, H., & Murakawa, H., 1982. In *The Galactic Centre*, AIP Conf. 83, eds G.R. Riegler & R.D. Blandford (New York: Am. Inst. of Physics), p. 48.

McGinn, M.T., Sellgren, K., Becklin, E.E., & Hall, D.N.B., 1989. ApJ, **338**, 824.
Menzies, J.W., 1990. In Proceedings of the ESO/CTIO Workshop on *Bulges of Galaxies*, eds B.J. Jarvis & D.M. Terndrup, p. 115.
Merrifield, M.R., 1991. AJ, **102**, 1335.
Minniti, D., White, S.D.M., Olszewski, E.W., & Hill, J.M., 1992. ApJ **393**, L47.
Mould, J.R., 1983. ApJ, **266**, 255.
Mulder, W.A., & Liem, B.T., 1986. A&A, **157**, 148.
Nakada, Y., Deguchi, S., Hashimoto, O., Izumiura, H., Onaka, T., Sekiguchi, K., & Yamamura, I., 1991. Nature, **353**, 140.
Nakada, Y., Onaka, T., Yamamura, I., Deguchi, D., Ukita, N., & Izumiura, H., 1993. PASJ, **45**, in press.
Pfenniger, D., & Norman, C., 1990. ApJ, **363**, 391.
Pfenniger, D., & Friedli, D., 1991. A&A, **252**, 75.
Raha, N., Sellwood, J.A., James, R.A., & Kahn, F.D., 1991. Nature, **352**, 411.
Rich, R.M., 1990. ApJ, **362**, 604.
Richstone, D.O., 1980. ApJ, **238**, 103.
Richstone, D.O., 1984. ApJ, **281**, 100.
Rieke, G.H., & Rieke, M.J., 1988. ApJL, **330**, L33.
Rix, H.-W., & White, S.D.M., 1992. MNRAS, **254**, 389.
Rix, H.-W., Franx, M., Fisher, D., & Illingworth, G.D., 1992. ApJL, **400**, L5.
Rowley, G., 1988. ApJ, **331**, 124.
Rybicki, G.R., 1987. In *IAU Symposium 127, Structure and Dynamics of Elliptical Galaxies*, ed. P.T. de Zeeuw (Dordrecht: Reidel), p. 397.
Satoh, C., 1980. PASJ, **32**, 41.
Schwarzschild, M., 1979. ApJ, **232**, 236.
Schwarzschild, M., 1993. ApJ, in press.
Sellgren, K., McGinn, M.T., Becklin, E.E., & Hall, D.N.B., 1990. ApJ, **359**, 112.
Sellwood, J.A., & Sanders, R.H., 1988. MNRAS, **233**, 611.
Sharples, R., Walker, A., & Cropper, M., 1990. MNRAS, **246**, 54.
Spaenhauer, A., Jones, B.F., & Whitford, A.E., 1992. AJ, **103**, 297.
Spergel, D.N., 1992. In *The Center, Bulge, and Disk of the Milky Way*, ed. L. Blitz (Dordrecht: Kluwer), p. 77.
Stark, A.A., 1977. ApJ, **213**, 368.
Stasińska, G. 1993. This volume.
te Lintel–Hekkert, P., Dejonghe, H.B., & Habing, H.J., 1991. Proc. Astron. Soc. Austr., **9**, 20.
te Lintel–Hekkert, P., Versteege–Hensel, H.A., Habing, H.J., & Wiertz, M., 1989. A&AS, **78**, 399.
te Lintel–Hekkert, P., Caswell, J.L., Habing, H.J., Norris, R.P., & Haynes, R.F., 1991. A&AS, **90**, 327.
Toomre, A., 1982. ApJ, **259**, 535.
Tyson, N.D., & Rich, R.M., 1991. ApJ, **362**, 547.
van Albada, G.D., 1985. In *IAU Symposium 106, The Milky Way Galaxy*, eds H. van Woerden, R.J. Allen, & W.B. Burton (Dordrecht: Reidel), p. 547.
van der Marel, R.P., 1991. MNRAS, **253**, 710.
van der Marel, R.P., & Franx, M., 1993. ApJ, in press.
van Langevelde, H.J., 1992. *PhD Thesis*, Rijksuniversiteit Leiden.
van Langevelde, H.J., Frail, D.A., Cordes, J.M., & Diamond, P.J., 1992a. ApJ, **396**, 686.
van Langevelde, H.J., Brown, A.G.A., Lindqvist, M., Habing, H.J., & de Zeeuw, P.T., 1992b. A&AL, **261**, L17.
Weinberg, M.D., 1992a. ApJ, **384**, 81.
Weinberg, M.D., 1992b. ApJL, **392**, L67.
Wesselink, Th., 1987. *PhD Thesis*, Nijmegen.
Whitelock, P.A., Feast, M.W., & Catchpole, R.M., 1991. MNRAS, **248**, 276.
Whitelock, P.A., & Catchpole, R.M., 1992. In *The Center, Bulge, and Disk of the Milky Way*, ed. L. Blitz, (Dordrecht: Kluwer), p. 103.
Winnberg, A., Baud, B., Matthews, H.E., Habing, H.J., & Olnon, F.M., 1985. ApJL, **291**, L45.
Yuan, C., 1984. ApJ, **281**, 600.

DISCUSSION

Habing: Firstly, it is dangerous to immediately identify regression lines with a rotation curve, because there is also the distribution of the stars along the line of sight to be taken into account. Secondly, the IRAS data processing was a highly complex process and definitely non-linear at points. So, IRAS data in confused regions like the galactic plane, are beset by selection effects. Therefore an analysis, like that by Weinberg, can only be done by somebody with a relatively naive mind. Which may be an advantage sometimes, so I don't disagree with that approach, but I think it should not be believed until it has been confirmed by a completely independent study.

de Zeeuw: I agree, as you saw from the viewgraph.

Gerhard: Perhaps part of the confusion comes from trying to fit all observations into one coherent bulge picture. It may turn out that the disk itself has some interesting structure, in the inner kpc, which may be different from the structure of the Bulge. So it may not be the best to combine measurements from close to the plane and those well above the plane.

de Zeeuw: I agree, but all the matter moves in the same potential, so it would be nice to make one coherent picture.

Whitelock: Comparing the Lindqvist OH/IR sample to that of te Lintel, which extends over the whole Bulge, it should be realised that the te Lintel sample is really somewhat biased towards the near side of the Bulge, due to sensitivity limitations in the Parkes telescope. I think that the Lindqvist et al sample does not have this bias.

de Zeeuw: Yes, but I don't think that particular bias can explain the very different rotational kinematics between the two samples. But we should keep it in mind.

KINEMATICS OF EXTRAGALACTIC BULGES: EVIDENCE THAT SOME BULGES ARE REALLY DISKS

JOHN KORMENDY*

Institute for Astronomy, University of Hawaii,
2680 Woodlawn Dr., Honolulu, HI 96822, USA

Abstract. Recent work on the dynamics of galaxy bulges has been dominated by two themes. (1) Bulges share the richness in kinematic structure that is currently being discovered in elliptical galaxies. This includes kinematic evidence for triaxiality and for accretion (counterrotating gas and stellar components). (2) The main subject of this paper is observational and theoretical evidence that some "bulges" are built secularly out of disk material. Many bulges show photometric and kinematic evidence for disklike dynamics. This includes (i) velocity dispersions σ much smaller than those predicted by the Faber-Jackson $\sigma - M_B$ correlation, (ii) rapid rotation $V(r)$ that implies V/σ values well above the "oblate line" describing rotationally flattened, isotropic spheroids in the V/σ – ellipticity diagram, and (iii) spiral structure dominating the $r^{1/4}$ part of the galaxy. In these galaxies, the steep, $r^{1/4}$-law central brightness profiles belong not to bulges but to disks. That is, some galaxy disks have central brightness profiles that are much steeper than the inward extrapolation of an exponential fit to the outer parts. These observations and n-body simulations of gas flow in nonaxisymmetric galaxies imply that high-central-concentration, flat components can be formed out of disk gas that is transported toward the center by bars and oval distortions. The n-body models suggest further that some "bulges" are built of disk stars heated in the axial direction by resonant scattering off of bars. These effects are signs that important secular evolution processes are at work in galaxy disks.

Key words: Galaxy Bulges – Galaxy Disks – Stellar Dynamics – Secular Evolution

1. Introduction

Observational and theoretical work on bulge dynamics is thriving; increasingly powerful tools provide us with much more information than we had a decade ago. Inevitably, bulge dynamics look more and more complicated. But two simple themes unify this work. One is well known: we see kinematic signatures of triaxiality and of galaxy accretion. This story is familiar from work on elliptical galaxies; I will review it only briefly (§ 2). The second theme is not well known; it will be the focus of this paper. A substantial body of evidence shows that some "bulges" are really disks: they have the steep, $r^{1/4}$-law brightness profiles that we associate with bulges, but they also have the "cold" (rotation-dominated) dynamics of disks. Numerical simulations suggest that they were built by various processes of secular evolution in disks, including inward gas transport by nonaxisymmetries in the potential and vertical heating by resonant scattering of stellar orbits off of bars. The importance of secular evolution driven by interactions with collective phenomena was emphasized by Kormendy (1982a, and references therein); §§ 3 – 7 bring these reviews up to date. Just as mergers showed that galaxies evolve not in isolation but in part through interactions, so the evidence for secular dynamical processes implies that internal evolution is more than the aging of stellar populations: the interactions of components can significantly change galaxy structure over a Hubble time.

* Visiting Astronomer at the Canada-France-Hawaii Telescope (CFHT), operated by the National Research Council of Canada, the Centre National de la Recherche Scientifique of France, and the University of Hawaii.

H. Dejonghe and H. J. Habing (eds.), Galactic Bulges, 209–228.

2. Stellar Dynamics of Bulges: Comparison With Elliptical Galaxies

2.1. The Dynamical Importance of Rotation

The "zeropoint" of this subject is well known (for reviews see Illingworth 1981; Kormendy 1982a; Binney 1982; Davies 1989; Kormendy and Djorgovski 1989, and de Zeeuw and Franx 1991). Work on elliptical galaxies was revolutionized by the discovery (Bertola and Capaccioli 1975; Illingworth 1977) that most bright ellipticals do not rotate significantly. Therefore their dynamics are controlled by velocity anisotropy, which implies that they are triaxial (Binney 1976, 1978a, b). Many observational signatures of triaxiality have been found. In contrast, bulges rotate essentially as rapidly as models of oblate spheroids that are isotropic and hence flattened only by rotation (e. g., Illingworth and Schechter 1982; Kormendy and Illingworth 1982; Kormendy 1982b); isotropic models are an excellent fit to the data (Jarvis and Freeman 1985a, b; Kent 1989). However, Davies *et al.* (1983) point out that most bulges are less luminous than the nonrotating ellipticals; low-luminosity ellipticals also rotate. It is therefore not clear whether bulges and ellipticals differ. The Sombrero Galaxy contains one of the few well-studied bulges that is as luminous as a typical nonrotating elliptical; it is an isotropic rotator, but so are a few bright ellipticals. Even now, too few high-luminosity bulges have been measured. We do not know whether virtually all bulges rotate rapidly or whether bulges and ellipticals show the same dependence of rotation on luminosity.

One complication (Kormendy and Djorgovski 1989) is that ellipticals can grow disks by accretion. If they originally did not rotate, they still will not rotate after becoming "bulges". They add noise to any intrinsic correlation between bulge rotation and luminosity.

Therefore we know of only modest dynamical differences between bulges and elliptical galaxies. Bulges lie nearly in the fundamental plane correlations for ellipticals (e. g., Kormendy 1985; Bender, Burstein, and Faber 1992). But there are some photometric differences (Hamabe and Kormendy 1987). Therefore a more quantitative comparison of bulges and elliptical galaxies may be profitable.

2.2. Kinematic (Sub)structure in Bulges and Elliptical Galaxies

Recent work on ellipticals has been dominated by continued attempts to measure the degree of velocity anisotropy and by discoveries of signatures of accretion. The latter include kinematically decoupled (misaligned) gaseous and stellar components. Reviews are given in Kormendy and Djorgovski (1989) and in de Zeeuw and Franx (1991). *This emerging richness in kinematic structure is shared by galaxy bulges.*

Kinematic evidence for accretion in disk galaxies includes the following. Gas that counterrotates with respect to the stars has been detected in NGC 4546 (Galletta 1987) and in NGC 2768, NGC 4379, and IC 4889 (Bertola, Buson, and Zeilinger 1992). We also have one spectacular case of an edge-on galaxy containing two stellar disks that counterrotate (NGC 4550: Rubin, Graham, and Kenney 1992; Rix *et al.* 1992). Bulges are less thoroughly studied than ellipticals; the early discovery of kinematically decoupled components in so many objects argues that bulge formation histories are as rich and complicated as those of elliptical galaxies.

Ellipticals are triaxial; are bulges triaxial, too? It is fashionable to suspect that they are. But *this is not an argument by analogy.* In ellipticals, triaxiality follows from the *unimportance* of rotation, which implies that velocity dispersion anisotropy is needed to account for the flattening. Once we realized that σ_z is smaller than the other two components, we had no reason to expect that $\sigma_r = \sigma_\phi$, either. But bulges rotate rapidly enough to be isotropic. If they are triaxial, this is not because they are like nonrotating ellipticals but rather because they are like bars.

What do observations tell us? It is too early to tell. In one galaxy, NGC 4845, kinematic evidence for noncircular streaming motions has been found and interpreted as a sign of triaxiality (Bertola, Rubin, and Zeilinger 1989; Gerhard, Vietri, and Kent 1989). From slit spectra at five position angles, Bertola *et al.* (1989) conclude that the rotation velocity at $7'' - 10''$ radius is smaller along the major axis than elsewhere in the bulge. A slight twist between bulge and disk isophotes is also seen. In the absence of complications, these observations imply that the bulge is triaxial with principal axes $a : b : c \simeq 1 : 0.75 : 0.5$. The above papers agree reasonably well on the implied ranges of b/a and c/a. However, there are significant uncertainties. NGC 4845 is almost edge-on (inclination $i \geq 75°$). This means that the major axis (PA = 78°) and neighboring (PA = 44° and 98°) slits are separated by only $2\overset{''}{.}8$ at $8''$ radius. How sure can we be that the velocities really differ when the deprojection corrections are factors of 2.9 and 1.8 for the two neighboring slits? Also, the galaxy is dusty; the brightness distribution clearly shows dust at radii of $\sim 5''$. Do we really see to the same depth along the line of sight at all three slit positions? Clearly this is a difficult object. Further such work is needed.

Bertola *et al.* (1989) note that NGC 4845 has a peanut-shaped bulge. Such bulges rotate particularly rapidly (§ 3); they may be related to or even formed by bars (§ 6). Peanut-shaped bulges are particularly likely to be triaxial.

3. Cylindrical Rotation in Box-Shaped Bulges

Kormendy and Illingworth (1982) found that in the box-shaped bulge of NGC 4565, rotation velocities remained almost constant with increasing height z above the disk plane to $z \simeq 30'' \simeq 2.7h^{-1}$ kpc (Hubble constant $H_0 = 50h$ km s^{-1} Mpc^{-1}). In contrast, rotation velocity decreased rapidly with increasing z in three galaxies with elliptical or disky-distorted bulge isophotes.

Cylindrical rotation is seen in all box-shaped bulges that have been observed. This includes the prototype, NGC 128 (Jarvis 1990; Bertola and Capaccioli 1977), which contains one of the most peanut-shaped bulges known (Sandage 1961; Jarvis 1990). In it, $V(z)$ is constant up to $z = 20'' = 8.5h^{-1}$ kpc. Three other box-shaped bulges have been measured and show cylindrical rotation, NGC 3079 (Shaw, Wilkinson, and Carter 1992), and NGC 1381 and NGC 7332 (Davies, Illingworth, and Kormendy 1993). Thus boxy bulges are particularly rapid rotators.

They also are more than just a curiosity. In a survey of all large, normal edge-on spiral and lenticular galaxies in the *Second Reference Catalog* (de Vaucouleurs, de Vaucouleurs, and Corwin 1976), Shaw (1987) found boxy bulges in $20 \pm 4\%$ of the objects. Therefore it is important to understand their origin.

Two mechanisms have been suggested. Most relevant to this paper is the suggestion that bars manufacture boxy bulges by vertical heating of the disk through resonant scattering of stellar orbits by the bar (see § 6). An alternative mechanism based on accretion has been suggested by Binney and Petrou (1985), Jarvis (1987), Whitmore and Bell (1988), Hernquist and Quinn (1988), and Statler (1988). If a bulge or elliptical galaxy accretes a dynamically cold object at an oblique angle, then differential precession will phase-mix the orbits until they produce a boxy or X-shaped structure embedded in the bulge. This must happen. But in general we cannot tell whether a particular boxy bulge originated in this or some other way. An accretion origin has been proposed for the boxy distortions in IC 4767 (Whitmore and Bell 1988) and IC 3370 (Jarvis 1987). IC 3370 is an elliptical; it clearly shows cylindrical rotation.

Therefore it is likely that at least two different processes make boxy bulges. Since boxy ellipticals generally do not rotate significantly, they are made in still a third way. Only a few, mostly low-luminosity boxy ellipticals appear related to boxy bulges. Three mechanisms of origin are an inconvenience. However, it is easy to identify anisotropic ellipticals by their slow rotation, and boxy bulges can be built by a bar only if there is a substantial disk. Therefore the critical question is: how common are boxy distortions made by accretion?

4. Some Bulges Are Really Disks

It is not generally realized that many galaxies contain central components that look like bulges but that have disk-like dynamics. This section reviews the evidence. A remarkably large number of papers at this meeting address this subject; it is clearly an idea whose time has come. As in previous papers (Kormendy 1982a, b; Kormendy and Illingworth 1983), I suggest that high-density disks are formed from ordinary disk gas that has been concentrated toward the center and turned into stars.

4.1. NGC 4736

The prototypical bulgelike disk is in the Sb galaxy NGC 4736 (Kormendy 1982a). It is illustrated in the *Hubble Atlas* (Sandage 1961). The central brightness profile (Fig. 1) is an $r^{1/4}$ law that reaches the high central brightness characteristic of a bulge (Boroson 1981). However, the $r^{1/4}$-law component shows a nuclear bar and spiral structure to within a few arcsec of the center (Fig. 2). Bars are disk phenomena. More importantly, it is not possible to make spiral structure in a bulge. Thus the morphology already shows that the $r^{1/4}$-law profile belongs to the disk. This is shown more quantitatively by the well-known $V_{\rm max}/\sigma - \epsilon$ diagram (Illingworth 1977; $V_{\rm max}$ = maximum rotation velocity; σ = mean velocity dispersion near the center; ϵ = ellipticity). Figure 3 shows that the "bulge" of NGC 4736 has an unusually large ratio of ordered to random velocities (the data are from Pellet and Simien 1982). It is well above the "oblate line" describing oblate spheroids with isotropic velocity distributions. Disks observed edge-on are near the oblate line; $\epsilon \gtrsim 0.8$ and $V/\sigma \gtrsim 2$. Seen more nearly face-on, they project to positions above the oblate line. Kormendy (1982a) therefore concluded that most of the light near the center is coming from a high-surface-brightness disk.

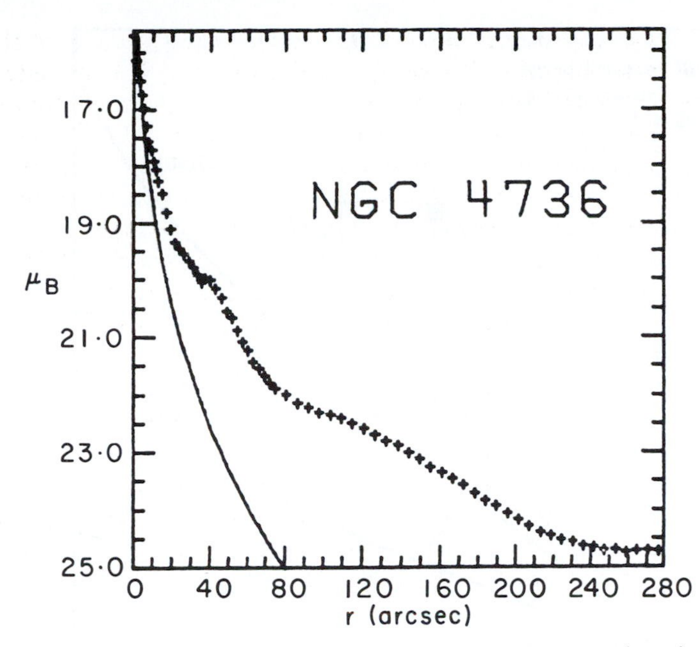

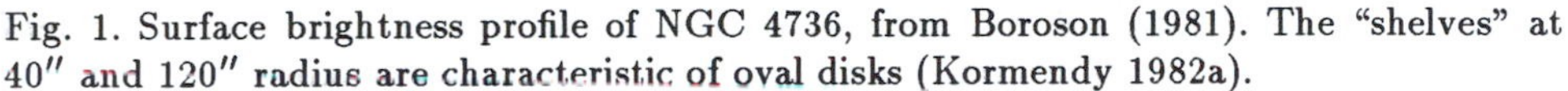
Fig. 1. Surface brightness profile of NGC 4736, from Boroson (1981). The "shelves" at 40″ and 120″ radius are characteristic of oval disks (Kormendy 1982a).

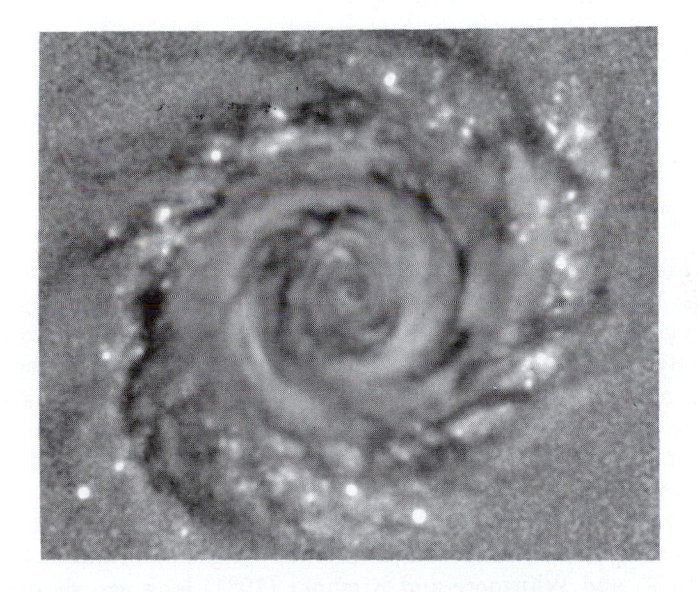

Fig. 2. CFHT image of NGC 4736 (106″ high). The radial brightness gradient has been removed: the image has been divided by a mask image with the brightness profile of the galaxy but exactly elliptical isophotes. The nuclear bar is elliptical and therefore also removed; it can be recognized by the spiral structure and dust morphology near the center.

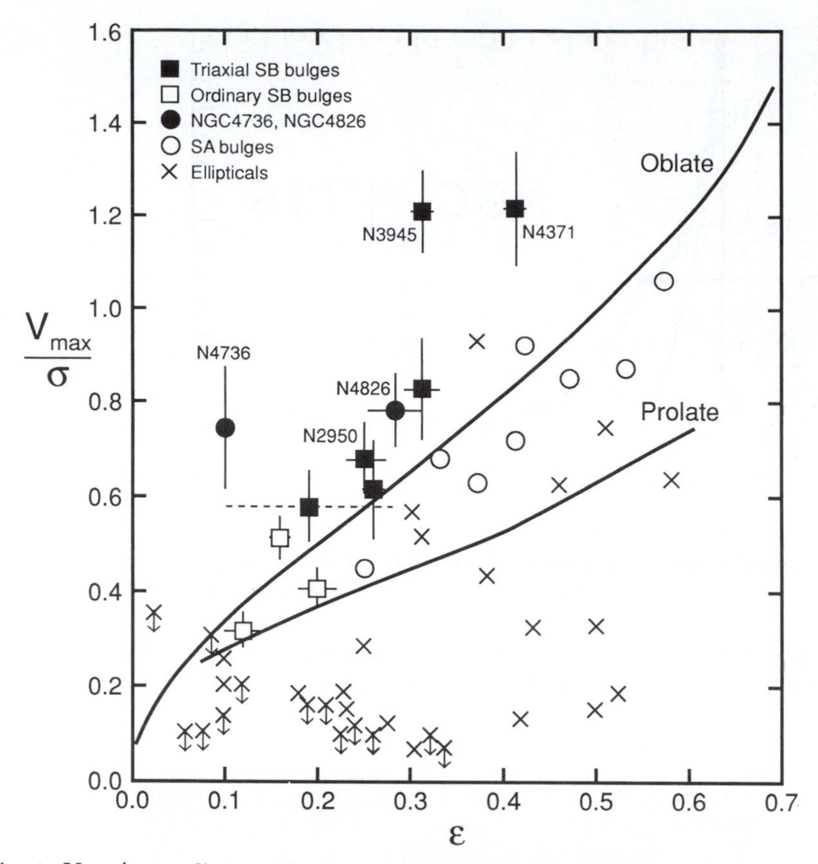

Fig. 3. $V_{\max}/\sigma - \epsilon$ diagram for various kinds of stellar systems (cf. Kormendy 1982a).

4.2. NGC 4826

NGC 4826 (Fig. 4) is another prototypical example (Kormendy 1993). Sandage (1961) calls it the earliest-type Sb galaxy in the *Hubble Atlas*; normally such objects contain a bulge that resembles an elliptical galaxy. Although partially obscured by dust, the central brightness profile is approximately an $r^{1/4}$ law. Also, the central brightness is normal for a bulge and much higher than in typical disks. But NGC 4826 does not have the velocity dispersion of a bulge, as would have been implied by the published $\sigma = 160 \pm 16$ km s^{-1} (see Whitmore, McElroy, and Tonry 1985). In fact, the central velocity dispersion is very low, $\sigma = 90 \pm 5$ km s^{-1} (Fig. 5).

Figure 6 shows the Faber-Jackson (1976) correlation between σ and luminosity. NGC 4826 is well below the scatter for normal bulges. Whitmore, Kirshner, and Schechter (1979) and Whitmore and Kirshner (1981) long ago showed that some bulges have smaller dispersions than ellipticals of the same M_B. Kormendy and Illingworth (1983) found that most of these are in barred galaxies. There were two prominent examples among unbarred galaxies, NGC 1172 (E/S0) and NGC 7457 (S0, see § 4.3). NGC 4826 is very like these.

Fig. 4. A V-band image of NGC 4826 taken with the CFHT. The scale is $0\farcs21$ pixel^{-1}; this panel is $210''$ wide. The Gaussian seeing dispersion radius is $\sigma_* = 0\farcs40$. Brightness and contrast have been adjusted to illustrate the dust disk; the central part of the bulge is saturated in this print. Compare the excellent photograph in the *Hubble Atlas*.

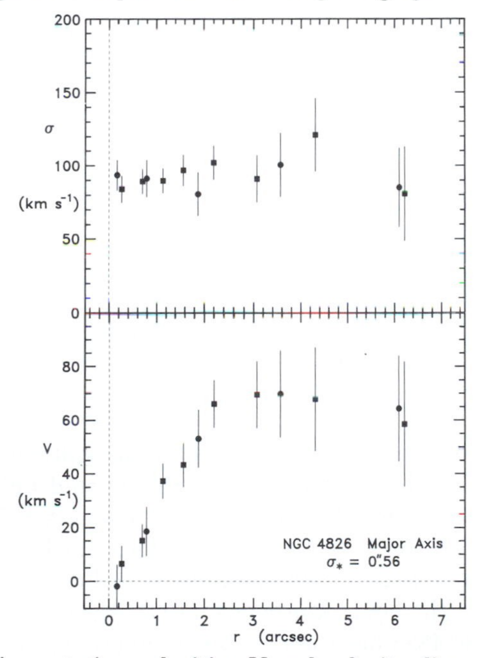

Fig. 5. Absorption-line rotation velocities V and velocity dispersions σ along the major axis of NGC 4826 (Kormendy 1993).

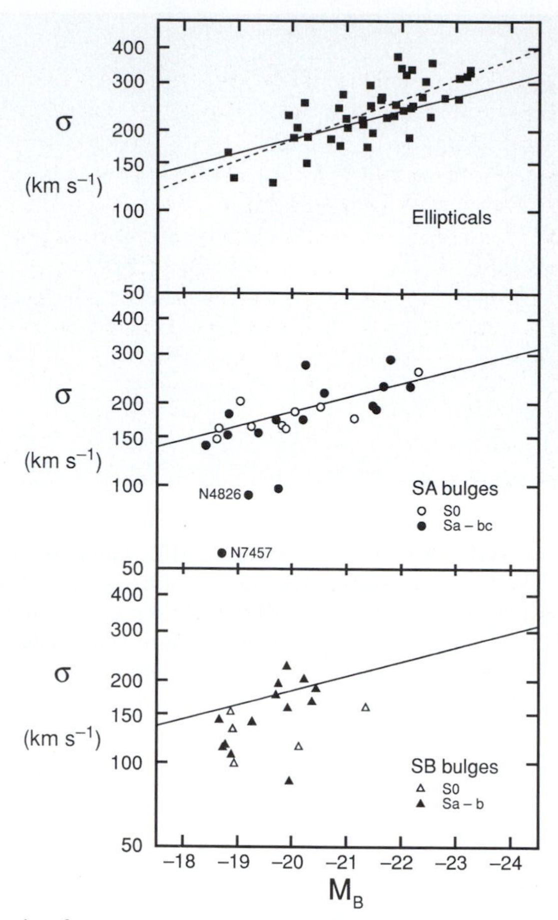

Fig. 6. Correlation between central velocity dispersion σ and absolute magnitude M_B for elliptical galaxies and for bulges of unbarred (SA) and barred (SB) disk galaxies. The solid line is a fit to the galaxies in the middle panel; the dashed line is a fit to the ellipticals. Except for the NGC 4826 point, this figure is from Kormendy and Illingworth (1983).

A small velocity dispersion is characteristic of disks. Kormendy and Illingworth (1983) and Kormendy (1982b) interpreted abnormally cold bulges as disk-like. A more definitive conclusion is provided by the $V/\sigma - \epsilon$ diagram (Fig. 3). Like NGC 4736, NGC 4826 is above the oblate line. Therefore much of the steep central brightness profile is coming from a cold component. A bulge may also contribute, but it does not dominate the light. Kormendy (1993) therefore concludes that the central disk light in NGC 4826 has the $r^{1/4}$-law brightness profile of a bulge.

Kormendy (1982b) found that many "bulges" of barred galaxies also are well above the oblate line in the $V/\sigma - \epsilon$ diagram. In particular, NGC 3945 and NGC 4371 are as dominated by rotation as NGC 4736. In all of these objects, the small σ (Fig. 6, bottom) and large V/σ (Fig. 3) show that the central components that we thought were bulges are really largely disk light.

4.3. NGC 7457

As a final example of a "bulge" that is really a disk, consider NGC 7457. This is a normal, unbarred S0 (*Hubble Atlas*) dominated by an exponential disk (Kormendy 1977). The "bulge" is faint, fractionally and in absolute luminosity ($M_B \simeq -18.5$). *Hubble Space Telescope* observations by Lauer *et al.* (1991) show that it has a steep brightness profile, a very high central surface brightness ($\mu_{0V} \lesssim 12.4$ V mag arcsec^{-2}), and an unresolved core. The limits on the core parameters are extreme, but they are in the range expected for such a low luminosity (Fig. 7). This "bulge" is enormously different from a normal disk; these typically have $\mu_{0V} \simeq 21$ V mag arcsec^{-2} (Freeman 1970). The rotation curve has not been measured well enough to allow us to plot the galaxy in the $V/\sigma - \epsilon$ diagram. But $\sigma = 65$ km s^{-1}, making this the coldest "bulge" in Fig. 6.

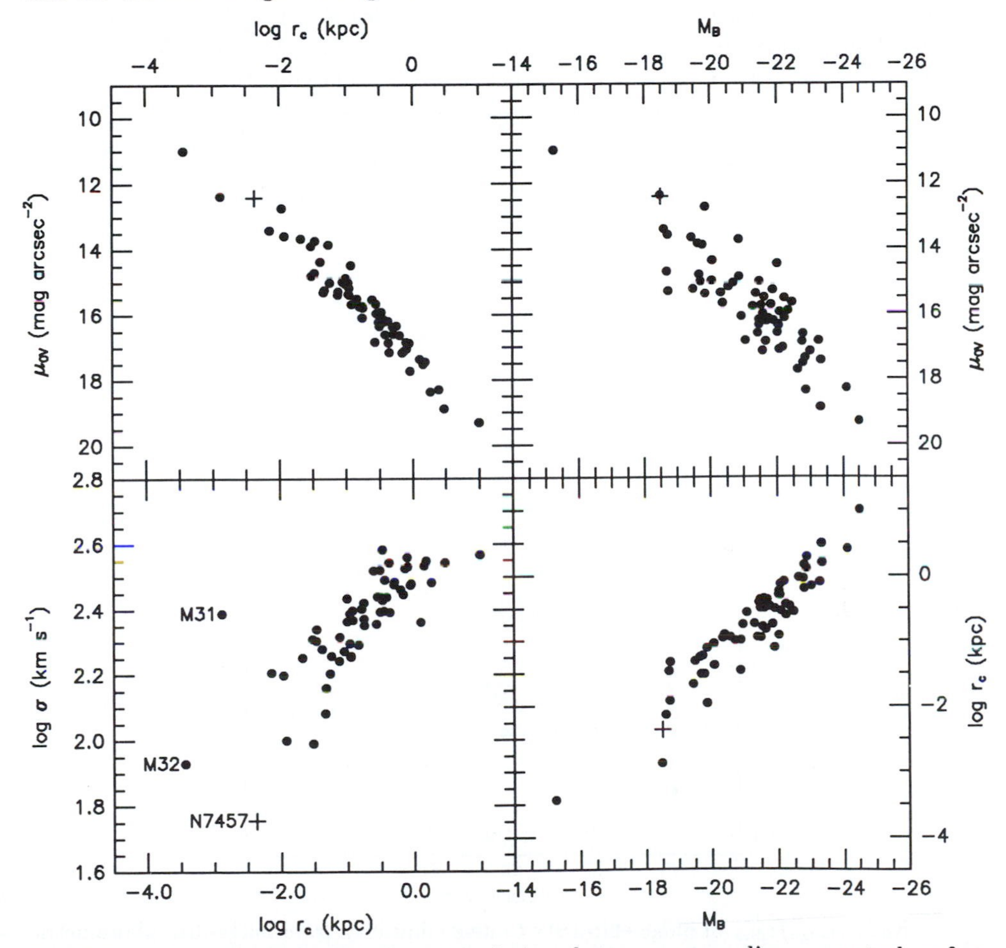

Fig. 7. Four of six fundamental plane correlations between core radius r_c, central surface brightness μ_{0V}, central velocity dispersion σ, and M_B. Approximate seeing corrections are from Kormendy (1987). NGC 7457 (Lauer *et al.* 1991) is shown by plus signs.

4.4. Many "Bulges" Look As Flat As Disks

If some "bulges" are really disks, then this should be evident in the distribution of bulge ellipticities. Figure 8 shows this effect in bulge-disk decompositions computed by Kent (1985, 1987, 1988). I use only the decompositions in which the ellipticity of the bulge was a free parameter. Also, I include only objects with disk ellipticities $\epsilon_{\rm disk} \geq 0.14$, since face-on objects have no leverage on the problem. Bulge-disk decompositions are uncertain, so interpretation should be cautious. However, in agreement with Kent, I conclude that many bulges look as flattened as their associated disks. Some look more flattened; these may be triaxial.

The median ratio $\epsilon_{\rm bulge}/\epsilon_{\rm disk}$ is smallest for Sa galaxies and increases toward later Hubble types. This agrees with other evidence which suggests that disklike "bulges" are more common at later Hubble types. Finally, it is interesting that the median $\epsilon_{\rm bulge}/\epsilon_{\rm disk}$ for S0 galaxies is similar to that for Scs, not Sas. Kinematically disklike bulges also are more common in S0s than in Sas. Similar effects led van den Bergh (1976b) to develop his "parallel sequence" classification.

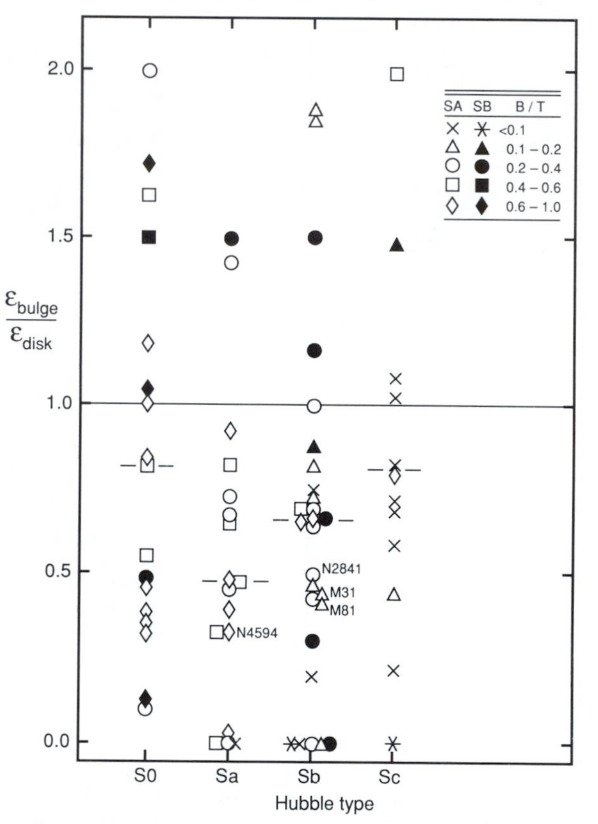

Fig. 8. Ratio $\epsilon_{\rm bulge}/\epsilon_{\rm disk}$ of bulge ellipticity to disk ellipticity, from bulge-disk photometric decompositions by Kent (1985, 1987, 1988). Different symbols encode bulge-to-total luminosity ratio B/T and the distinction between barred and unbarred galaxies (see key). Horizontal tics are drawn at the median $\epsilon_{\rm bulge}/\epsilon_{\rm disk}$ for each Hubble type.

4.5. Nuclear Disks

"Nuclei" are central star clusters that are dynamically distinct from the bulge (see Kormendy and Djorgovski 1989 for a review). *Some nuclei are disks.* A particularly clear example is NGC 4594 (the Sombrero Galaxy): the nuclear isophotes are very flattened (Burkhead 1986, 1991; Kormendy 1988b), and even the observed spectrum, which is a composite of the bulge and nucleus, implies a velocity dispersion 181 ± 6 km s^{-1} at $r \simeq 3''.7$ that is smaller than the velocity dispersion 240 ± 4 km s^{-1} of the bulge (Kormendy 1988b; Jarvis and Dubath 1988). Another possible example is the nucleus of M 31: after subtraction of the superposed bulge spectrum, this is colder at $r \gtrsim 2''$ than the bulge (Kormendy 1988a, but contrast Dressler and Richstone 1988). Good spatial resolution is required to see a kinematic signature. More commonly, we recognize nuclear disks only through their effects on isophote shapes. Disky distortions in the central few arcsec of many ellipticals suggest that nuclear disks are common (Nieto *et al.* 1991; Scorza 1993). They are another example of high-density disk material near galaxy centers.

4.6. Triaxial Disklike "Bulges" In Barred Galaxies

In the discussion of the $V_{\max}/\sigma - \epsilon$ diagram, I noted that bulges of barred galaxies tend to be more dominated by rotation than bulges of unbarred galaxies. That is, they are more disk-like. Also, it has been known for a long time that many of them are triaxial. Examples include the SB0 galaxies NGC 1291 and NGC 1543 (de Vaucouleurs 1975; Jarvis *et al.* 1988) and NGC 2950 (Kormendy 1981). In all of these, the isophotes clearly show nuclear bars that have position angles different from the main bar and from the outer disk. Other examples are discussed in Kormendy (1979b) and in Buta (1986a, b, 1990). Sometimes the inner bar is a nucleus distinct from the bulge, sometimes the whole "bulge" in an SB galaxy is triaxial. Large $V_{\max}/\sigma$ values imply that these exceptionally triaxial "bulges" are dynamically like bars (Kormendy 1983; Kent and Glaudell 1989; Kent 1990) and not like triaxial giant ellipticals.

4.7. "Bulges" Made Of Population I Material

A suggestive clue to the origin of bulgelike disks is provided by their stellar content. Many of them contain or even are dominated by Population I material. For example, SB galaxies frequently contain nuclear hot spots of young stars and gas. Well known examples include NGC 1097 (Hummel, van der Hulst, and Keel 1987), NGC 4314 (*Hubble Atlas*), and NGC 4321 (Arsenault *et al.* 1988); general references are Sérsic and Pastoriza (1965), Alloin and Kunth (1979), and Buta and Crocker (1991). The same is true of "bulges" in some oval galaxies, including the prototype NGC 4736. Bulges are most likely to contain substantial Population I material at low luminosities and late Hubble types (Bica and Alloin 1987; Frogel 1992).

Even S0 bulges can contain molecular gas and star formation (e.g., at this meeting: NGC 4710, Wrobel and Kenney 1992, 1993; see many papers in Combes and Casoli 1991, but especially Sofue 1991). Also, a modest fraction contain older starbursts that give them A-type integrated spectra (e.g., Gallagher, Faber, and

Balick 1975; Burstein 1979; Sparke, Kormendy, and Spinrad 1980; Sil'chenko 1993), although some of the young stars may result from accretion events. And central dust disks are very common in early-type galaxies, both S0s ("$S0_3$" objects in the *Hubble Atlas*) and ellipticals (see Kormendy and Djorgovski 1989 for a review).

Young stars are not always present: in early-type galaxies, $r^{1/4}$-law disks are usually made of old stars. However, the above observations show that when gas is present, it knows how to find the center and it likes to make stars there.

4.8. Conclusions

This section leads to two observational conclusions. First (Fig. 9): Besides true exponential disks and Freeman (1970) Type II exponentials, *some disks have steeper density profiles near the center than the inward extrapolation of an exponential fitted at large r. Disks can even have $r^{1/4}$-law central brightness profiles, in which case they are indistinguishable from bulges on the basis of density or density gradient alone.* They can often be recognized by V_{max}/σ values that are larger than normal for isotropic spheroids of the observed ϵ (however, see § 6). I do not mean to imply that there is no bulge at all in all of these galaxies; at least at early Hubble types, the high-density disk material has probably been added to a preexisting bulge. But in extreme cases, the disk dominates the projected density.

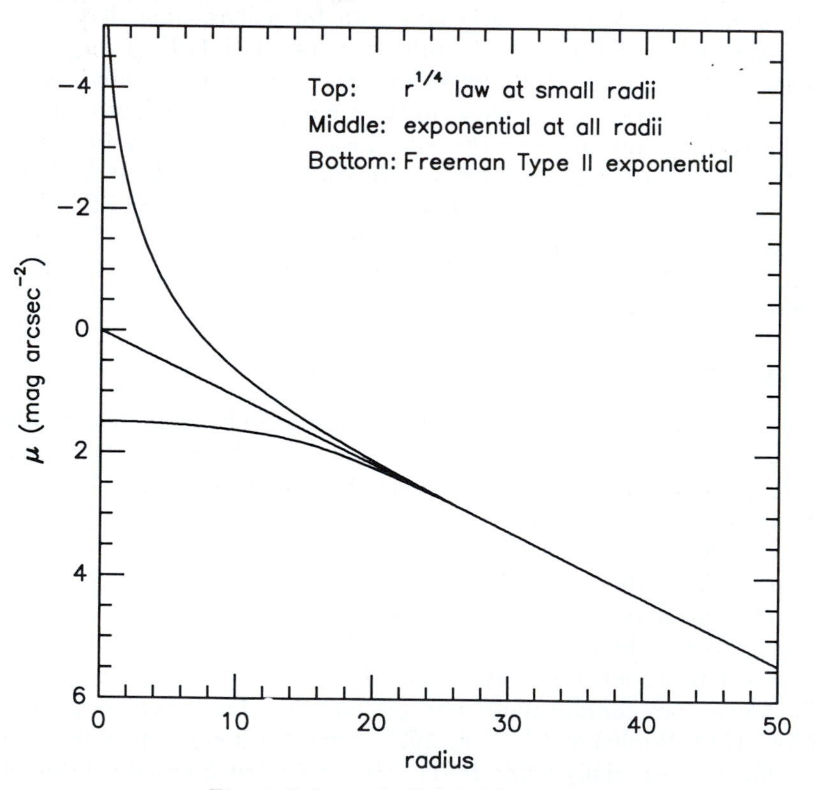

Fig. 9. Schematic disk brightness profiles.

Second (Fig. 10): If we want B/T to measure the true bulge-to-total luminosity ratio, i.e., the fraction of the total light that is contributed by an ellipsoidal component that is more-or-less like an elliptical galaxy, then *the true distribution of bulge-to-total luminosity ratios is skewed toward smaller values than those we derive from a blind decomposition of luminosity distributions into $r^{1/4}$-law and exponential parts.* At all Hubble types, this effect is most common in barred and oval galaxies. Otherwise, it is smallest at early Hubble types, where most bulges are like ellipticals (except in some S0s). At type Sb, there are already some galaxies in which most of the "bulge" is really disk material, although others (e.g., M 31 and M 81) contain true bulges. By type Sc, I do not believe that any galaxies contain true bulges.

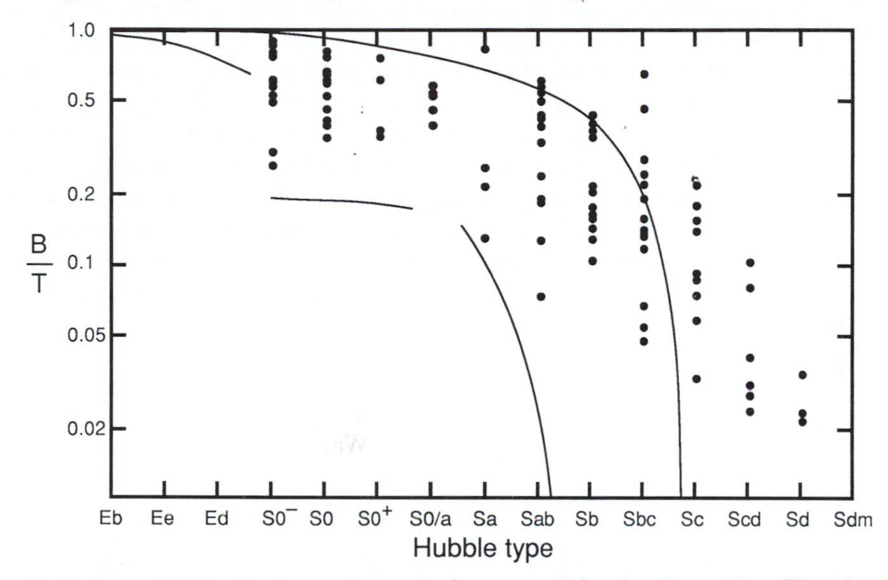

Fig. 10. Schematic distributions of true bulge-to-total luminosity ratios. This figure is adapted from Fig. 2 in Simien and de Vaucouleurs (1986); Eb, Ee, and Ed refer to ellipticals with boxy, elliptical, and disky isophotes, respectively. The points are values measured by decomposing profiles into $r^{1/4}$-law and exponential parts. The curves estimate how the boundaries of the B/T distributions change if we include disk-like "bulges" with disks. More accurate B/T distributions should ultimately be based on dynamical data.

5. Building "Bulges" By Secular Evolution. I. Making Bulgelike Disks By Bar-Driven Inward Gas Transport

A unifying hypothesis for the origin of bulgelike disks emerges from discussions by a number of authors (e.g., van den Bergh 1976a; Kormendy 1982a, b, 1988b; Gallagher, Goad, and Mould 1982; O'Connell 1983; Kormendy and Illingworth 1983; Kormendy and Djorgovski 1989; Kormendy and McClure 1993). The suggestion is that *the central concentration of galaxy disks can be increased dramatically by inward gas transport and subsequent star formation. Bars and oval disks are particularly efficient engines for this process* (see Kormendy 1982a; Prendergast 1983; Combes 1991 for reviews).

This idea also is a natural consequence of the hypothesis that nuclear activity in galaxies results when black holes are fueled by infalling gas. If gas can reach the black hole, it may form stars along the way when the density gets high enough in the gravitational funnel. This may even be a necessary step in the formation of nuclear black holes, since core collapse times in giant ellipticals are long, while nuclei can evolve more rapidly (Kormendy 1988c).

Shlosman and Begelman (1987, 1989) and Shlosman, Frank, and Begelman (1989) take this idea one step further and suggest that hierarchical bar formation through inward gas transport also solves the problem of how to get fuel to small radii where the main bar no longer affects the gravitational potential. They suggest that a bar transports gas inward to radii much smaller than its length; at this point the central concentration has increased enough to make the gas self-gravitating; it becomes unstable to the formation of a new and shorter bar, and the process repeats itself with residual gas. It is not clear that all of this actually happens, nor is it clear that this is the dominant transport process (contrast Gunn 1979). But it is interesting to note that the nuclear bars discussed in the previous section are predicted by the Shlosman and Begelman mechanism. *Space Telescope* observations should tell us whether there exist additional levels of bar-within-bar hierarchy.

There has recently been a resurgence of theoretical work on the building of nuclear disks and bulges by secular processes. For example, Hasan and Norman (1990), Norman and Hasan (1990), and Hasan, Pfenniger, and Norman (1993) point out that bars may be destroyed when the central mass concentration is increased sufficiently. Related papers include Duschl (1988a, b) and Pfenniger and Norman (1990). Many papers at this meeting also discuss aspects of this picture (Gerhard 1993; Hasan and Norman 1993; Sellwood 1993; Wada and Habe 1993).

6. Building "Bulges" By Secular Evolution. II. Making Box-Shaped Bulges By Bar-Driven Vertical Disk Heating

The previous section suggests that bars increase the central mass concentration of disks, but it does not explain how to make bulges that are thick in the axial direction. This section reviews a heating process that may allow bars to manufacture box-shaped bulges out of disks. This has clearly captured the interest of the n-body modeling community; it has been discussed repeatedly at this meeting.

The idea is this: at radii where the vertical oscillation frequency ν_z of disk stars is in resonance with the pattern speed Ω_B of the bar, the vertical motions are amplified by the bar. The most important such resonance is vertical inner Lindblad resonance (ILR), $\Omega_B = \Omega - \nu_z/2$ (Ω is the angular velocity of rotation about the center). At vertical ILR, the z oscillations look periodic to the bar, so perturbations accumulate quickly. The effect is analogous to azimuthal disk heating at planar ILR by bars and spiral arms. The result is that the disk thickens. Many authors have developed this picture; they show that the result is a rapidly rotating box-shaped "bulge" (Combes and Sanders 1981; Pfenniger 1984, 1985; Combes *et al.* 1990; Pfenniger and Norman 1990; Friedli and Pfenniger 1990; Pfenniger and Friedli 1991; Raha *et al.* 1991). A number of papers at this meeting further develop the theme (Friedli and Udry 1993; Hasan and Norman 1993; Pfenniger 1993; Sellwood 1993).

The process is attractive because it is specific (we can simulate it rigorously) and because it seems to work. The heating must happen. It also agrees in important ways with the observations: the resulting box-shaped bulges rotate cylindrically. But we do not yet know whether this process is sufficient in practice or whether it is the main process making boxy bulges. I am especially concerned about the extreme view that boxy bulges are bars seen side-on:

(i) Edge-on bars are very flat, and in some cases they coexist with a boxy bulge. One example is NGC 4762 (*Hubble Atlas*; Wakamatsu and Hamabe 1984).

(ii) If boxy bulges are side-on bars, then the longest major axes of boxy bulges should equal the lengths of bars in face-on galaxies.

(iii) The distributions of luminosity and B/T ratio for boxy bulges should equal the analogous distributions for bars in face-on galaxies.

Predictions (ii) and (iii) can be checked observationally. I suspect that all three points will be problems for the idea that boxy bulges are side-on bars. But I know of no observation that conflicts with the idea that boxy bulges are manufactured by vertical heating of the centrally concentrated disks discussed in §§ 4 and 5.

7. Conclusion: Our Galaxy

The main conclusion of this paper is that some "bulges" are disk-like in their dynamics and origin. That is, disks can have central brightness profiles that are much steeper than exponentials; in extreme cases, they are as well fitted by $r^{1/4}$ laws as are ellipticals. Then true bulge-to-total light ratios are smaller than we think. One consequence is that we should be cautious in interpreting evidence for bulge triaxiality; we may be seeing disk effects.

Although the details are far from clear, we believe that all this results from one or more processes of secular dynamical evolution. Bars and oval disks are particularly likely engines, but dissipation and gas infall can happen without them, and a variety of processes may operate. The importance of secular evolution by the interaction of galaxy components has been emphasized by Kormendy (1979a, b, 1981, 1982a).

The above results are particularly relevant to this meeting because our Galaxy has probably been affected. It is an Sbc (de Vaucouleurs and Pence 1978), so it is unlikely to contain a pristine, pure bulge (Fig. 10). It is barred. The beautiful COBE photograph of the Galaxy (Mather *et al.* 1990; see the Bahcall Committee report for a color version) shows that its bulge is box-shaped. Massive molecular clouds live near the center. We should be on the lookout for young stars. The age of the bulge has been the most controversial subject discussed at this meeting; surely the stars at large z are old, but does the bulge contain young stars near the disk plane? *If external galaxies are any guide, it is likely that our Galaxy has had its bulge augmented by high-density disk material and that active star formation continues near the center.* Were this not so, our Galaxy would be quite unusual. When we discuss the stellar population, the dynamics, and the evolution of the Galactic bulge, we should ask: *Are some effects that we see properties not of a true bulge but rather of a high central concentration of disk material that may have been heated by instabilities and resonance effects?* I believe that it will be important to look for the effects discussed in §§ 4 – 6 in our own front yard.

Acknowledgements

This talk was prepared at the Landessternwarte Königstuhl, Heidelberg, Germany; it is a pleasure to thank I. Appenzeller and R. Bender for their hospitality.

References

Alloin, D., and Kunth, D.: 1979, *Astr. Ap.* **71**, 335
Arsenault, R., Boulesteix, J., Georgelin, Y., and Roy, J.-R.: 1988, *Astr. Ap.* **200**, 29
Bender, R., Burstein, D., and Faber, S. M.: 1992, *Ap. J.* **399**, 462
Bertola, F., Buson, L. M., and Zeilinger, W. W.: 1992, *Ap. J.* **401**, L79
Bertola, F., Rubin, V. C., and Zeilinger, W. W.: 1989, *Ap. J.* **345** L29
Bertola, F., and Capaccioli, M.: 1975, *Ap. J.* **200**, 439
Bertola, F., and Capaccioli, M.: 1977, *Ap. J.* **211**, 697
Bica, E., and Alloin, D.: 1987, *Asr. Ap. Suppl.* **70**, 281
Binney, J.: 1976, *M. N. R. A. S.* **177**, 19
Binney, J.: 1978a, *M. N. R. A. S.* **183**, 501
Binney, J.: 1978b, *Comments Ap.* **8**, 27
Binney, J.: 1982, in *Morphology and Dynamics of Galaxies, Twelfth Advanced Course of the Swiss Society of Astronomy and Astrophysics*, eds. L. Martinet and M. Mayor, Geneva Observatory: Sauverny, 1
Binney, J., and Petrou, M.: 1985, *M. N. R. A. S.* **214**, 449
Boroson, T.: 1981, *Ap. J. Suppl.* **46**, 177
Burkhead, M. S.: 1986, *A. J.* **91**, 777
Burkhead, M. S.: 1991, *A. J.* **102**, 893
Burstein, D.: 1979, *Ap. J.* **232**, 74
Buta, R.: 1986a, *Ap. J. Suppl.* **61**, 609
Buta, R.: 1986b, *Ap. J. Suppl.* **61**, 631
Buta, R.: 1990, *Ap. J.* **351**, 62
Buta, R., and Crocker, D. A.: 1991, *A. J.* **102**, 1715
Combes, F.: 1991, in *IAU Symposium 146, Dynamics of Galaxies and Their Molecular Cloud Distributions*, ed. F. Combes and F. Casoli, Kluwer: Dordrecht, 255
Combes, F., and Casoli, F., eds.: 1991, *IAU Symposium 146, Dynamics of Galaxies and Their Molecular Cloud Distributions*, Kluwer: Dordrecht
Combes, F., Debbasch, F., Friedli, D., and Pfenniger, D.: 1990, *Astr. Ap.* **233**, 82
Combes, F., and Sanders, R. H.: 1981, *Astr. Ap.* **96**, 164
Davies, R. L.: 1989, in *The World of Galaxies*, eds. H. G. Corwin and L. Bottinelli, Springer-Verlag: New York, 312
Davies, R. L., Efstathiou, G., Fall, S. M., Illingworth, G., and Schechter, P. L.: 1983, *Ap. J.* **266**, 41
Davies, R. L., Illingworth, G., and Kormendy, J.: 1993, in preparation
de Vaucouleurs, G.: 1975, *Ap. J. Suppl.* **29**, 193
de Vaucouleurs, G., de Vaucouleurs, A., and Corwin, H. G.: 1976, *Second Reference Catalogue of Bright Galaxies*, Univ. of Texas Press: Austin
de Vaucouleurs, G., and Pence, W. D.: 1978, *A. J.* **83**, 1163
de Zeeuw, T., and Franx, M.: 1991, *Ann. Rev. Astr. Ap.* **29**, 239
Dressler, A., and Richstone, D. O.: 1988, *Ap. J.* **324**, 701
Duschl, W. J.: 1988a, *Astr. Ap.* **194**, 33
Duschl, W. J.: 1988b, *Astr. Ap.* **194**, 43
Faber, S. M., and Jackson, R. E.: 1976, *Ap. J.* **204**, 668
Freeman, K. C.: 1970, *Ap. J.* **160**, 811
Friedli, D., and Pfenniger, D.: 1990, in *ESO Workshop on Bulges of Galaxies*, eds. B. Jarvis and D. M. Terndrup, ESO: Garching, 265
Friedli, D., and Udry, S.: 1993, in *IAU Symposium 153, Galactic Bulges*, eds. H. Habing and H. Dejonghe, Kluwer: Dordrecht, in press
Frogel, J. A.: 1992, in *IAU Symposium 149, The Stellar Populations of Galaxies*, eds. B. Barbuy and A. Renzini, Kluwer: Dordrecht, 245

Gallagher, J. S., Faber, S. M., and Balick, B.: 1975, *Ap. J.* **202**, 7
Gallagher, J. S., Goad, J. W., and Mould, J.: 1982, *Ap. J.* **263**, 101
Galletta, G.: 1987, *Ap. J.* **318**, 531
Gerhard, O. E.: 1993, in *IAU Symposium 153, Galactic Bulges*, eds. H. Habing and H. Dejonghe, Kluwer: Dordrecht, in press
Gerhard, O. E., Vietri, M., and Kent, S. M.: 1989, *Ap. J.* **345**, L33
Gunn, J. E.: 1979, in *Active Galactic Nuclei*, eds. C. Hazard and S. Mitton, Cambridge Univ. Press: Cambridge, 213
Hamabe, M., and Kormendy, J.: 1987, in *IAU Symposium 127, Structure and Dynamics of Elliptical Galaxies*, ed. T. de Zeeuw, Reidel: Dordrecht: Reidel, 379
Hasan, H., and Norman, C.: 1990, *Ap. J.* **361**, 69
Hasan, H., Pfenniger, D., and Norman, C.: 1993, *Ap. J.* **409**, in press
Hasan, H., and Norman, C.: 1993, in *IAU Symposium 153, Galactic Bulges*, eds. H. Habing and H. Dejonghe, Kluwer: Dordrecht, in press
Hernquist, L., and Quinn, P. J.: 1988, *Ap. J.* **331**, 682
Hummel, E., van der Hulst, J. M., and Keel, W. C.: 1987, *Astr. Ap.* **172**, 32
Illingworth, G.: 1977, *Ap. J.* **218**, L43
Illingworth, G.: 1981, in *The Structure and Evolution of Normal Galaxies*, eds. S. M. Fall and D. Lynden-Bell, Cambridge Univ. Press: Cambridge, 27
Illingworth, G., and Schechter, P. L.: 1982, *Ap. J.* **256**, 481
Jarvis, B.: 1987, *A. J.*, **94**, 30
Jarvis, B.: 1990, in *Dynamics and Interactions of Galaxies*, ed. R. Wielen, Springer-Verlag: New York, 416
Jarvis, B. J., and Dubath, P.: 1988, *Astr. Ap.* **201**, L33
Jarvis, B. J., Dubath, P., Martinet, L., and Bacon, R.: 1988, *Astr. Ap. Suppl.* **74**, 513
Jarvis, B. J., and Freeman, K. C.: 1985a, *Ap. J.* **295**, 314
Jarvis, B. J., and Freeman, K. C.: 1985b, *Ap. J.* **295**, 324
Kent, S. M.: 1985, *Ap. J. Suppl.* **59**, 115
Kent, S. M.: 1987, *A. J.* **93**, 816
Kent, S. M.: 1988, *A. J.* **96**, 514
Kent, S. M.: 1989, *A. J.* **97**, 1614
Kent, S. M.: 1990, *A. J.* **100**, 377
Kent, S. M., and Glaudell, G.: 1989, *A. J.* **98**, 1588
Kormendy, J.: 1977, *Ap. J.* **217**, 406
Kormendy, J.: 1979a, in *Photometry, Kinematics and Dynamics of Galaxies*, ed. D. S. Evans, Dept. of Astronomy, Univ. of Texas at Austin: Austin, 341
Kormendy, J.: 1979b, *Ap. J.* **227**, 714
Kormendy, J.: 1981, in *The Structure and Evolution of Normal Galaxies*, eds. S. M. Fall and D. Lynden-Bell, Cambridge Univ. Press: Cambridge, 85
Kormendy, J.: 1982a, in *Morphology and Dynamics of Galaxies, Twelfth Advanced Course of the Swiss Society of Astronomy and Astrophysics*, eds. L. Martinet and M. Mayor, Geneva Observatory: Sauverny, 113
Kormendy, J.: 1982b, *Ap. J.* **257**, 75
Kormendy, J.: 1983, *Ap. J.* **275**, 529
Kormendy, J.: 1985, *Ap. J.* **295**, 73
Kormendy, J.: 1987, in *Nearly Normal Galaxies: From the Planck Time to the Present*, ed. S. M. Faber, Springer-Verlag: New York, 163
Kormendy, J.: 1988a, *Ap. J.* **325**, 128
Kormendy, J.: 1988b, *Ap. J.* **335**, 40
Kormendy, J.: 1988c, in *Supermassive Black Holes*, ed. M. Kafatos, Cambridge Univ. Press: Cambridge, 219
Kormendy, J.: 1993, in preparation
Kormendy, J., and Djorgovski, S.: 1989, *Ann. Rev. Astr. Ap.* **27**, 235
Kormendy, J., and Illingworth, G.: 1982, *Ap. J.* **256**, 460
Kormendy, J., and Illingworth, G.: 1983, *Ap. J.* **265**, 632
Kormendy, J., and McClure, R. D.: 1993, *A. J.*, in press
Lauer, T. R., *et al.* : 1991, *Ap. J.* **369**, L41

Mather, J. C., *et al.* 1990, in *Observatories in Earth Orbit and Beyond*, ed. Y. Kondo, Kluwer: Dordrecht, 9
Nieto, J.-L., Bender, R., Arnaud, J., and Surma, P.: 1991, *Astr. Ap.* **244**, L25
Norman, C., and Hasan, H.: 1990, in *Dynamics and Interactions of Galaxies*, ed. R. Wielen, Springer-Verlag: New York, 479
O'Connell, R. W.: 1983, *Ap. J.* **267**, 80
Pellet, A., and Simien, F.: 1982, *Astr. Ap.* **106**, 214
Pfenniger, D.: 1984, *Astr. Ap.* **134**, 373
Pfenniger, D.: 1985, *Astr. Ap.* **150**, 112
Pfenniger, D.: 1993, in *IAU Symposium 153, Galactic Bulges*, eds. H. Habing and H. Dejonghe, Kluwer: Dordrecht, in press
Pfenniger, D., and Friedli, D.: 1991, *Astr. Ap.* **252**, 75
Pfenniger, D., and Norman, C.: 1990, *Ap. J.* **363**, 391
Prendergast, K. H.: 1983, in *IAU Symposium 100, Internal Kinematics and Dynamics of Galaxies*, ed. E. Athanassoula, Reidel: Dordrecht, 215
Raha, N., Sellwood, J. A., James, R. A., and Kahn, F. D.: 1991, *Nature* **352**, 411
Rix, H.-W., Franx, M., Fisher, D., and Illingworth, G.: 1992, *Ap. J.* **400**, L5
Rubin, V. C., Graham, J. A., and Kenney, J. D. P.: 1992, *Ap. J.* **394**, L9
Sandage, A.: 1961, *The Hubble Atlas of Galaxies*, Carnegie Institution of Washington: Washington
Scorza, C.: 1993, in *ESO/EIPC Workshop, Structure, Dynamics, and Chemical Evolution of Early-Type Galaxies*, Elba, 25 – 30 May, 1992
Sellwood, J. A.: 1993, in *IAU Symposium 153, Galactic Bulges*, eds. H. Habing and H. Dejonghe, Kluwer: Dordrecht, in press
Sérsic, J. L., and Pastoriza, M.: 1965, *Publ. A. S. P.* **77**, 287
Shaw, M. A.: 1987, *M. N. R. A. S.* **229**, 691
Shaw, M. A., Wilkinson, A., and Carter, D.: 1992, *Astr. Ap.*, in press
Shlosman, I., and Begelman, M. C.: 1987, *Nature* **329**, 810
Shlosman, I., and Begelman, M. C.: 1989, *Ap. J.* **341**, 685
Shlosman, I., Frank, J., and Begelman, M. C.: 1989, *Nature* **338**, 45
Sil'chenko, O. K.: 1993, in *IAU Symposium 153, Galactic Bulges*, eds. H. Habing and H. Dejonghe, Kluwer: Dordrecht, in press
Simien, F., and de Vaucouleurs, G.: 1986, *Ap. J* **302**, 564
Sofue, Y.: 1991, in *IAU Symposium 146, Dynamics of Galaxies and Their Molecular Cloud Distributions*, eds. F. Combes and F. Casoli, Kluwer: Dordrecht, 287
Sparke, L. S., Kormendy, J., and Spinrad, H.: 1980, *Ap. J.* **235**, 755
Statler, T. S.: 1988, *Ap. J.* **331**, 71
van den Bergh, S.: 1976a, *Ap. J.* **203**, 764
van den Bergh, S.: 1976b, *Ap. J.* **206**, 883
Wada, K., and Habe, A.: 1993, in *IAU Symposium 153, Galactic Bulges*, eds. H. Habing and H. Dejonghe, Kluwer: Dordrecht, in press
Wakamatsu, K.-I., and Hamabe, M.: 1984, *Ap. J. Suppl.* **56**, 283
Whitmore, B. C., and Bell, M.: 1988, *Ap. J.* **324**, 741
Whitmore, B. C., and Kirshner, R. P.: 1981, *Ap. J.* **250**, 43
Whitmore, B. C., Kirshner, R. P., and Schechter, P. L.: 1979, *Ap. J.* **234**, 68
Whitmore, B. C., McElroy, D. B., and Tonry, J. L.: 1985, *Ap. J. Suppl.* **59**, 1
Wrobel, J. M., and Kenney, J. D. P.: 1992, *Ap. J.* **399**, 94
Wrobel, J. M., and Kenney, J. D. P.: 1993, in *IAU Symposium 153, Galactic Bulges*, eds. H. Habing and H. Dejonghe, Kluwer: Dordrecht, in press

DISCUSSION

Statler: You say there might not be bulges in some galaxies, they're really just the disks. What do you mean by a disk? When I think of a disk, I think of something that's very flat. When I look at the side of NGC 4826, at least to the eye and unless the gas distribution is very asymmetric, it looks like it has a thick bulge that is poking up above the back side of the dust lane.

Kormendy: In NGC 4826, you can't tell what the apparent axial ratio of the disk is, because the exposure is too short. It would require looking at ellipticity as a function of radius out where the disk is, which is not evident on the slide you are referring to. But, I was not trying to say that all the material is disk material or that this object has no bulge whatsoever. You also asked, what's a disk anyway? That is a question to which none of us has formulated a rigorous answer. But I can give you some qualitative idea of what the picture in my mind is. If you look at the V/σ–diagram for edge-on objects, which as you say are if they are not triaxial somewhere around the ridge-line, disks live up where the ratio is much larger than the most flattened ellipticals (E6). So if you see an object of E9, I would call that a disk. If you see an E3, call it a spheroid. But the dividing line is only as sharp as nature makes it and we should not be surprised if there is not a sharp boundary.

Statler: If you're looking at the V/σ–diagram, you have to be very careful of interpreting $V/\sigma > 1$ as being a disk signature (with respect to the oblate isotropic line). Isotropy is of course just a middle case and there is nothing preventing you from having a thickened ellipsoidal object that is slightly tangential, or radially cold and vertically hot. Of course there are formation problems associated with that. But if it is radially cold, and you want to make the bulges out of vertical heating by Pfenniger, Norman or Sellwood's mechanisms, bars are radially hot rather than radially cold. So I don't really see how you would get a disk-like low velocity dispersion object from this bar mechanism which makes things radially very hot.

Kormendy: On the second point: I wasn't trying to suggest that the vertical heating produced that bar, or that it was the only way to produce a thickened object. I was mainly bringing that up in connection with the box shaped objects, and the connection there does not seem too bad, although I do have some worries with it.

Tendrup: We (Davies *et.al.*) are completing a survey of 60 or 80 inner regions of spiral galaxies in *JHK*, where the effects of dust are minimized. We have a large number of galaxies for which, although the profile changes from exponential to identically $r^{1/4}$ within a certain radius, there is no change in anything like color or ellipticity. Just as though you have an elliptical galaxy stuck in a disk. You could decompose them, but they don't look like what we would normally think of as a bulge oldish population and a disk with somewhat younger population. On the other hand, there are galaxies that can be separated quite well into bulge and disk components. We don't have enough data to do statistics on this, but there are several examples. As a particular example, when you do K-band imaging, a lot of Sb galaxies have a lot more triaxial signatures in them, than you would see in the optical, since you can see right through at K.

Kormendy: So you see at K some bulges that are as flattened as disks. These are very welcome observations, because it is clear that this game should be played further into the IR where dust and young stars mess up the story much less.

Renzini: Is it possible to attempt a quantitative estimate of the amount of star formation of the bulges for two cases in particular, the Galactic bulge and that of M31?

Kormendy: In M31, there are only two observations (that I know about), that point to this kind of phenomenon. M31 is a classic example of a relatively ordinary elliptical-galaxy-like bulge, but it is true that there are clear signs that the dust distribution that belongs to the disk, extends inside the bulge. Not very much light needs to be invoked, but the nucleus is almost negligible in terms of its fractional light contribution. So M31 happens to be a galaxy which has this disease, clearly in the central 2 arcsec, but that almost doesn't count in terms of mass. In our own Galaxy, I would love to know the answer and I have been listening very carefully to see whether elements of this are coming out in this meeting. But one of the things that is very likely to happen, is that this phenomenon is a strong function of z-distance. Much of the work we have seen, such as on Baade's window, is already at moderately large z-distances. So we want observations much nearer to the centre. Perhaps the OH/IR stars are an indication, but then the ages of those stars have been debated here, so we just don't have all the answers yet.

John KORMENDY

Francesco BERTOLA

PHOTOMETRIC PROPERTIES OF EXTRAGALACTIC BULGES

FRANCESCO BERTOLA
Department of Astronomy, University of Padova

1. Introduction

The "unresolved nuclear region", as Hubble (1926) indicated the bulge of spiral galaxies, became a physically distinct component with the introduction by Baade (1944) of the concept of stellar populations.

Since then, the role of the bulge itself and of its luminosity with respect to that of the disk is crucial in the context of understanding the Hubble sequence and the formation history of galaxies. For instance, it has been proposed to regard the Hubble sequence as a sequence of spheroidal luminosities (Meisels and Ostriker 1984) or as a sequence of decreasing bulge to disk ratios (van der Bergh 1976). Concerning the formation processes, in addition to the classical collapse picture (Eggen, Lynden-Bell, Sandage, 1963) where the bulge stars are formed before the formation of the disk, other possibilities are suggested. Bulges, with the exception of the very small ones characteristic of late-type spirals, could have been formed by merging of small objects (Larson 1990) or by two galaxies (Rix and White 1990). In any case bulges should have been formed before the disk, as indicated by their lower metallicity, so that, as stated by Binney and May (1986), disk galaxies probably started as undressed spheroids.

It is therefore clear that the detailed study of bulges, both from the kinematic, photometric and chemical abundance point of view, will offer interesting clues on their role in galaxy formation.

2. The bulge-disk decomposition

The most direct information that can be derived from the photometric study of disk galaxies is the determination of the photometric parameters of the disk and bulge component, disregarding the presence of superimposed features such as bars, lenses, spiral arms and rings. Such parameters are the effective radius r_e of the bulge and the corresponding surface brightness μ_e, the ellipticity of the bulge isophotes, the central surface brightness of the disk and its scale length α and the ellipticity of the disk isophotes. The luminosity profile of the bulge is supposed to follow the $r^{1/4}$ law, while in the disk the surface brightness decreases exponentially. Additional information is represented by possible trends of the ellipticity and of the position angle of the major axis as a function of the distance from the center.

After the pioneering work of de Vaucouleurs (1959) several methods of decomposing the bulge and the disk component have been developed. They are reviewed by Simien (1989) and Capaccioli and Caon (1992). The modern techniques make use of the full two dimensional images obtained with CCD detectors and can be described as follows. The image is cleaned using masking techniques, removing disturbances such as spiral arms, dust lanes and superimposed star images. Then a

H. Dejonghe and H. J. Habing (eds.), Galactic Bulges, 231–241.

smooth model of the disk is constructed using the parameters derived in the outer parts of the image and subtracted from the image of the galaxy. What remains is the bulge component, from which a smooth model is also constructed. The model galaxy, obtained by adding the two components, is then compared with the original. This procedure is repeated several times adjusting the input parameters of the models until the residuals (galaxy minus model galaxy) are minimized.

The bulge to disk ratio (B/D), namely the ratio between the luminosity of the disk and that of the bulge, derived with the above methods is an important parameter characterizing disk galaxies .

One of the most interesting points is to establish how the B/D ratio is correlated with the morphological type, as suggested by the original Hubble criteria of classification, and, if this correlation is present, which are the separated roles of the disk and the bulge.

In the morphological approach of van der Bergh (1976) *S*0 galaxies are considered as forming a sequence of decreasing B/D parallel to that of spirals, thus they are removed from being a transition between ellipticals and spirals. This point has been contradicted by the photometric decomposition of Simien and de Vaucouleurs (1986) who find a correlation between B/D and the morphological type, B/D being highest for *S*0 galaxies. The decreasing B/D for late-type galaxies is mainly due to a decreasing luminosity of the bulge. Similar results were also found by Kent (1985). However the analysis of a large sample of galaxies led Kodaira *et al* (1986) to the result that the mean absolute magnitude of disks and bulges is nearly constant over the entire sample of morphological types. They concluded that the Hubble type does not show a tight correlation with either the spheroid magnitude, the disk magnitude or the B/D ratio.

The analysis of a sample of 35 spirals and 19 *S*0 galaxies in the Virgo Cluster (Bertola and D'Onofrio 1992) shows a slight increase of the absolute magnitude of the disk while the luminosity of the bulge decreases from *S*0 to *Sc* type. In this way the D/B ratio varies from 0.5 to 2, a much smaller range that the one given by Simien and de Vaucouleurs (1986) (Fig.1). These discrepancies could be due to errors in the decomposition, different classification criteria and incompleteness of the samples. Further studies should clarify the point.

3. The detection of hidden disks

One of the major results of the photometric studies of early type galaxies in recent years has been the discovery of faint disks relative to the spheroidal component and therefore almost hidden in it. Galaxies with such disks were previously classified as ellipticals, they are often called "disky" ellipticals in contrast with "boxy" ellipticals, where no trace of disk is present and at the same time the outer isophotes show departures from the ellipticity towards squared shapes.

The hidden disks are usually found by representing the departures from ellipticity of the isophotes by means of a Fourier analysis. If the coefficient a_4 is positive, it implies a pointed structure of the isophotes, indicating the presence of the disk. This method has been extensively used by Bender *et al* (1988,1989), who established that disky and boxy ellipticals are characterized by different physical properties.

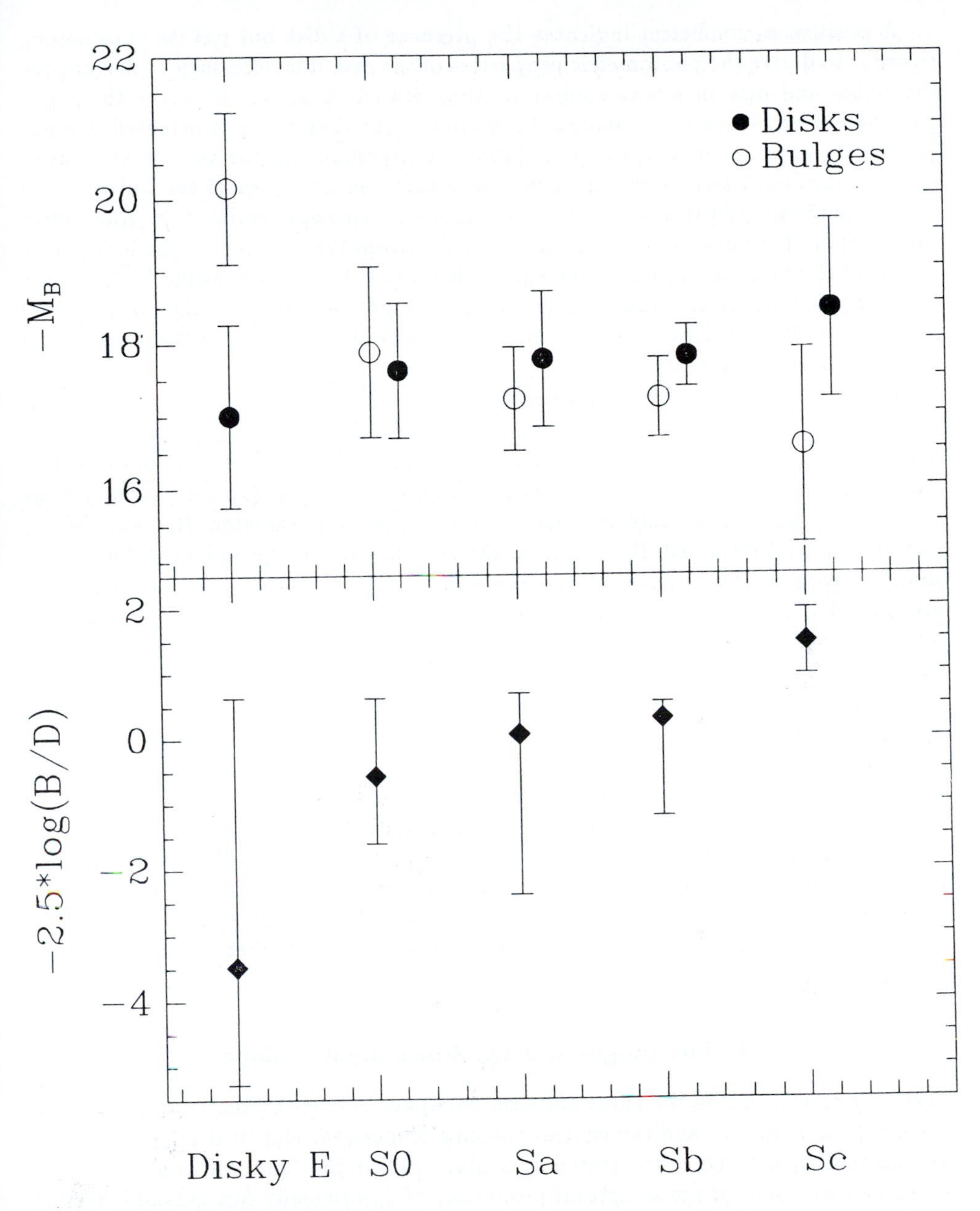

Fig. 1. Absolute magnitudes of disks and bulges and B/D ratios for a sample of *S*0 and spiral galaxies in the Virgo Cluster (Bertola and D'Onofrio 1992). The available data on "disky" ellipticals are also plotted.

A positive a_4 coefficient indicates the presence of a disk but not its parameters. In order to derive the photometric properties of the disk it is necessary to decompose the bulge and disk in a way similar to that described above. However the input parameters necessary to construct the model of the disk to be subtracted are not so easy to select. Scorza and Bender (1990) in decomposing the galaxy NGC 3610 selected the input parameters in such a way that the isophotes of the bulge would appear perfectly elliptical after the subtraction of an exponential disk. Also, after subtraction of the disk component the deviation from the $r^{1/4}$ law on the luminosity profile of the bulge are largely reduced. In this way it has been possible to find disks whose consistence is between 15% to 2% of the luminosity of the bulge. Simien and Michard (1990) developed a similar method, which does not imply the assumption that the disk is exponential.

Rix and White (1990) have studied the detectability of the hidden disks in elliptical galaxies, which is a function of the D/B ratio and of the viewing angles. Face on disks are generally more difficult to detect than edge on ones and their photometric signatures reside mainly on the deviations from the $r^{1/4}$ law. On the contrary edge on disks produce noticeably positive values of the a_4 parameter. Rix and White estimate that 50% of all disks with a D/B<0.25 can not be detected by photometric means. Disks in elliptical galaxies can be detected also spectroscopically by studying the stellar line profiles, as they are produced by the two components. Rix and White (1992) were able to derive in this way the luminosity profiles of the disk and the bulge. However, the most appropriate way to decompose disk and bulge is to use both photometric and kinematical data. This method has been successfully applied by Cinzano and van der Marel (1992) to NGC 2974. They were not obliged to make assumptions on the perfect ellipticity of the bulge isophotes due to the presence of the kinematical constraints. In Fig.1 mean absolute magnitudes of the disk and the bulge and the mean B/D ratio are plotted also for a sample of seven "disky" ellipticals present in the literature (Scorza and Bender 1990; Capaccioli *et al.* 1991; Vader and Vigroux 1991; Scorza and Bender 1992; Cinzano and van der Marel 1992). The mean D/B ratio of 0.04 is produced by a slightly lower luminosity of the disk than in $S0$ and spiral galaxies and by a much higher luminosity of the bulge.

4. The bulges and the fundamental plane

It is well known that in the three dimensional space defined by the two photometric parameters μ_e and r_e and the central velocity dispersion, elliptical galaxies show a strong tendency to be concentrated in a plane called the "fundamental plane". In a recent discussion of the structural properties of dynamically hot galaxies Bender, Burstein and Faber (1992) included, in addition to ellipticals, bulges of disk galaxies. They find that bulges share common properties with ellipticals of intermediate luminosity (-20.5< M_B <-18.5), both looking edge-on and face-on at the fundamental plane. There is a tendency for the representative points of the bulges to lie slightly below the plane, but it is not clear whether this might be a genuinely lower M/L than in ellipticals. In the edge on view of the fundamental plane bulges and ellipticals of intermediate luminosity occupy the central position while at the two

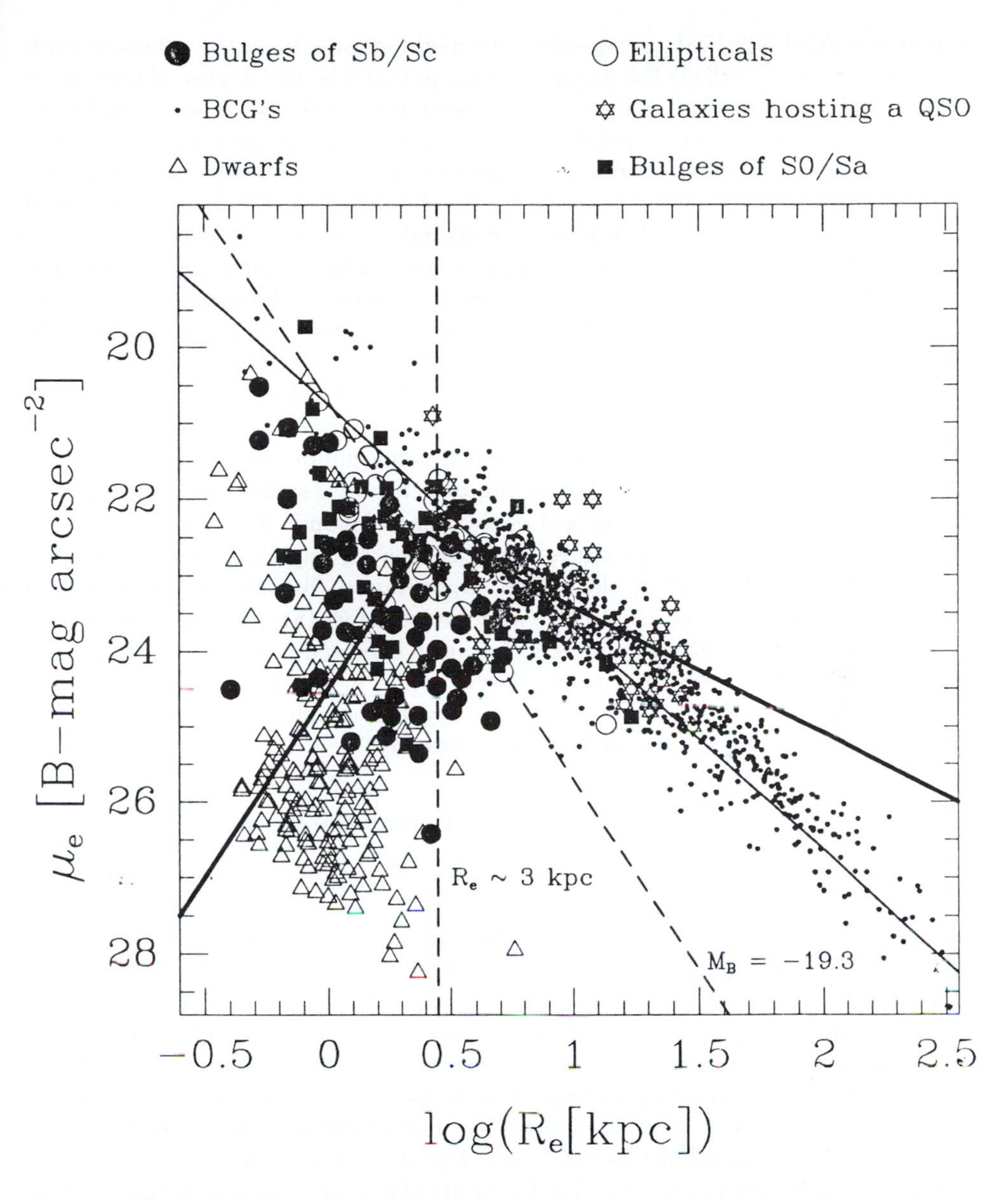

Fig. 2. Elliptical galaxies and bulges in the μ_e,Re plane according to Capaccioli *et al.* (1992). The thick line represents the relationship by Binggelli *et al.* (1984).

extreme are faint ellipticals on one side and dwarf ellipticals on the other. According to Bender *et al.* (1992) the discriminating property is the degree of anisotropy in the velocity distribution. Bulges and intermediate ellipticals are associated since they both are isotropic rotators (Davies *et al.* 1983). Since giant ellipticals tend to be "boxy" and intermediate luminosity ellipticals tend to be "disky", it is suggested that the distinction between these two kinds of galaxies is due to the absence or presence of a disk. Such a disk is not recognizable in a visual inspection of the ellipticals of intermediate luminosity, characterized by an higher bulge to disk ratio than ordinary *S*0's and spirals to which they are related. The same mixing of intermediate luminosity ellipticals and bulges is present also when the fundamental plane is projected into the plane defined by the two photometric parameters μ_e, r_e (Fig.2) (Capaccioli *et al.* 1992) with a tendency of disky ellipticals to concentrate toward the high luminosity lines.

Recent numerical simulations seem to cast some doubts on the interpretation that disky isophotes generally indicates the presence of a stellar disk. By studying elliptical galaxies as the result of a dissipationless collapse Stiavelli *et al.* (1991) found that the isophotal shape varies with the viewing angles, giving to the same galaxies either the "disky" or the "boxy" appearance. Similar results have been obtained by Governato *et al.* (1992), who investigated the properties of merger remnants from two spherical galaxies and from a disk and a spherical galaxy. The isophotes in the first case tend to be "boxy" while in the second case there is a marked tendency to be "disky". However they are unable to reproduce objects with high degrees of diskyness, as are often observed.

5. The box- and peanut-shaped bulges

*S*0 and spiral galaxies, when seen on edge, often reveal bulges whose shape is box like or even peanut like, namely with a depression of the isophotes along the minor axis. This phenomenon has been known since long time and lists of galaxies exhibiting such peculiar bulges were published by Jarvis (1986), Shaw (1987) and Souza and dos Anjos (1987). Shaw (1987) finds that 20% of disk galaxies exhibit a box or peanut-shaped bulge. This percentage could be even higher taking into account that the peculiarity is not easily recognizable in late-type spirals. According to de Souza and dos Anjos (1987) this frequency, at least for *S*0 galaxies, is consistent with the idea that all of them are barred galaxies as invoked by some recent numerical simulations. Jarvis (1986) has studied the occurence of box and peanut-shaped bulges in galaxies of different Hubble type, concluding that there is a strong tendency for them to occur in *S*0 and *Sab* to *Sb* types. He also finds that 63% have peanut-shaped bulges, compared to box-shaped ones.

Shaw (1992) has recently estimated that the extra light, producing the box/peanut shape, after the subtraction of a model bulge, is 5-15% of that in the bulge as a whole. This luminosity excess appears greater in later type galaxies, where the bulges are less prominent. Among the different mechanisms proposed to explain the formation of box/peanut bulges, such as external torques (May *et al.* 1985), merging (Binney and Petron 1985) and the presence of a bar (Combes *et al.* 1990), the latter seems to be the most promising one. By means of three-dimension N-body

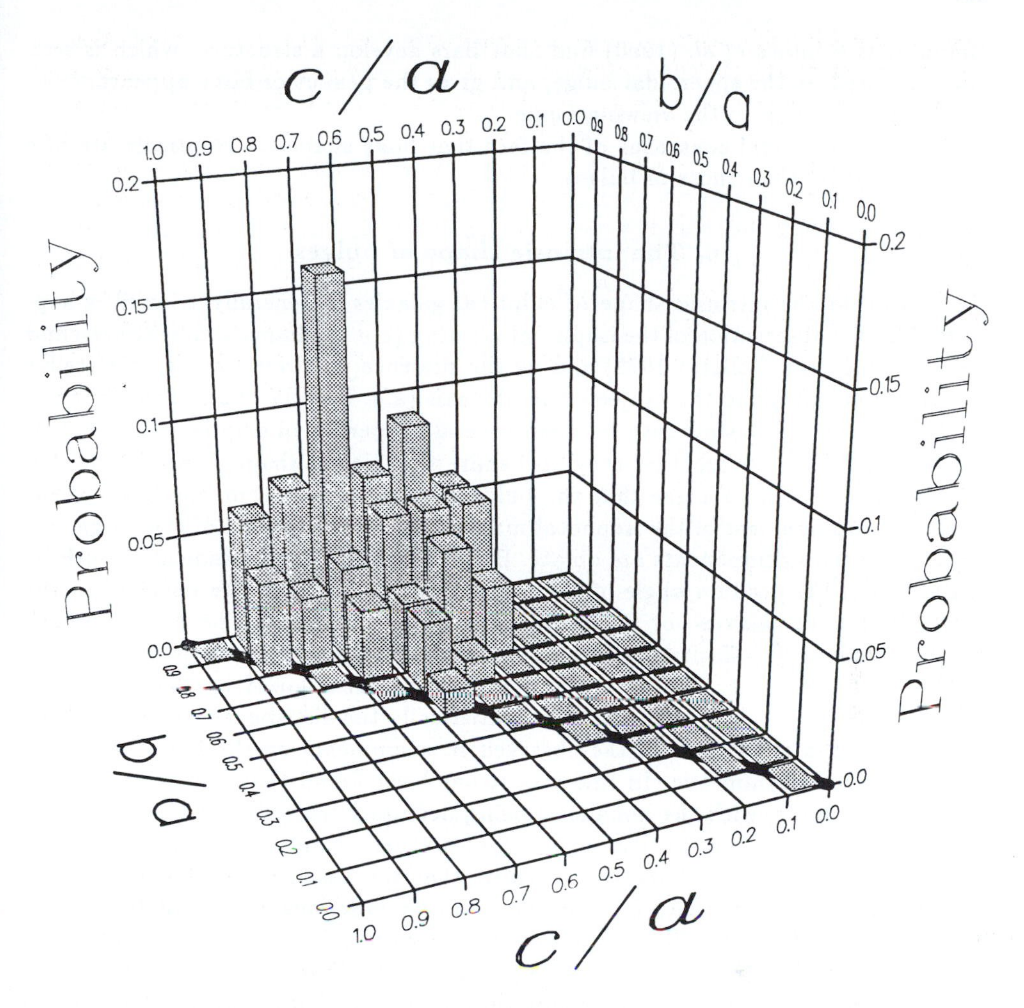

Fig. 3. The distribution function of the intrinsic axial ratios of bulges derived by Bertola *et al.* (1991) on the assumption that disks are circular.

simulations Combes *et al.* (1990) find that bars develop a structure, which is seen superimposed on the spheroidal bulge, and gives the peanut or boxy appearance to the bulge according to the viewing angle.

There is a general consensus of the fact that boxy shapes in ellipticals are of a different nature than those of bulges.

6. The intrinsic shape of bulges

The fact that the intrinsic shape of elliptical galaxies is generally triaxial is supported by the observation of the isophotal twisting (*e.g.* Williams and Schwarzschild 1979, Bertola and Galletta 1979) and by the presence, in several cases, of a stellar velocity gradient along the projected minor axis (*e.g.* Bertola *et al.* 1989, Franx *et al.* 1989). Due to the similarity between several properties of ellipticals and bulges one is tempted to ask whether they also share the triaxial shape. Indeed Lindblad (1956) was the first to argue that the bulge of M31 is triaxial on the basis of the observed misalignment of the isophotal major axis of the disk and of the bulge, not expected if both components are oblate. Bulge triaxiality can be indicated both by variations of the position angle of the major axis within the bulge itself, as in the case of elliptical galaxies, or by the bulge disk misalignment. The first approach has been followed by Zaritsky and Le (1986) who find major axis twisting in all 11 bulges of their sample. The second approach has been followed by Bertola, Vietri and Zeilinger (1991) who tried to derive statistically the distribution function of the axial ratios b/a and c/a from the observed misalignments on the hypothesis that the disks are axisymmetric. In this way it has been found that half of the bulges are close to oblate, with the remaining being definitely triaxial (Fig.3). However, a subsequent investigation by Fasano *et al.* (1992) based on the distribution of the apparent axial ratios of disks has revealed that the assumption of axisymmetry is not correct. In fact broad intrinsic distribution functions peaked at b/a in the range $0.8 \div 0.9$ for both $S0$s and spirals were derived (Fig.4). Similar results were also found by Huizinga and van Albada (1992) and by Franx and de Zeeuw (1992). The latter authors, however, point out that these ellipticities are too large when considering the small scatter present in the Tully-Fisher relation.

The fact that disks are triaxial modifies the distribution of Fig.3, in the sense that bulges tend to be more triaxial if they are elongated along the same direction as the disk, while they are more oblate if the two directions are perpendicular.

7. Concluding remarks

The photometric studies of the bulges pose several questions. Here are a few:

i) Do $S0$ galaxies represent a transitional stage in the Hubble classification scheme or do they form a sequence parallel to that of spirals in terms of B/D ratio ? According to Simien and de Vaucouleurs (1986) they represent a transitional stage. However the data of Fig. 1 suggest that they do not behave differently from *Sa* and *Sb* galaxies.

ii) Do disky ellipticals constitute the natural continuation of the $S0$s and spirals ? According to the study of their properties in the fundamental plane Bender

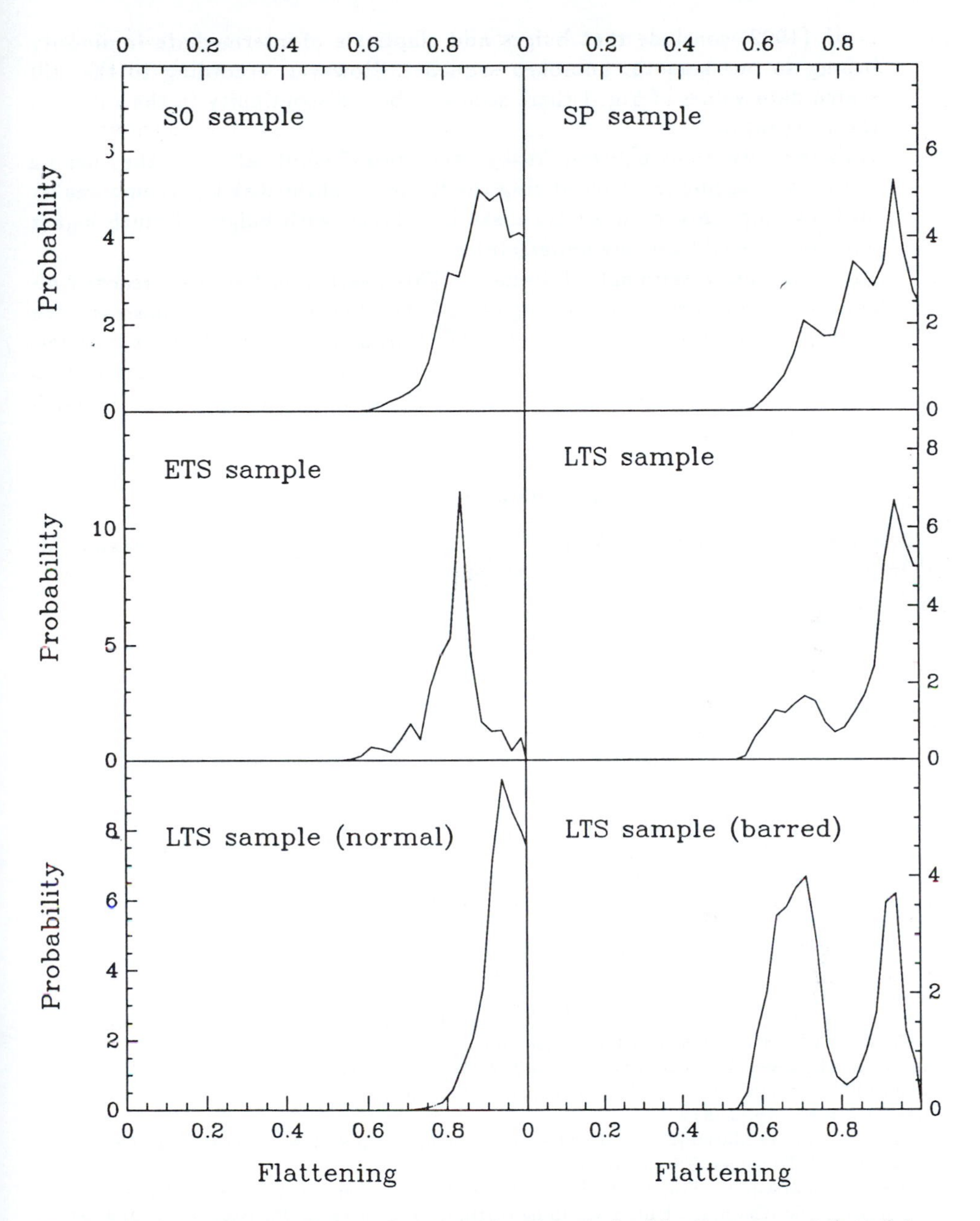

Fig. 4. The flattening distribution functions for disks derived by Fasano *et al.* (1992) for a sample of *S*0 and spirals galaxies (top panels). The spirals are subdivided into early-type and late-type in the middle panels. Late-type spirals are subdivided into normal and barred in the bottom panels. Notice the effect of the bar.

et al. (1992) conclude that bulges and ellipticals of intermediate luminosity belong to one smooth, unbroken sequence. However, according to the still scarce data values of Fig. 1 there seems to be a discontinuity in the values of the B/D ratio.

iii) Is there a dichotomy between "disky" and "boxy" ellipticals or do they belong to the same sequence ? Considering the tendency of the disk to become weaker in disky ellipticals, even weaker disks in galaxies with bulges of much higher luminosity would become undetectable.

iiii) Are all bulges rotationally flattened ? This conclusion has been reached almost ten years ago by Davies *et al.* (1983). However, we now possess both dynamical evidence (Bertola *et al.* 1990; Gerhard *et al.* 1990) and photometric evidence (Bertola *et al.* 1991) that a fraction of bulges are triaxial. This suggests the presence of a degree of velocity anisotropy, which would be worth to investigating.

Acknowledgements

It is a pleasure to thank P. Amico, C. Bender, M. Capaccioli, P. Cinzano, M. D'Onofrio, A. Pizzella and C. Scorza. for useful discussion. J. Pesce was so kind to read the manuscript.

References

Baade, W.: 1944, *Ap. J.* **100**, 137
Bender, R., Burstein, D., Faber, S.M.: 1992, *Ap. J. in press*
Bender, R., Döbereiner, S., Möllenhoff, C.: 1988, *Astron. & Astrophys. suppl. ser.* **74**, 385
Bender, R., Surma, P., Döbereiner, S., Möllenhoff, C., Madejsky, R.: 1989, *Astron. & Astrophys.* **217**, 35
Bertola, F., Capaccioli, M., Galletta, G., Rampazzo, R.: 1989, *Astron. & Astrophy.* **192**, 24
Bertola, F., D'Onofrio M.: 1992, *in preparation* ,
Bertola, F., Galletta, G.: 1979, *Astron. & Astrophys.* **77**, 363
Bertola, F., Rubin, V.C., Zeilinger, W.W.: 1989, *Ap. J. Letters* **345**, L29
Bertola, F., Vietri, M., Zeilinger, W.W.: 1991, *Ap. J.* **374**, L13
Binggelli, B., Sandage, A., Tarenghi, M.: 1984, *Astron. J.* **89**, 64
Binney, J., May, A.: 1986, *M.N.R.A.S.* **218**, 743
Binney, J., Petrou, M.: 1985, *M.N.R.A.S* **214**, 449
Capaccioli, M., Caon, N.: 1992, *Morphological and Physical Classification of galaxies*, Longo G., Capaccioli M., Busarello G.: Kluwer Academic
Capaccioli, M., Caon, N., D'Onofrio, M.: 1992, *M.N.R.A.S in press*
Capaccioli, M., Vietri, M., Held, E., Lorenz, H.: 1991, *Ap. J.* **371**, 535
Carter, D.: 1987, *Ap. J.* **312**, 514
Cinzano, P., van der Marel, R.P.: 1992, *Structure, Dynamics and chemical evolution of early-type galaxies.*, J. Danziger: ESO Garching
Combes, F., Debbash, F., Friedli, D., Pfenniger, D.: 1990, *Astron. & Astrophys.* **233**, 82
Davies, R.L., Efstathion, G., Fall, S.M., Illingworth, G.D., Schechter, P.: 1983, *Ap.J.* **266**, 41
de Souza, R.E., dos Anjos: 1987, *Astron. & Astrophys. suppl.* **70**, 465
de Vaucouleurs, G.: 1958, *Ap. J.* **128**, 465
Eggen, O.J., Lunden-Bell, D., Sandage, A.R.: 1963, *Ap. J.* **136**, 748
Fasano, M., Amico, P., Bertola, F., Vio, R., Zeilinger, W.W.: 1992, *M.N.R.A.S. in press*
Franx, M., de Zeeuw, P.T.: 1992, *Ap. J. in press*
Franx, M., Illingworth, G., Heckman, T.: 1989, *Ap. J.* **344**, 613
Governato, F., Reduzzi, L., Rampazzo, R.: 1992, *M.N.R.A.S in press*
Huizinga, J.E., van Albada, T.S.: 1992, *M.N.R.A.S in press*

Hubble, E.P.: 1926, *Ap. J.* **64**, 321
Jarvis, B.J.: 1986, *A.J.* **91**, 65
Kent, S.M.: 1985, *Ap. J. Suppl.* **59**, 115
Kodaira, K., Watenabe, M., Okamura, S.: 1986, *Ap. J. Suppl.* **62**, 703
Lindblad, B.: 1956, *Stockholm Obs. Ann.* **Vol.19**,N.2
May, A., van Albada, T.S., Norman, C.A.: 1985, *M.N.R.A.S.* **214**, 131
Meisels, A., Ostriker, J.P.: 1984, *A. J.* **89**, 1451
Rix, H.-W., White, D.M.: 1990, *Ap. J.* **362**, 52
Rix, H.-W., White, D.M.: 1992, *M.N.R.A.S* **254**, 389
Scorza, C., Bender, R.: 1990, *Astron. & Astrophys.* **235**, 49
Scorza, C., Bender, R.: 1992, *Morphological and Physical Classification of galaxies*, Longo G., Capaccioli M., Busarello G.: Kluwer Academies, 389
Shaw, M.A.: 1987, *M.N.R.A.S* **229**, 691
Shaw, M.A.: 1992 *M.N.R.A.S in press*
Simien, F.: 1989, *The World of Galaxies*, Corwin H.G. and Bottinelli L.: Springer Verlag, 293
Simien, F., de Vaucouleurs, G.: 1986, *Ap. J.* **302**, 564
Stiavelli, M., Londrillo, P., Messina, A.: 1991, *M.N.R.A.S* **251**, 57p
van den Bergh, S.: 1976, *Ap. J.* **206**, 883
Williams, T.B., Schwarzschild, M.: 1979, *Ap. J.* **227**, 56
Zaritsky, D., Lo, K.Y.: 1986, *Ap. J.* **303**, 56

DISCUSSION

Sellwood: I was delighted to hear that the fraction of box shaped bulges is slowly rising as observers look more carefully. But I must caution about pointing to the coincidence of 30% fraction of box bulges, compared to 30% of galaxies that are barred. Because when we see a galaxy edge on, the bar can be aligned at any random angle to the line of sight. Therefore we would expect the fraction of box bulges, if they were made from bars, to be less than the fraction of bars. We should also check whether the dimensions of these two things are comparable, before we really start being confident that this is an explanation for the peanut phenomenon.

Rix: I have a comment on your analysis of the shape of the disks. It seems you find that the disks in late type spirals and barred systems, deviate more strongly from axisymmetric. Both of those systems, more or less by definition for the late type spirals and for bars, are strong two armed spirals. Also, you inferred the triaxiality from the lack of systems with unit axis ratio. Is it clear to you in those cases that the axis ratios in those case do not only reflect the spiral arms, but have something to do with the underlying disk?

Bertola: We say no at this point and as I said, we took the data from the literature and also a sample of 100 galaxies, which we analyzed. We tried to analyze the influence of the spiral arms. What we found was that the disk is dominant and there is not so much an effect from the spiral arms. Of course in the inner regions this would not be the case and they would cause a twisting of the isophotes.

MARIJN FRANX

WHAT IS THE CONNECTION BETWEEN ELLIPTICALS AND BULGES ?

MARIJN FRANX
Harvard-Smithsonian Center for Astrophysics, 60 Garden Street, Cambridge, MA 02138

Abstract. The structure of bulges and ellipticals, and their relation to galaxy halos are reviewed. Since many ellipticals contain faint disks, the qualitative distinction between bulges and ellipticals is more accurately described as a quantitative variation in Bulge/Disk ratio. The exception may be the brightest ellipticals, which are usually Bright(est) Cluster Members. The available evidence suggests that the spheroid properties are determined by more than halo properties alone. This is clearest for systems with low B/D ratio, where bulge velocity dispersions are similar to disk velocity dispersions. Constraints on the stellar formation scenarios are discussed.

Key words: elliptical galaxies - bulges - dark halos - formation - stellar populations

1. Introduction

The idea that galaxies can be dissected into different components is fundamental to almost all classification schemes. As an example, one of the important characteristics of the Hubble sequence (Hubble 1936) is that the bulge to disk ratio increases towards early types. The underlying assumption is that the different components by themselves are very similar between galaxies, but that their relative masses within galaxies differ, and determine the Hubble type.

We would currently distinguish four different components in galaxies: the bulge or spheroid, the disk, the interstellar medium, and the dark matter halo. These four components may be present in almost all galaxies. Probably all galaxy types contain significant amounts of gas, from ellipticals which contain hot gas observed in X-rays, to late-type spirals containing large amounts of cool H I. Spiral galaxies definitely have large, massive halos, but the situation is not completely clear for ellipticals. This will be discussed below. Even disk components may be present in almost all ellipticals, except possibly, the brightest (which are in many cases also the brightest cluster members). Thus, the general separation of galaxies into these components appears to be very successful.

The next questions that one can ask is how the components compare in detail, and, more fundamentally, how they were formed. Here we focus on the properties of spheroids, (ellipticals and bulges), their relations to halos, and constraints on their formation.

2. How simple are Spheroids ?

In the 1960's and early 70's, it was believed that spheroids were simple systems. Mass was thought to be the main parameter; the flattening was thought to be due to rotation. The stellar populations were modeled with single metallicity, single age components. The formation of spheroids was thought to have taken place in the very early universe.

H. Dejonghe and H. J. Habing (eds.), Galactic Bulges, 243–260.

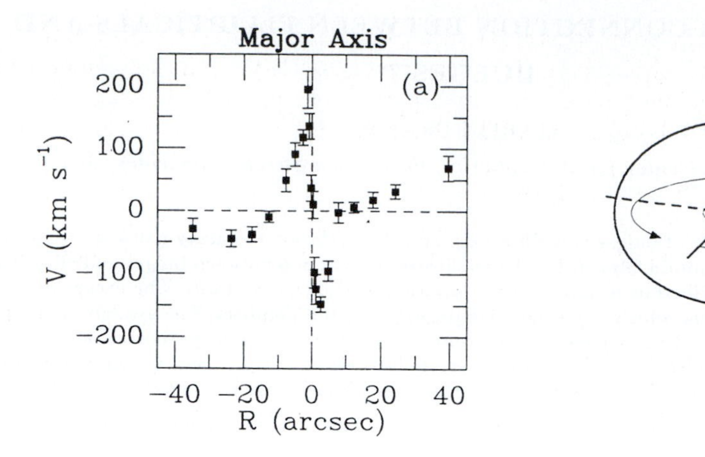

Fig. 1. The complex kinematics of IC 1459. (a) the rotational velocity along the major axis. The inner parts appear to counter-rotate compared to the outer parts. This is schematically drawn in (b). A more detailed analysis has shown that the central parts can be decomposed into two components: one hot component, slowly rotating like the outer parts, and one cold component, rapidly counter rotating with respect to the outer parts. The cold component may not contribute more than 20% of the light, but dominates the kinematics (Franx & Illingworth 1988).

2.1. KINEMATICS

New observational evidence collected over the last 2 decades has challenged this common wisdom. The conference proceedings of IAU 100 (Athanassoula 1983), and IAU 127 (de Zeeuw 1987) contain much of the new material. As an extreme example, figure 1 shows the kinematics of IC 1459, a normal, southern elliptical. This galaxy appears perfectly normal on the sky, but the kinematics show that the inner parts are counter-rotating with respect to the outer parts. A detailed spectroscopic analysis showed that the central part consists of two components, a slowly rotating component related to the bulk of the galaxy, and a rapidly, counter-rotating component, which may contribute about 20% of the light (Franx and Illingworth 1988).

Many other such systems have been found (Bender 1988, Jedrzejewski & Schechter 1988, Franx et al 1989), including systems like NGC 4406, in which the angular momentum of the central regions is perpendicular to the angular momentum of the outer parts. The overall statistics are that about 25% of observed ellipticals show such kinematically distinct nuclei.

The formation of such systems is not known with certainty. It is likely that a merger, or accretion of gas or stars played a role. The phenomenon is possibly related to star burst galaxies, in which comparable amounts of CO are observed near the central parts, and which are thought to be triggered by mergers.

There is no reason to assume that the kinematically distinct cores were formed only recently. The two body relaxation time is so large, that the subcomponents will persist for many Hubble times. It is very well possible that the subcomponents were formed as part of the generic formation process of ellipticals. The fact that the stellar populations of the galaxies with such nuclei are normal is consistent with this hypothesis.

The relatively high rate of kinematically distinct nuclei in ellipticals poses the question whether similar nuclei exist in bulges. Some bulges in nearby galaxies (like M31, NGC 4594) contain rapidly rotating subcomponents (e.g., Dressler 1984, Kormendy 1988) , but until now, no counter rotating subcomponents have been found. A survey of the kinematics of S0 galaxies has until now not produced a single such case (Fisher et al 1993).

However, Rubin has recently discovered the case of NGC 4550 in a similar survey (Rubin et al 1992). Her data showed clear evidence for two counter rotating components in this S0 galaxy. Whereas the author thought that this galaxy might be the first case of a kinematically distinct nucleus in a bulge, this turns out not to be the case. A multicomponent analysis by Rix et al (1992, see also this volume), shows that the two counter-rotating components reside in two co-spatial, cold, disks. The two disks have comparable exponential luminosity profiles, and are like normal disks. The bulge of the galaxy is not rotating at all. It is likely that the counter rotating disk was formed by the infall of counter rotating gas. Either the bulge was not rotating before the accretion took place, or during the accretion process bulge stars were added to produce a non rotating bulge.

2.2. PHOTOMETRY

In addition, there is photometric evidence for substructure in ellipticals. Carter (1978, 1986) detected non-elliptical isophotes in ellipticals, and later surveys showed large numbers of ellipticals with such deviations (e.g., Bender et al 1988). A significant number of those deviations are "disk-like", i.e., they can be explained by the presence of a disk. The deviations are expressed in a parameter c_4, which characterizes the residual amplitude after the ellipsefit in the $\cos 4\theta$ harmonical. Positive c_4 indicates disky isophotes, negative c_4 indicates boxy isophotes.

The value of c_4 lies usually in the range of $-0.04 \leq c_4 \leq 0.04$. This might suggest that any disk contributes only several percents of the light. However, an analysis by Rix and White (1990) showed that this is a serious underestimate of the required disk fraction. Because of projection effects, disks produce very small c_4 terms when the galaxy is seen face-on. When the galaxies are seen edge-on, they tend to be classified as S0, and not elliptical, because of the disk. They concluded that a large fraction of ellipticals may contain disks contributing 20 % of the light.

This result is consistent with the work of others, who had speculated that many ellipticals were face-on S0's (i.e., they have disks), e.g. Capaccioli (1986), van den Bergh (1990). Complete surveys of ellipticals and S0s are needed to put strong constraints on the fraction of the light in such disks. It is interesting to note, that two independent surveys both conclude that many, if not most, low luminosity ellipticals contain disks (Capaccioli et al 1992, Rix 1993).

We conclude that both the kinematics and the photometry show that most ellipticals have substructure. This significantly narrows the "gap" between ellipticals and S0's, and indicates that the formation of disks is an even more common aspect of galaxy formation than thought before. Truly disk-less ellipticals may be exceedingly rare, and those may be mainly bright(est) cluster members.

3. Can we distinguish between Ellipticals and Bulges ?

This issue has been reviewed extensively by Kormendy (1982), Illingworth (1983), Barnes and White (1984). Here a brief summary is given.

3.1. STRUCTURE

The fact by itself that it has taken more than 10 years of CCD photometry studies to decide whether or not ellipticals have substantial disks demonstrates the photometric similarity of bulges and ellipticals. Most bulges follow an $r^{1/4}$ law reasonably well, although some exceptions have been reported (see, e.g. Kormendy 1982). These exceptions are generally systems with low Bulge/Disk ratio. Infrared observations would be useful to minimize the influence of dust on the observed profiles.

An early analysis by Barnes and White (1984) of the data by Kent (1984) showed that the distribution of structural parameters (effective radii and surface brightness) were very similar. This important result puts tight constraints on the formation of spheroids.

3.2. KINEMATICS

The kinematics of bulges and ellipticals give a somewhat more confusing result. The rotational support of ellipticals correlates rather strongly with luminosity, in the sense that bright ellipticals have low rotation, and faint ellipticals (around L_*), rotate rather rapidly (Davies et al 1983). No such trend has been observed for bulges, all bulges have been found to be rotating rapidly (e.g., Kormendy and Illingworth 1982). This is partly or fully due to the fact that it is hard to find truly bright bulges, but it remains somewhat uncomfortable that no slowly rotating bulges have been found. The only known exception is the bulge of NGC 4550, which is faint, and not rotating (Rix et al 1992). As discussed above, this galaxy has two counter rotating disk components, and no rotation is detected in the central bulge region.

The interpretation that luminosity is the driving parameter has become more uncertain with the discovery that many of the low luminosity ellipticals have disks, and that rotation correlates with isophotal shapes (Bender 1988a). Thus, when we compare low luminosity ellipticals with low luminosity bulges, we may be comparing similar galaxies at different inclinations, and not different galaxy types. It is very hard to separate the contribution of the disk to the apparent rotation, and it may be that the rotation of the spheroid in inclined systems has been overestimated.

Another complication is the result that some extremely low luminosity ellipticals show no significant rotation. Bender & Nieto (1990) found that low surface brightness dwarfs show little rotation, demonstrating that more parameters than luminosity alone determine the rotation of galaxies.

It has been established that elliptical galaxies satisfy relations like the Faber Jackson relation, the Fundamental Plane, and the $D_n - \sigma$ relation (see de Zeeuw and Franx 1991, and references therein). These are basically relations between structural parameters (effective radius, surface brightness), and velocity dispersion. Early studies gave conflicting results as to the question whether bulges follow the same relations as ellipticals. Recent results appear to indicate that this is indeed the case for the Fundamental Plane, within reasonable accuracy (Dressler 1987, Bender et al 1992). The question whether the accuracy is high enough to allow the use of bulges as distance indicators is not completely resolved.

The origin of these relations is not quite clear. The fundamental plane relation implies that Mass to Light ratio (M/L) varies slowly with luminosity (Faber et al 1987). It is not known with certainty whether the M/L ratio is entirely determined by the stellar population, or whether dark matter plays a role. It is not known either whether the metallicity (and thereby the M/L) of bulges is determined by the luminosity of the bulge, or by the total luminosity. Thus the application of these relations to bulge systems is somewhat uncertain. It appears, however, that the M/L ratios of bulges are roughly the same as those of ellipticals.

3.3. STATISTICAL RELATIONS

The above suggests that it is virtually impossible to distinguish between bulges and ellipticals by a study of the spheroidal component alone. Furthermore, many ellipticals turn out to have weak disks. This suggests that the main distinction between bulges and ellipticals is Bulge to Disk ratio (B/D). The question whether differences exist between bulges and ellipticals can be rephrased into the question whether the B/D ratio can be determined from properties of the spheroid. In the following, we address the question whether such distinction can be made in a statistical sense.

3.3.1. Luminosity Function

Dressler and Sandage (1983) suggested that the luminosity function of bulges and ellipticals differ. In the faint regime, bulges are more prevalent, and at the high end, ellipticals dominate. A problem with this kind of comparison is that many faint bulges reside in luminous disks, and are included in magnitude selected samples because of the bright disk. However, a detailed study of the luminosity function of Virgo galaxies by types has confirmed that ellipticals are more abundant at the high luminosity end (Binggeli, Sandage, and Tammann 1988).

Schechter & Dressler (1987) determined B/D ratios for a complete field sample. The result is given in Figure 2. The scatter in B/D ratio is larger than the systematic trend; but the median points do indicate that the B/D ratio increases slightly with luminosity. Since the B/D ratio is plotted against total luminosity, not bulge luminosity, this data supports strongly the notion that faint bulges have generally lower B/D ratios.

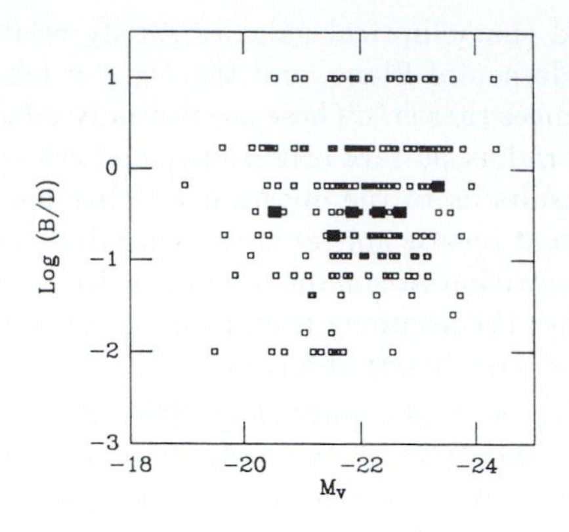

Fig. 2. The Bulge/Disk ratio plotted against total luminosity, based on the data from Schechter & Dressler (1987). The scatter dominates the relation. The filled symbols are medians of equal sized subsamples. There is a weak relation, in the sense that brighter galaxies have somewhat larger B/D ratios.

3.3.2. Environment

Dressler (1980) and Schechter & Dressler (1987) also found that the B/D ratio depends on environment, in the sense that galaxies have higher B/D ratios at higher densities. This interesting result is not well understood. It can be partly due to the fact that disk galaxies in dense environments have low starformation rates, and thus the disk M/L ratio may be systematically higher. It is just as well possible, however, that processes at early times caused the bulge to be larger in higher densities, possibly at the cost of part of the disk. One might think of interactions with other galaxies, accretion of smaller galaxies, etc. For discussions of possible effects, see Dressler (1980), Kent (1981), Schechter & Dressler (1987).

4. Relation between spheroid and halo

One of the most fundamental structural questions one can ask in the comparison of bulges and ellipticals is whether or not both are surrounded by a halo. We know that many field spirals have halos, which contribute 50% or more of the mass in the luminous region, and which continue to large radii (e.g., Broeils 1992, and references therein). Halos are no doubt the dominant component in spiral galaxies, and the formation of spirals is currently thought to be driven by the collapse of the halos. The luminous matter may not be important at all for the collapse of the proto galaxy.

An obvious question therefore is whether ellipticals, and S0's, have halos. If the answer is negative, then the process by which these galaxies formed must have been dramatically different, or they must have lost their halo later.

4.1. DO ELLIPTICALS HAVE HALOS ?

One of the easiest ways to determine the mass profile of a galaxy is by measuring the rotation curve from a gas disk. Unfortunately, there are not many ellipticals with regular gas disks. The number of known gas disks in truly luminous ellipticals is virtually zero. Thus the information is rather sparse. Kent (1990) and de Zeeuw (1992) reviewed the available data, and concluded that the H1 data is generally consistent with the hypothesis that ellipticals have halos, but the evidence is far from overwhelming. A complicating factor is that ellipticals may have triaxial shapes (e.g., Franx et al 1991), and that gas orbits may be non-circular.

It is much more difficult to use the stellar kinematics to determine the density profile of elliptical galaxies. The reason is that the velocity dispersions can be anisotropic, and that introduces an extra degree of freedom (Binney & Mamon 1982). This makes it impossible to determine the density profile with the same accuracy as density profiles for spirals. Nevertheless, early results by Efstathiou et al (1982) indicated that the M/L ratio of NGC 5813 increases at large radii.

One way to solve the ambiguity is to measure velocity dispersions at large distances from the center (e.g., Saglia et al 1993). Unfortunately, this becomes extremely hard, as the surface brightness decreases rapidly with radius beyond 1 effective radius ($I \propto r^{-2}$).

Possibly the best tracer to use is the X-ray emitting hot gas, which is quite ubiquitous in luminous ellipticals (e.g., Forman et al 1985, Trinchieri et al 1986). Surface brightness and temperature measurements are necessary to obtain good density profiles. Only very few galaxies observed with the Einstein satellite had their temperature profiles measured, and thus the mass determinations were still rather uncertain. It is to be expected that the ROSAT satellite will produce accurate temperature measurements in the near future.

A more indirect argument is based on the fact that rich clusters appear to have rather high mass to light ratios. When one compares poor groups to rich clusters, the overall luminous baryonic matter to dark matter ratio appears to be roughly constant (Blumenthal et al 1984). Since the morphological mixes in groups and clusters vary widely with their density, this can be taken as proof that the ratio of dark matter per luminous baryonic matter does not change systematically with morphological type.

4.1.1. Do we expect Ellipticals to have halos ?

Another constraint on the mass of ellipticals or S0's follows from the dynamics of clusters. Merritt (1988) analyzed the effect of tidal stripping in the center of the Coma cluster. He concluded that the ratio of matter bound to galaxies to the total matter $M_{gal}/M_{tot} \leq 0.15$. The result implies that for individual galaxies $M_{gal} < 5M_{lum}$, where M_{gal} is the total galaxy mass. This implies that the early type galaxies in the centers of clusters may have lost a significant part of their halos to the cluster. The dark matter content of galaxies may obviously be influenced by their environment.

4.2. HOW ARE THE HALOS RELATED TO THE LUMINOUS PARTS ?

In the following, it is assumed that early type galaxies do have halos. It is then natural to ask, how the halos of galaxies are related to the luminous parts, and whether any systematic relationships can be found. The best observed parameter of a halo of a spiral galaxy is its circular velocity. We use the circular velocity of the H I gas as the circular velocity of the halo. For elliptical galaxies, no similar observable exists, but we can estimate circular velocities from observations of velocity dispersion profiles.

Van der Marel (1991) calculated axisymmetric models to fit observed photometry and velocity dispersions of a sample of 37 ellipticals. These models calculations were repeated by the author, with the inclusion of a dark halo, which produced a flat rotation curve. The circular velocity of the halo was taken to be the maximum of the rotation curve. The models are limited in the sense that $\sigma_z = \sigma_r$ at all places. Whereas this assumption makes it possible to easily calculate models, it is not required by any means that this assumption holds. Full, three integral models are needed for a more extensive exploration of parameter space, but that is too complex for the current purposes. As an alternative, the models were "corrected" for anisotropy as if they were spherical. That is, apparent velocity dispersion profiles were calculated for a spherical model, under different assumptions for the anisotropy $\beta(r) = 1 - \sigma_T^2(r)/\sigma_r^2(r)$. Then "correction" functions $cor(r,\beta) = \sigma_{obs}(r,\beta)/\sigma_{obs}(r, isotropic)$ were evaluated, and applied to the axisymmetric models. These corrections should be taken only as an exploration of parameter space, not as exact answers.

4.2.1. The Tully-Fisher relation for ellipticals

As a result, we were able to calculate circular velocities for elliptical galaxies. The first question one can ask, is whether ellipticals fall onto the same relation as spirals in the Tully-Fisher diagram, which relates luminosity and circular velocity. Figure 3 shows the relation for our sample of ellipticals, compared to the sample of Pierce and Tully (1988) on spiral galaxies in Virgo and Ursa Major. Figure 3a shows the maximum circular velocities of ellipticals under the assumption that ellipticals do not have halos. It is clear, that spirals and ellipticals are well separated in this diagram, by about 0.7 mag. The circular velocities used in Figure 3b were derived under the assumption that ellipticals have dark halos, and that they are radially anisotropic in the outer parts $[\beta = r^2/(r^2 + r_e^2)]$. Clearly the correspondence is better, the difference in magnitude is reduced to 0.3 when Pierce and Tully's relation is used, and -0.1 for the relation by Willick (1990). The circular velocities for the last models are reduced, because the halo helps to support the galaxy, and because of the anisotropy. Both effects reduce the circular velocity by $\approx$ 10% and $\approx$ 5%, respectively.

Ellipticals satisfy Pierce and Tully's relation exactly if the circular velocities of ellipticals are related to the central velocity dispersions by $v_{cir}/\sigma_{cen}(obs) = 1.27$. The anisotropic halo models discussed above produce an average ratio of $v_{cir}/\sigma_{cen}(obs) = 1.38$. Thus the circular velocities from the models are 10 % higher

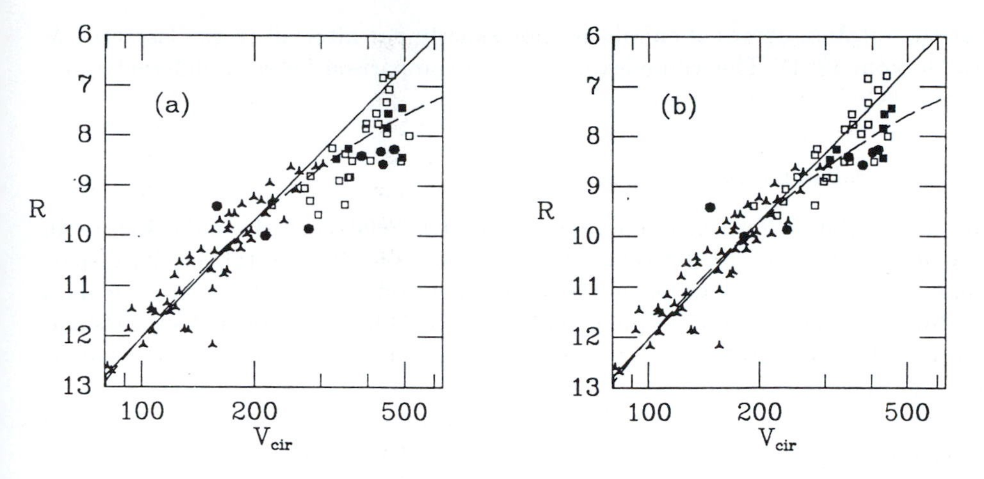

Fig. 3. The R band Tully Fisher relation for spirals in Virgo and Ursa Major (triangles), and ellipticals (boxes and filled symbols). Ellipticals have been rescaled to Virgo distance. Ellipticals in Virgo and Fornax have filled symbols. The drawn line is the Tully-Fisher relation by Pierce and Tully (1988), the dashed line is the relation given by Willick (1990). (a) The ellipticals are assumed not to have halos, and the circular velocity is the maximum circular velocity. (b) The ellipticals are assumed to have halos, and have anisotropic velocity dispersions. The circular velocities for these models are constant in the outer parts. The offset between ellipticals and spirals is rather small for these models.

than the circular velocities expected if ellipticals satisfy the Pierce and Tully's relation exactly.

It is important to note, however, that there is very little overlap in the circular velocities. Most ellipticals have large circular velocities ($>$ 250 km/s), whereas most spirals have $v_c < 250$ km/s. Thus a relation with a different shape will give a different offset. It is obvious that in the area of overlap, the offset between spirals and ellipticals is not very large in Fig. 3b. The offset reported here is smaller than those reported by Blumenthal et al (1984) and Hernquist and Quinn (1988).

In hindsight, it may not be a surprise that an offset remains between spirals and ellipticals. A residual is expected from the fact that spirals are still forming stars. Even if the baryonic content of a galaxy is uniquely related to the circular velocity, we do not expect star forming galaxies to have exactly the same luminosity as galaxies without star formation. After all, the star forming galaxies will change luminosity with time in a different way as the "dead" galaxies, and we would have been living in a very special epoch if the luminosities would have been a unique function of circular velocity, independent of galaxy type. More models are needed to investigate whether star formation alone can produce the observed differences. Furthermore, there is no obvious reason that the baryonic content is uniquely related to the circular velocity; as a matter of fact, several authors have attempted to explain the Hubble sequence by variations of halo density (and v_c) at constant baryonic mass (e.g., Blumenthal et al 1984, Zurek et al 1988). Finally, it may be that the rotation curves of spirals are not exactly flat after all (e.g., Casertano & van Gorkom 1991). This complicates a direct comparison between different types of galaxies.

4.2.2. The relation between v_c and σ_{cen}

The next question to ask, is how the spheroid is related to the halo. Figure 4a presents a plot of the circular velocity of the halo versus the central velocity dispersion of the bulge. The data for spirals were taken from Kent (1986, 1988), the data for ellipticals were taken from the models described above. Many of the bulges and ellipticals fall along the line $v_c = 1.38\sigma_{cen}$, with a large scatter below the line at low circular velocities.

The main parameter responsible for the scatter below the line may very well be bulge/disk ratio. Figure 4b presents the ratio of bulge light to total light ratio (B/T) versus σ_{cen}/v_c. The ellipticals are taken to have $B/T = 1$, and the B/T ratios for spirals were taken from Kent (1986, 1988). Note that similar diagrams were published by Whitmore et al (1979), and Whitmore and Kirshner (1981). The left most point is the extrapolated ratio of σ_{cen}/v_c for a pure disk, taken from the work of Bottema (see Bottema et al 1991, and references therein). There is an obvious trend, in the sense that galaxies with B/T < 0.2 have a low σ_{cen}/v_c. The ratio of σ_{cen}/v_c for these galaxies is comparable to the mean of σ_{cen}/v_c for pure disks. At larger values of B/T there may be a weak trend. It is encouraging that the *observed* values for bulges with $B/T > 0.5$ agrees well with the *model value* for ellipticals.

It is not clear how much of the scatter is observational, and more and better data would be useful. The data are certainly consistent with the idea that σ_{cen}/v_c drops systematically with decreasing bulge to disk ratio. This is not required by stellar dynamics, but must be imposed by the formation process (e.g., Faber 1982).

5. Stellar Populations

Recent reviews of this subject can be found in other papers in this conference proceedings, and in the proceedings of IAU Symp 149 (Barbuy and Renzini 1992). Here a short review of some topics of interest is given.

5.1. AGES

The most relevant quantity one would like to determine of a stellar population, is probably its age. It is notoriously difficult, however, to separate the effects of age differences from the effects of metallicity differences. Broad band colors allow no separation (see, e.g., the results of Arimoto and Yoshii 1987). It had been hoped that absorption line strengths could provide a way to separate age effects and metallicity effects, but even that turns out to be very difficult (e.g., Peletier, 1989, Worthey et al 1992). Detailed analyses of wide band spectrophotometric data find evidence for intermediate age populations (e.g., Pickles, 1985).

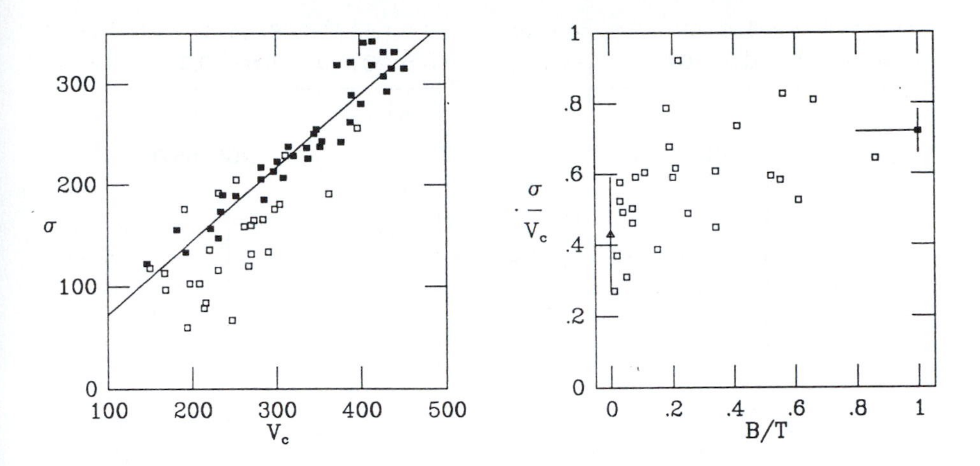

Fig. 4. (a) The central velocity dispersion σ_{cen} against halo circular velocity v_c. Open symbols are bulges, closed symbols are ellipticals. The circular velocities for ellipticals are the model values from figure 3b. (b) The ratio of σ_{cen}/v_c against the ratio of Bulge to Total light (B/T). The triangle at left is valid for pure disks, the square at right for ellipticals. Note that systems with low B/T have σ_{cen}/v_c almost equal to σ_{cen}/v_c for disks.

A complicating problem is that elliptical galaxies must contain mixes of stellar populations. In a closed box model of chemical enrichment, it is obvious that low metallicity stars form before high metallicity stars do. Thus, one expects a large range of metallicities in ellipticals, ranging from $[Fe/H] = -\infty$ to $[Fe/H] = 1$ (?). Observations of the galactic bulge have confirmed this prediction. Rich (1988) found a range in metallicities of a factor of 100 in the galactic bulge.

Arimoto and Yoshii (1987) constructed models of populations with a large intrinsic range in metallicity, and calculated broad band colors. They showed that some results change quite drastically when a range in metallicities is used, in contrast to a single metallicity. The author is not aware of independent follow up studies along these lines. It appears that any analysis based on a single metallicity model is somewhat uncertain until a more complete distribution of metallicities has been used.

5.2. COLOR MAGNITUDE RELATION

The best known relation is probably the color magnitude relation (Sandage and Visnavathan 1978). The brighter the galaxy, the redder. Since the absorption line strengths increase too with magnitude, this is generally interpreted as a correlation of metallicity with magnitude. Sandage and Visnavathan (1978) concluded that S0's follow the same relation as ellipticals, and that their populations must be very much alike. Larson et al (1980) found that the spread around the relation depends on galaxy type and environment. The S0's and ellipticals in the field have generally higher scatter than ellipticals in rich clusters. Ellipticals in the field are somewhat

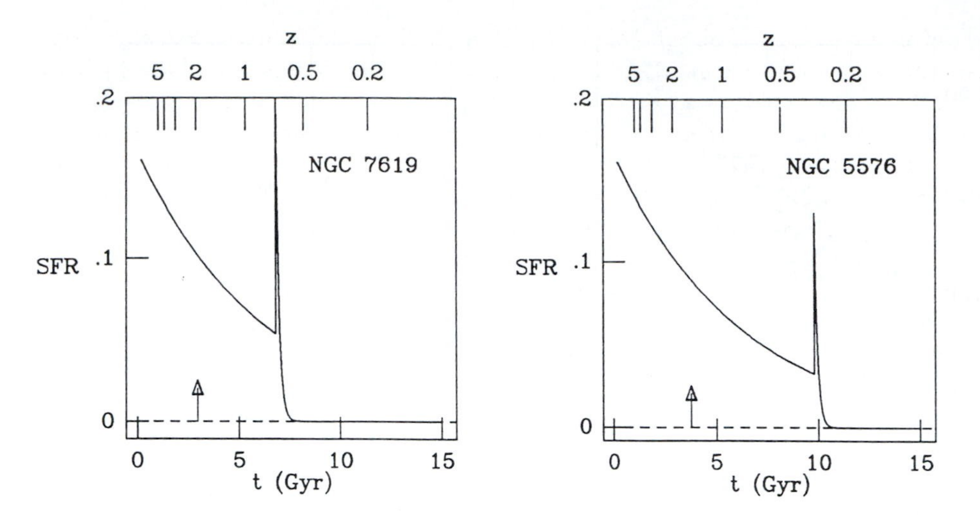

Fig. 5. The star formation histories for ellipticals as modeled by Schweizer & Seitzer (1992). The galaxies are assumed to evolve like an Sb galaxy, until they merge, and quickly exhaust their gas supply. The two galaxies span a typical range in merger age. Note that merging occurs very late for these systems ($0.3 < z < 0.7$, $q_0 = 1/2$). The arrows indicate the mean stellar age. The mean stellar is very uniform at t=3 Gyr equivalent to $z \approx 2$.

bluer than cluster ellipticals. These results indicate that star formation histories probably depend on the environment. More systematic surveys would obviously be useful.

5.3. THE SPREAD IN MEAN AGES

The color magnitude relation implies that the scatter in color is quite low, given the magnitude. This can be used to obtain a constraint on the spread in ages. Schweizer and Seitzer (1992) analyzed a sample of field ellipticals. They used deep photometry, and overall colors of galaxies to try to estimate the star formation history of galaxies. Their observed scatter in $U - V$ at a given absolute magnitude was 0.07 mag. Their (merger) models assumed that the star formation decays exponentially in the progenitor galaxies, until a merger occurs, in which part of the remaining gas is quickly converted into stars, and the rest is expelled from the galaxy. A schematic representation of such a formation scenario is given in Figure 5a and b. The mergers occurred about 8 ± 2 Gyr ago, and the universe is taken to be 15 Gyr old. This relatively late "assembly" implies that we can observe these mergers at redshifts between 0.3 and 0.7. The mean stellar age implied by their model is much higher, 12 $\pm$ 0.5 Gyr, corresponding to a redshift of 1.9. This illustrates clearly that the "epoch of star formation" is not necessarily the same as the "epoch of galaxy assembly", and constraints on the mean ages are not necessarily constraints on the last major event in a galaxies lifetime.

The same spread in colors can be modeled in a completely different way. If it is assumed that the galaxy forms in a very short star burst, then the spread implies a spread in stellar age of about 4 Gyr (FWHM). This model implies that enormous star bursts occur in a redshift range of 1.4 to ∞, if the bursts start at very high redshift. It may equally well be that metallicity variations add to the scatter (e.g., Bertola et al 1993), or small bursts of starformation at very recent times. It is clear that different models, which reproduce the data equally well, produce very different predictions in redshift space. One may therefore hope that observations of high redshift galaxies will help greatly to solve the ambiguity.

A recent study of Coma and Virgo by Bower et al (1992) produced a spread in the color magnitude diagram of about 0.04 in $U - V$, about half the spread found by Schweizer & Seitzer (1992) for their field sample. The result implies that the formation history of these cluster galaxies was more uniform. The same ambiguities remain in the modeling of this data.

6. Summary and Discussion

The similarity of ellipticals and bulges is striking. Many ellipticals may actually have faint disks, demonstrating that disk formation is an even more common aspect of galaxy formation than previously thought. The exception is formed by very bright elliptical galaxies, which are usually the brightest, or among the brightest galaxies in groups and clusters. There is no strong evidence for disks in these galaxies. Thus, if one defines ellipticals narrowly as galaxies without disks, many, if not most of the ellipticals should be redefined as S0's, and we are left with Bright Cluster Members as "true ellipticals". However, the renaming of (faint) ellipticals into S0's does not explain very much, and is certainly not favored by the author.

Thus, whereas disk formation may be a common process in the formation of ellipticals, or, generally speaking, high Bulge/Disk ratio systems, bulge formation may be part of the disk formation in low B/D systems. At the lowest B/D ratios, the bulge velocity dispersions are possibly coupled to the disk velocity dispersions, and the bulges may have formed as part of the formation of the disks.

The available data is consistent with the hypothesis that high B/D systems have halos, but the evidence is still scarce. An exploration of models indicates that ellipticals may follow more or less the same relation between luminous mass and halo circular velocity, but this result remains model dependent. The ratio of bulge velocity dispersion to halo circular velocity decreases with decreasing B/D ratio. This must be imposed by the formation process.

The question to be answered at this stage is what determines the Bulge/Disk ratio of galaxies, and the correlation between B/D ratio and environment and luminosity. It is obvious that baryonic processes (like dissipation, star formation, reheating, etc.) play an important role. Dissipationless collapse would produce rather similar halos, and without the systematic changes that are observed in the baryonic components (e.g., Barnes and Efstathiou 1987). Nevertheless, since the dark matter contributes most of the mass, it must play an important role in determining the structure of the galaxies. Possibly the most essential question that can be asked at this moment is how the luminous parts of galaxies are related to their halos. The current evidence suggests that early type systems have slightly higher circular velocities for the same red luminosity. The difference may only be small, however, (15%

?), and is partly or fully caused by different M/L ratios, and systematic changes in the shapes of rotation curve with mass.

Zurek et al (1988), and Katz (1992) speculated that the Hubble type was driven by the properties of the individual (dark matter) clumps before assembly. A higher differentiation of clumps before assembly can lead to higher cooling rates, and higher star formation rates before the assembly. During the assembly (or "merger") the more differentiated clumps can loose more angular momentum against the dark halo material. This explains why the specific angular momentum decreases with earlier Hubble type. The fact that early type galaxies may have higher halo densities (as demonstrated by their higher circular velocities) is in agreement with this hypothesis.

6.1. FUTURE WORK

The numerical simulations of galaxy formation in CDM type universes are maturing at this very moment. Previously, questions about details of the baryonic component of galaxies could not be addressed, but now, with the inclusion of hydrodynamics, such questions can actually be studied (e.g., Katz et al 1992). It appears as if much progress can be made along these lines. The current results are very encouraging, but, of course, one cannot claim to have understood all properties at this moment. Feedback processes like heat injection from supernovae in the ISM, and the role of nuclear activity have not been fully explored. The total amount of energy released by both phenomena is very large, and must be important in the overall galaxy formation process.

On the observational side, there is a need for systematic studies of the dark matter components of galaxies, and their relation with the luminous parts. It is important to try to understand the relation between galaxy morphology/structure versus halo structure. Unfortunately, such studies are very difficult for galaxies other than late type spirals, and they will require large investments of telescope time. Studies of systematic changes of galaxy properties with environment (or mean density) are extremely useful. It should not come as a surprise, if the galaxy formation process is significantly influenced by the underlying, large scale density distribution (e.g., Katz et al, 1992).

Secondly, direct observations of galaxies at high redshift are becoming possible, which enables us to observe some of the processes that are speculated to occur. An interesting example are the high redshift radio galaxies, where star formation appears to be coupled to the nuclear activity (e.g., Chambers et al 1990). Another example is the detection of evolution of galaxies in rich clusters (" the Butcher–Oemler effect", Butcher & Oemler 1984). Studies of such galaxies may provide a glimpse of the formation process in working.

Acknowledgements

It is a pleasure to thank the organizers of the conference for their support. Discussions with Steve Kent, Neal Katz, Hans-Walter Rix, and Paul Schechter are gratefully acknowledged.

Tim de Zeeuw and Konrad Kuijken commented on the manuscript. This research has been funded by Hubble Fellowship grant HF-1016.01.91A.

References

Arimoto, N., & Yoshii, Y. 1987, *AA*, **173**, 23.

Athanassoula, E., 1983, IAU Symposium 100 "Internal Kinematics and Dynamics of Galaxies" (Dordrecht: Reidel).

Barnes, J., & White, S. D. M., 1984, *MNRAS*, **211**, 253.

Barnes, J., & Efstathiou, G., 1987, *ApJ*, **319**, 575.

Barbuy, B., & Renzini, A., IAU symposium 149 "The Stellar Populations of Galaxies", (Dordrecht: Kluwer).

Bender, R., 1988a, *AA*, **193**, L7.

Bender, R., 1988b, *AA*, **202**, L5.

Bender, R., Burstein, D., & Faber, S. M., 1992, *ApJ*, **399**, 462.

Bender, R., Döbereiner, S., & Möllenhof, C., 1988, *AASup*, **74**, 385.

Bender, R., & Nieto, J. L., 1990, *AA*, **239**, 97.

Bertola, F., Burstein, D., & Buson, L., 1993, *ApJ*, **403**, 573.

van den Bergh, S., 1990, *ApJ*, **348**, 57.

Binggeli, B., Sandage, A., & Tammann, G. A., 1988, *ARAA*, **26**, 509.

Binney, J. J., & Manon, G. A., 1982, *MNRAS*, **200**, 361.

Blumenthal, G. R., Faber, S. M., Primack, J. R., & Rees, M. J., 1984, *Nature*, **311**, 517.

Bottema, R., van der Kruit, B. & Valentijn, E., 1991, *AA*, **247**, 357.

Bower, R. G., Lucey, J. R., & Ellis, R. S., 1992, *MNRAS*, **254**, 601.

Broeils, A. H., 1992, PhD Thesis, University of Groningen.

Butcher, H., & Oemler, A., 1984, *ApJ*, **285**, 426.

Carter, D., 1978, *MNRAS*, **182**, 797.

Carter, D., 1987, *ApJ*, **312**, 514.

Capaccioli, M., 1987, in IAU symposium 127, "Structure and Dynamics of Elliptical Galaxies", ed. T. de Zeeuw (Dordrecht: Reidel), p47.

Capaccioli, M., Caon, N., & D'Onofrio, M. 1992, *MNRAS*, **259**, 323.

Casertano, S., & van Gorkom, J. H., 1991, *AJ*, **101**, 1231.

Chambers, K. C., Miley, G. K., & van Breugel, W. J. M., 1990, *ApJ*, **363**, 21.

de Zeeuw, P. T., 1987, IAU symposium 127, "Structure and Dynamics of Elliptical Galaxies", (Dordrecht: Reidel).

de Zeeuw, P. T., 1992, in "Morphological and Physical Classification of Galaxies", ed. Busarello et al (Dordrecht: Kluwer).

de Zeeuw, P. T., & Franx, M, 1991, ARAA, **29**, 239.

Davies, R. L., Efstathiou, G., Fall, S. M., Illingworth, G. D., & Schechter, P. L., 1983, *ApJ*, **266**, 41.

Dressler, A., 1980, *ApJ*, **236**, 351.

Dressler, A., 1984, *ApJ*, **286**, 97.

Dressler, A., 1987, *ApJ*, **317**, 1.

Dressler, A., & Sandage, A. 1983, *ApJ*, **265**, 664.

Efstathiou, G., Ellis, R. S., & Carter, D., 1982, *MNRAS*, **201**, 975.

Faber, S. M., 1982, in "Astrophysical Cosmology", ed. H.A. Brück et al (Vatican: Pontificia Academia Scientiarum), p219.

Faber, S. M., Dressler, A., Davies, R. L., Burstein, D., Lynden-Bell, D., Terlevich, R. J., & Wegner, G., 1987, in "Nearly Normal Galaxies", ed. S. M. Faber (New York: Springer), p. 175

Forman, W., Jones, C., & Tucker, W., 1985, *ApJ*, **293**, 102.

Fisher, D., Illingworth, G. D., & Franx, M., 1993, in preparation.

Franx, M., & Illingworth, G. D., 1988, *ApJ (Letters)*, **327**, 55.

Franx, M., Illingworth, G. D., & Heckman, T., 1989, *ApJ*, **344**, 613.

Franx, M., Illingworth, G. D., & de Zeeuw, P. T., 1991, *ApJ*, **383**, 112.

Hernquist, L., & Quinn, P. J., 1987, *ApJ*, **312**, 1.

Hubble, E., 1936, "Realm of Nebulae" (New Haven: Yale University Press).

Illingworth, G. D., 1983, in IAU Symposium 100, "Internal Kinematics and Dynamics of Galaxies", ed. E. Athanassoula (Dordrecht: Reidel), p257.

Jedrzejewski, R., & Schechter, P. L., 1988, *ApJ (Letters)*, **330**, 87.
Katz, N., 1992, *PASP*, **104**, 852.
Katz, N., Hernquist, L., & Weinberg, D. H., 1992, *ApJ (Letters)*, **399**, 109.
Kent, S. M., 1981, *ApJ*, **245**, 805.
Kent, S. M., 1984, *ApJSup.*, **56**, 105.
Kent, S. M., 1986, *AJ*, **91**, 1301.
Kent, S. M., 1988, *AJ*, **96**, 514.
Kent, S. M., 1990, in ASP conf ser. 10, "Evolution of the Universe of Galaxies", ed. R. G. Kron (Provo: Brigham Young Univ.), p109.
Kormendy, J., & Illingworth, G. D., 1982, *ApJ*, **256**, 460.
Kormendy, J., 1988, *ApJ*, **335**, 40.
Kormendy, J., 1982, in "Morphology and Dynamics of Galaxies", ed. L. Martinet & M. Mayor (Sauverny: Geneva Obs.), p113.
Larson,, R. B., Tinsley, B. M., & Caldwell, C. N., 1980, *ApJ*, **237**, 692.
Merritt, D. 1988, in ASP conf. ser. 5, "The Minnesota Lectures on Clusters of Galaxies and Large Scale Structure", ed. J. M. Dickey (Provo: Brigham Young Univ.), 175.
Peletier, R. F. P., 1989, PhD Thesis, University of Groningen.
Pierce, M.J., & Tully, R.B., 1988, *ApJ*, **330**, 579.
Pickles, A. J., 1985, *ApJ*, **296**, 340.
Rich, R. M., 1988, *AJ*, **95**, 828.
Rix, H. W., Franx, M., Fisher, D., & Illingworth, G., 1992, *ApJ (Letters)*, **400**, 5.
Rix, H. W., & White, S. D. M., 1990, *ApJ*, **362**, 52.
Rubin, V. C., Graham, J. A., & Kenney, J. D. P., 1992, *ApJ (Letters)*, **394**, 9.
Sandage, A., & Visnavathan, N., 1978, *ApJ*, **225**, 742.
Saglia, R. P., Bertin, G., Bertola, F., Danziger, J., Dejonghe, H., Sadler, E. M., Stiavelli, M., de Zeeuw, P. T., & Zeilinger, W. W., 1993, *ApJ*, **403**, 573.
Schechter, P. L., & Dressler, A., 1987, *AJ*, **94**, 563.
Schweizer, F., & Seitzer, P. 1992, *AJ*, **104**, 1039.
Trinchieri, G., Fabbiano, G., & Canizares, C. R., 1986, *ApJ*, **310**, 637.
van der Marel, R. P. 1992, *MNRAS*, **253**, 710.
Whitmore, B. C., Kirshner, R. P., & Schechter, P. L., 1979, *ApJ*, **234**, 68.
Whitmore, B. C. & Kirshner, R. P., 1981, *ApJ*, **250**, 43.
Willick, J.A., 1990, *ApJ (Letters)*, **351**, L5.
Worthey, G., Faber, S. M., & Gonzales, J. J., 1992, *ApJ*, **398**, 69.
Zurek, W. H.,Quinn, P. J., & Salmon, J. K., 1988, *ApJ*, **330**, 519.

DISCUSSION

Ferguson: The exercise with the colours is an interesting one, but has yet to be subjected to the sort of tests that the Yoshii and Arimoto models have been, with lots of different colours and in predicting the C–M diagram and so on. It seems that if the simple chemical evolution models work, they are the Occam's razor solution and the merger models need to be very convincing to argue that they are a significant factor in elliptical formation in general.

Franx: You can separate two things, merging and star formation, and they are not necessarily the same thing of course. My impression is that it will be very hard to distinguish between those two options in the nearby galaxies. When star formation has stopped for 5 Gyr, it becomes very hard to say very much about what happened before that, and if you look in redshift space those differences become irrelevant.

Gerhard: A comment on the determination of the halo in ellipticals: I don't think that the anisotropy problem is really such a big one. I know it has been around for 10 years and it is an effect, but once you go out into the regimes where you actually expect some effect of the halo, I think that the effect of tangential anisotropy, which is one that could mimic dark matter, relative to isotropy is only 20% or so. When it comes to these differences, I don't think you would be bold enough to interpret this as a heavy halo anyway. The biggest problem is to go out to large enough radii, where you can collect sufficient photons, which is the topic of an ESO key project.

Franx: I think you are right and we can improve on what has been done. We want to settle more than just the question whether elliptical galaxies are very large; we want to determine what their circular velocities are. It still remains to be seen how good you can do that with kinematic data from the light of the stars.

Gerhard: You were showing some rather interesting cases for ellipticals and bulges with sub–structure, and, yes, you can argue that you wouldn't see that structure in a lot of cases, and we could say that in a third of cases you would definitely not see it, but isn't it none the less true that the majority of them are rather boring in a sense?

Franx: Yes, I completely agree with you that we only see this structure in a small fraction of ellipticals. The most amazing aspect is though, that all the other properties of these ellipticals appear to be very normal, so it suggests but doesn't prove in any way, that this kind of sub–structure that we are seeing is one of the many outcomes of the formation process.

Renzini: The recent paper by Bowers, Lucy and Ellis, does much better than Schweizer *et.al.*: they have a dispersion in colour of only 0.03 magnitude for Virgo and Coma ellipticals, for a given central velocity dispersion. This allows them to set a lower limit to the age of these elliptical galaxies' stellar populations, which is at least 13 Gyr for over 90% of the light we see today. So, at this point it is important to get it through to people that when you compare the so called fast formation process for ellipticals to a slow formation process, we have the two time scales which are at this point constrained to be less than just a few Gyr. That has not been, in clusters, a continuous formation of ellipticals at a nearly constant rate over the last 15 Gyr, as a result of merging.

Franx: It is well known effect, that when we look at clusters of galaxies at intermediate redshifts, they appear to be full with old galaxies. I showed a cluster at a redshift of only 0.2, but when we make a colour map of that cluster, it is indeed full of old galaxies, but it is well known that even at these redshifts we see some indications of blue galaxies. This is of course the Butcher-Oemler effect, which has been observed by many other people as well. It actually turns out that it is possible to derive rotation curves for these galaxies and in the cluster I showed, the largest blue galaxy is rotating very rapidly. So what this galaxy probably is, is an S0 galaxy in transformation. So we shouldn't lose sight that, despite the constraints of newer data, even at very low redshifts we do see evidence of star formation and galaxy transformation.

Renzini: Okay, one can always find weird objects, but one must quantify statements instead of making an example of one case, no matter how many such weird cases there are. I wanted to make a second point, you have used so far only the dispersion in the C-σ relation, there is the trend to explain as well. Galaxies with higher central velocity dispersion, which means a deeper potential well, are redder and this to me means you have to form the stars within that particular potential. You don't merge already formed stellar systems. You merge, if you do merge, gas and then you make the stars.

Franx: The CDM simulations (in the literature) show that the depth of a perturbation at redshift zero is closely correlated with the depth of the perturbations of which they are formed from at a redshift of 1.0, so there is a very good correlation between the halos before they merge and after they merge. Which is probably only saying that big clumps form even bigger clumps.

The artist, L. Debrouwere (first in front row), enjoying a rare moment of rest during Habing's concluding words

Colin NORMAN

KEN FREEMAN
'92

SUMMARY: ACHIEVEMENTS AND OPEN QUESTIONS

K.C. FREEMAN
Mount Stromlo and Siding Spring Observatories
Weston Creek PO, ACT 2611, Australia

1. Introduction

Without attempting to cover all of the topics discussed at this meeting, I plan to talk briefly about some of the major achievements and ideas that came up here, and then mention a few of the important open questions and some work that might be done to answer them.

2. Kent's Model

A good galactic model for the inner Galaxy is needed to interpret the kinematics and light distribution of various populations. Kent's (1992) simple model, based on the Spacelab 2.4μ surface photometry, has a radial density distribution $\rho(r) \propto r^{-1.8}$ for $r < 900$ pc and steepens to $\rho(r) \propto r^{-3.7}$ for larger r. The model is axisymmetric, isotropic and flattened, with axial ratio 0.6 and mass-to-light ratio $M/L(2.2\mu) = 1$, and it accounts nicely for the kinematics of a wide range of bulge tracers (OH/IR stars, K giants, the integrated bulge light, planetary nebulae and M giants) between about 1 and 1500 pc from the galactic center.

The model shows that the high apparent rotational velocities observed for HI and CO within about 1 kpc of the galactic center are unlikely to represent the true circular velocity in this region (as most people have long suspected anyway). Detailed analysis of gas motions in the inner few hundred parsecs indicates that there is a central bar, with pattern speed $\Omega_p = 63$ km s^{-1} kpc^{-1} (Gerhard). Corotation is at 2.4 kpc from the center, and the inferred density distribution is in excellent agreement with Kent's model. There is now much direct evidence for the presence of a central bar, from the distributions of several different bulge tracers (but not from the distribution of the RR Lyrae stars). Analysis of the COBE data should help define the parameters of this bar (de Zeeuw).

The true radial density distribution of the bulge is difficult to establish directly, because it is inevitably confused with gradients in the abundance and age of the stellar population (King, Rich). Kent's model is derived from infrared surface photometry and agrees well with the radial density distribution of Miras (Whitelock). However the other M giants and RR Lyrae stars in the inner bulge show a steeper radial gradient.

H. Dejonghe and H. J. Habing (eds.), Galactic Bulges, 263–269.

3. Kinematics of OH/IR Sources and SiO Masers

The (longitude, velocity) distributions are similar for the various kinds of bulge objects in late stages of stellar evolution (planetary nebulae, Miras, OH/IR stars, SiO masers), and there seems to be agreement that at least a fair fraction of these objects are very old (ages > 10 Gyr). The typical mean rotation of the bulge is about 10 km s^{-1} $degree^{-1}$. The kinematics of the SiO masers and OH/IR sources indicate that the velocity dispersion and mean rotation of the bulge decrease slowly with increasing galactic latitude for $|b| > 3^\circ$. The velocity dispersion is $101 - 3.6\ |b|$ km s^{-1}, and the mean angular velocity is $15.6 - 1.23\ |b|$ km s^{-1} $degree^{-1}$ (Izumiura), so the rotation of the galactic bulge may be somewhat different from the cylindrical behaviour seen in the rotation of other boxy bulges.

Habing and Dejonghe presented a two-integral quadratic programming model to represent the distribution and kinematics of the OH/IR stars. The model gives the distribution of stellar orbits in the (angular momentum, energy) plane. The morphology of the model in this plane is interesting: qualitatively, it is like the two-integral distribution function models that have been constructed for boxy bulges (eg Rowley, 1988). For the future, the two-integral distribution function could also be derived for the inner bulge by applying Hunter's (1992) new inversion technique to Kent's model (de Zeeuw), and it will be interesting to see how well these two approaches agree.

The kinematics of the inner OH/IR stars (Lindqvist *et al*) show that the mean velocity reaches about 150 km s^{-1} at radius $R = 0^\circ.5$. This velocity is close to the circular velocity for the Kent model, and suggests the presence of an inner disk which may be somewhat younger (Whitford). The brightest bulge objects at the K-band, which would include the youngest and most metal-rich, show the most flattening.

For the future, we look forward to the new VLA / Australia Telescope survey of the bulge OH/IR stars, and offer every encouragement for proper motion studies of the SiO and OH masers.

4. The bulge - halo transition

As King pointed out in his opening talk, we do not yet know how the galactic bulge is related to the other components of the Galaxy (thin disk, thick disk, halo). Many authors in the past have regarded the bulge and the metal-poor halo as together making up the galactic spheroid. However, in a late type galaxy like ours, the concept of the spheroid is itself not well defined (eg Wyse and Gilmore, 1988), and King suggests that this terminology should be avoided. Some results presented at this meeting indicate that the metal-poor halo and the bulge are probably not closely related.

The second parameter effect on the morphology of the horizontal branch, as seen in the halo globular clusters, is also shown by the blue horizontal branch stars (BHB) of the metal-poor halo. The ratio of RR Lyrae stars to BHB stars increases with increasing galactocentric distance, from the edge of the bulge ($l = 0^{\circ}$, $b = -10^{\circ}$) out to about 25 kpc from the center , and the BHB becomes less blue (Preston). For the globular clusters, the second parameter effect is now believed to be due to age variations with galactocentric radius: the outer metal-poor halo formed later and over a longer period of time.

In Baade's window, the horizontal branch is predominantly red. There appears to be a marked transition in HB morphology between halo and bulge, from the very blue HB just outside the bulge (ie inner halo) to the relatively red HB in Baade's window (ie bulge). Lee (1992) proposed that the RR Lyrae stars in Baade's window may be the oldest objects in the Galaxy. Renzini argued on other grounds that bulges are on average older than the metal-poor halos.

Preston estimated that the mass of the metal-poor halo stars ([Fe/H] < -1) within 1 kpc of the galactic center is about 3.10^8 $M_{\odot}$; this is only a few percent of the mass of the bulge. Morrison and Harding studied the stellar population in a bulge field at $l = -10^{\circ}$, $b = -10^{\circ}$. In this field, the peak of the chemical abundance distribution is at [Fe/H] $\approx$ -0.8. The metal-poor stars with [Fe/H] < -1 have halo kinematics, while the more metal-rich stars with [Fe/H] > -1 have typical bulge (or thick disk ?) kinematics. Their halo model indicates that only a small fraction of the Baade's window giants belong to the metal-poor halo, supporting the view that the bulge is *not* just the inner part of the halo.

For the future, it would be very interesting to know more about the kinematics of the intermediate abundance 47 Tuc-like population that pervades the outer bulge and the thick disk out to galactocentric distances R of at least 12 kpc. Not much is known about the kinematics of this population in the inner few kpc of the Galaxy.

5. Stellar Proper Motions in the Bulge

Proper motions are now becoming available for stars in the bulge. This is an important advance for understanding the dynamics of the bulge region, because proper motions and radial velocities give direct information about the rotation and kinematic isotropy of the bulge (eg Rich 1990). Proper motion data presented at this meeting (Minniti, Terndrup *et al*) indicate that the bulge is kinematically roughly isotropic in Baade's window and in a field at R = 1.6 kpc, with a velocity dispersion of about 110 km s^{-1}. There is a hint of tangential anisotropy for the more metal rich stars with [Fe/H] > 0.3. The stellar bulge as defined by the K giants is clearly rotating.

We look forward to more proper motion data for bulge stars, and to some absolute proper motions in the future.

6. The Abundance Gradient and Abundance Distribution

The observed abundance distribution for the inner bulge extends over the range $-1 < [Fe/H] < +1$, and is similar to that for the closed box model. Tyson reported on the radial abundance distribution along the minor axis of the bulge. The mean abundance appears to be roughly constant out to about 1.2 kpc from the center (near the edge of the COBE bulge) and then drops abruptly. This is an important observation, and we look forward to better statistics which will soon be available. The origin of the high mean abundance within the COBE bulge is not yet clear. Is it the product of early dissipational collapse and chemical evolution, or does it result from bar-driven starbursts in the inner bulge (see §9) ? The element ratios observed in the K giants of the bulge indicate moderate enhancement of the α-elements and enhancement of [Eu/Fe], suggesting a significant contribution from SN II to the chemical evolution.

Balcells found evidence for color gradients in the bulges of some external galaxies, particularly in the more luminous systems.

7. Structure and Content of other Bulges

Bulges appear to be dynamically close to isotropic oblate rotators, although the kinematics and light distributions show that some are clearly triaxial. Some bulges are boxshaped, like the bulge of our Galaxy; the incidence of boxiness in bulges is about 30 percent. The origin of this boxiness (vertical heating of the disk, oblique accretion, dissipative collapse ?) is not yet fully understood.

The high V/σ values and relatively low velocity dispersions of some apparent bulges, particularly in the later-type spirals, indicate that they are likely to be features in the disk (Kormendy). This suggests that secular evolution has taken place in these disks, possibly through bar formation.

Bulges and intermediate luminosity ellipticals show similar global properties, such as the color-magnitude and metallicity-luminosity relations (Bertola).

8. The Hot Stellar Component in the Bulge of M31

Studies of the hot stellar component in the bulge of M31 and in the ellipticals M32 and NGC 1399 (Ferguson) indicate that the incidence of classic post asymptotic giant branch stars falls as the abundance [Fe/H] increases (as predicted by Greggio and Renzini, 1990). This suggests that the incidence of planetary nebulae may be biased away from higher [Fe/H] values.

9. Bar Formation and Destruction, and the Bulge

Bar formation and destruction may be very important for the formation of bulges in disk galaxies, as discussed at this meeting by Hasan, Norman, Pfenniger and Sellwood. The sequence of events may proceed in this way:

1. Vertical resonances associated with bar formation pump disk stars into a box-shaped bar/bulge (here bar/bulge is just used to mean a barlike or triaxial bulge).
2. A bar is easily destroyed by a small central mass that is only a few percent of the total bar mass. This central mass affects the orbital structure of the bar, and transforms the bar into a roughly axisymmetric bulge-like system.
3. A bar in a disk galaxy drives a gas inflow which leads to a rapid aggregation of mass at the center of the bar (eg Wada's talk at this meeting). In this sense the bar provides the source of its own destruction. Central starbursts are associated with this gas flow, and these star-bursts may contribute significantly to the chemical enrichment of the inner bar/bulge.
4. This cycle of events may happen more than once: infall of cold gas into the inner regions of the disk may allow a bar to form again.

In this picture, normal bulges form from disk matter through the dynamical effects associated with bar formation and destruction. The accretion of a large (~ 10 percent) satellite can also lead to the formation of a bulge from disk matter. This may be appropriate to the formation of the larger bulges like that of the Sombrero galaxy NGC 4594.

If this picture of bulge formation through bar formation is correct, then we need to understand (as for any bulge formation picture) why so many late type disk galaxies (Sc and Sd) show no evidence for a bulge. Is it possible that these late type systems are stabilized against bar formation by their dark matter ? This seems unlikely, because late type barred galaxies are common enough. Or is the vertical pumping by the bar in these very flat systems less efficient over the lifetime of the bar ?

10. Some questions

10.1 THE AGE OF THE BULGE

To understand the sequence of events that led to the formation of the bulge, we need to know the relative ages of the metal-poor halo, the galactic bulge and the old disk in the inner Galaxy. It is important to have a definitive non-kinematical measurement of the age of the bulge. In the inner Galaxy, stellar kinematics may discriminate between populations older or younger than a few Gyr, but are unlikely to be reliable age indicators for older populations. Some age estimators that could be useful for the bulge include:

- Main sequence turnoff colors in Baade's window (Baum, King).

- The magnitude difference between the giant branch clump and the main sequence turnoff.
- Properties of the HB and RR Lyrae stars (eg Lee 1992).
- Limits on turnoff masses estimated from Preston's bulge Algol stars.
- Strömgren photometry of the evolved F/G star population in the bulge.

10.2 DISK-BULGE COEXISTENCE

Is there an old disk coexistent with the bulge, as in some ellipticals ? Or is there a Kormendy (1977) hole in the inner disk ? This question is relevant to the idea of bulge formation *via* bars. How could we tell ? For external galaxies, we would need detailed modelling to estimate the expected kinematical properties of stars along lines of sight passing through the inner disk and bulge. Then infrared photometric and spectroscopic observations of the inner regions of some edge-on bulges, may help to answer this question. For the inner regions of our Galaxy, studies of OH/IR sources are important for this problem (see §3). More proper motion work in low-latitude galactic windows would also be useful.

10.3 BAR/BULGES

We need to know more about the observational structure, kinematics and metallicity distribution of other small bar/bulges like that of NGC 4565 in order to interpret the observations of the galactic bulge. Realistic dynamical models of rotating bars would be very helpful for interpreting the observed structure and kinematics of small bar/bulges.

10.4 DYNAMICAL COMPARISON OF OUR BULGE WITH OTHERS

In other bulges, the kinematics are measured from integrated light (ie K giants) and planetary nebulae (for a few nearby systems like NGC 3115 and the Sombrero). It would be useful to exploit these techniques for the bulge of our Galaxy, in order to facilitate direct comparison of our bulge with others. Integrated light observations of the galactic bulge are straightforward in regions away from high interstellar absorption (eg Freeman *et al* 1988) but have not been used much so far. More extensive surveys for planetary nebulae in the galactic bulge, for detailed kinematical studies, would be easy and useful.

11. Conclusion

Bulges are interesting because they contain an important part of the history of galaxies like the Milky Way, and we need to know how the bulges fit in to the overall picture of galactic formation and evolution. Bulges may be the seeds for galaxy formation (at least for those galaxies that have bulges), or they may form later through dynamical effects (bar formation, accretion) acting on the disk. This makes the age determination of the galactic bulge particularly interesting. In the future, if we want to understand the events leading to the formation of bulges, then studies of galaxies at high redshift will surely be very important.

References

Freeman, K., de Vaucouleurs, G., de Vaucouleurs, A., Wainscoat, R. 1988. ApJ, 325, 563.
Greggio, L. & Renzini, A. 1990. ApJ, 364, 35.
Hunter, C. 1992. Presentation at Heidelberg Workshop: *Internal Dynamics of Ellipticals and Bulges.*
Kent, S. 1992. ApJ, 387, 181.
Kormendy, J. 1977. ApJ, 217, 406.
Lee, Y-W. 1992. Presentation at 11th Santa Cruz Workshop: *The Globular Cluster - Galaxy Connection.*
Rich, R.M. 1990. ApJ, 362, 604.
Rowley, G. 1988. ApJ, 331, 124.
Wyse, R. & Gilmore, G. 1988. AJ, 95, 1404.

DISCUSSION

Sellwood: Several different groups arguing on several different grounds, have all come to the conclusion that our galaxy is barred. What is reassuring about this is that they all put the angle of the bar in the same quadrant, although there is disagreement about the exact angle.

King: Are you bothered by the fact that the bar does not seem to show its signature in the stellar kinematics?

Freeman: Yes I am, but I'm still confused about what we are really looking, with the highly evolved objects.

Rich: I remain worried about the fact that we see these correlations between abundance and kinematics, and it seems hard to get them with the bar thickening mechanisms. Also, the correlations exist through the solar region of metallicity and above. So we are talking about the majority of stars in the Bulge, not just separating out halo and extremely metal rich.

Balcells: A basic test of the bar model for peanut shaped bulges, I think, is quite simple. If we place them on a V/σ v's flattening diagram, the behavior of oblate spheroids and bars should be quite different.

Michael FEAST

PAPERS

3D GAS DYNAMICS IN TRIAXIAL SYSTEMS

D. FRIEDLI and S. UDRY
Geneva Observatory, CH-1290 Sauverny, Switzerland

1. INTRODUCTION

Depending on the nature of the various components (stars, gas) present in triaxial stellar systems (elliptical galaxies, bulges and bars), the dynamics is expected to be rather different. The stars are collisionless, dissipationless, and dynamically hot; they are mainly trapped by quasi-periodic or chaotic orbits. On the contrary, the gas is collisional, dissipational, and dynamically cold; the cold or warm gas ($\lesssim 10^4$ K) is a powerful orbital tracer, however shocks prevent it from following self-crossing orbits. The hot gas ($\gtrsim 10^6$ K) is influenced by "repulsive" pressure forces which prevent in close encounters the flow from being strongly shocked; it rather follows chaotic trajectories. By means of fully self-consistent **3D** simulations with stars and gas using PM (Pfenniger & Friedli 1992) and SPH (Friedli & Benz 1992) techniques, we investigate the response of gaseous components in the following situations: 1) slow or fast pattern speed Ω_p, 2) direct or retrograde gas motion with respect to the stars, and 3) warm or hot gas temperature T. Initial parameters and final characteristics of each runs are reported in Table I.

TABLE I

For all models, the total mass $M_t = M_g + M_* = 2 \cdot 10^{11}$ $M_\odot$, with M_g the gas mass and M_* the star mass. The number of gas and star particles are $N_g = 10^4$ and $N_* = 10^5$ respectively.

	Direct warm gas		Retrograde warm gas		Hot gas	
Rotation	Slow	Fast	Slow	Fast	Slow	Fast
Figures	1a-b	2a-b	–	3a-b	4a-b	–
t_{fig} [Myr]	400	1600	300	1200	1000	2000
Ω_{p} [kms^{-1}kpc^{-1}]	5	28	7	22	5	20
T_{beg} [K] (center)	5000	5000	5000	5000	$2.5 \cdot 10^5$	$2.5 \cdot 10^5$
T_{end} [K] (center)	5000	5000	5000	5000	$7 \cdot 10^6$	$3 \cdot 10^6$
M_g/M_t	0.10	0.05	0.01	0.05	0.05	0.05

2. MAIN RESULTS

1) The warm gas is a tracer of the galaxies' gravitational potential and of its related resonances. Due to shocks, it cannot remain on self-crossing trajectories including (a) quasi-periodic or chaotic orbits, (b) periodic orbit families with loops, (c) different periodic orbit families devoid of loops but having mutually crossing orbits.
2) The warm gas is sensitive to vertical resonances associated with triaxial shapes. It can leave the galactic plane to follow **3D** orbit families and can be found at large distances in z ($\gtrsim 1$ kpc). Its detection well above the galactic plane in nearly edge-on spiral galaxies will be a reliable observational test of the presence of a bar.

H. Dejonghe and H. J. Habing (eds.), Galactic Bulges, 273–274.

3) *Warm direct gas.* (a) Slow Ω_p: the gas will follow the x_2 family (A, associated to the ILR), more circular than the x_1 (B) family. Spiral arm features can be noticed. As predicted by Udry (1991), no vertical resonance appears on this family (Fig. 1). (b) Fast Ω_p: the gas will follow the x_1 family; Ω_p is too high to allow the presence of an ILR. Near the center, the gas can leave the galactic plane to follow the ABAN orbit family (2/1 vertical resonance; Pfenniger & Friedli 1991; Fig. 2).
Warm retrograde gas. (a) Slow Ω_p: at a given energy interval the gas will be trapped by one of the branches of the anomalous orbit families ANO_x or ANO_y (1/1 vertical resonance; Pfenniger & Friedli 1991) bifurcating from the retrograde plane x_4 (R) family which is depopulated. (b) Fast Ω_p: the gas will be trapped by one of the branches of the anomalous orbit families ANO_x (Fig. 3). This can explain the present distribution of inclined retrograde gas observed in some SB0 galaxies.
Hot gas. The contribution of the pressure forces is high and dominates the dynamics. The gravitational potential contributes only to the general (triaxial) shape of the gas density (Fig. 4) and hot gas is not sensible to the detail of the orbital structure.

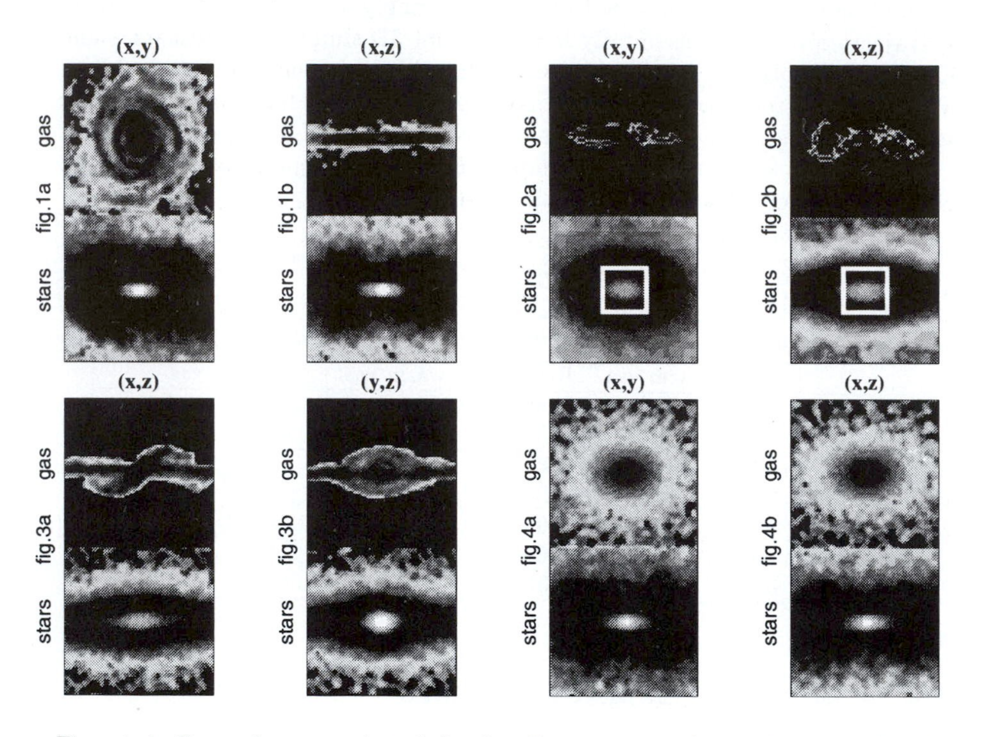

Figs. 1–4. Gas and star projected density. Frames are 20 kpc wide except for the warm direct gas where they are 6 kpc wide (corresponding to the square in the star panel).

References

Friedli, D., Benz, W., 1992, A&A, in press
Pfenniger, D., Friedli, D., 1991, A&A 252, 75
Pfenniger, D., Friedli, D., 1992, A&A, (submitted)
Udry, S., 1991, A&A, 245, 99

DYNAMICS OF THE GALACTIC BULGE FROM GAS MOTIONS

Ortwin E. Gerhard
Landessternwarte, Königstuhl, D-6900 Heidelberg, Germany

James Binney
Theoretical Physics, Keble Road, Oxford OX1 3NP, U.K.

ABSTRACT: Observations of cold gas in the inner galactic disk show a clumpy, highly asymmetric distribution, with large non-circular velocities. Recent work has shown that the flow of gas in the inner few kpc is dominated by a rapidly rotating bar, with corotation at $R \simeq 2.4\,\mathrm{kpc}$ and oriented at an angle of $\theta_{\mathrm{incl}} = 16 \pm 2°$ with the line-of-sight to the Galactic Center. From the kinematical model the gravitational potential in the inner Galaxy can be determined. Gas falls inwards through the bar's inner Lindblad resonance at a rate of $\sim 0.1\,\mathrm{M}_\odot/\,\mathrm{yr}$; this suggests that in episodic star formation significant mass and angular momentum may be added to the inner bulge.

1. Cold Gas in the Inner Galaxy

The central 500 pc of the Galaxy contain of order 3% of the total Galactic luminous mass, 3% of the cold gas component, and 10% of all starforming activity (e.g., Güsten 1989). Only a small fraction of the cold gas in the Galactic Center (GC) region is in the form of neutral HI, measured at 21 cm (Burton & Liszt 1978, 1983; Sinha 1979; Braunsfurth & Rohlfs 1981); the total HI mass inside $\sim 1.5\,\mathrm{kpc}$ is $1 \times 10^7\,\mathrm{M}_\odot$. The dominant component is the cold molecular gas observed in mm-emission lines of molecules such as ^{12}CO, ^{13}CO, CS (Bania 1977, Sanders *et al.* 1984, Heiligman 1987, Dame *et al.* 1987, Bally *et al.* 1988). The inferred gas densities range from $\sim 200\,\mathrm{cm}^{-3}$ to $\sim 3 \times 10^4\,\mathrm{cm}^{-3}$, the total mass is $\sim 10^8\,\mathrm{M}_\odot$, and the surface density of the inner layer is $\sim 300\,\mathrm{M}_\odot/\,\mathrm{pc}^2$.

It has long been known that the distribution of the cold gas in the (l, v)-diagram is not well described by models in which gas moves on circular orbits in an axisymmetric potential. At least four aspects support this view:

(i) Its clumpiness: mm-observations detect structure from the smallest resolved scales ($\sim 1' \sim 3\,\mathrm{pc}$) to objects like Sgr B, with a mass $\simeq 5 \times 10^6\,\mathrm{M}_\odot$

H. Dejonghe and H. J. Habing (eds.), *Galactic Bulges*, 275–282.

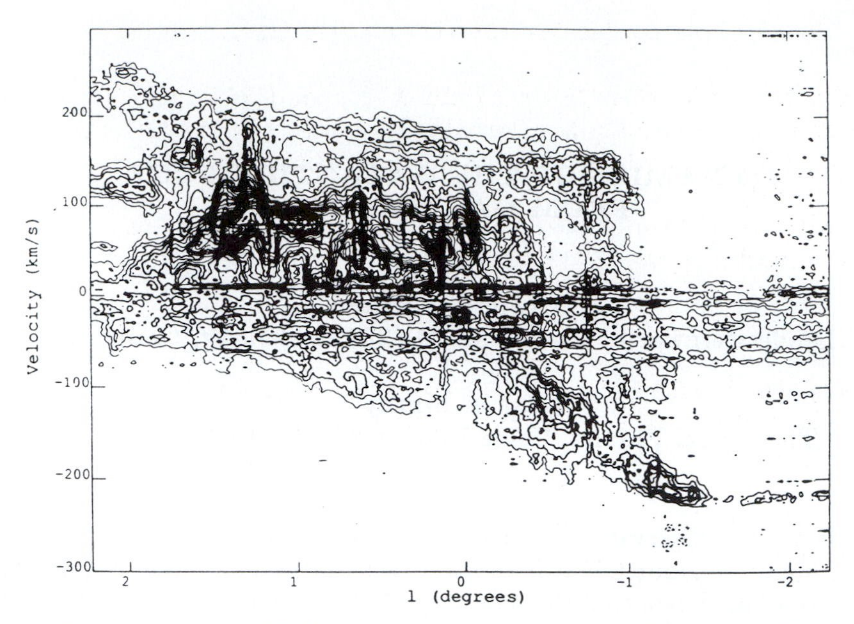

Figure 1. (l, v) digram of ^{12}CO $J = 1 \rightarrow 0$ averaged over $|b| < 0.1°$. The contours (spaced at intervals of 1 K) show a striking parallelogram. (From Binney *et al.* 1991).

(Stark *et al.* 1991). About one third of the GC molecular gas is in such giant cloud complexes (Bally *et al.* 1988; henceforth B88).

(ii) Its asymmetry: About three quarters of the ^{13}CO and CS emission come from positive longitudes and three quarters from material at positive velocities (B88). Part of the longitude asymmetry can be explained as a perspective effect (see below), and part of both asymmetries is caused by the one-sided distribution of the small number of giant cloud complexes. However, an additional overall asymmetry of either hydrodynamic or gravitational origin may be indicated.

(iii) Out-of-plane material: Only about 70% of the GC molecular gas appears to have settled to an equilibrium configuration in the Galactic plane (B88). According to Liszt & Burton (1980), most aspects of the HI distribution somewhat further out can be modelled in terms of *inclined* (by $\sim 13°$) elliptical streamlines. Whether a similar but weaker tilt can be seen in some of the molecular gas is controversial (Heiligman 1987, B88).

(iv) Non-circular and 'forbidden' motions: Known for a long time, these are perhaps most clearly seen in the 'molecular parallelogram' in the (l, v) diagram of Fig. 1 (reproduced from Binney *et al.* 1991). Essentially, the deviations from circular velocities in this structure are as large as the alleged circular velocities themselves. As many papers have noted, either expansion involving large (perhaps improbable) amounts of energy, or a non-axisymmetric potential, is required to generate the observed velocities.

2. Kinematic Model of Galactic Center Gas

Recently, Binney, Gerhard, Stark, Bally & Uchida (1991) have argued that the complex distribution and kinematics of cold gas in the GC can best be understood in terms of motions in a rapidly rotating barred potential. The key to the Binney *et al.* model is the 'molecular parallelogram' referred to above, with its strikingly vertical edges in the (l, v)-diagram. We argue that this feature arises from gas on a narrow range of closed orbits in a rotating barred potential.

Given the extreme inhomogeneity of the ISM and its wide range of scales, it is not immediately clear whether hydrodynamic modelling should concentrate on the ISM's particle nature, or on its smooth fluid aspects – see Binney & Gerhard (1992). However, all hydrodynamic simulations of gas flows in gravitational potentials, using either sticky particles (Schwarz 1981, 1984; Habe & Ikeuchi 1985) or smooth fluids (Sanders & Huntley 1976; van Albada 1985; Mulder & Liem 1986; Athanassoula 1992), show that, if a quasi-equilibrium flow is established, it is generally a good approximation to think of such a flow in terms of the closed ballistic orbits in the underlying gravitational potential. Only near resonances or where closed orbits intersect, do hydrodynamic forces significantly change this picture.

In a barred potential, gas inside corotation follows the main prograde x_1 family until these orbits becomes self-intersecting near the inner Lindblad resonance (ILR). When the inflowing gas reaches the highest-energy self-intersecting x_1-orbit (the 'cusped orbit' in Fig. 2), the hydrodynamic simlations show that a shock forms, which causes it to switch to the closed 'x_2' orbits – another prograde family lying deeper in the potential well, but elongated along the short axis of the potential. Essentially, gas reaching pericentre on the 'cusped' orbit crashes into gas at apocentre on the largest embedded x_2 orbit, producing a spray of material, which in turn causes a shock at the far side of the 'cusped' orbit. From there, material drifts through steadily deeper-lying x_2 orbits, which are not affected by the shock. It is this shock which Binney *et al.* argue gives rise to the observed parallelogram.

The sequence of closed x_1 and x_2 orbits and their projection into the (l, v)-diagram for an observer near the long axis of a rotating bar is shown in Fig. 2. Requiring that the (l, v) trace of the 'cusped' orbit resemble the observed parallelogram determines that observers near the Sun must view the GC bar at a narrowly constrained angle around $\theta_{\rm incl} = 16°$. By contrast, only a weak limit on the bar's axial ratio q can be obtained, as this does not influence the orbit shapes near the ILR much. If we model the inner density distribution as a prolate ellipsoid, then $q \gtrsim 0.75$.

With the same viewing angle, it is also possible to account for the rapid fall-off in the HI terminal velocity observed near $l = 2°$ and the subsequent slower decline out to $l = 12°$. For a given scale of the parallelogram and bulge mass profile, the HI data determine the pattern speed of the bar such

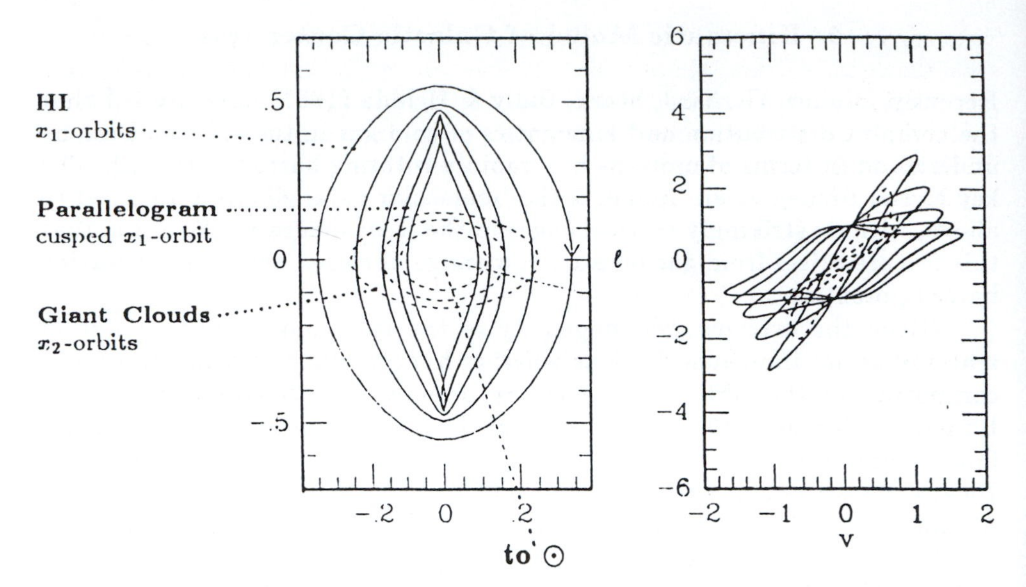

Figure 2. Closed orbits in a rotating barred potential and their projection into the (l, v)-diagram for an observer near the long axis of the bar. (From Gerhard 1992, after Binney *et al.* 1991).

that corotation is at $R_{\rm corot} \simeq 2.4 \pm 0.4$ kpc. In the region $R \lesssim 1.2$ kpc, where the HI terminal-velocity curve decreases outwards, the circular velocity is, in fact, rising.

The strong points of the Binney *et al.* model are that it can fit the (l, v) distribution and the associated large non-circular motions of the molecular parallelogram and the HI terminal curve. Further, that it predicts a perspective asymmetry in the gas distribution which goes in the same sense as that observed, and in the sense expected from the bulge infrared photometry (Blitz & Spergel 1991, COBE). The model naturally accounts for the absence of cold gas between $R \sim 1.5$ kpc and $R \sim 3.5$ kpc, since in the neighbourhood of corotation no stable closed orbits exist on which cold gas could settle. The molecular ring at $R \sim 3.5$ kpc may be associated with gas accumulating near the bar's outer Lindblad resonance. Finally, the giant molecular clouds at the GC, such as Sgr B and Sgr C, are observed in the region of the (l, v)-diagram predicted to be occupied by x_2-orbits inside the ILR .

However, it also appears that some essential ingredients are still missing from the model. On the one hand, the observed lopsidedness of the GC molecular gas is probably too great to be accounted for just in terms of perspective effects, and it remains to be demonstrated whether these can be supplemented and enhanced by hydrodynamic clumping when the self-gravity of the gas is included. The interpretation of the molecular parallelogram in terms of the transition from x_1 to x_2 orbits is only qualitatively born out by simple particle-hydrodynamic simulations (Jenkins 1991); this issue may also be connected with the choice of 'correct' description for the interstellar gas. Finally, there is no explanation yet for the three-dimensional gas motions, in particular the tilt in the HI gas distribution. While gas dynamical simulations show that some gas in peanut-bulge potentials may settle on non-planar closed orbits (Friedli & Benz 1992, Friedli & Udry, this conference), no obvious explanation of the HI tilt has yet emerged.

3. The Gravitational Potential in the Galactic Bulge Region

The model of the Galactic Center HI and molecular gas kinematics just summarized assumed a rotating prolate mass distribution with axis ratio 0.75 and density profile (Binney *et al.* 1991)

$$\rho(a) = 53 \left(\frac{a}{100\,\mathrm{pc}}\right)^{-1.75} \mathrm{M}_\odot\,\mathrm{pc}^{-3}, \tag{1}$$

rotating with a pattern speed of $\omega_p = 63\,\mathrm{km\,s^{-1}\,kpc^{-1}}$. Here a is a prolate-spheroidal radius. As the gravitational potential in the inner Galaxy may arise from a superposition of several components, it may be useful to specify the monopole and quadrupole parts in the kinematic model separately. The spherical part of the density in eq. (1) is (Gerhard 1992)

$$\rho_0(r) = \rho_s \left(\frac{r}{r_s}\right)^{-7/4} = 37 \left(\frac{r}{100\,\mathrm{pc}}\right)^{-7/4} \frac{\mathrm{M}_\odot}{\mathrm{pc}^3}, \tag{2}$$

corresponding (up to an additivive constant) to the potential

$$\Phi_0(r) = \frac{64\pi}{5} G\rho_s r_s^2 \left(\frac{r}{r_s}\right)^{1/4} = 6.4 \times 10^4 \left(\frac{\mathrm{km}}{\mathrm{s}}\right)^2 \left(\frac{r}{100\,\mathrm{pc}}\right)^{1/4}. \tag{3}$$

This spherically averaged density profile is in very good agreement with that derived recently by Kent (1992). He constructed an oblate-isotropic model of the inner bulge, which is based on InfraRed Telescope photometry from the Spacelab Shuttle mission (Kent *et al.* 1992). With a density normalisation of $41.5\,\mathrm{M}_\odot/\mathrm{pc}^3$ and a slightly different exponent of 1.85 in eq. (1), this model accounts for a variety of *stellar* kinematics in the bulge.

From the IRT photometry, the inner bulge has ellipticity $\epsilon = 0.39$. If one assumes an oblate-spheroidal mass distribution, such flattening increases the circular velocity by a factor of 1.09 over that in the spherical model with the same spherically averaged $\rho_0(r)$ (Binney & Tremaine 1987; eq. (2-91)). Thus in the plane we expect

$$v_c(R) = 1.09 \left(\frac{16\pi}{5} G\rho_s r_s^2\right)^{1/2} \left(\frac{R}{r_s}\right)^{1/8} = 137 \left(\frac{R}{100\,\mathrm{pc}}\right)^{1/8} \frac{\mathrm{km}}{\mathrm{s}}. \tag{4}$$

Because the fit of the GC bar model to the kinematic data is not sensitive to the bar's axis ratio, the amplitude of the quadrupole term in the plane is uncertain. For the rotating density distribution of eq. (1) it is

$$\Phi_2 \sim -1000 \left(\frac{\mathrm{km}}{\mathrm{s}}\right)^2 \left(\frac{r}{100\,\mathrm{pc}}\right)^{1/4} \cos\left[2(\phi - \phi_0 - \omega_p t)\right], \tag{5}$$

where the pattern speed is $\omega_p = 63\,\mathrm{km\,s^{-1}\,kpc^{-1}}$ and $\phi_0 = 16°$ at the present time, $t = 0$.

One question that still needs to be answered is whether the dominant part of the quadrupole potential measured in the gas motions comes from the inner bulge or disk. Recent COBE near-infrared observations show several qualitative features predicted by Blitz & Spergel (1991) on the basis of earlier 2.4 μ data of Matsumoto *et al.* (1982), suggesting a bar-like triaxial shape for the Galactic bulge. Also several observations of tracer populations have shown asymmetries in l with respect to the Sun-Galactic Center line, consistent with a barred distribution with its nearer end at $l > 0$ (Nakada *et al.* 1991, Whitelock & Catchpole 1992).

However, the large figure rotation rate ($R_{\mathrm{corot}} \simeq 2.4\,\mathrm{kpc}$) inferred by Binney *et al.* from the gas motions may suggest that the rapidly rotating component is more akin to a (possibly thick) disk-like bar than to a triaxial bulge-spheroid. For example, the peanut bulge of NGC 4565 has an estimated minimum corotation radius of $\sim$ 15 kpc (Gerhard & Vietri 1986). It is not known whether the inner stellar disk of the Galaxy is barred. However, while the mass fraction of cold gas in the inner 500 pc is only $\sim$ 5% and it therefore does not substantially influence the azimuthally averaged rotation curve, it may nevertheless make a significant contribution to the potential's quadrupole moment, because it is flattened to the plane and moves on highly elongated orbits. Based on observations of external barred galaxies (e.g., Kormendy 1982) it would not be surprising if the gravitational potential in the inner Galaxy were made up of non-trivial contributions from several components.

4. Gas Infall and Bulge Evolution

In this rapidly rotating barred potential, molecular gas is moving inwards through the shock associated with the cusped orbit. We may roughly estimate the mass inflow time-scale by noting that the gas cannot stay on the cusped orbit for much more than one dynamical time, 2×10^7 yr. The mass in the parallelogram has been estimated to be $\sim 2 \times 10^6\,\mathrm{M}_\odot$ (Bania 1977), so the the mass inflow rate through the ILR is approximately $0.1\,\mathrm{M}_\odot/$ yr.

Whether or not the gas then settles on the inner x_2-orbits, its highly clumped nature must lead to further angular momentum loss to the stars in the bulge/bar. Since the orbital speed of the clouds on x_2 orbits is faster than the pattern velocity of the bar, orbital angular momentum is lost (i) by gravitational torques from the rotating quadrupole, and (ii) by *dynamical friction* against the bulge stars. The latter process occurs because the clouds are very massive, and it alone causes gas inflow to yet smaller radii at a rate comparable to that estimated above (Stark *et al.* 1991). Thus the present population of Galactic centre clouds must be transient, and for the next Gyr material will be accreting onto the GC at an average rate of $\sim 0.01 - 0.1\,M_\odot/$ yr.

An accretion rate of $0.1\,M_\odot/$ yr, if typical, would amount to of order one third of the total bulge mass inside 500 pc over a Hubble time. Since the typical circular velocity in the bulge is $\sim 180\,\mathrm{km\,s^{-1}}$ and the internal streaming velocity in the oblate-isotropic case is only $\sim 65\,\mathrm{km\,s^{-1}}$ (estimated from Kent's model, 1992), such continuous accretion would also transfer to the bulge stars much of their present angular momentum content. While some of this angular momentum may be transported outwards again, in winds or through the rotating barred potential, some fraction may remain locked in internal streaming motions.

These would be remarkable consequences. On their way to the Galactic Center, the GC clouds are presumably making stars. As they sink in the potential well, they may then dump their remaining mass onto the GC, causing a period of enhanced star formation. One may speculate that, integrated over the age of the Galaxy, a significant part of the stellar mass in the inner bulge region might have been formed in this way, and perhaps been scattered out of the galactic plane into the bulge by the vertical instability processes discussed by Combes *et al.* (1990) and Raha *et al.* (1991). The agent for these processes is a rotating bar, which either scatters resonant stars out of the plane or becomes itself unstable to a bending mode. Not enough is yet known about these processes to say whether they can occur continually or periodically, and how they interact with the expected gas flow patterns in the disk.

There is some evidence for an intermediate-age metal rich population in the Galactic bulge (e.g., Harmon & Gilmore 1988), and also for dissipative mechanisms in the formation of the bulge (Rich 1990; Minniti *et al.* 1992): these authors show that the velocity dispersion of K-giants in some bulge fields is higher for metal-poor stars than for metal-rich stars. But even the metal-richer stars have ages of at least 5 Gyr, so this part of the Galactic bulge cannot have been formed recently.

Athanassoula, E., 1992. *Astr. Ap.* , in press.
Bally, J., Stark, A.A., Wilson, R.W., Henkel, C., 1988. *Ap. J.* **324**, 223.
Bania, T.M., 1977. *Ap. J.* **216**, 381.
Binney, J.J., Gerhard, O.E., 1992. In: Proceedings of 3rd Maryland Astrophysics Conference *Back to the Galaxy*, ed. Blitz, L., in press.
Binney, J.J., Gerhard, O.E., Stark, A.A., Bally, J., Uchida, K.I., 1991. *Mon.*

Not. R. astr. Soc. **252**, 210.
Binney, J.J. & Tremaine, S.D., 1987. *Galactic Dynamics*, Princeton University Press, Princeton.
Blitz, L., Spergel, D.N., 1991. *Ap. J.* **379**, 631.
Braunsfurth, E., Rohlfs, K., 1981. *Astr. Ap. Supp.* **44**, 437.
Burton, W.B., Liszt, H.S., 1978. *Ap. J.* **225**, 815.
Burton, W.B., Liszt, H.S., 1983. *Astr. Ap. Supp.* **52**, 63.
Combes, F., Debbasch, F., Friedli, D., Pfenniger, D., 1990. *Astr. Ap.* **233**, 82.
Dame, T.M., Ungerechts, H., Cohen, R.S., de Geus, E.J., Grenier, I.A., May, J., Murphy, D.C., Nyman, L.-A., and Thaddeus, P., 1987. *Ap. J.* **322**, 706.
Friedli, D., Benz, W, 1992. *Astr. Ap.* , in press.
Gerhard, O.E., 1992. *Reviews in Modern Astronomy* **5**, 174.
Gerhard, O.E., Vietri, M., 1986. *Mon. Not. R. astr. Soc.* **223**, 377.
Güsten, R., 1989. In: IAU Symposium 136, *The Center of the Galaxy*, ed. Morris, M., (Kluwer, Dordrecht), p. 89.
Habe, A., Ikeuchi, S., 1985. *Ap. J.* **289**, 540.
Harmon, R., Gilmore, G., 1988. *Mon. Not. R. astr. Soc.* **235**, 1025.
Heiligman, G.M., 1987. *Ap. J.* **314**, 747.
Jenkins, A.R., 1992. PhD Thesis, Oxford University.
Kent, S.M., 1992. *Ap. J.* **387**, 181.
Kent, S.M., Mink, D., Fazio, G., Koch, D., Melnick, G., Tardiff, A., Maxson, C., 1992. *Ap. J. Supp.* **78**, 403.
Kormendy, J., 1982. *Ap. J.* **257**, 75.
Liszt, H.S., Burton, W.B., 1980. *Ap. J.* **236**, 779.
Matsumoto, T. *et al.* ., 1982. In: *The Galactic Center*, AIP Conf. No. 83, eds. Riegler, G.R., Blandford, R.D. (AIP, New York) p. 48.
Minniti, D., White, S.D.M., Olszewski, E., Hill, J., Irwin, M., 1992. In: IAU Symposium 149, ed. Barbuy, B. (Kluwer, Dordrecht).
Mulder, W.A., Liem, B.T., 1986. *Astr. Ap.* **157**, 148.
Nakada, Y., Deguchi, S., Hashimoto, O., Izumiura, H., Onaka, T., Sekiguchi, K., Yamamura, I., 1991. *Nature* **353**, 140.
Raha, N., Sellwood, J.A., James, R.A., Kahn, F.D., 1991. *Nature* **352**, 411.
Rich, R.M., 1990. *Ap. J.* **362**, 604.
Sanders, D.B., Solomon, P.M., Scoville, N.Z., 1984. *Ap. J.* **276**, 182.
Sanders, R.H., Huntley, J.M., 1976. *Ap. J.* **209**, 53.
Schwarz, M.P., 1981. *Ap. J.* **247**, 77.
Schwarz, M.P., 1984. *Mon. Not. R. astr. Soc.* **209**, 93.
Sinha, R.P., 1979. *Astr. Ap. Supp.* **37**, 403.
Stark, A.A., Gerhard, O.E., Binney, J.J., Bally, J., 1991. *Mon. Not. R. astr. Soc.* **248**, 14P.
van Alabada, G.D., 1985. *Astr. Ap.* **142**, 491.
Whitelock, P., Catchpole, R., 1992. In: *Large Scale Distribution of Gas and Dust in the Galaxy*, ed. Blitz, L. (Kluwer, Dordrecht).

THE BULGE OF THE MILKY WAY AND COSMIC RAYS

F.Jansen[1], K.-P.Wenzel[1], D.O'Sullivan[2], A.Thompson[2]
1.Space Sci. Dept. of ESA, ESTEC, Noordwijk, NL
2.Dublin Inst.for Adv.Studies, Dublin, Ireland

ABSTRACT. The propagation of cosmic ray protons and anti-protons from the inner Galaxy via the galactic halo to the Sun supplies a good agreement with the observed cosmic ray gradient and is in the order of the measured anti-proton flux. Ultra heavy cosmic ray nuclei may have the same origin.

1. Introduction

The near-infrared image of the Milky Way obtained by DIRBE on the satellite COBE lead to the question of the contribution from cosmic ray sources within the "peanut-shaped" bulge to the cosmic ray flux at the Sun. The trend of the preliminary results from the DUBLIN-ESTEC experiment on LDEF satellite (O'Sullivan et al., 1992) is a confirmation of earlier measurements from Ariel 6 and HEAO-3. Therefore cosmic ray nuclei with charge greater than 60 disagree with the predictions of the standard source and propagation models of cosmic rays. For instance in the **L**eaky **B**ox **M**odel such ultra heavy cosmic ray nuclei are a factor of two less abundant than observed (Fowler et al. 1987). The observed anti-protons are a factor of three over abundant than predicted in the LBM (Tan and Ng, 1983). In both cases the mean thickness traversed by cosmic rays before reaching the Sun position in the Galaxy was taken to be 7 gcm^{-2} as measured for ^{10}Be cosmic ray nuclei. Moreover the gamma-ray bulge detected by COS-B mission indicated the existence of a cosmic ray gradient between the inner region of the Galaxy and the Sun of about 1,5 (Strong et al., 1988). This cosmic ray gradient disagrees with one of the basic assumptions in the LBM that there is a homogeneous distribution of cosmic rays in the Galaxy. A significant mixing of cosmic rays in the galactic halo must occur , because the observations at 408 MHz show in the inner Galaxy many spurs out of the galactic plane (Sofue, 1988).

2. The halo diffusion model and the results

The aim of the proposed **H**alo **D**iffusion **M**odel for cosmic ray protons and antiprotons by Halm et al. (1992a,b) was to build a model under conditions as close as possible to those described in the introduction. A diffusion equation for the cosmic ray density N was solved

$$-\frac{D(E)}{r}\frac{\partial^2}{\partial r^2}(rN(r,E)) + \frac{\partial}{\partial E}(I(E)N(r,E)) + \Sigma(E)N(r,E) = q(E)\,e^{-(r/R_0)^2}$$

H. Dejonghe and H. J. Habing (eds.), Galactic Bulges, 283–284.

(From left to right the terms for diffusion, ionization, annihilation and the source term, E is the total energy, the radial coordinate r is directed along the path of a cosmic ray particle with R_0 as the cosmic ray source parameter.). The calculated and the observed cosmic ray gradient are in agreement if $r/R_0 \approx 1{,}05$ (r = 10 kpc for the position of the Sun.). The gradient is less dependent on the diffusion constant D_0 in D(E). With $D_0 \leq 5\ 10^{28}$ cm^2 s^{-1} (Bykov and Toptygin, 1990) and from the equation

$$n = 2D_0x_0/(R_0mv)$$

the gas particle density n traversed by relativistic cosmic rays is smaller than 10^{-2} cm^{-3} (m proton rest mass, v velocity of the cosmic rays). This value agrees with the observation mentioned above, that the cosmic ray quickly leave the region of the inner Galaxy and propagate mainly via the halo to the Sun. With $D_0 = 5\ 10^{28}$ cm^2 s^{-1} the flux of anti-protons, which are produced during the propagation by collisions of cosmic ray protons with the interstellar gas is in the order of the LBM. In contrast to the protons the anti-proton flux strongly depends on D_0. For instance a value of $5\ 10^{27}$ cm^2 s^{-1} gives about three times more anti-protons.
Because it was shown by Halm et al. (1992a) for antiprotons that the LBM and the HDM solutions are similar for $r/R_0 \approx 1$ and because of the increasing anti-proton flux with decreasing diffusion constant it may be possible to produce an ultraheavy cosmic ray abundance in the HDM which is similar to, or greater than, that in the LBM.

References

Bykov, A.M.; Toptygin, I.N.: 1990, *Proc. 21th Intern. Conf. Cosmic Rays* **3**, 307

Fowler, P.H.; Walker, R.N.F.; Masheder, M.R.W.; Moses, R.T.; Worley, A.; Gay, A.M.: 1987, *Astrophys. J.* **314**, 739

Halm, I.; Jansen, F.; de Niem, D.: 1992a, *Astron. Astrophys.*, in press

Halm, I.; Jansen, F.; de Niem, D.: 1992b, *Astron. Astrophys.*, subm.

O 'Sullivan, D.; Thompson, A.; Bosch, J.; Keegan, R.; Wenzel, K.-P.; Jansen, F.; Domingo, C.: 1992, *COSPAR Washington*, in prep.

Sofue, Y.: 1988, *Publ. Astron. Soc. Japan* **40**, 567

Strong, A.W.; Bloemen, J.B.G.M., Dame, T.M., Grenier, I.A.; Hermsen, W.; Lebrun, F.; Nyman, L.-Å.; Pollock, A.M.T.; Thaddeus, P.:1988, *Astron. Astrophys.* **207**, 1

Tan,L.C.; Ng, L.C.: 1983, *J. Phys. G: Nucl. Phys.* **9**, 227

RAPID GAS FUELING IN A BARRED POTENTIAL BY SELF-GRAVITATIONAL INSTABILITY

Keiichi WADA *and* **Asao HABE**
Department of Physics, Hokkaido University, Sappro 060 Japan

Recent *IRAS* surveys have revealed that large infrared luminosities are originated from active star forming regions in galaxies (Soifer *et al.* 1986). Such starburst regions are frequently located in galactic central regions and CO observations indicate that these regions contain a large amount of molecular gas (e.g. Ishizuki *et al.* 1990). However, the triggering mechanism for starbursts and the mechanism of the high mass supply rate of gas into a galactic center are still unclear.

It is suggested that tidal encounters of galaxies remove angular momentum of gas and trigger rapid gas accretion and starburst. Noguchi(1988) has shown by computer simulations that galaxy-galaxy interactions induce a stellar bar, and gas loses its angular momentum and accumulates to a galactic center. In his numerical simulation, non-axisymmetric potential of a stellar bar plays an important role in the accretion of gas. However, it is not obvious as to whether or not gas accretes into a nuclear region within a few hundred pc only by the effects of stellar bar.

There were a number of studies concerning gas dynamics in rotating bar potential to find the mechanism which forms and maintains galactic spiral arms (e.g. Matsuda *et al.* 1987). These numerical studies revealed that oval gas rings are formed near the two *ILR*. The rings are generally elongated and the major radius of the ring nearly coincides with the radius at outer *ILR*. Although a large fraction of gas inside the CR radius is accumulated into the ring, gas cannot accrete to the galactic center beyond inner *ILR*. However, self-gravity of gas is not taken into account in these simulations.

If a pattern speed of the weak barred potential induced by galaxy-galaxy encounters, which has suitable for *ILR*s, some fraction of gas in the galaxy is rapidly accumulated into the near proximity of *ILR* radius. In this case, if self-gravity of the accumulated gas is taken into account, the gas becomes gravitationally unstable and it is expected that a rapid gas supply into the center of the galaxy would occur.

We investigate the dynamics of self-gravitating gas in the weak barred potential by 2-D numerical simulation, and show that for the initial gas mass ratio to stellar mass grater than 10%, a central elongated gas ring becomes unstable and collapses after gas accumulates to form the elongated

H. Dejonghe and H. J. Habing (eds.), Galactic Bulges, 285–286.

gas ring near the *ILR* radius(Wada and Habe 1992). To analyze this rapid and complex process, we made a video of this simulation. The video clearly shows that the formation of dense gas clumps at both sides of the elongated ring triggeres instability of the ring: The dense clumps move and collide with each other on a highly elongated orbit; After frequent collisions, these gas clumps merge and finally form a central dense core. Instead of the video, we show here the time evolution of the velocity of gas in the ring (**Fig.1**).

References

Ishizuki, S. *et al.* 1990. *Nature*, **344**,224 .

Matsuda, T. *et al.* 1987. *Mon. Not. R. astr. Soc.*, **229**, 295.

Noguchi,M.,1988. *Astr. Astrophys.*, **203**,259.

Soifer, B. T.,1984. *Astrophys. J.*, **283**,L1.

Wada,K. & Habe,A., 1992. *Mon. Not. R. astr. Soc.*, **258**,82.

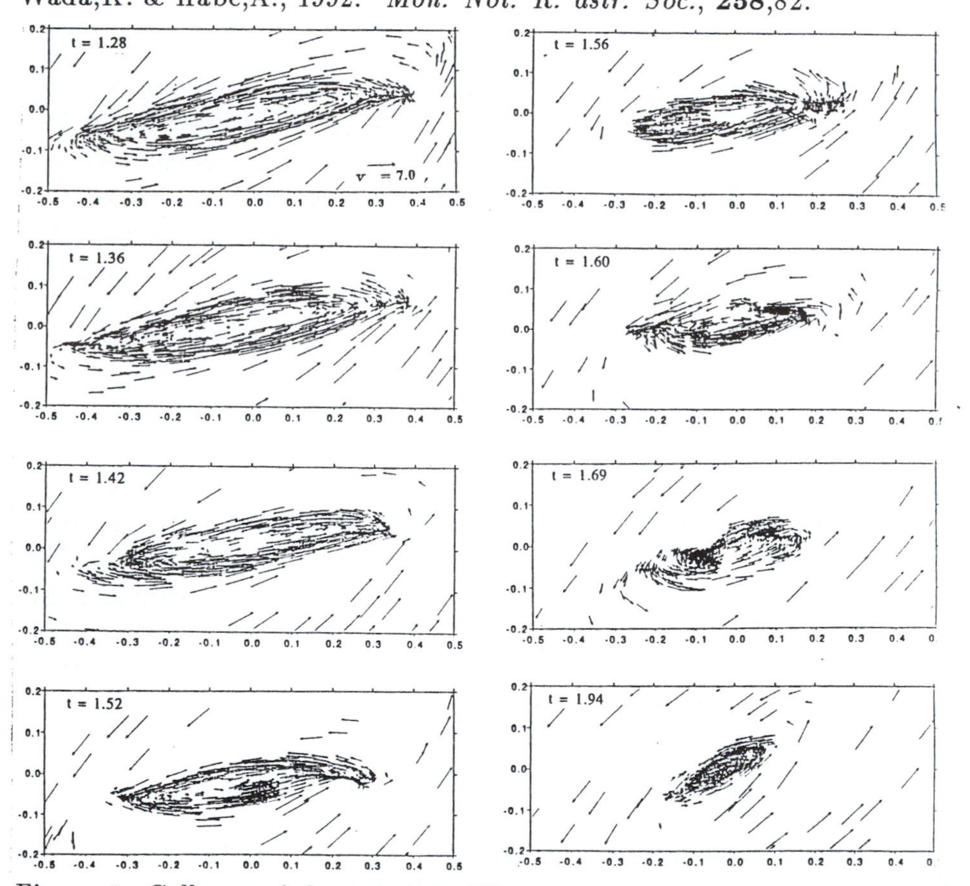

Figure 1: Collapse of the gas ring. The arows represent the velocity vector of the gas. The unit of the time is $3.5 \times 10^8 yr$.

MgI TRIPLET LINES IN COMPOSITE SYSTEMS

B. BARBUY, R.E. DE SOUZA, S. DOS ANJOS
IAG-USP, CP 9639, 01065-970 São Paulo, Brazil

ABSTRACT. We use information contained in CMDs of metal-rich globular clusters, in order to compute a grid synthetic spectra for 10 evolutionary stages. These synthetic spectra are added up to reproduce the composite spectra of S0/E galaxies.

1. Introduction

Renzini (this symposium) pointed out that stellar evolutionary tracks for metallicities above solar are presently difficult to compute for a lack of information on the helium abundance, besides the problem of dealing with very high opacities, suggesting that observed colour magnitude diagrams (CMDs) of metal-rich globular clusters are a better tool.

In the present work, in order to build composite spectra for metal-rich populations contained in bulges of S0/E galaxies, we have used the information contained in CMDs of the metal-rich bulge clusters NGC 6553 (Ortolani et al 1990), NGC 6528 (Ortolani et al 1992a) and Terzan 1 (Ortolani et al 1992b). Some characteristics of these clusters are a fainter red giant branch (RGB) tip, very red horizontal branch (HB), and the fact that the brighter giants are fainter by $>$ 1 magnitude than metal-poor ones.

Further information on the clusters were provided by a high resolution analysis of a star in NGC 6553, where [M/H] = -0.2 was found, and the relative location of CMDs of different metallicities as shown in Bica et al (1991); from this work we concluded that the shift in colour is due to increasing opacities, i.e., it is essentially due to metallicity (and not temperature).

We adopt then the number of stars in each evolutionary stage from counts in CMDs of NGC 6553 and their respective temperatures are derived as described in Barbuy et al (1992). A library of 10 spectra was created for different evolutionary stages. The MgI triplet region at $\lambda\lambda$ 490-540 nm was computed, where atomic plus MgH,C_2,CN and TiO molecular lines were included.

The synthetic spectra were then added up weighted by their luminosities and relative number of stars. These composite spectra are then fitted to observed spectra of lenticular galaxies. In Fig. 1 the matches, using synthetic spectra for [M/H] = 0.0, are shown for the galaxies NGC 4565 and NGC 4594. The observations are described in de Souza et al (this symposium). We see a main disagreement in the intensity of strong lines in the range $\lambda\lambda$ 523-538 nm, which will be further investigated.

References

Barbuy, B., Castro, S., Ortolani, S., Bica, E.: 1992, A&A 259, 607
Bica, E., Barbuy, B., Ortolani, S.: 1991, ApJ 382, L15
Ortolani, S., Barbuy, B., Bica, E.: 1990, A&A 236, 362
Ortolani, S., Bica, E., Barbuy, B.: 1992a, A&AS 92, 441
Ortolani, S., Bica, E., Barbuy, B.: 1992b, A&A, in press

H. Dejonghe and H. J. Habing (eds.), Galactic Bulges, 287–288.

Figure 1 - Fit of synthetic spectra (solid lines) computed with [M/H] = 0.0 to observed spectra (dotted lines) of NGC 4565 and NGC 4594.

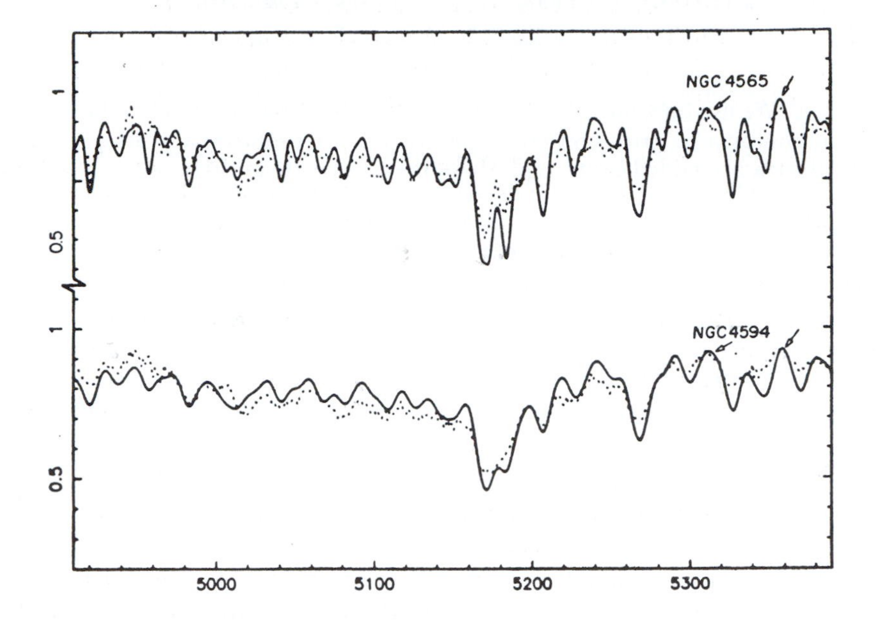

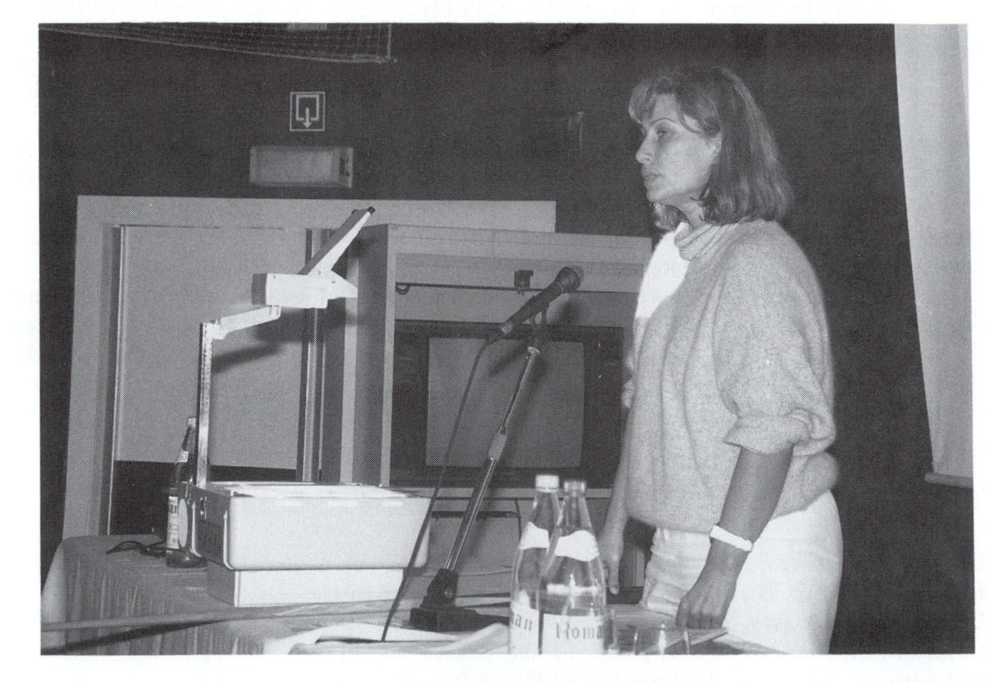

B. Barbuy

HUBBLE SPACE TELESCOPE OBSERVATIONS OF BAADE'S WINDOW

W. A. BAUM, *Univ. of Washington, Seattle, WA 98195, USA*
R. M. LIGHT, *Univ. of California, Santa Cruz, CA 95064, USA*
J. HOLTZMAN, D. HUNTER, T. KREIDL, *Lowell Obs., Flagstaff, AZ 86001, USA*
E. J. O'NEIL, JR., *Kitt Peak Nat. Obs., Tucson, AZ 85726, USA*
E. J. GROTH, *Princeton Univ., Princeton, NJ 08544, USA*

This is a status report on a continuing program using the Hubble Space Telescope (HST) Wide–Field Camera (WFC) to probe the stellar population of the Galactic bulge to fainter magnitudes. We seek the mean age of the stars and the initial mass function (IMF). Galactic bulge stars offer the *only* opportunity to investigate the IMF of a super metal–rich population. They are 100 times closer than the next nearest sample.

On 18 August 1991, two long HST–WFC exposures (1600^s and 2000^s) were made with the F555W filter and two (2000^s each) with the F785LP filter. The WFC is the f/12.9 mode of the "WF/PC" (Westphal 1982; MacKenty 1992), and it covers 2.5 ×2.5 arcmin at an image scale of 0.10″ per pixel. Our WFC field is located at $18^h03^m10^s$, -29°51′43″ (2000). It was selected to lie in the apparently most transparent part of Baade's Window, 11.4′ NNW from the imbedded globular star cluster NGC 6522, and 3.8° from the direction to the Galactic center. We should expect A_V for the HST-WFC field to be less than the average for Baade's Window.

After standard WF/PC reduction procedures (Lauer 1989) were applied to the images, the F555W set and the F785LP set were each stacked (co–added), and DAOPHOT was used for stellar photometry. Instrumental magnitudes in the F555W and F785LP passbands were transformed to the Johnson–Cousins V, I system using WF/PC calibration data published by Harris *et al.* (1991).

Our color–magnitude diagram in Figure 1 is based on profile–fitting photometry. This CM diagram is limited at the bright end by the onset of CCD saturation at $V \sim 18$ mag. At the faint end, it ceases to be meaningful below $V \sim 23.5$ mag, where photometric errors are rising steeply. Terndrup's (1988) CM diagram, which bottomed at $V \sim 20$ mag, can be merged with the top of ours, and they join well if we assume Terndrup's $V - I$ to include a few hundredths of a magnitude more reddening than ours. His field was located 3.1′ NW of NGC 6522, or about 8′ south of our HST field.

Contamination by the foreground disk population was discussed by Terndrup. In our case, it would average only about 10 stars per 0.25-mag bin over the range of our data. That is of no importance below $V \sim 20$ mag in Figure 1, but it is a significant contributor around 18th and 19th magnitude, and it masks the bulge turnoff.

By slicing our CM diagram into $\Delta V = 0.25$–mag bins and finding the median value of $V - I$ in each bin, we obtain the ridge–line plotted in Figure 2. Theoretical isochrones, such as those of Green, Demarque, and King (1987) fit the main sequence (i.e., the $20.0 < V < 23.5$ mag portion of this ridge–line) impressively well, but the fit is sensitive to any zero–point error in the magnitude scales and to the validity of the models. Moreover, the mean age of the population is not uniquely determined by the fit, because for each choice of mean age, one can find associated values of A_V and R_0 that permit an acceptable fit. Each choice must thus be judged by the plausibility of those associated values. Taking abundances $Z = 0.04$ and $Y = 0.30$, we find, for example,

H. Dejonghe and H. J. Habing (eds.), Galactic Bulges, 289–290.

that a 10 Gyr isochrone fits if $A_V \approx 1.2$ mag and $R_0 \approx 7.4$ kpc. By Buonanno's (1986) criterion, the turnoff in this case should fall at $V_t \approx 19.8$. For each +1 Gyr in assumed mean age, the corresponding differentials are $\Delta A_V \sim -0.02$ mag, $\Delta R_0 \sim -0.06$ kpc, and $\Delta V_t \sim +0.08$ mag. Although the present results appear to favor a scenario in which star formation in the bulge continued longer than in the halo, the uncertainties are large, and a 15 Gyr mean age cannot be ruled out.

To produce a V luminosity function, star counts were made in 0.25–mag bins, and a very large number of simulated stars with appropriate PSFs were inserted, one at a time, to derive completeness corrections as a function of magnitude. (The simulated stars also served to check against systematic errors in the photometry.) Preliminary analysis yields a corrected luminosity function for the main sequence in Baade's Window which does *not* differ much from that in the solar neighborhood.

References

Buonanno, R. 1986. *Mem.Soc.Astron.Ital.* **57**, 333.

Green, E. M., Demarque, P., and King, C. R. 1987. *The Revised Yale Isochrones and Luminosity Functions* (New Haven: Yale Univ. Observatory).

MacKenty, J. W. 1992. *HST Wide–Field/Planetary Camera Instrument Handbook* (Baltimore: Space Telescope Science Institute).

Harris, H. C., Baum, W. A., Hunter, D. A., and Kreidl, T. J. 1991. *Astron.J.* **101**, 677.

Lauer, T. R. 1989. *Publ.Astron.Soc.Pac.* **101**, 445.

Terndrup, D. M. 1988. *Astron.J.* **96**, 884.

Westphal, J. A. 1982. In *The Space Telescope Observatory* NASA CP-2244, ed. D. N. B. Hall (Washington: NASA Scientific and Technical Information Branch), 28.

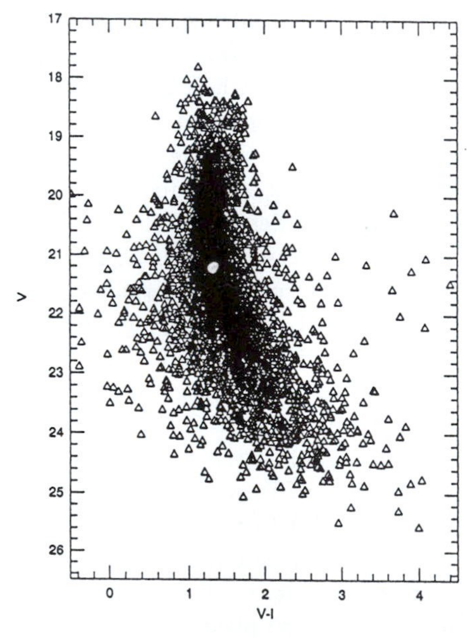

Fig 1. CM diagram of Baade's Window.

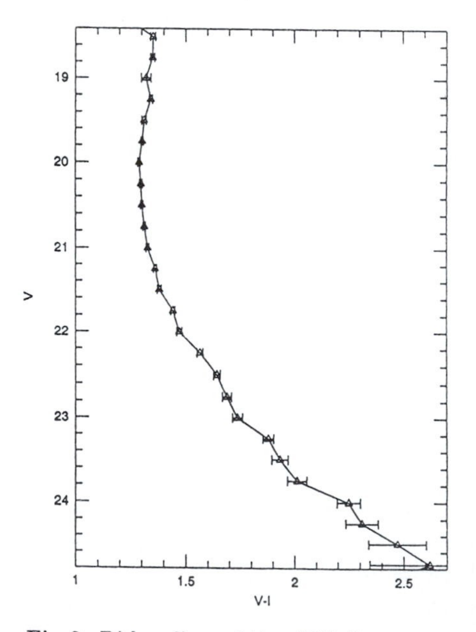

Fig 2. Ridge–line of the CM diagram.

Long-period variable AGB stars in a field towards the galactic bulge *

J.A.D.L. Blommaert[1], A.G.A. Brown[1],
H.J. Habing[1], W.E.C.J. van der Veen[2] and Y.K. Ng[1].

1: Sterrewacht Leiden, P.O. Box 9513, 2300 RA Leiden, The Netherlands
2: Columbia University, Dpt. of Astronomy, 538 West 120th St., New York, NY 10027, USA

We study two different samples of long-period variable Asymptotic Giant Branch (AGB) stars in a field of low and homogeneous extinction towards the Galactic bulge, the Palomar-Groningen field Nr. 3. The samples were selected to study the evolution of the late phases on the AGB. One sample consists of 486 variables (mostly Miras) optically detected and studied by Plaut (1971) and by Wesselink (1987). The other sample is selected from the IRAS Point Source Catalogue and consists of 239 sources. We made additional infrared measurements between 1.2 and 13 μm for a large fraction of both samples. This information was used to identify the IRAS sources and derive the apparent bolometric magnitudes. The samples of Miras and variable IRAS sources have a similar apparent bolometric magnitude distributions, but are displaced by an amount significantly less than expected from the Mira period-luminosity relation (Feast *et al.* 1989; 0.3 magnitudes as opposed to 0.6 magnitudes). The surface density distribution along the minor axis of the bulge is the same for both samples. We conclude that both samples have evolved from the same parent population and that they represent different evolutionary stages on the AGB. The IRAS sources with the longer periods (on average 450 days (Whitelock *et al.* 1991) versus on average 250 days for the optical sample) are the further evolved objects. As the IRAS sources have higher mass loss rates we conclude that mass loss increases during the late stages of the evolution. However, we find indications that in some stars the mass loss process has been interrupted for some time; mass loss could be an intermittent process although its overall rate increases in time.

The Miras and the IRAS sources in the bulge have a very similar spatial distribution which also agrees with that found by Blanco (1988) for late type M giants (Fig. 1). The distribution differs considerably from that of the metal-poor RR Lyrae stars. From model calculations in the literature we estimate that the population of long-period variable AGB stars is more than 10 Gyrs old and originates from stars with a Main Sequence mass of about 1 $M_\odot$. Unlike Harmon and Gilmore (1988) who estimate the ages

* Based on observations collected at the European Southern Observatory (La Silla, Chile) as part of the ESO Key Programme "Stellar Evolution in the Bulge".

H. Dejonghe and H. J. Habing (eds.), Galactic Bulges, 291–292.

of the IRAS variables to be less than 5 Gyr, we conclude that the ages of the AGB stars are comparable to those of the metal-poor globular cluster population (16 Gyrs).

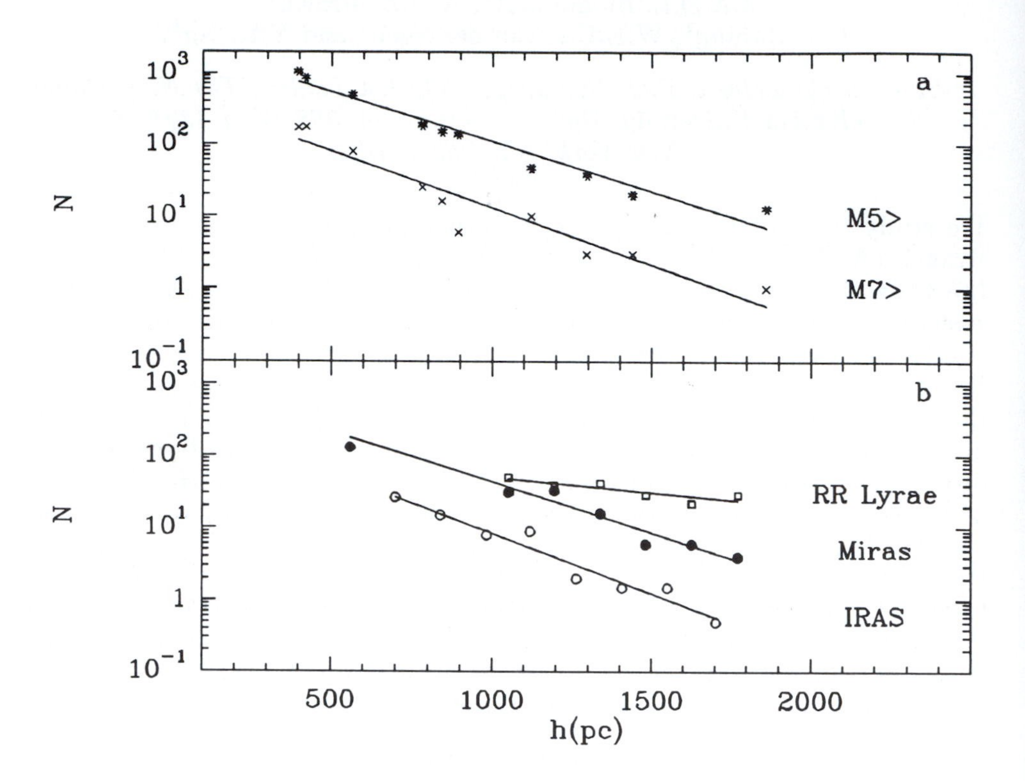

Figure 1. *# a: The projected number density of bulge giants along the $l = 0°$ axis (Blanco 1988). # b: The same but now for the RR Lyrae, Miras and IRAS sources (number of sources per 2 square degrees). $R_o = 8$ kpc is assumed.*

References

Blanco, V.M.: 1988, *Astron. J.* **95**, 1400

Harmon, R., Gilmore, G.: 1988, *Monthly Notices Roy. Astron. Soc.* **235**, 1025

Feast, M.W., Glass, I.A., Whitelock, P.A., Catchpole, R.M.: 1989, *Monthly Notices Roy. Astron. Soc.* **241**, 375

Plaut, L.: 1971, *Astron. Astrophys. Supplem. Series* **4**, 75

Whitelock, P.A., Feast, M.W., Catchpole, R.M.: 1991, *Monthly Notices Roy. Astron. Soc.* **248**, 276

Wesselink, T.J.H.: 1987, Ph.D. Thesis, Catholic University of Nijmegen

Preliminary results of the Two Micron Galactic Survey

X. Calbet, T. Mahoney, P. Hammersley, F. Garzón and M. Selby

Instituto de Astrofísica de Canarias, E-38200 La Laguna, Tenerife (Spain)

Abstract

We present the first results of the Two-Micron Galactic Survey (TMGS). The sources are concentrated on the Galactic plane, the width of the distribution increasing between 0° and 30° longitude. Several obscuration lanes are seen, showing that K-band absorption can be high and patchy.

The aim of the TMGS is to map large areas of the Galactic plane and bulge (Mahoney *et al.* 1990). The TMGS, which was started in 1988, uses the 1.5 m Carlos Sánchez Telescope of the Observatorio del Teide (Tenerife). It uses the ICSTM Seven Channel Near-Infrared Camera, which has 7×15 arcsec^2 detectors aligned in declination and a standard K filter. The data are obtained by making drift scans, typically two hours long, across the Galactic plane. The limiting magnitude of $\approx +10.5$ and is complete to $\approx +9.8$. The positional accuracy is currently 1 arcsec in RA and 12 arcsec in declination.

From the data reduced so far, which corresponds to the first four years of the project, we have detected 400 000 sources and covered an area of 240 deg^2. The map shown in Figure 1 is of a region centred on $l = 30°, b = 0°$. A histogram of source counts and superimposed plot of $K-[60]$ against RA is given in Figure 2. Correlating this survey with IRAS data, it can be seen that objects at the centre of the plane or in one of these obscuration zones are more reddened.

References

Mahoney, T., Selby, M., Garzón, F., Hammersley, P. and Calbet, X 1990, in B. Jarvis and D. Terndrup (eds.) *Bulges of Galaxies*. ESO/CTIO Workshop held at La Serena, Chile 16–19 January 1990.

H. Dejonghe and H. J. Habing (eds.), Galactic Bulges, 293–294.

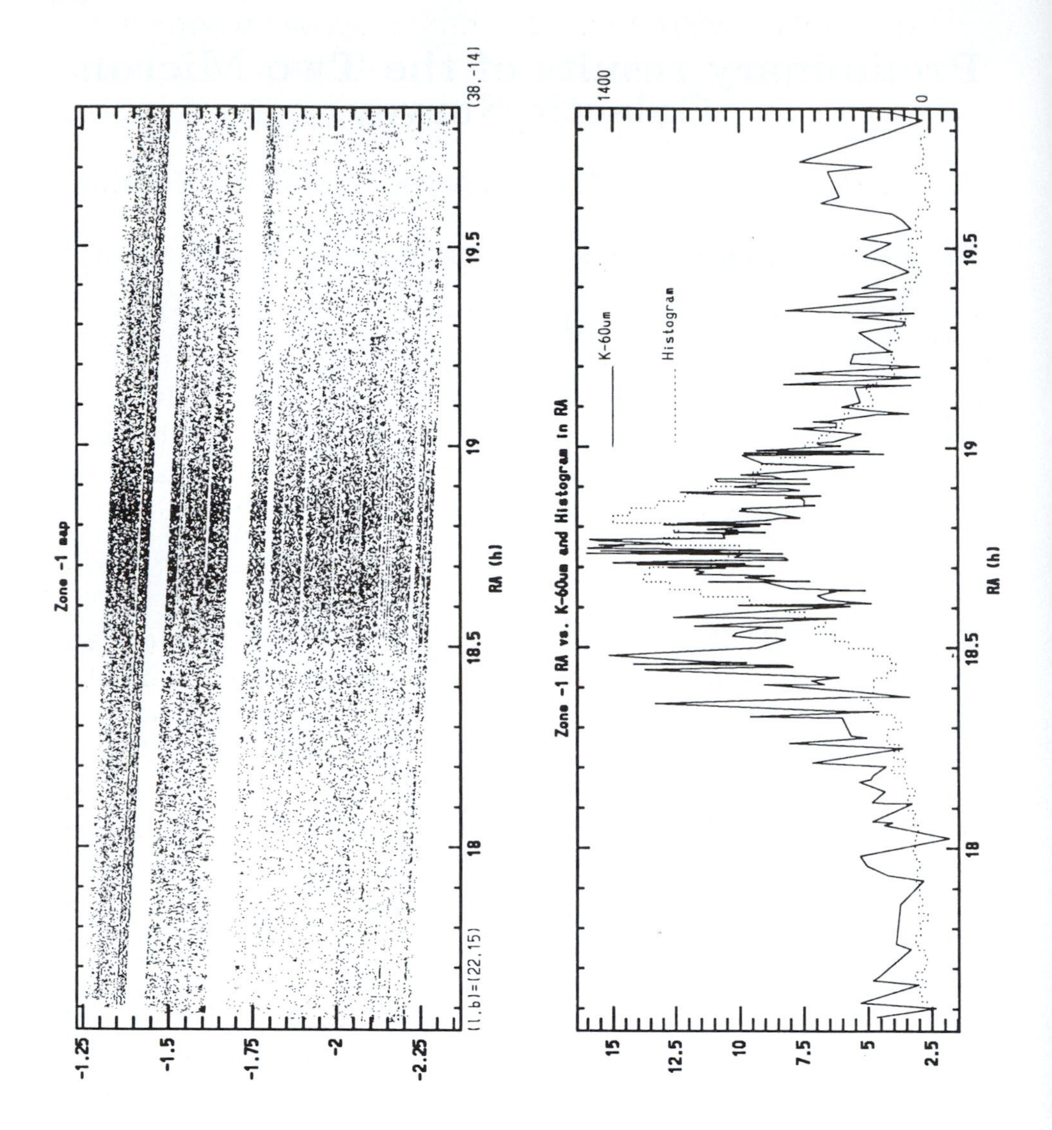
Zone -1 map
DEC (deg)
RA (h)
(l,b)=(22,15)
(38,-14)
Zone -1 RA vs. K-60um and Histogram in RA
K-60um
Histogram
K-60um or N. Stars
RA (h)
1400
0

Figure 1

METALLICITY OF TWO STARS IN BAADE'S WINDOW *

S. CASTRO[1], B. BARBUY[1], T. RICHTLER[2]
1 IAG-USP, São Paulo, Brazil; 2 Bonn Univ., Germany

ABSTRACT. Low resolution spectra of two stars of the Baade Window were obtained with the 1.5m telescope at ESO. Their temperatures and gravities are obtained by using their location in colour magnitude diagrams of bulge clusters, and through synthetic spectra fitting, their metallicities are derived.

1. Introdution

Whitford & Rich (1983) and Rich (1988, 1992) described the stellar populations of the galactic bulge, in particular that of the Baade Window (BW). The bulge contains a distinct stellar population relative to other components of the Galaxy, including RR Lyrae, M giants, OH/IR stars and Miras. The metallicity of bulge stars in the BW range from 1/10 to 10 times solar, according to Rich.
Further studies on the metallicity of BW's stars are of great interest, and this poster addresses the study of two BW stars for which we have low resolution spectra.

2. Observations and Stellar Parameters

The observations were obtained by T. Richtler at the 1.5m telescope of the *European Southern Observatory* - ESO, at La Silla (Chile). The *Image Dissector Scanner* - IDS was used as detector. Mid-resolution spectra of $\Delta\lambda \approx 5$ Å were obtained in the wavelength range $\lambda\lambda$ 400 - 620 nm for the stars 2 and 9 of the Baade Window's field near NGC 6528 according to the identifications by van den Bergh (1971). The BV colours from van den Bergh (1971) are used to derive the temperature, using the method presented in Barbuy et al. (1992). The reddening E(B-V) = 0.58 indicated by van den Bergh & Younger (1979) was adopted, pointing out that this is a mean between the values E(B-V) = 0.55 found for the direction of the globular cluster NGC 6528 by Ortolani, Bica & Barbuy (1992, OBB92), and the mean values E(V-I) = 0.70 given by Walker & Mack (1986) and adopted by Sadler (1992) for the Baade Window.
Below are shown their colours and derived parameters. For star 9, it could be ascent RGB or RGB tip which in (B-V) mixes either the hotter stars (see OBB92).

star	V	$(B-V)$	$T_{ef}(K)$	$\log g$
2	15.66	2.05	4000	0.71
9	16.03	1.99	4100, 3500	0.95

* Observations collected at the European Southern Observatory, La Silla, Chile

H. Dejonghe and H. J. Habing (eds.), Galactic Bulges, 295–296.

3. Derivation of Metallicity

The MgI triplet plus continua at 489.7-495.8 and 530.3-536.6 nm are used here: we computed synthetic spectra for the region $\lambda\lambda$ 490-540 nm, which are compared to the observed spectra.
Model atmospheres are interpolated in the grid by Gustafsson et al. (1975), using the same assumptions used in the calculations of the models.
In figure 1 are shown the synthetic spectrum, computed with [M/H]=0.0, fitted to star 2 in the range $\lambda\lambda$ 505-516 nm. The star 9 proved to be a cool M giant of metallicity also approximately solar, where TiO bands dominate.

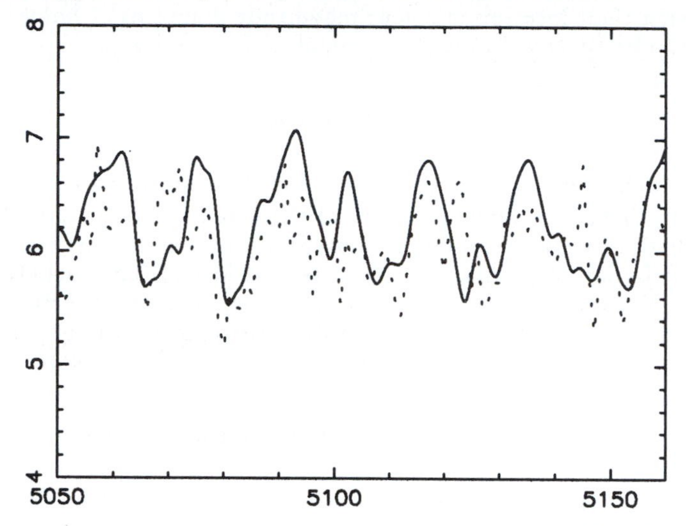

Fig. 1 - Synthetic spectrum (–) of star 2, computed with [M/H]=0.0

References

Barbuy, B., Castro, S., Ortolani, S., Bica, E.: 1992, A&A 259, 607
Bica, E., Barbuy, B., Ortolani, S.: 1991, ApJ Lett. 382, L15
Gustafsson, B., Bell, R.A., Eriksson, K., Nordlund, A.:1975, A&A 42, 407
Ortolani, S., Bica, E., Barbuy, B.: 1992, A&AS 92, 441
Schmidt-Kaler, A.: 1982, in Landolt-Börnstein, group VI, vol. 2, subvol. b (Stars and star clusters)
Rich, M.: 1988, AJ, 95, 828
Rich, M.: 1992, in *The Stellar Populations of Galaxies*, IAU Symp. 149, eds. B. Barbuy, A. Renzini, Kluwer Academic Publishers, p. 29
Sadler, E.: 1992, in *The Stellar Populations of Galaxies*, IAU Symp. 149, eds. B. Barbuy, A. Renzini, Kluwer Academic Publishers, p. 41
van den Bergh, S.: 1971, AJ 76, 1082
Walker, A.R., Mack, P.: 1986, MNRAS, 220, 69
Whitford, A., Rich, M.: 1988, ApJ 274, 723

THE BULGE / HALO INTERFACE

ROTATIONAL KINEMATICS FROM [Fe/H] = –3.0 TO SOLAR

PAUL HARDING and HEATHER MORRISON

Observatories of the Carnegie Institution of Washington, Pasadena, CA, USA and Kitt Peak National Observatory, National Optical Astronomy Observatories, Tucson, AZ, USA

Abstract. Kinematics and abundances have been determined for a large sample of K giants in a field at ~2 kpc from the center to study the relationship between the bulge and the halo. There are two different kinematic groups evident. The metal-poor stars have low rotation and high velocity dispersion, and so can be associated with the halo. The metal-richer stars have intermediate rotation. Our data suggest that the bulge and halo could be separate kinematical entities.

Key words: Kinematics - K giants - Halo

The term "spheroid" is used to describe an amalgam of properties of the halo and bulge of both our Galaxy and other spiral galaxies. However, if the halo and bulge are not related, attempting to combine their properties into one category may confuse rather than provide any coherent picture.

In order to determine if the halo and bulge can be considered as distinct populations towards the center of our Galaxy, we have surveyed a field at $l = 350, b = -10$, to examine the dependence of rotational kinematics on abundance. Along this line of sight velocities of stars close to the tangent point are dominated by rotation. The minimum distance is ~2 kpc from the Galactic center. Accurate colors and magnitudes using Washington photometry have been measured for ~250000 stars in the 2 square degree field. The reddening has also been measured to be both low (E(B–V) = 0.07) and uniform over the whole field. Washington photometry provides good estimates of abundance (our photometric errors translate to an error of 0.3 in [Fe/H]) and temperature for the ~5000 K giants in the field. Distances can then be estimated to an accuracy of 25 – 30% for the giants. The mean abundance of the whole sample is [Fe/H] = –0.8, and only 30% of stars have [Fe/H] < –1.

Radial velocities have been measured for a subset of the K giants sampled uniformly on abundance and chosen to be close to the tangent point. Figure 1 shows the plot of line of sight velocity versus abundance. These giants separate clearly into two kinematic groups. (Note that our [Fe/H] errors will make the transition seem more gradual than it really is). Stars with [Fe/H] < –1 have a mean line-of-sight velocity of –37 ±8 km/s and dispersion $\sigma_{los} = 107 \pm 6$ km/s. Those with [Fe/H] > –1 have a mean velocity of –82±8 km/s and $\sigma_{los} = 67 \pm 6$ km/s. (Mean velocities are negative because the field is at negative longitude.)

Our distance estimates for each star allow us to calculate rotational kinematics directly following the procedure of Morrison et al (1990). We solve for V_{rot} (the mean rotational velocity about the galactic center) and σ_ϕ (the velocity dispersion of the azimuthal component). The change in kinematics at [Fe/H] = –1 is still apparent. For the metal-poor group ([Fe/H]<–1.0) V_{rot} = 44 ±12 km/s and $\sigma_\phi = 107 \pm 11$ km/s. The errors quoted include the effect of distance errors. The abundance errors, which will cause the more numerous stars with [Fe/H] >–1.0 to spill over into this group, mean that this is probably an over-estimate by 10–20 km/s. The kinematics

H. Dejonghe and H. J. Habing (eds.), Galactic Bulges, 297–298.

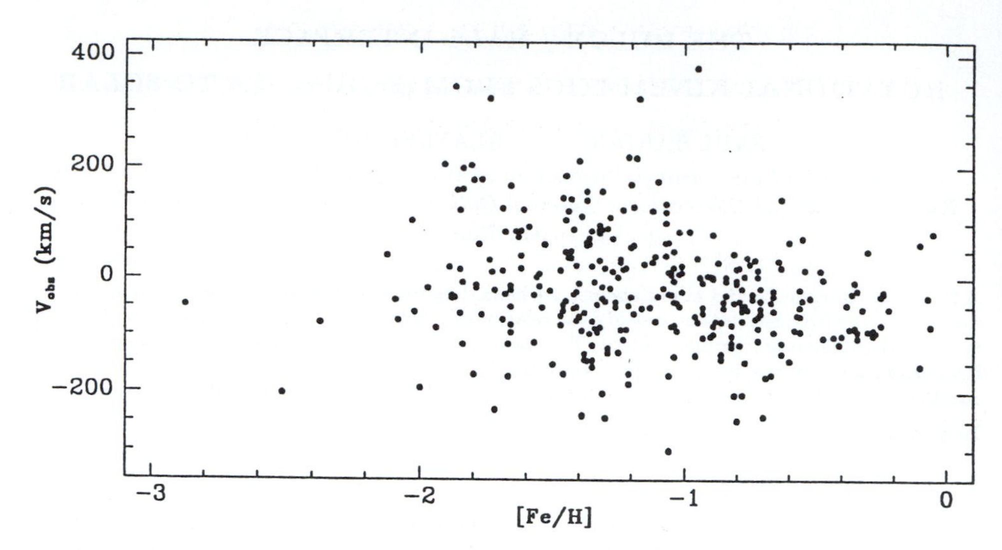

Fig. 1. The observed velocity with respect to the LSR, plotted against [Fe/H] from the Washington photometry. A change in the kinematics is seen at [Fe/H]=–1.0.

of this group are remarkably similar to those of the halo in the solar neighborhood and thus it seems appropriate to associate these stars with the halo.

The stars with [Fe/H] > -1 have $V_{rot} = 112 \pm 12$ km/s and $\sigma_{\phi} = 60 \pm 12$ km/s. These measures are consistent with the kinematics of the Mira variables (Menzies 1990), planetary nebulae (Kinman et al 1988), and OH/IR stars (Dejonghe, this volume, te Lintel-Hekkert et al 1991) at similar l and b but show higher mean velocity than the sample of Minitti et al (1992).

Our data are consistent, within the errors, with an abrupt change in kinematics near [Fe/H] $= -1.0$, which would imply that the bulge and halo are different entities, possibly having completely different origins. We plan to measure more accurate abundances from the stellar spectra, and this will provide a stronger test of this hypothesis.

Acknowledgements

We would like to thank George Preston and Leonard Searle for many useful discussions, and the OCIW TAC for the generous allocations of time at Las Campanas that made this project possible.

References

Kinman, T.D., Feast, M.W. and Lasker, B.M. 1988, AJ, 95, 804

Menzies, J.W. 1990, In "Bulges of Galaxies", ESO/CTIO Workshop, La Serena, Chile, ed. B.J. Jarvis and D.M. Terndrup

Minitti, D., White, S.D.M., Olszewski, E.W. and Hill, J.M. 1992, ApJ, 393, L47

Morrison, H.L., Flynn, C.M. and Freeman, K.C. 1990, AJ, 100, 1191

te Lintel-Hekkert, P., Dejonghe, H., and Habing, H.J. 1991, Proc ASA, 9, 20

DETECTABILITY OF BULGE STARS *

M.HERNANDEZ-PAJARES, R.CUBARSI and J.SANZ-SUBIRANA
Departament de Matemàtica Aplicada i Telemàtica
and
J.M.JUAN-ZORNOZA
Departament de Física Aplicada
ETSETB, Universitat Politècnica de Catalunya,
P.O.Box 30002, E-08080 Barcelona, Spain
First author's e-mail: matmhp@mat.upc.es

Abstract. Classification algorithms based on neural network techniques are applied to study if the bulge stars can be differentiated from other stars belonging to other Galactic components. A synthetic sample is build as a mixture of four components, thin disk, thick disk, halo and bulge, according to some stellar system models and considering observational errors.

Key words: Galactic Bulge - Stellar Classification - Neural Networks

In the recent literature some authors consider directly or indirectly, the possible existence of bulge stars in the solar vicinity. Rich 92 suggests that a central bar might scatter bulge stars, thus providing an explanation for the high velocity metal rich stars in the solar neighbourhood. By other hand Feast et al. 92 propose the existence of a *generalized thick disk* component, which could be regarded as the Local Bulge type population. This idea is suggested from the study of Mira variables and M-giants local stars. These considerations lead to the problem of the detectability of possible local bulge stars: Which star attributes are necessaries to know, to separate the different components in an efficient way? Can the bulge be detected within a local sample? Which is the possible predominant population that could hide a bulge component? To give first answers to the last questions we follow a synthetic approach within two steeps:

First we generate a four component local sample, with thin disk (t), thick disk (T), halo (h) and bulge (b) stars. We suppose, as Nemec et al. 92 do, that the sample has a kinematic bias that allows a strong presence of spheroidal component stars; for instance a proportion of $t : T : h = 2 : 2 : 1$. We introduce a small *contamination* of bulge stars in this composition with a sample that contains 1000 stars with 400 belonging to each disk component, 150 to the halo and 50 stars to the bulge. The real heliocentric distances have been taken lesser than 2.5 kpc with heights above the galactic plane lesser than 0.5 kpc. The sample contains 1000 stars and 8 attributes: 3-D positions (r, l, b), 3-D velocities (U, V, W), metallicities $([Fe/H])$ and ages. The details with the different distributions adopted for the thin disk, thick disk and halo are explained in Hernández et al. 92. The main features of the local bulge are based in the model of Cubarsí et al. 92 (see Hernández 92).

Secondly, the synthetic catalogue is analyzed with Principal Component Analysis (PCA) and with the Self Organizing Map algorithm (SOM). PCA is a known multivariate technique (see for instance Murtagh et al. 87) that looks for the maximum variance directions and gives us a quantitative idea of which attributes are more discriminant for the sample: in this case the circular rotation component, V,

* This work has been supported by the D.G.I.C.I.T. of Spain under Grant No. PB90-0478

H. Dejonghe and H. J. Habing (eds.), Galactic Bulges, 299–300.

377	5	0	0
23	347	9	13
0	19	139	0
0	29	2	37

TABLE I

The confusion matrix between the thin disk (1), thick disk (2), halo (3) and bulge (4) is given. The number in the i-th row and j-th column indicates the stars that really belong to the population j and appear in a centroid with predominant component i.

the metallicity and the age with the 24% of the total variance. By other hand SOM is an unsupervised neural classifier (see for details Kohonen 90) that has been applied to astronomical data in Hernández et al. 91. It is an efficient way to arrange the stars of the sample in subgroups in terms of the proximity in the characteristic space formed by the m-attributes considered, in our case $m = 8$. The basic aim is finding a smaller set of centroids that provides a good approximation of the original set S of n stars with m attributes, encoded as "vectors" $x \in S$. However, the main advantage of the algorithm is that it also arranges the centroids so that the associated mapping from S to A maps the topology of the set S in a least distorting way. Usually A is a bidimensional set of indexes where proximity between them means similarity between the global properties of the associated groups of stars. The SOM of the sample, after 500000 iterations and with a size of 5x5=25 centroids, has been calculated, and a good discrimination has been obtained by this neural classifier. The bulge is concentrated in two centroids where appears mixed with thick disk stars. Practically all centroids contain a predominant population with more than 50% of the respective stars. The purity of the classification can be also appreciated looking at the confusion matrix given in Table I. From this table the recognition error is 100 stars = 10%. It is important to note that the main source of confusion happens between the bulge and the thick disk. This component could hide the possible local bulge in observational samples with few bulge stars. Such result is compatible with the identification of local bulge stars with a generalized thick disk component (Feast et al. 92). By other hand, no confusion appears between the thin disk, and the halo and bulge components.

References

R. Cubarsí and M. Hernández-Pajares, IAU Symp. 153 on Galactic Bulges, 1992.

M.W. Feast, P.A. Whitelock, and R. Sarples, *The Stellar Populations of Galaxies*, Barbuy and Renzini, eds., IAU Symp. 149, Kluwer Acad. Press, Dordrecht, 1992, p. 77.

M. Hernández-Pajares, to be published, 1992.

M. Hernández-Pajares, R. Cubarsí, and E. Monte, in Heck and Murtagh, Eds., Astronomy from Large Databases II, Eur. Southern Obs., 1992 (proc. , Haguenau, France, 14-16 Sept. 1992).

M. Hernández-Pajares and E. Monte, *Artificial Neural Networks*, Lecture Notes in Computer Science, vol. 540, Springer-Verlag, Berlin, 1991, p. 422.

T. Kohonen, Proceedings of the IEEE **78** (1990), no. 9, 1464.

F. Murtagh and A. Heck, *Multivariate data analysis*, D.Reidel Publishing Comp., Dordrecht, 1987.

J.M. Nemec and A.F.L. Nemec, *The Stellar Populations of Galaxies*, Barbuy and Renzini, eds., IAU Symp. 149, Kluwer Acad. Press, Dordrecht, 1992, p. 103.

R.M. Rich, *The Stellar Populations of Galaxies*, Barbuy and Renzini, eds., IAU Symp. 149, Kluwer Acad. Press, Dordrecht, 1992, p. 29.

The Bulge-Halo Transition Region

RODRIGO IBATA and GERARD GILMORE

Institute of Astronomy, Madingley Road, Cambridge, England CB3 OHA

Abstract.

The Baade's Window bulge is a metal-rich, rotationally supported entity, whereas the solar neighbourhood halo is metal-poor and non-rotating; what is the connection between these two spheroid components?

We present results of a spectroscopic survey of 847 K-giants from 3 low absorption windows at (5,-12), (-5,-12) and (-25,-12). This is a seven times larger sample than has been previously studied in this region.

Key words: Galaxy - Structure - Bulge - Halo

The bulge is in several ways very different from the halo: it is predominantly rotationally supported, it is metal enriched, non-axisymmetric and exhibits a steep density distribution; in contrast the halo is pressure supported, metal poor, axisymmetric and has a shallower density distribution (Gilmore 1989).

We wish to determine whether these two components are indeed separate entities or just different regions of a spheroid whose attributes vary in some smooth way from the centre of the galaxy to beyond the solar neighbourhood. This, apart from letting us discover the Galaxy's appearance, may well shed light on the nature of its formation, and the physical processes involved in the formation of spiral galaxies in general.

There have been many studies investigating the inner bulge of the Galaxy (mostly Baade's Window), the minor axis (Blanco et al) and the halo. However, surprisingly few data have been published on the intermediate region between the bulge and the halo. It is interesting to point out that it is this spheroid region that is easiest to observe in external galaxies (Kormendy et al 1982), where halos are too dim and inner bulges too heavily obscured.

Three low absorption regions of approximately half a degree in diameter at (5,-12), (-5,-12) and (-25,-12) were selected for observation. UK Schmidt-Telescope plates of these regions were scanned using the APM facility at Cambridge, giving a complete sample for $14 \leq m_V \leq 16$, $1.1 < (B-V) < 2.0$. This selction window was chosen so as to find K-giants in the galactic bulge region.

Spectra of 847 candidate K-giants were observed with the AUTOFIB + IPCS configuration at the AAT. Radial velocities were calculated by crosscorrelation with standard radial velocity star spectra. Figure 1 shows the resulting radial velocity distributions.

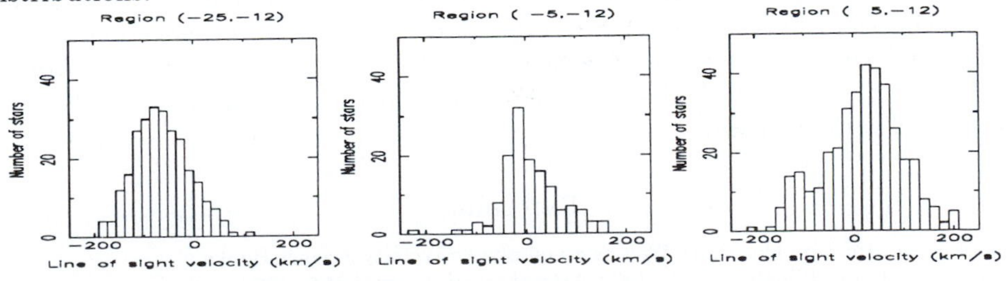

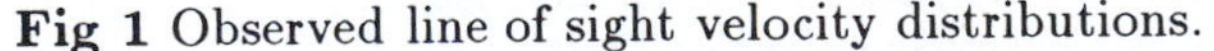
Fig 1 Observed line of sight velocity distributions.

H. Dejonghe and H. J. Habing (eds.), Galactic Bulges, 301–302.

To investigate the contribution of the disk, halo and thick disk and the effect of our selection criteria on the velocity distribution, we constructed a galaxy model which calculates the expected number of stars down a line of sight and the velocity distribution associated with these.

For the bulge we assume a gaussian velocity distribution along each of the familiar (u,v,w) axes. The total line of sight velocity distribution is calculated by summing the velocity distributions projected onto the line of sight of the stars found in each distance bin:

$$K(v) = \omega \int_0^\infty \sum_{\substack{spectral\\classes}} \left(\sum_{m=m_1}^{m_2} \Phi[M,S] D_S(r) r^2 \frac{e^{-\frac{(v-\bar{v}P_L)^2}{2\sigma_L}}}{\sqrt{2\pi}\sigma_L} \right) dr$$

where v is velocity, Φ is the luminosity function, $D_S(r)$ is the relative density of spectral class S, ω is the solid angle viewed, r is the distance along the line of sight, P_L is a projection of v onto the line of sight, C is stellar colour and $R(r)$ is intestellar reddening.

The integrand is put to zero at those values of r where the apparent colour lies outside the observational colour bounds: $(B-V)_1 \leq C - R(r) \leq (B-V)_2$.

The course we take is to insert into the model all the necessary relations and parameters pertaining to the disk, thick disk and halo. For the bulge we assume a luminosity function and colour-magnitude relation, and guess functional forms for the bulge geometry and velocity distribution.

Assuming the model to be correct, we compute the most likely set of bulge geometry and velocity distribution parameters using the maximum likelihood method given our observational data. We minimise the model's likelihood function using an amoeba-simplex routine (Press et al 1989). The robustness of any resultant parameter set is then tested by 'bootstrapping'.

We are in the process of examining our model parameter space, and as yet have no firm preference for a particular set of bulge parameters. However, it is clear that the data is inconsistent with a galaxy model being either axisymmetric or having a single bulge component with a density distribution of scale length 200-400 kpc.

1. Conclusions

We have reduced a large sample of K-giant stars away from the minor axis in the region 1.8-3.7 kpc from the galactic centre. We hope this data set will offer a significant contribution to the understanding of the halo-bulge connection. A star-count and kinematic model has been written to understand our results. Work is currently in progress to find which sets of bulge parameters are most likely to be consistent with the data.

References

Gilmore, G., 1990 In The Milky Way as a Galaxy: University Science Books
Kormendy, J. and Illingworth, G., 1982 *Astrophysical Journal* **256**, 460
Press et al., 1989 Numerical Recipes: Cambridge University Press

SIO MASER SURVEY OF THE BULGE IRAS SOURCES

H. IZUMIURA AND T. ONO
Department of Astronomy and Earth Sciences, Tokyo Gakugei University
4-1-1 Nukui-kita, Koganei, Tokyo 184, Japan

I. YAMAMURA, K. OKUMURA, AND T. ONAKA
Department of Astronomy, Faculty of Science, University of Tokyo
Bunkyo-ku, Tokyo 113, Japan

S. DEGUCHI AND N. UKITA
Nobeyama Radio Observatory, National Astronomical Observatory
Mimamimaki, Minamisaku, Nagano 384-13, Japan

O. HASHIMOTO
Department of Applied Physics, Seikei University
3-3-1 Kichijouji-kita, Musashino, Tokyo 180, Japan

AND

Y. NAKADA
Kiso Observatory, Faculty of Science, University of Tokyo
Mitake-mura, Kiso-gun, Nagano, 397-01, Japan

ABSTRACT. SiO maser emission from the Bulge IRAS sources has been searched by the v=1, J=1−0 and v=2, J=1−0 transitions to investigate the kinematics of the Galactic Bulge, resulting in a sample of 124 line-of-sight velocities. The rotation velocity, velocity dispersion, and velocity offset at $l = 0°$ for the sample are found to be 9.3±1.4 km s^{-1} deg^{-1}, 75.8 $^{+6.6}_{-5.9}$ km s^{-1}, and −18.2±9.7 km s^{-1}, respectively (80% confidence interval). Furthermore we find trends for the rotation velocity and velocity dispersion to decrease with distance from the galactic plane. These trends are supported by a larger sample constructed by incorporating other available velocity data on the Bulge IRAS sources. The rotation velocity and velocity dispersion are expressed as $15.6-1.23\times|b(\mathrm{deg})|$ km s^{-1} deg^{-1} and $101-3.6\times$ $|b(\mathrm{deg})|$ km s^{-1} , respectively. The implications of the observed quantities are discussed.

1. Introduction

The Galactic Bulge is a unique system for which we can investigate the kinematical properties in detail. So far several attempts were made to examine the motion of the Bulge at optical wavelengths (Menzies 1990 and references therein, Minniti et al. 1992). Most of them were, however, made toward some restricted regions because interstellar extinction hampers homogeneous sampling toward the Bulge. The success in IRAS mission changed this situation dramatically. As shown by Habing et al. (1985) it has become possible to extract the Bugle source candidates from IRAS Point Source Catalog with color and flux constraints (the Bulge IRAS sources). They are most likely mass losing AGB stars, hence SiO maser sources. SiO maser emission is one of the best tools to study stellar kinematics in distant and obscured regions

H. Dejonghe and H. J. Habing (eds.), Galactic Bulges, 303–308.

like the Bulge (cf. Lindqvist et al. 1991). We have started a survey of SiO masers from the color-selected Bulge IRAS sources with the Nobeyama 45-m telescope (cf. Nakada et al. 1992).

2. Observations and Survey Results

The observations were made with the 45-m telescope of Nobeyama Radio Observatory from January 1990 to June 1992 (still continuing). We used a 4-K cooled SIS mixer receiver with T_{sys} around 200 K (SSB, Ta*) at the elevation angle of 25°. The beam size of the telescope is 39"(HPBW) at 43 GHz. We observed two SiO maser lines simultaneously: SiO v=1, J=1−0 (43.122027 GHz) and v=2, J=1−0 (42.820539 GHz). This dual line observation assures firm detection of the emission like conspicuous twin-peaks of OH masers. The velocity coverage in V_{LSR} is from -300 km s^{-1} to $+300$ km s^{-1}. The velocity resolution is 0.28 km s^{-1} at the observing frequencies. Pointing accuracy was better than 10", which is accurate enough for our observing beam size of 39".

The Bulge IRAS sources are extracted from the IRAS Point Source Catalog with the conditions of F12<10Jy, 0.0≤log(F25/F12)≤0.2, $-15° \leq l \leq 15°$, and $3° \leq |b| \leq 15°$. The upper limit to the 12μm flux and color range were determined so that the sources were distributed on the sky to delineate clearly the shape of the Bulge (cf. Habing et al. 1985) and by consulting the study by Whitelock et al. (1991). The region near the galactic plane was excluded because of the contamination and confusion by disk sources. We included sources which did not satisfy the above criteria but were assigned to the Bulge members by Whitelock et al. (1991). To the present 186 "Bulge" sources were observed at Nobeyama and the SiO maser emission has been detected in 124 (67%) of them. The distribution of the detected sources on the sky is shown in Figure 1 in the galactic coordinates. They seem to concentrate rather into the four strips, $4° \leq |b| \leq 5°$ and $7° \leq |b| \leq 8°$. This is because we gave a priority to these areas in extending our survey down to sources having 12μm fluxes smaller than 5Jy. The detection rate exceeds about 70% for the sources which satisfy the above color criterion but have 12μm flux greater than 10Jy (32 out of 44, 73%). The velocities between SiO masers and OH masers (te Lintel et al. 1991) agree well within ± 3 km s^{-1} for commonly observed sources. These results are in good agreement with those by Jewell et al. (1991) on the sources in the solar neighbourhood and by Lindqvist et al. (1991) on the OH/IR stars close to the galactic center.

3. Kinematical Properties of the Bulge

We here show some results on the Bulge kinematics. We assume that the Local Standard of Rest moves with respect to the Galactic Center in a circular oribit at a velocity of 220 km s^{-1}. Velocities given here are refered to the Galactic Standard of Rest.

We first show radial velocities of the Bulge IRAS sources as a function of galactic longitude (l–v diagram, Figure 2). Clear evidence is found for the rotation of the Bulge. We obtained the rotation velocity to be 9.3 ± 1.4 km s^{-1} deg^{-1} (80% confidence interval) from a linear regression. This corresponds to 67.4 ± 10.4 km s^{-1}

kpc^{-1} assuming Ro=8.0 kpc. The residual velocity dispersion is $75.8^{+6.6}_{-5.9}$ km s^{-1} (80% confidence interval). Minkowski (1964) first pointed out that the Bulge of our Galaxy slowly rotates in the same direction as the disk component does through the analysis of planetary nebulae near the Galactic Center region. Since then several works are made on the motion of the Bulge (Menzies 1990 and references therein, Minniti et al. 1992, Nakada et al. 1992). They are consistent with the present results on both the direction and speed of rotation of about 10 km s^{-1} deg^{-1}.

In Figure 1 a velocity offset at l = 0° of −18.2±9.7 km s^{-1} (80% confidence interval) is found. Similar offsets are observed for Baade's Window K-giants (Rich 1990) and planetary nebulae in the vicinity of the galactic center (Kinman et al. 1988). Some possible reasons can be thought of immediately:

- It is suffered from statistical fluctuations,
- Our survey is still too shallow to obtain a l–v diagram symmetric with regard to the origin,
- There is another component about the motion of the Local Standard of Rest.

The second reason suggests a systematic velocity structure in the Bulge. It requires some relation between the distances to the sources and the radial velocities. Nakada et al. (1991) and Blitz and Spargel (1991) suggest existence of a bar-like bulge in the center of our Galaxy. In such a bar-like system, systematic streaming motion can be expected (cf. Freeman 1966). More observations in the $l \leq 0°$ region with higher sensitivity are necessary to discriminate one from these possibilities.

Second, we divided our survey region into areas of 1 degree width in $|b|$ and examined the dependence of the kinematical parameters on the distance from the galactic plane. We have found that the rotation velocity tends to decrease with distance. For the velocity dispersion a similar trend is discernible but less conclusive. To improve the statistics we have incorporated other available data on the radial velocities of the Bulge IRAS sources, i.e., the OH maser survey by te Lintel et al. (1991). The number of the sample is increased from 124 to 183. With the improved statistics the trends became more evident (Figure 3,4). The rotation velocity and velocity dispersion can be expressed as $15.6-1.23\times|b(\mathrm{deg})|$ km s^{-1} deg^{-1} and $101-3.6\times|b(\mathrm{deg})|$ km s^{-1}, respectively. Two possibilities are considered:

- Both quantities really depend on $|b|$,
- The contamination by disk sources decreases with distance from the galactic plane.

The former can be anticipated for kinematic studies, and if real, it is found observationally for the first time. The velocity dispersion is suggested to decline with distance from the galactic center (Minniti et al. 1992), which favours the former possibility. We expect the disk source contamination is low, however, we still cannot rule out completely the latter possibility at present. We are now exmamining influence by the contamination.

4. Summary

1). SiO maser emission has been searched for v=1 J=1-0 and v=2 J=1-0 transitions simultaneously toward 186 Bulge IRAS sources and detected in 124 sources.

The detection rate is 67% for the present sample and exceeds 70 % for the bright sources. Further it was found that the SiO maser emission gave the stellar systemic velocities as accurately as OH masers. Thus the SiO masers have been proved to be one of the best tools to probe line-of-sight velocities of the Bulge IRAS sources.

2). The mean rotation speed and velocity dispersion of the Bulge were obtained to be 9.3 $\pm$1.4 km s^{-1} deg^{-1} (67.4 km s^{-1} kpc^{-1}, Ro=8.0 kpc) and 75.8 $^{+6.6}_{-5.9}$ km s^{-1}, respectively.

3). The velocity offset at $l = 0°$ is -18.2 ± 9.7 km s^{-1}. This may suggest existence of a systematic velocity structure in the Bulge or that of an additional velocity component in the motion of the Local Standard of Rest. However, statistical fluctuations cannot be ruled out.

4). Both the rotaion velocity and velocity dispersion tend to decrease with distance from the galactic plane, and are expressed as $15.6-1.23\times|b(\mathrm{deg})|$ km s^{-1} deg^{-1} and $101-3.6\times|b(\mathrm{deg})|$ km s^{-1} , respectively, for the sample including OH data.

Our survey is still on its way. We will be able to discuss the above issues with greater confidence and to understand the origin and kinematics of the Bulge in conjunction with the rest of the Galaxy in near future.

Acknowledgements

The authors express their hearty thanks to the staff of the Nobeyama 45-m telescope. This work is partly supported by Grant-in-Aid from the Ministry of Education, Science and Culture, Japan (No.03640244C).

References

Blitz, L. & Spargel, D. 1991, Ap.J., 379, 631

Freeman, K. C. 1966, M.N.R.A.S., 134, 1

Habing, H., Olnon, F. M., Chester, T., Gillett, F., Rowan-Robinson, M., & Neugebauer, G. 1985, A.A. 152, L1

IRAS, Point Source Catalog 1985, US Government Publication Office

Jewell, P. R., Snyder, L. E., Walmsley, C. M., Wilson, T. L., & Gensheimer, P. D. 1991, A.A., 242, 211

Kinman, T. D., Feast, M. W., & Lasker, B. M. 1988, A.J., 95, 804

Lindqvist, M., Ukita, N., Winnberg, A., & Johansson, L. E. B. 1991, A.A., 250, 431

Menzies 1990, in *BULGES of GALAXIES*, ESO-CTIO Workshop, eds. B. J. Jarvis & D. M. Terndrup, p.115

Minkowski 1964, P.A.S.P., 76, 197

Minniti, D., White, S. D. M., Olszewski, E. W., & Hill, J. M. 1992, Ap.J. Letters, 393, L47

Nakada, Y., Deguchi, S., Hashimoto, O., Izumiura, H., Onaka, T., Sekiguchi, K., & Yamamura, I. 1991, Nature, 353, 140

Nakada, Y., Onaka, T., Yamamura, I., Deguchi, S., Ukita, N., & Izumiura, H. 1992, P.A.S.J. in press

Rich 1990, Ap.J. 362, 604

te Lintel Hekkert, P., Caswell, J. L., Habing, H. J., Haynes, R. F., & Norris, R. P. 1991, A.A. Suppl., 90, 327

Whitelock, P. A., Feast, M. W., & Catchpole, R. M. 1991, M.N.R.A.S. 248, 276

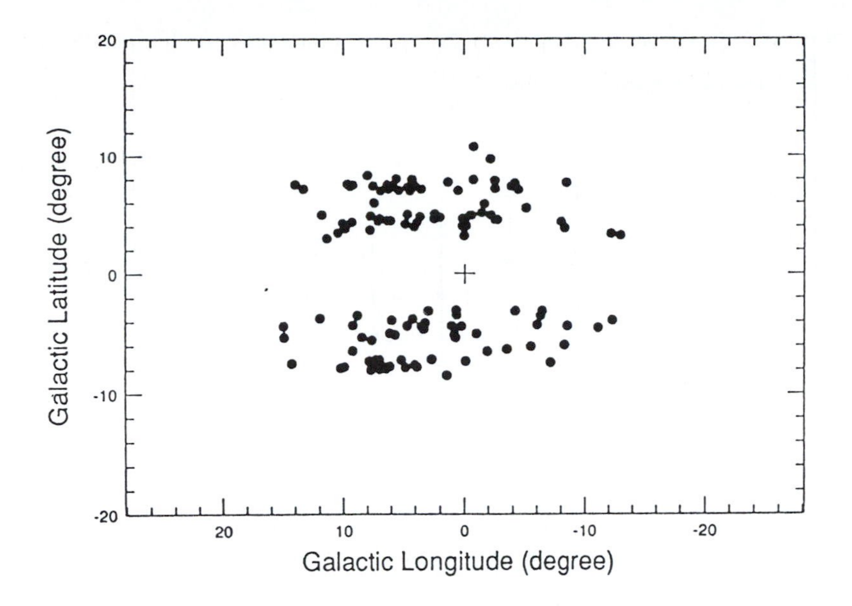

Figure 1: The distribution of 124 Bulge IRAS sources associated with SiO maser emission. The region of $|b| < 3°$ is excluded in our survey.

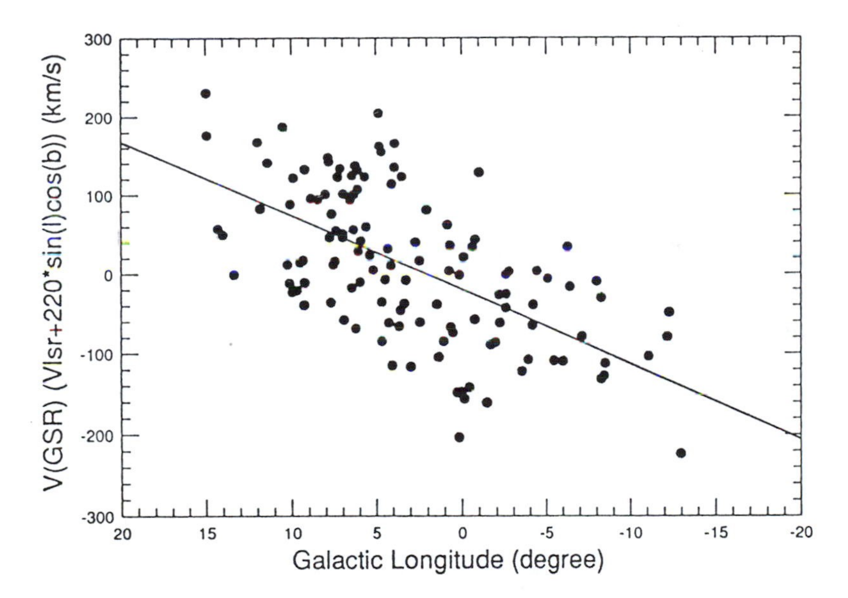

Figure 2: The radial velocities of the Bulge IRAS sources as a function of galactic longitude. A linear regression gives the rotation velocity, velocity dispersion, and velocity offset at $l = 0°$ to be 9.3 ± 1.4 km s^{-1} deg^{-1}, $75.8^{+6.6}_{-5.9}$ km s^{-1}, and -18.2 ± 9.7 km s^{-1}, respectively.

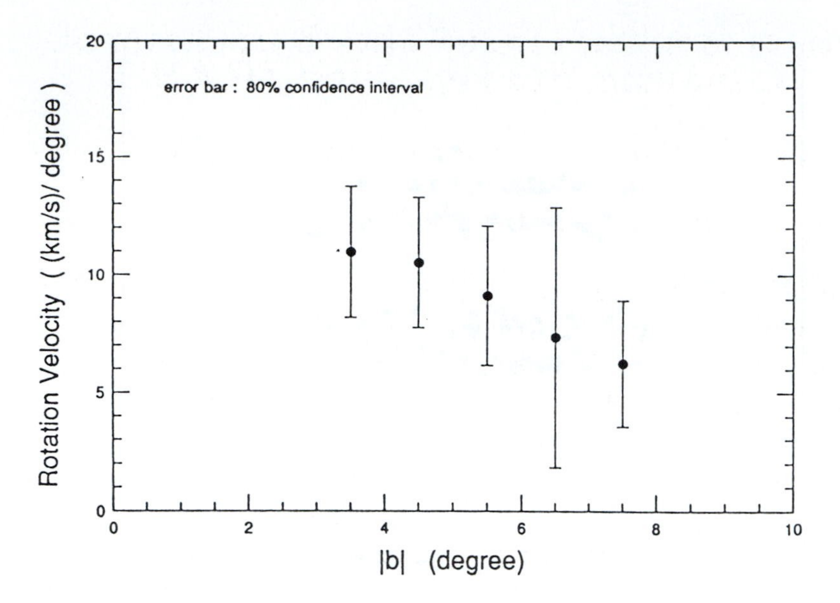

Figure 3: The dependence of rotation velocity on galactic latitude for a sample of 186 sources in which OH data are also included. The error bars express 80% confidence intervals. The least squares fit to a straight line gives a slope of -1.23 km s^{-1} deg^{-2}.

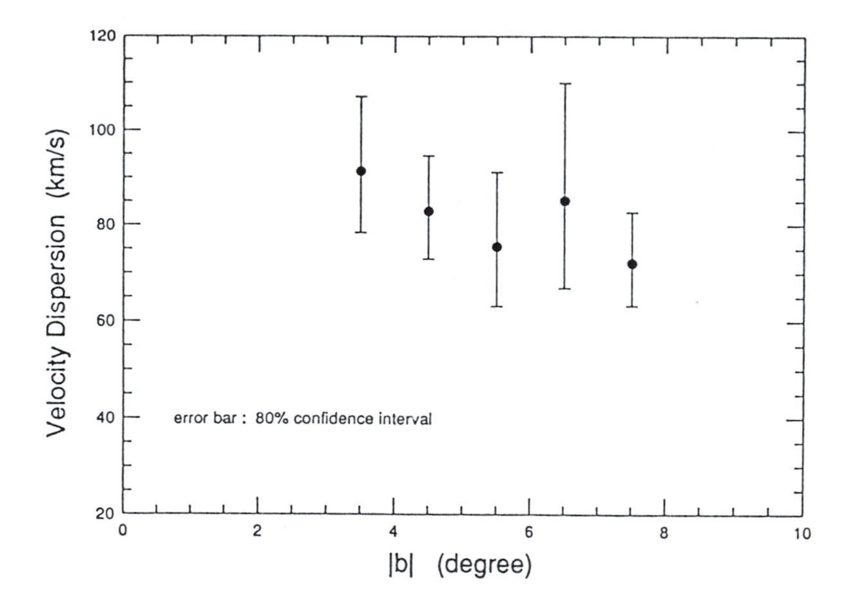

Figure 4: Same as in Figure 3 but for velocity dispersion. The slope is found to be -3.6 km s^{-1} deg^{-1}.

Cross-correlation of the Two Micron Galactic Survey with IRAS

T. Mahoney, F. Garzón, X. Calbet, P. Hammersley, M. J. Selby

Instituto de Astrofísica de Canarias, E-38200 La Laguna, Tenerife (Spain)

The Two Micron Galactic Survey (TMGS) is described by Calbet *et al.* (this symposium) and Mahoney *et al.* (1990). It has been cross-correlated with the IRAS point source catalogue (IRPS) using a 15-arcsec search radius, and preliminary results are presented here.

There were many one-to-one correspondences, but source confusion was severe close to the plane. At the limiting magnitude of the TMGS ($m_K = +10.5^{\rm m}$) and with a galactocentric distance of 8 kpc and extinction of 0.0002 mag $\rm pc^{-1}$ a typical M3 III giant with $M_K = -5.5^{\rm m}$ will have an apparent magnitude of $m_K = +10.6^{\rm m}$. So the TMGS should see such objects almost as far as the Galactic centre. The IRPS sources examined correspond to three regions scanned by the TMGS. These zones are roughly 1 degree wide in declination and 2–3 hours long in right ascension. They are centred at $\delta = -1^\circ, -22^\circ$ and -30° (approximately) and are centred in RA on the Galactic plane.

Colour-colour plots and histograms of source distributions of the IRPS data alone for each of the three regions were compared with similar plots and histograms for sources correlated with the TMGS. For IRAS-only data, the Galactic plane is seen at $\delta = -1^\circ$. The bulge is evident at $l = -22^\circ$ and dominates at the Galactic centre. In 12–25–65 μm diagrams all sources above the $F_\nu(12\mu{\rm m}) = F_\nu(25\mu{\rm m})$ line are stellar, whereas galaxies and "Galactic components" lie to the left of the $F_\nu(25\mu{\rm m}) = F_\nu(60\mu{\rm m})$ line. Stellar sources are of three types: photospheres, with $F_\nu(12\mu{\rm m}) > 3F_\nu(25\mu{\rm m})$, circumstellar dust shells (CDS), with $F_\nu(25\mu{\rm m}) < F_\nu(12\mu{\rm m}) < 3F_\nu(25\mu{\rm m})$, and "disk and bulge stars", which extend the stellar region of the colour-colour diagram below the $F_\nu(12\mu{\rm m}) = F_\nu(25\mu{\rm m})$ line. Nearly all source types are represented where all the sources in the zones are included. Closer in towards the Galactic plane, however, only the redder CDS's remain, along with the Galactic components. Photospheres become much scarcer.

For cross-correlation we used high-quality IRAS sources for $\mid b \mid < 15^\circ$ and applied two criteria: (i) 6.0 Jy $< F_\nu(12\mu{\rm m}) <$ 0.6 Jy, and (ii) the redness parameter $R = F_\nu(25\mu{\rm m})/F_\nu(12\mu{\rm m}) > 0.7$ (Habing 1986).

H. Dejonghe and H. J. Habing (eds.), Galactic Bulges, 309–310.

The plane is still evident at $\delta = -1^\circ$, but the bulge is clearly seen at $\delta = -22^\circ$. Source confusion, particularly in the IRAS data, is the principal reason for the poor showing of the bulge in the $\delta = -30^\circ$ zone. It seems, nevertheless, that the TMGS is seeing into the bulge. New 12–25–60 μm c-c plots were made for the remaining sources. CDS's cluster about the -0.1 abscissa value, although this is not so evident for the -30° zone, due, probably, to confusion problems. In common with Habing, we find with cross-correlation that the plane and bulge become more pronounced as the two criteria are applied, hence confirming that the TMGS is detecting bulge sources.

References

Habing, H 1986, in F. P. Israel (ed.), *Light on Dark Matter.* Proceedings of the First IRAS Conference, Noordwijk, The Netherlands. Dordrecht: Reidel.

Mahoney, T., Selby, M. J., Garzón, F., Hammersley, P. and Calbet, X. 1990, in B. J. Jarvis and D. M. Terndrup (eds.), *Bulges of galaxies.* ESO/CTIO workshop, La Serena, Chile.

Microlensing Towards the Galactic Bulge

ANDRZEJ UDALSKI, MICHAL SZYMANSKI, JANUSZ KALUZNY, MARCIN KUBIAK
Warsaw University, Astronomical Observatory, Al. Ujazdowskie 4, PL-60-286, Warszawa, Poland

MARIO MATEO, GEORGE PRESTON, WOJCIECH KRZEMINSKI
Carnegie Observatories, 813 Santa Barbara Street, Pasadena CA, 91101, USA

and

BOHDAN PACZYNSKI
Princeton University Observatory, 124 Peyton Hall, Princeton, NJ, 08544, USA

Abstract. We describe an ongoing survey to search for dark matter via lensing events of stars in the Galactic Bulge. The principal properties of the survey are described, and some preliminary results are shown for newly-discovered variables. We discuss some of the projects related to the study of the Galactic Bulge that can be addressed using these data, and describe the future plans for the survey over the coming few years.

Key words: methods: observational – surveys – stars: imaging

Galaxy rotation curves provide convincing evidence for the existence massive halos of unseen material. Paczynski (1986, ApJ, 304, 1) first suggested this dark matter can be detected via gravitational lensing effects if composed of objects more massive than about $10^{-8}M_{\odot}$. More recently, Paczynski (1991, ApJL, 371, L63) and Griest, et al. (1991, ApJL, 372, L79) noted the significant benefits of searching for microlensing events towards the Galactic Bulge. These include a) the existence of a minimum lensing rate due to low-mass disk stars, b) the presence of a multitude of background sources at a common distance in many different Bulge fields, and c) a reasonably high lensing rate due to putative halo objects. The obvious practical difficulty with this experiment is that only about one star in 10^6 is expected to be 'strongly' lensed (i.e., Δmag ≥ 0.34) at any given time. Thus, millions of stars must be regularly monitored to detect a significant number of events.

We have embarked on a project to conduct a microlensing survey towards the Galactic Bulge. The 1992 observations have been acquired using a 2048×2048 Ford (Loral) CCD on the 1m telescope at Las Campanas Observatory. Fourteen fields were monitored in the V and I bands, though most were obtained in I. Galaxy clusters have been monitored for SNe when the Bulge is not visible or the seeing is worse than 1.5-1.7 arcsec. The primary survey field is Baade's Window (BW; $l = 1°, b = -4°$); in addition, fields on either side of the Galactic center ($l \sim \pm5°, b \sim -3°$) were also monitored. Over 1100 I-band images were obtained on $\sim$ 45 useful nights between 12 Apr and 10 Aug 1992; an additional 200 V-band images were also acquired.

All analysis was done in near-real-time at the telescope, the photometry being performed using a modified version of the DoPHOT program developed by Paul Schechter. We were able to reduce about 150,000 stars per hour using two Sparc 2 computers at LCO. The Bulge frames typically contain about 120,000-180,000 measurable stars. Though the reductions are optimized for repeat measurements

H. Dejonghe and H. J. Habing (eds.), Galactic Bulges, 311–312.

of crowded fields, the quantity and quality of the stellar photometry is nonetheless critically dependent on the seeing. The median seeing for acceptable frames is approximately 1.2 arcsec.

The Figure below shows (uncalibrated) I-band light curves of four *newly discovered* variables in our survey. Two of these are RR Lyr stars, while the others are different sorts of eclipsing binaries. The Algol variable (star 108954) has a short duty cycle, yet was easily identified with our photometry. The systematic identification of intrinsic variables is important to a) eliminate them from consideration in our survey for lensing events, b) as a source of study in their own right, and c) as a check on the quality of the photometry and our ability to detect variables in our crowded fields.

The 1992 observations were obtained as a pilot project for a more ambitious future survey. We have a) demonstrated the feasibility of monitoring 10^6 stars nightly, b) developed software and techniques to analyze these data in near-real-time, and c) identified many of the intrinsic variable in the Bulge. In the future we plan to pursue a number of projects on the Bulge using the existing data. These include a survey of the age distribution of the Bulge from deep, co-added VI frames of each field and from analysis of the photometric properties of the Algols (Preston, this volume). Also, we plan to study the structure of the Bulge by comparing the distribution of brightnesses of RR Lyr stars in different regions. Our second major goal is to observe more stars, on more nights, with faster and more complete software. We plan to achieve this by applying faster hardware to the survey, constructing a multi-CCD camera, and possibly building a dedicated telescope for the project. Our immediate goals for 1993 are to extend the current survey to more stars in other Bulge fields, and improve our ability to do real-time analysis of the incoming data.

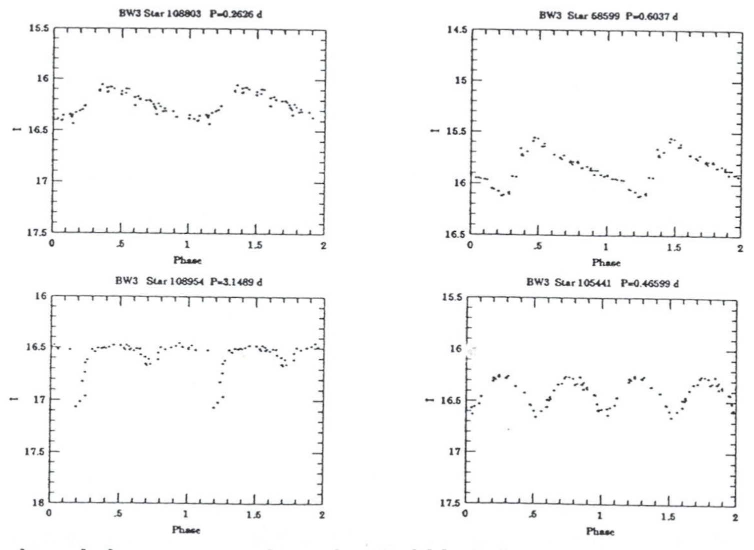

MM acknowledges support through a Hubble Fellowship from STScI under contract to NASA (HF-1007.01-90A). BP acknowledges support from the NSF through grant AST-9023775.

TIO BANDS IN COMPOSITE SYSTEMS

A. MILONE, B. BARBUY
IAG-USP, CP 9638, CEP 01065-970, São Paulo, Brazil

ABSTRACT. TiO bands at λ 620 nm are synthesized. The behaviour of these bands as a function of stellar parameters is studied. Application to composite spectra of one galactic bulge globular cluster and to bulges of elliptical galaxies is also carried out. TiO bands may be useful metallicity indicators.

1. Behaviour of TiO Bands

TiO bands in the red and near-infrared regions of the early-type galaxies spectra are strong absorption features(e.g. Bica & Alloin 1986). A detailed study of strong spectral features is an important link between stellar spectroscopy and population synthesis in composite systems.

In order to study the behaviour of TiO bands in stars as function of metallicity, effective temperature and gravity, we have computed TiO and total synthetic spectra at $\lambda\lambda$ 614.5 - 627.5 nm, in steps of 0.02 Å and with FWHM=0.8 Å using the code by Barbuy(1989), for a grid of the atmospheric parameters: T_{eff} = 4000, 4500, 5000, 5500 K, log g = 0.0, 1.0, 2.0, 3.0, 4.0 , 4.5, 5.0, and [M/H] = -3.0, -2.0, -1.0, 0.0, +0.5 dex; in the same mode employed by Barbuy *et al.*(1992b) for the study of the Mg indices. The used model atmospheres are interpolated in the grids of models by Bell *et al.*(1976) and by B. Gustafsson(p.c.). The atomic data base consists of lines by Moore *et al.*(1966), whose the oscillator strengths are obtained by fitting the computed lines to the solar spectrum. The molecular data base consist of the line identification: by J.G. Phillips(p.c.) for TiO α ($C^3\Delta - X^3\Delta$), $\gamma(A^3\Phi - X^3\Delta)$, and γ prime ($A^3\Pi - X^3\Delta$) systems; by Davis & Phillips(1963) for CN red system($A^2\Pi - X^2\Sigma$); and by Phillips & Davis(1968) for C_2 Swan system($A^3\Pi - X^3\Pi$). The molecular oscillator strengths are calculated using the formulae by Kovacs(1969).

We have measured TiO and total absorptions, F(TiO) and F(6145-6275 Å)respectively, and analyzed its dependence on metallicity, effetive temperature and gravity. The logarithm of TiO absorption is strongly dependent on the metallicity for [M/H] $\geq$ -1.0 dex. A similar dependence exist on θ_{eff} in giants and dwarfs with [M/H] $\geq$ 0.0 dex. There is practically no correlation between F(TiO) and gravity. Similar results are obtained when we have analyzed the relationship between log[F(6145-6275 Å)] and metallicity, temperature and gravity. The linear correlation between log[F(6145-6275 Å)] and [M/H] become stronger and extend to all metallicities(see Figure 1).

2. Synthetic Grid and Composite Spectra

We have used the information contained in the BV colour-magnitude diagram of the galactic globular cluster NGC 6553, whose metallicity is nearly solar: [M/H] = -0.2 dex by Barbuy *et al.*(1992a).

We have computed a grid of synthetic spectra at $\lambda\lambda$ 614 - 648 nm for ten different

H. Dejonghe and H. J. Habing (eds.), Galactic Bulges, 313–314.

stages of evolution observed in the NGC 6553 for the solar metallicity only. The metodology is the same applied by de Souza *et al.*(1992).
We have built a composite spectrum, which represent a sum of the individual synthetic spectra according to the relative number of star in each evolutionary stage and the corresponding flux contribution. The resulting composite spectrum was convolved with FWHM from 3.2 until 9.2 Å, in order to adjust to the spectra of composite systems.
We have fitted convolved composite spectra to the globular cluster G1 Bica(1988) spectrum, corresponding to metal-rich globular clusters, and the NGC 4936 elliptical galaxy(E0).

Acknowledgements. The calculations and reductions were carried out with the VAX 8530 of the IAG-USP. Financial support by FAPESP is acknowledged.

References

Barbuy B.:1989, Ap&SS, **157**, 111
Barbuy B., Castro S., Ortolani, S., Bica, E.:1992a, A&A,**259**, 607
Barbuy B., Erdelyi-Mendes M., Milone A.:1992b, A&AS, **93**, 235
Bell R.A., Eriksson K., Gustafsson B., Nordlund A.:1976, A&AS,**23**, 37
Bica E., Alloin D.:1986, A&A, **162**, 21
Bica, E.:1988, A&A, **195**, 76
Davis S.P., Phillips J.G.:1963,"The Red System of CN Molecule", University of California Press, Berkeley
Kovacs, I.:1969, "Rotational Structure in the Spectra of Diatomic Molecules", Adam Helger Ed. , London
Moore C.E., Minnaert M.G., Houtgast J.: 1966, NBS Monograph n° 61, Washington
Phillips J.G., Davis S.P.:1968, "The Swan System of the C_2 Molecule", University of California Press, Berkeley
de Souza R.E., Barbuy B., dos Anjos S.:1992, AJ (submitted)

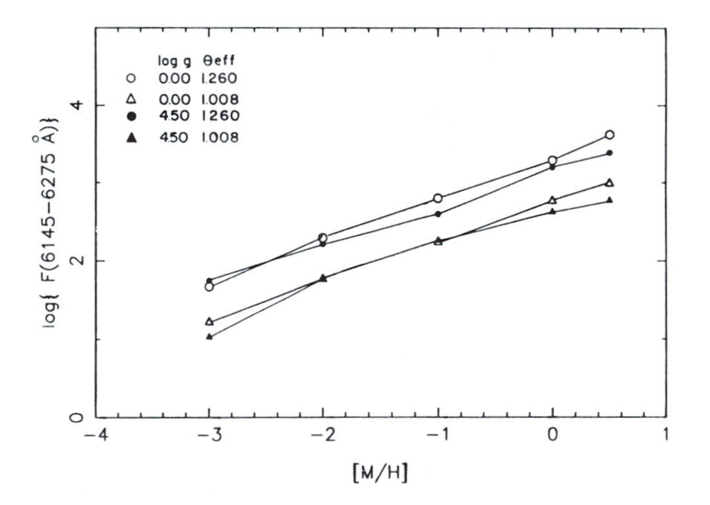

Figure 1 - Logarithm of total absorption as a function of metallicity

KINEMATICS OF THE GALACTIC BULGE: THE VELOCITY ELLIPSOID

Dante Minniti
Steward Observatory, Univ. of Arizona, Tucson AZ 85721, U.S.A.

Abstract: We investigate the shape of the velocity ellipsoid in the Galactic bulge by combining radial velocities and proper motions. We find that the velocity ellipsoid is nearly isotropic.

The Galactic bulge is the only one where we can actually measure the 3-D kinematics. Because of the lack of 3-D information in distant elliptical galaxies and bulges of spirals, the degree of anisotropy has been just an assumption in every kinematic study of bulges and elliptical galaxies (*e.g.* Binney and Tremaine 1987, 'Galactic Dynamics', Princeton Univ. Press). With accurate determinations of proper motions towards different bulge fields, we would be in position to constrain that parameter for the first time.

Cudworth (1986, A.J. 92, 348, referred to as C86) measured proper motions in the field of the globular cluster M22 ($l = 9.9, b = -7.6$). The foreground contamination by bulge stars is important in this low latitude field at a projected distance of 1.6 kpc from the Galactic center, as can be seen in C86 color magnitude diagrams. About 200 stars were found to be nonmembers of M22 on the basis of proper motions. By doing appropriate color and magnitude cuts, we select 81 of them which lie in the region of the CMD occupied by the bulge giant branch. Minniti et al. (1992, Ap.J. 393, L47) have measured radial velocities for 78 spectroscopically confirmed K giants in a field at $l = 8, b = 7$, most of which are expected to be bulge members. This field lies in a different quadrant, but at about the same projected distance from the Galactic center as the M22 field. Thus, it is not unreasonable to combine the two data sets to study the 3-D kinematics of the bulge. None of our conclusions rely on the absolute proper motions, we focus here on the relative proper motions, which give the dispersions σ_l and σ_b.

The proper motions reveal that the velocity ellipsoid projected on the plane of the sky is nearly isotropic, with slight elongation towards the Galactic center. If we assume the distance to the Galactic center to be $R_0 = 7.5$ kpc, both the radial velocity dispersion σ_r and the two proper motion dispersions σ_l and σ_b are similar, consistent with isotropy. The Figure below shows the distribution of radial velocities and proper motions in the l,b directions, scaled assuming $R_0 = 7.5$ kpc. Recent work by Spaenhauer et al. (1992, A.J. 103, 297) in Baade's window shows agreement with our results.

H. Dejonghe and H. J. Habing (eds.), Galactic Bulges, 315–316.

We divide the proper motion sample into 2 subsamples with colors redder and bluer than $(B-V)_0 = 1.4$, corresponding roughly to metal rich and metal poor giants, respectively. Such a division is preliminary, and must be confirmed by more accurate metallicity determinations. However, we find that the metal rich stars tend to have smaller σ_l and σ_b than the metal poor stars. The radial velocities show the same effect, the metal poor subsample has a larger velocity dispersion than the metal rich one. Also, the metal rich stars have a shift in the mean μ_l with respect to the metal poor stars. This shift is in the direction of galactic rotation, again evidence that the metal rich subsample is rotating more rapidly. It is important to note that these effects are small (only $\sim$2 σ), and their reality should be checked with more extensive data samples.

To analyze the data, Minniti & White (1993, in preparation) derive the geometry involved in the cases with spherical and cylindrical symmetry. We model the bulge density and velocity dispersion laws, and perform Monte-Carlo simulations to reproduce the data. We find that a new and independent measure of the distance to the Galactic center can be obtained from combining the proper motions with radial velocities. A distance of R_0 = 7-8 kpc is found from the C86 proper motions, the uncertainty is given only by the small size of the sample.

In summary, we find that the bulge velocity ellipsoid is nearly isotropic, supporting the Minniti et al. (1992) conclusion that the Galactic bulge is flattened by rotation and not by velocity anisotropies. The shape of the velocity ellipsoid of the Galactic bulge is different than that in the other components of the Galaxy (thin disk, thick disk and halo), that have been measured locally. There are trends of the kinematics with metallicity, in the sense that the metal rich subsamples have larger rotation and smaller velocity dispersions. A very important byproduct of a larger proper motion study in the bulge will be the determination of an accurate distance to the Galactic center R_0. This distance scale is absolutely essential to determine the age of the Galactic bulge: for R_0 = 7 and 8.5 kpc, population models give a difference in ages of 4 Gyr, the bulge being younger for the larger R_0.

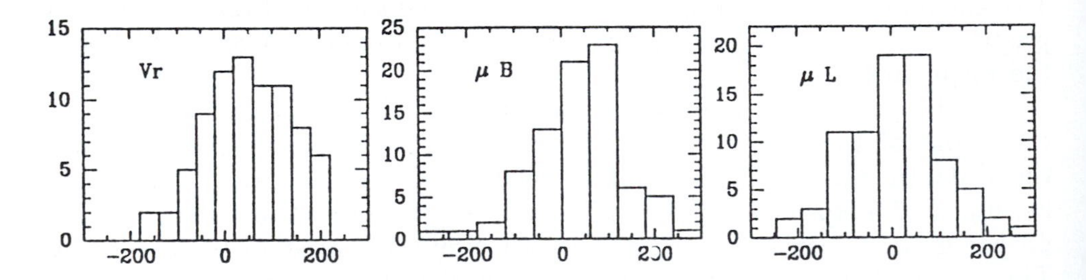

COMPARING M AND K GIANTS IN THE BULGE

HEATHER MORRISON and PAUL HARDING

Observatories of the Carnegie Institution of Washington, Pasadena, CA, and Kitt Peak National Observatory, National Optical Astronomy Observatories, Tucson, AZ, USA

and

PETER MCGREGOR

Mt Stromlo and Siding Spring Observatories, The Australian National University, ACT, Australia

Abstract. Kinematics and abundances of samples of K and M giants in a field at the edge of the bulge are compared. Despite a higher mean abundance, the M giants have the same kinematics as the metal-richer K giants.

Key words: Kinematics - M giants - K giants

While studies of M giants in the galactic bulge (see Frogel *et al* 1990 and references therein) have contributed a great deal to our understanding of the bulge, there still remain some puzzling discrepancies. For example, Frogel *et al* find evidence for an abundance gradient along the minor axis, but find a very narrow metallicity dispersion within a given field, in contrast with results for K giants in the same fields. This may be due to the use of M giants as tracers: it is difficult to measure [Fe/H] for late M giants, both because of the lack of good model atmospheres for these very cool stars and because we lack empirical calibrators (most globular clusters have [Fe/H] well below solar).

Harding and Morrison (this volume) describe a survey of a field at the edge of the bulge ($l = 350, b = -10$) where the mean abundance has dropped to [Fe/H] $\simeq$ -0.8. A comparison of the K and M giants in this field with those in fields closer to the galactic center is useful because we do not need to extrapolate to measure abundances in our field. Different methods of selection have been used for M giants in the Blanco surveys used by Frogel *et al* and our survey. Both have abundance selection effects: Blanco identified M giants by the presence of strong TiO bands on grism spectra, while we used T1–T2 color (very close to Cousins R–I) of stars from our CCD scans, which is less sensitive to abundance but still not a pure estimate of temperature for M giants.

Abundance has a strong effect on magnitude of M giants in an optical color-magnitude diagram. Thus comparison of the positions of stars in the T1 vs. T1–T2 diagram with the loci of globular cluster giant branches (placed at the distance of the bulge) gives a rough estimate of our M giant abundances. In our field almost all the M giants are fainter than the 47 Tuc locus, and thus have [Fe/H] significantly higher than -0.75. JHK photometry, obtained at Siding Spring Observatory for 75 of the M giants, confirms that their mean abundance lies between -0.75 and solar. Since the K giants in our field have a mean [Fe/H] of ~ -0.8, this means that the M giants have a higher mean abundance than the K giants, as we would expect from stellar evolution. Interestingly, the evidence for this effect is much less strong in Baade's window (Frogel *et al* 1990).

If the K and M giants have different mean abundances, are their kinematics

H. Dejonghe and H. J. Habing (eds.), Galactic Bulges, 317–318.

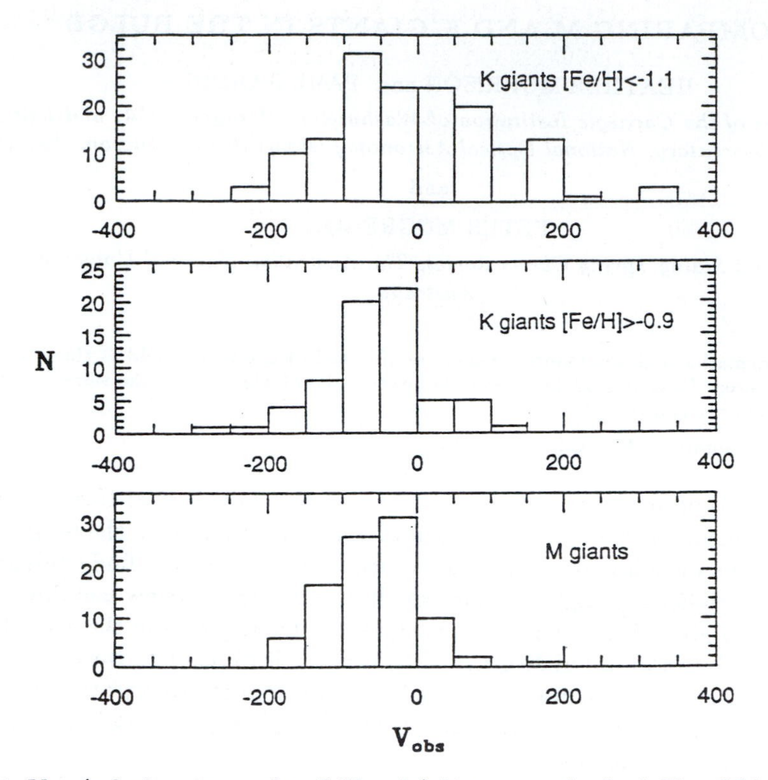

Fig. 1. V_{obs} (velocity corrected to LSR only) histograms for both K and M giants. It can be seen that the M giant velocities are very similar to those of the metal-rich K stars, suggesting that their rotational kinematics are also similar.

different too? The rotational kinematics of the M giants are harder to measure, because we need an accurate distance in order to calculate V_ϕ, and this requires more accurate abundance estimates than we presently have. However, it is possible to simply compare the line-of-sight radial velocities of the K and M giants in the field. Figure 1 shows that the kinematics of the M giants are unlike that of the metal-poor K giants but very similar to the metal-rich K giants. Thus, at this distance from the galactic center, both M giants and metal-rich K giants seem to be drawn from the same population.

Acknowledgements

HLM and PH would like to thank George Preston and Leonard Searle for helpful discussions, and the OCIW and MSSSO TACs for the allocations of time that made this project possible.

References

Frogel, J.A., Terndrup, D.M., Blanco, V.M. and Whitford, A.E. 1990, ApJ, 353, 494

PG3, A field in the Bulge of Our Galaxy: Description of a Galactic Model

Y.K. Ng[1,*], G. Bertelli[2,3], C. Chiosi[3]
A. Bressan[4], H. Habing[1] and R.S. Le Poole[1]

[1] *Leiden Observatory, Leiden, the Netherlands*
[2] *National Council of Research, Rome, Italy*
[3] *Department of Astronomy, Padua, Italy*
[4] *Astronomical Observatory, Padua, Italy*

Abstract

Preliminary results are shown of our study on the metallicities and ages of the stellar populations present in different components along the line of sight to field #3 of the Palomar-Groningen Variable Star Survey (PG3) with synthetic Hertzsprung-Russell Diagrams (HRDs).

Model Building

Wesselink (1987) re-examined the variable stars in PG3 (Larsson-Leander 1959; $l = 0^o$, $b = -10^o$) by measuring B_J and R Schmidt plates on the Leiden Astroscan (de Vries 1986). Wesselink's work is extended to all stars in PG3 (Ng et al. 1990, 1992). A CMD has been obtained for a great many stars stars in an area sized $3\overset{\circ}{.}5 \times 3\overset{\circ}{.}5$ centered around $l = 1^o$ and $b = -11^o$. A first analysis (Ng et al. 1992) with synthetic CMDs (Chiosi et al. 1988) showed that the contamination with the foreground stars is significant. With a more elaborated version of the CMD synthesis tool with a continuous metallicity coverage the analysis is continued. Four datasets from the latest models computed (Bertelli et al. 1990; Bressan et al. 1192; Alongi et al. 1992) with Z={0.0004,0.001,0.008,0.02} are the basis for the continuous Z synthetic HRD generator. Inputs to start the synthesis engine are

1) the slope for the powerlaw IMF (we used the Salpeter value);
2) the time interval for and the shape of the star formation rate assumed;
3) the range in Z (for simplicity a linear increase in time is assumed).

The B,R magnitudes for the synthetic stars are transformed into the photographic B_J,R magnitudes (Blair and Gilmore 1983). Synthetic HRDs are generated and 'observed' through an HRD Software Telescope which contains a two component galactic model: a spheroidal component (age: 17-15 Gyr, Z=0.0004-0.0156) and a disk component (age: 15-8 Gyr, Z=0.009-0.02, scalelength 3.5 kpc and scaleheight 200 pc). Realistic features of the HRD Software Telescope are the reddening of the synthetic star, the inclusion of apparent (unresolved) photometric binaries and the simulation of photometric errors as a function of magnitude for each passband.

---

* The research of Y.K. Ng is supported under grant 782-373-040 by the Netherlands Foundation for Research in Astronomy (ASTRON) which receives its funds from the Netherlands Organization for Scientific Research (NWO).

H. Dejonghe and H. J. Habing (eds.), Galactic Bulges, 319–321.

Preliminary Results and Discussion

Synthetic stars, with a limiting magnitude $B_{J,lim} = 19^m$, are generated for a two component (spheroid, disk) Galaxy with a ratio Spheroid : Disk = 3 : 4 . We adopted R_0 = 8 kpc for the distance to the Galactic Centre and a mean reddening $E(B_J-R)=0\overset{m}{.}15 \pm 0\overset{m}{.}05$ (Wesselink, 1987). The reddening for a synthetic star with $|z| < 90$ pc is assumed proportional to its distance from the Sun. In the figures 1 and 2 we show the results of star counts in different magnitude and colour bins. The effects due to photometric errors and unresolved apparent binaries are not considered.

Figure 1 shows the synthetic color distribution for the individual, combined and and the observed color distribution (solid line=combined, dashed line=spheroid, dot dashed line=disk, circles=observed).

Figure 2 shows the synthetic starcounts for the combined and individual components together with the observed starcounts (solid line=combined, dashed line=spheroid, dot dashed line=disk, circles=observed).

The color distribution of the stars in PG3, figure 1, is primarily governed by the scaleheight of the disk main sequence stars. The evolved stars require an additional contribution by the spheroidal component. The adopted metallicity and age dispersion for the spheroidal and disk component determines the width of the color distribution. Figure 2 shows that the color equations used (Blair and Gilmore 1982) aren't optimal because in each figure the shape of the *synthetic* starcounts resembles more closely the observed data in the next figure. Bessel (1986) pointed out that cubic transformations should be used, because the extended red tail of the Kron-Cousins R band differs from the photographic R band. In figure 2 the peak near $B_J=16\overset{m}{.}5$ in the observational data in the slice for $B_J-R=[1.2;1.4]$ is a signature of the horizontal branch from the spheroidal component and it holds information about the metallicity and the history of star formation. The simulated starcounts are one magnitude fainter than the observed data. Uncertainties in the theoretical models and the assumed metallicity range can result in a 0.5 magnitude shift while the shift from the color transformation is less than 0.5 magnitude. The reddening tends to shift the simulated starcounts to fainter magnitudes because the adopted value is regarded as a lower limit. This result suggests that the distance to the Galactic Centre is smaller than the 8 kpc adopted. Some future improvements on the galactic model description and a check on the theoretical evolution models are required before any definitive conclusions can be made.

References

Alongi, M. et al., 1992, Astron. Astrophys Suppl. Ser., submitted

Bertelli, G. et al., 1990, Astron. Astrophys. Suppl. Ser. **85**, 845-853

Bessell, M.S., 1986, P.A.S.P. **98**, 1303-1311

Blair, M. and Gilmore, G., 1982, P.A.S.P. **94**, 742

Bressan, A. et al., 1992, in preparation

Chiosi, C. et al., 1988, Astron. Astrophys. **196**, 84

Larsson-Leander, G., 1959, Proc. IAU symposium **7**, 22

Ng, Y.K. et al., 1990, Proc. ESO/CTIO Workshop on 'Bulges of Galaxies', B.J. Jarvis and D.M. Terndrup (eds.), 89-92

Ng, Y,K. et al., 1992, Proc. IAU symposium 149, 'Stellar Populations of Galaxies', B. Barbuy and A. Renzini (eds.), 462

Vries, de C., 1986, Ph.D. thesis, Leiden University, the Netherlands

Wesselink, T.J.H., 1987, Ph.D. thesis, Nijmegen University, the Netherlands

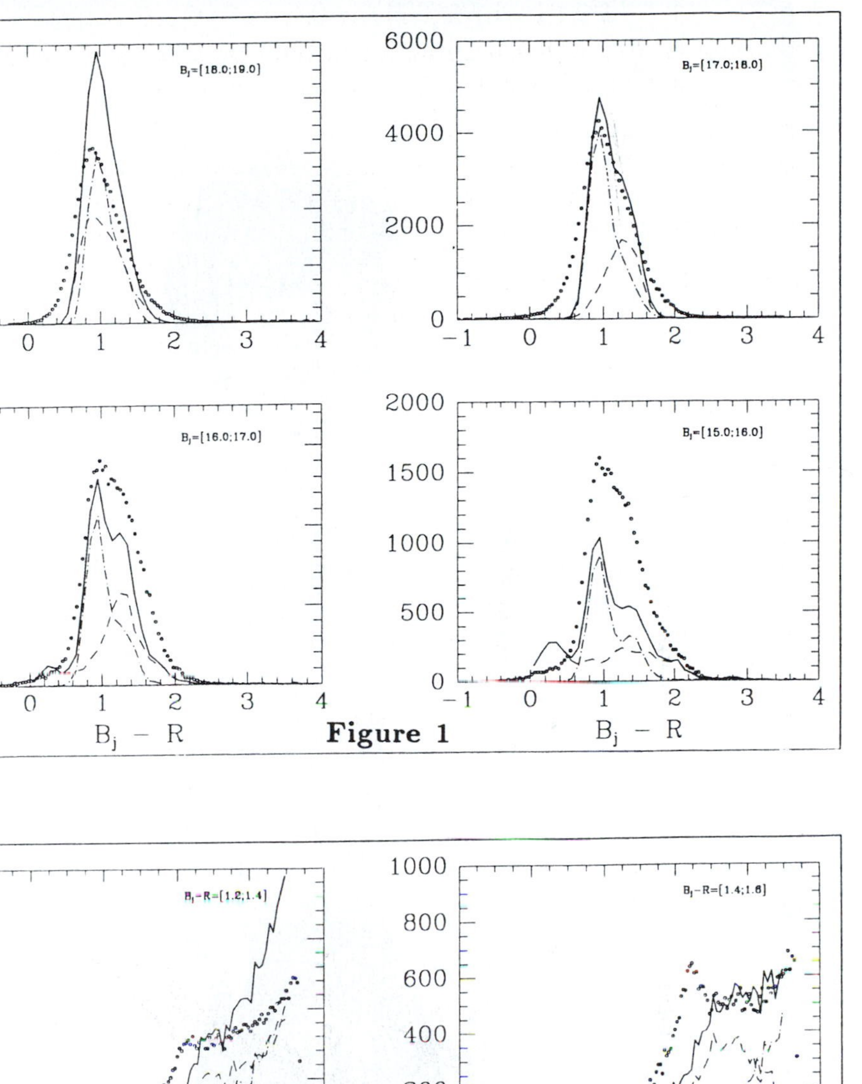

Figure 1

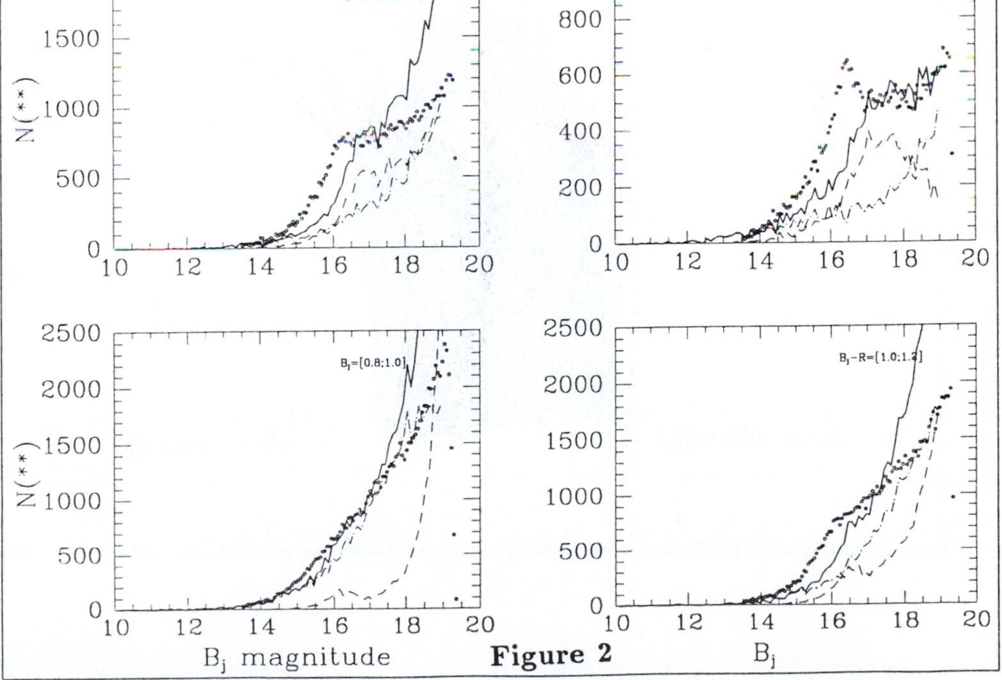

Figure 2

Ortwin GERHARD

STAR CLUSTERS AS TRACERS OF GALACTIC SUBSYSTEMS

B. BARBUY[1], E. BICA[2], S. ORTOLANI[3]
1 IAG-USP, S. Paulo, 2 UFRGS, Porto Alegre, Brazil; 3 U. Padova, Italy

ABSTRACT. We have obtained CCD BVRI colour-magnitude diagrams for a series of disk globular clusters, improving parameters and detecting a new one: Lyngå 7. Using the magnitude difference between turn-off and horizontal branch Δ(TO-HB) as an age discriminator, and their spatial distribution we compare old disk open clusters, young halo globular clusters, and metal-rich disk globular clusters, obtaining clues to the Galaxy formation process.

1. Discussion

Some star clusters previously catalogued as open clusters have been recently classified as globular clusters from colour magnitude (CMD) analyses, e.g. Ruprecht 106 (Buonanno et al 1990) and Lyngå 7 (Ortolani et al 1992a). They arise a major interest because they might present age and metallicity values of a possible intermediate population between the young disk and the old spheroidal subsystems of the Galaxy. We have improved the sample of metal-rich globular clusters using CCD photometry. These clusters are: NGC 6553 (Ortolani et al 1990), NGC 6528 (Ortolani et al 1992b), Terzan 1 (Ortolani et al 1992c), as well as Lyngå 7. The analyses of Colour Magnitude Diagrams (CMDs) allowed us to better derive reddening, metallicity, position in the Galaxy, and age estimates for the clusters, also based on Δ(TO-HB) values. In Table 1 are given the relevant parameters to the present discussion. Also included in the Table are: (i) the young halo globular clusters Pal 12 (Gratton & Ortolani 1989) and Rup 106 (Buonanno et al. 1989); (ii) some of the oldest open clusters in the Galaxy (Demarque et al 1992); (iv) the classical metal-rich globular cluster 47 Tuc, which seems to be a transient cluster between the halo and disk globular cluster systems.

The spatial distribution of these clusters in the scale height z(kpc) as a function of the projected distance from the Galactic center Rxy(kpc), assuming a distance from the Galactic center from the Sun of 8.8 kpc, is given in Figure 1. The disk system studied by Armandroff (1989) are also shown. Lyngå 7 is at the edge of the disk subsystem of metal-rich globular clusters in the Galaxy, very close to the plane. The scale height of $z \approx 0.8$ kpc for the Galactic disk corresponds to the old open clusters. From the absolute ages and ΔV(TO-HB) estimates (Table 1), one can conclude that while the halo was still forming clusters at moderate metallicities like Pal 12 and Rup 106, the metal-rich subsystem of disk globular clusters with scale height $z \approx 1$ kpc was forming Lyngå 7. This epoch almost overlaps with the formation of old disk clusters of about solar metallicity, like NGC 6791.

Lyngå 7 appears to be the young tail of the disk metal-rich globular clusters.

The connections in time, metallicity and spatial distributions give important clues to the collapse steps of the Galaxy, and consequently to the stellar subsystems formation.

H. Dejonghe and H. J. Habing (eds.), Galactic Bulges, 323–324.

References

Armandroff, T.E.: 1989, AJ 97, 375
Buonanno et al.: 1980, AJ 100, 1911
Demarque, P., Green, E., Guenther, D.: 1992, AJ 103, 151
Gratton, R., Ortolani, S.: 1989, A&AS 73, 137
Ortolani, S., Barbuy, B., Bica, E.: 1990, A&A, 236, 362
Ortolani, S., Barbuy, B., Bica, E.: 1992a, A&A, submitted
Ortolani, S., Bica, E., Barbuy, B.: 1992b, A&AS, 92, 441
Ortolani, S., Bica, E., Barbuy, B.: 1992c, A&A, in press

Table 1 - Basic parameters for the studied clusters

Cluster	ΔV(TO − HB)	z(kpc)	R_{xy}(kpc)	[M/H]
NGC188	2.75	0.6	8.1	0.0
NGC6791	2.95	0.8	8.3	≥ 0.0
M67	2.20	0.4	8.3	0.0
NGC6528	3.3	0.6	0.5	+0.12
NGC6553	3.3	0.3	3.9	−0.29
Terzan 1	− − −	0.1	4.3	+0.24
Lynga 7	3.05	0.3	4.7	−0.4
Rup 106	3.15	4.1	17.1	−1.09
Pal 12	3.10	13.1	6.2	−1.14
47 Tuc	3.60	3.2	7.4	−0.71

Figure 1 - Scale height z(kpc) vs. distance from the Galactic center in the plane R_{xy}(kpc)

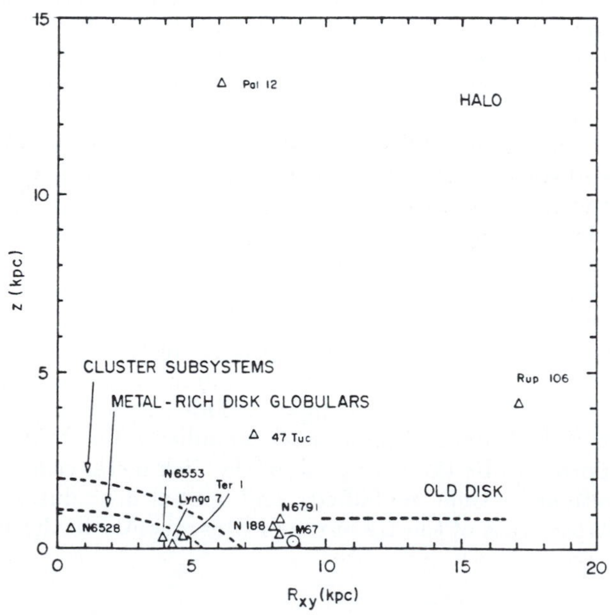

Stellar Population Synthesis in the Bulge of our Galaxy

ANNIE C. ROBIN

Observatoire de Besançon, BP 1615, F-25010 Besançon cedex, France

Abstract. Using a model of population synthesis well tuned to reproduce the disc and halo populations of the Galaxy, we show how to set constrains on the bulge population parameters using magnitude and colour distribution in 4 directions near the bulge. This first analysis result in a bulge with a disc-like LF but a density law much steeper in the center.

Key words: Milky Way Galaxy - galaxy modeling - stellar populations

1. Introduction

Connecting observable distributions in the Galaxy to the main processes they come from is basically a multivariate problem for which we have developed a synthetic approach of Galaxy modeling (hereafter referred to as the Besançon model, Robin and Crézé, 1986, Bienaymé et al., 1987). The relation between the bulge, the disc, the thick disc and the halo should be addressed using all kinds of observational parameters (photometric as well as astrometric). In this first attempt we compare magnitude and colour distributions in the visible and the near-infrared in 4 directions close to the galactic bulge in order to put constraints on its stellar content.

2. Galaxy model

Our specific approach is based on assumptions directly derived from current pictures of galactic evolution. Observational parameters like velocities, space distribution and metallicities are all driven by the age parameter. Age distributions come from a specific evolution model (Rocca-Volmerange et al., 1981), while the disc density laws have been self consistently constrained with the whole potential of the galaxy using Boltzmann and Poisson equations and an age-velocity dispersion relation (Bienaymé et al., 1987).

3. Comparison with bulge data

Three directions at longitude 10° and latitude around −10° ,−20° and −30° have been observed by Rodgers & Harding (1989) (hereafter RH) giving a good sample to study the gradient of populations near the bulge. While Ruelas-Mayorga & Teague (hereafter RT) produces K and J-K star counts in the Baade window to K=13.5.

4. Results

By comparing model predictions with RT data towards the Baade Window and RH data at three latitudes we found that :

The *luminosity function* most appropriate for the bulge population is similar to the disc one and quite different from the globular cluster LF for the stars of absolute magnitude between -1 and 4. However this property of the bulge is no more true at latitudes larger than 20°-25° (ex: fig. 2) where a M3 LF is more suitable. This is probably related to the specific process of chemical evolution in the bulge. The

H. Dejonghe and H. J. Habing (eds.), Galactic Bulges, 325–326.

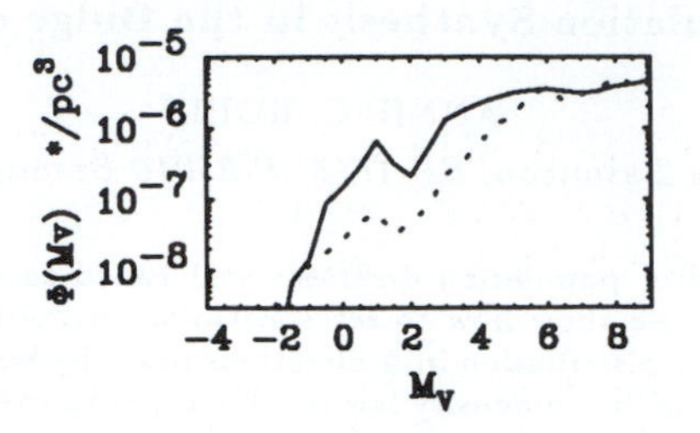

Fig. 1. Two possible bulge luminosity functions. Solid line: disc LF normalized to 0.2% of the disc; Dashed line: M3 LF.

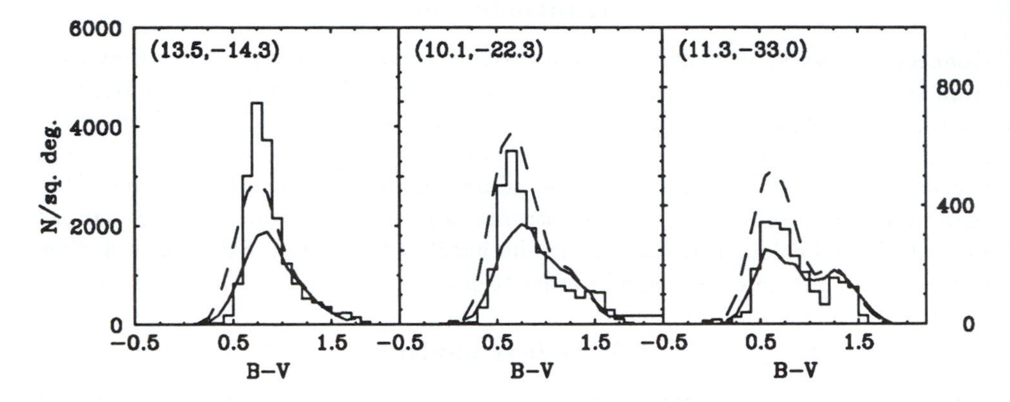

Fig. 2. B-V distribution at three latitudes for stars with $18 < V < 19$, compared with model predictions assuming two bulge LF. Solid line: M3 LF, dashed line: disc-like LF.

overall process of evolution may result in a luminosity function closer to the disc one than to the halo one.

The *bulge density law* is well fitted by a power law with an exponent of 3.8 at latitude 4° (Baade window) while a smaller value (3.1, as the halo) is suitable for fields at b≥ 10°.

This first analysis result in a bulge with a disc-like LF but a density law much steeper in the center. In the near future we plan to apply this model of population synthesis to proper motion and radial velocity distributions to contrain the kinematics of the bulge, hence to undertand more about its scenario of formation.

References

Bienaymé, O., Robin, A.C., Crézé, M., 1987, A&A 180, 94

Robin, A.C., Crézé, M. 1986, A&A 157, 71

Rocca-Volmerange et al., 1981, A&A 104, 177.

Rodgers, A.W., Harding, P., 1989, A.J., 97, 1036 (RH)

Ruelas-Mayorga, R.A., Teague, P.F., 1992, A&A suppl. 93, 61 (RT)

Yoshii, Y., Rodgers, A.W., 1989, A.J., 98, 853

CCD BVRI PHOTOMETRY OF STARS IN THE GALACTIC BULGE

A. RUELAS-MAYORGA[1], S. WEST[2] and A. PEIMBERT-TORRES[1]
1 Instituto de Astronomía. UNAM. México.
2 University of Christchurch. New Zealand.

ABSTRACT. We present preliminary results from BVRI CCD photometry of stars in the galactic bulge. The importance of disc contamination is shown. Several statistical criteria to separate disc from bulge stars are used.

1. INTRODUCTION

The nuclear bulge of the Galaxy contains a stellar population that does not conform to the old ideas we had about its origin and physical characteristics.
In Ruelas-Mayorga and Teague (1992a) the stars of the bulge of our galaxy are characterized as follows:
i) They appear mainly at faint magnitudes, show
ii) Low values for their CO index, which might indicate an associated low value for their metallicity, and
iii) They appear to have a Luminosity Function similar to that of globular clusters. Further studies of the stellar population in the galactic bulge are needed to solve discrepancies between published results.

2. RESULTS

The observations were obtained using a CCD and filters B, V, R and I. Standard observation and reduction techniques were used. Here we present a preliminary set of results (for a 2' x 3' field) based on magnitude versus colour and colour-colour diagrams. A brief discussion in terms of a stellar distribution model of the galaxy (Ruelas-Mayorga, 1991a) is also given.
The V vs B-V and the V vs V-I diagrams (Figures 1 and 2) show the same character as those of Terndrup's (1988); both the red and blue sequences identified by him are seen here. There are very red stars in this field with V-I in excess of 4.0. The V-R vs R-I two-colour diagram shows a linear elongation almost parallel to the reddening vector (Johnson, 1968). This may not be produced by depth effects in the galactic bulge since the E(R-I) ~ 2.3 would imply a diameter of the order 20 kpc. We must conclude that very red stars are present in this field. Differential reddening is ruled out due to the small size of the field under study.

H. Dejonghe and H. J. Habing (eds.), Galactic Bulges, 327–328.

According to the Ruelas-Mayorga (1991a) model of the galaxy, 79 % of the stars (110) observed down to V=18 mag must belong to the disc, and the rest (21 % = 29) to the bulge of our galaxy. Model predictions in this direction produce the correct proportions if V-I = 2.0 is used as separator; the bulge is redder and the disc is bluer than this value. We propose the 29 redmost (V-I ≥ 2.0) stars to be the bulge stars.

3. CONCLUSIONS

From these preliminary results we conclude the following:

i) There is a substantial number of disc stars in the low absorption windows through which the galactic bulge is usually studied. Allowance for their presence must always be made.

ii) In agreement with the results found in the bright IR (see Ruelas-Mayorga and Teague, 1992c); in the bright visual magnitude range (V≤+18) bulge sources must have on the average red colours (V-I ≥ +2.0) which, after dereddening [$(V\text{-}I)_0 \geq +1.21$] imply spectral types later than G8 III.

4. REFERENCES

Johnson, H.L., 1968, *Stars and Stellar Systems: Nebulae and Interstellar Matter.*, **vol. VII**,ed. B.M.Middlehurst and H.L. Aller, 167.

Ruelas-Mayorga, R.A., 1991a, Rev. Mex. Astron. Astrofis., **22**, 27.

Ruelas-Mayorga, R.A. and Teague, P.F., 1992a, A &AS, **93**, 61.

Ruelas-Mayorga, R.A. and Teague, P.F., 1992c, A & A, In press.

Terndrup, D.M., 1988, AJ, **96**, 884.

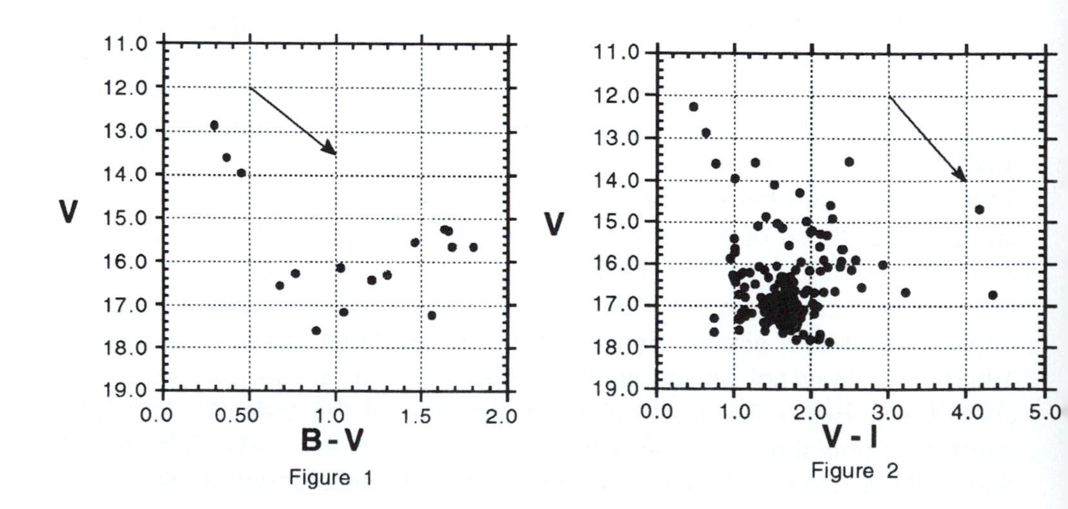

Figure 1

Figure 2

THE IR STELLAR POPULATION AROUND THE GALACTIC CENTRE

A. RUELAS-MAYORGA[1] and P. TEAGUE[2]
1 Instituto de Astronomía. UNAM. México.
2 The Niels Bohr Institute, Copenhagen ø. Denmark.

ABSTRACT. We present K-counts for a number of stars with K ≤ +9.0 in 3 clear regions near the galactic centre. They are located approximately at l~0.0 and b ~ -3.5, -4.0 and -4.5. Their Cumulative Counts Functions (CCF's) are formed, and with the aid of an exponential disc model for the Galaxy are separated into disc and bulge CCF. As for the case in Baade's Window (BW) (Ruelas-Mayorga and Teague, 1992a) the disc is dominant at bright magnitudes whereas the bulge dominates at fainter ones. The slope of the bulge CCF is steeper than that for the disc in all cases corroborating the result obtained for BW. The two colour JHK and the K vs J-K diagrams show that the stellar population in these areas is similar to that studied in BW by us and by Frogel and Whitford (1987). An average E(J-K) ~0.42 mag for the reddening is obtained. At K ≤ +9.0, the disc may be accounted for by those sources with J-K ≤ +1.6 whereas the bulge population presents values for J-K in excess of +1.6.

1. INTRODUCTION

In Ruelas-Mayorga and Teague (1992a,b) we present our infrared study for Baade's Window (BW). In this paper we study three clear windows in the general direction of the galactic centre. These regions are located at galactic coordinates (0.0, -3.5), (0.0, -4.0) and (0.0, -4.5); they are chosen from a plate of this area in which it is obvious they also enjoy low extinction and therefore a study of their stellar populations similar to that made for BW may be performed.

2. OBSERVATIONS

The observations were obtained scanning a single detector over the chosen area of the sky. Only the K-filter was used for reductions.

3. ANALYSIS

From the paper chart records of the observations the respective Cumulative Counts Functions (CCF's) (Number of sources per square degree down to a given K-magnitude) for each region are formed. Using the Ruelas-Mayorga (1991a) model of the galaxy a disc prediction for each region is obtained; simple subtraction yields a bulge estimate for

H. Dejonghe and H. J. Habing (eds.), Galactic Bulges, 329–330.

each position. The model also predicts colour distributions for the disc, which, when subtracted from the observed colour distribution, yield a colour distribution for the bulge sources. The main results of this analysis are summarised in the next section.

4. SUMMARY

i) Scanning observations in the IR (2.2 μm) have been performed for three clear windows (R1: (0.0, -3.5), R2: (0.0,-4.0) and R3: (0.0,-4.5)) in the general direction of the galactic centre.
ii) After correction for confusion, the CCF for each region is formed.
iii) An exponential disc simulation indicates that there is a fair agreement with the observed CCF at bright magnitudes. This fact suggests that the disc stellar population dominates the bright sections of the CCF and also lends further support to the equivalent finding for BW.
iv) The disc and bulge appear to be equally important at magnitudes in the range $+8.0 \leq K \leq +9.5$. The bulge component becomes definitely the dominant contributor at fainter magnitudes.
v) A mean CCF determination for the bulge is obtained from the observed and the predicted disc CCF's and the bulge CCF for the BW area. The slopes of the mean bulge CCF's are always steeper than those of the disc CCF's.
vi) All the stars lie along the reddening line at a mean E(J-K) ~0.42.
vii) The colour-magnitude (K vs. (J-K)) diagram shows that sources lie above and to redder colours than the GB of 47 Tuc when properly reddened. The photometric results suggest that the stars of our photometric sample are more metal rich than those in the 47 Tuc GB.
viii) The mean reddening value for R1, R2 and R3 (E(J-K) ~0.42) is larger than that for BW (E(J-K)~0.27) and is obtained by assuming that the stellar mix in R1, R2 and R3 is the same as that in the BW region.
ix) A theoretical prediction of the distribution of disc stars for $K \leq +9.0$ with respect to J-K, shows that most of the objects in our photometric sample with $(J-K) \leq +1.6$ may be accounted for as disc members.
x) The histrogram of number of sources versus (J-K) shows that the majority of the stars with $(J-K) \geq +1.8$ and $K \leq +9.0$ may be thought as real bulge members.

5. REFERENCES

Frogel, J.A. and Whitford, A.E., 1987, ApJ, **320**, 199.
Ruelas-Mayorga, R.A., 1991a, Rev. Mex. Astron. Astrofis., **22**, 27.
Ruelas-Mayorga, R.A. and Teague, P.F., 1992a, A&AS., **93**, 61.
Ruelas-Mayorga, R.A. and Teague, P.F., 1992b, A&AS, in press.
Terndrup, D.M., 1988, AJ, **96**, 884.

STUDIES OF THE GALACTIC BULGES USING THE POST-THEORETICAL MASS METHOD

DORU MARIAN SURAN and NEDELIA ANTONIA POPESCU
Astronomical Institute of Romanian Academy, Str. Cutitul de Argint, No.5, 75212 Bucharest 28, ROMANIA

ABSTRACT. The results of a new method, Post Theoretical Mass Method (PTM) are presented in order to investigate radial pulsating stars (RPS) in galaxies. Further implications for stellar and galactic astrophysics are also discussed.
Key Words: radial pulsating stars, galaxies-components.

1. Introduction

In the present work we try to determine the properties of different galactic components using RPS as tracers. Detailed calculations are made for disk (Cepheids in our Galaxy) and halo (RR Lyraes in two dE satellites of M31), in order to determine the chemical structure (radial and transverse gradients) of galaxies. Because bulges are transient component between disk and halo, this method can be very atractive to investigate them.

2. Observations

In the CEPHEIDS program we investigate stars from the galactic disk (3kpc≤d≤15kpc) with low period (1d≤P≤10d) of intermediate mass (3MO≤M≤11MO) in the 2-nd and 3-rd crossing of the instability band. We have 10 Cepheids for which was determined (P, <B>-<V>). Also the data are supplemented by photometrical and spectroscopical results for more than 50 Cepheids.

In the RR LYRAES program we investigate 10 stars from the halo of two dE's (NGC 185,147) with periods 0.3d≤P≤0.8d (RRab) with amplitudes AV≥1, no M31 contaminations, for which we obtain (P, Ag).

3. Theory

Our method is based on the concept of PTM (Suran 1985) and was largely dicussed in Suran (1991) (see his eq. [15]) where the third relation is used in the form of six-parameters relation (P-L-M-Te-Y-Z).

H. Dejonghe and H. J. Habing (eds.), Galactic Bulges, 331–332.

Solutions of PTM method implie:

$$\left.\begin{array}{l}[P, \langle B\rangle-\langle V\rangle] \quad (\mathrm{Cep}) \\ {[P, \quad A_g \quad](\mathrm{RR\ Lyr})\ (l, \kappa, \mathrm{eos}, H_{puls})(a_0, .., f)}\end{array}\right\} \Longrightarrow [(M, L, T_e), (\langle Y\rangle, Z)] \qquad (1)$$

4. Calibrations

In the CEPHEIDS program we use the (P-L-M) relation in the form of Becker, Iben Tuggle (1977). The temperature relation is in the form:

$$\log T_e = 3.886-0.175\ (\langle B\rangle_0-\langle V\rangle_0)\ ,\ E_{B-V}=E_J \qquad (2)$$

with an error limit $\Delta Te \leq \pm\ 100^0$K. We use $\langle Y\rangle=0.28$.

In the RR LYRAES program we use the full 6-parameters relation (Suran 1992):

$$\log L = 0.81\log M-5.97+0.595\log P+2.07\log T_e-0.05[Fe/H] \qquad (3)$$

For the determination of temperatures we use the relation:

$$\log T_e = 0.05A_g+3.774 \qquad (4)$$

with a limiting error of $\Delta Te \leq \pm\ 200^0$K. We use $\langle Y\rangle=0.23$.

5. Results

For the CEPHEIDS program we obtain two parameters linear relations (Suran 1985) and a true chemical gradient for Cepheids in the disk of the Galaxy of: -0.06/kpc.

In the RR LYRAES program we obtain estimations of the chemical structure of the two dE's ([Fe/H]=-1.77 for NGC185 and respectively -1.31 for NGC 147).

References

Becker, S. A., Iben, I., Tuggle, R. S. 1977, *Ap. J.*, **218**, 533;

Suran, M. D. 1985, *Topics in Astronomy and Astrophys.*, **1**, 1;

Suran, M. D. 1992, in *Proceedings ofESO/EIPC Workshop on Early Type Galaxies,* Marciana Marina (in press).

A STUDY OF THE ABUNDANCE DISTRIBUTIONS ALONG THE MINOR AXIS OF THE GALACTIC BULGE

NEIL D. TYSON
Princeton University Observatory, Princeton, New Jersey 08544 USA

R. MICHAEL RICH
Department of Astronomy, Columbia University, New York 10027 USA

ABSTRACT

We derive the heavy element abundances for hundreds of K-giants in seven windows of low extinction, along or near the minor axis of the Galactic bulge. By using the recently-calibrated Washington photometric filter system, the distribution function in [Fe/H] is determined for each field. Within 8° of the Galactic center (~ 1 kpc) our data are consistent with no gradient in the distribution of [Fe/H], which may hint to a dissipationless collapse, and/or sufficient mixing during the star-forming epoch when Fe was produced in the bulge. The mean abundance over this region is between two and five times solar. The form of these distributions is well-fitted by the simple (closed box) model of chemical evolution where the bulge is self-enriched by processing its original gas content to completion. Beyond 8° from the Galactic center, our data show that the mean of the abundance distributions drops precipitously. This is consistent with the notion that the inner bulge is chemically distinct from the halo.

1. OBSERVATIONS

We used the recently calibrated Washington photometric system (Geisler, et al. 1991) to obtain photometric abundances for hundreds of K-giants in seven low extinction windows to the Galactic bulge. The data were obtained on the 0.9m telescope at CTIO. Abundances accurate to ±0.25 dex in [Fe/H] were derived for each K-giant using the metallicity-sensitive indices of the Washington system. (For a recent review of the Washington system see Tyson 1991).

2. ABUNDANCE DISTRIBUTIONS

The shape of the stellar abundance distribution holds valuable information about the history of chemical enrichment in a system. The distributions for the inner fields (i.e. $b = -2.7°, -4°, -6°$) bear a remarkable resemblance to the gaussian-convolved abundance distribution of the closed box, simple model of chemical evolution (Searle & Zinn 1978, also Rich 1990). Fig. 1 displays the abundance distribution for the inner-most field of this study. Error bars represent the typical errors in precision for a single observation derived from the photometric uncertainties. There is the

H. Dejonghe and H. J. Habing (eds.), Galactic Bulges, 333–335.

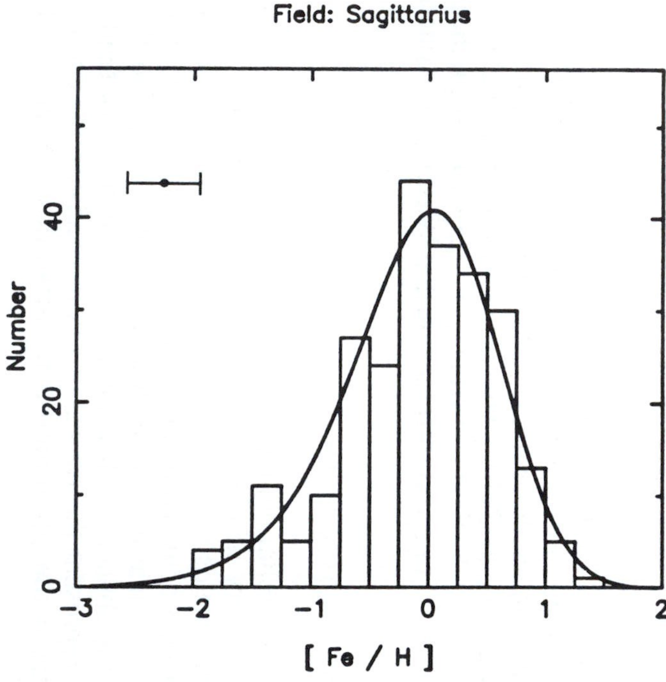

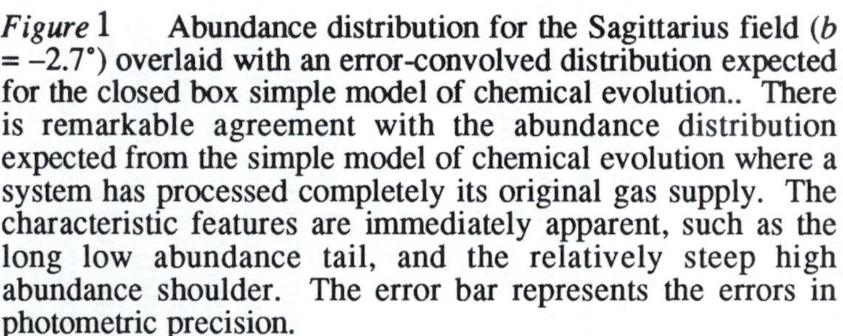

Figure 1 Abundance distribution for the Sagittarius field ($b = -2.7°$) overlaid with an error-convolved distribution expected for the closed box simple model of chemical evolution.. There is remarkable agreement with the abundance distribution expected from the simple model of chemical evolution where a system has processed completely its original gas supply. The characteristic features are immediately apparent, such as the long low abundance tail, and the relatively steep high abundance shoulder. The error bar represents the errors in photometric precision.

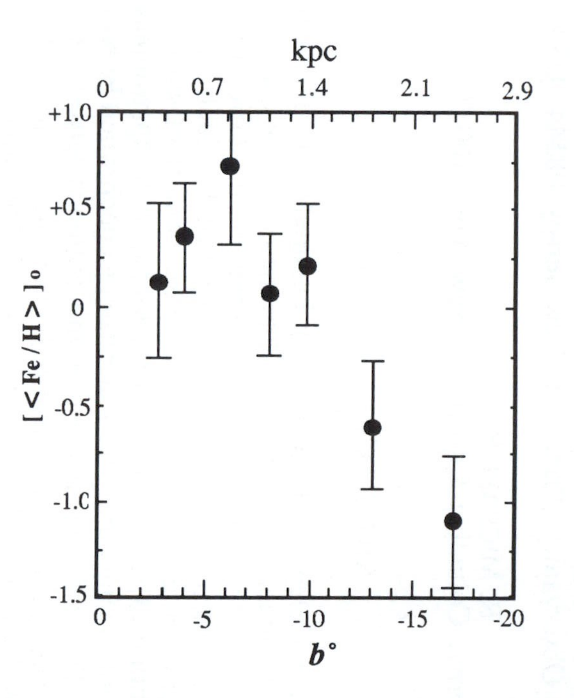

Figure 2. Galactic latitude dependence of the mean in the abundance distributions. Plotted is the error-corrected log of the mean abundance at all latitudes. The mean abundance does not show a trend downward until beyond ~ 1 kpc. The similarity among the inner latitudes in their mean abundance suggests strongly that the bulge, within about 1 kpc, followed common evolution. The similarity and shape of the abundance distributions for the inner fields suggests a common evolution described by the closed box simple model of chemical evolution.

characteristic extended tail toward lower abundances and the relatively steep shoulder at high abundance. Consequently, our data suggest that the bulge abundance distributions, out to $b = -6°$, are consistent with a fully mixed, closed box, simple model of chemical evolution that has processed its gas to completion.

3. DISCUSSION

To supplement interpretation of the abundance distributions we can look at the run of the log mean abundance, [<Fe/H>], as a function of Galactic latitude. This constitutes the first direct measurement of the latitude dependence of [Fe/H] in the Galactic bulge. Fig. 2 presents these data. The errors are large, but two convincing features are immediately apparent: [1] The inner latitudes stay at roughly constant mean abundance [Fe/H] $\approx 0.4 \pm 0.3$ ($r < 1$ kpc), which suggests that the bulge, within about 1 kpc, is a chemically well-mixed volume. There is a small trend for [<Fe/H>] to drop within $b = -6°$, but given the relatively large errors, it is not clear whether much should be made of this feature. The fact that the abundance distributions for the three inner latitudes show good agreement with the analytic form of the abundance distribution from the closed box simple model suggests strongly that the inner bulge is self-enriched. [2] Outside of the inner 1 kpc, the mean abundance begin a precipitous 1.5 dex drop to latitude $b = -17.3°$. This drop is remarkable when we consider that it occurs over just 1 kpc. The 1σ error bars in Fig. 2 are computed from the photometric *and* zero-point errors. The statistical uncertainty in the mean abundance ($\sigma / N^{-1/2}$) for each field is considerably smaller.

4. FUTURE

One of the most useful aspects of the Washington system is its ability to flag interesting stars of unusually high (or low) abundance for follow-up spectroscopy. As emphasized by D. Terndrup at this conference, the idea of a kinematic dependence on abundance in the Galactic bulge is an exciting topic that demands much further study. In collaboration with D. N. Spergel (Princeton), we plan to obtain medium resolution spectra of the highest and lowest abundance stars from the abundance distributions of this study. Spergel has developed computer models of the Galactic bulge that allow one to derive observable quantities from segregated orbit families in phase space. We expect this approach to be a powerful analytic tool that will enable interpretive leverage on abundance and velocity distributions that would not otherwise be possible.

REFERENCES

Geisler, Clariá, J. & Minniti, D. 1991, AJ 102, 1836

Rich, R. M. 1990, in *Bulges of Galaxies*, ed. Jarvis & Terndrup (Garching: ESO) p. 65

Searle, L. & Zinn R. 1978, ApJ 225, 357.

Tyson, N. 1991, in *Precision Photometry: Astrophysics of the Galaxy*, ed. Philip, Upgren, & Janes, (Schenectady: L. Davis Press), p. 193

Alex RUELAS
Kutrowner '92

The UV and Optical Reddening Law to the Galactic Bulge and CNO Abundances in Bulge Planetary Nebulae

N.A. WALTON and M.J. BARLOW
Department of Physics & Astronomy, University College London, Gower Street, London WC1E 6BT, UK

and

R.E.S. CLEGG
Royal Greenwich Observatory, Madingley Road, Cambridge CB3 0EZ, UK

Abstract. An analysis of the differential *ultraviolet* extinction towards four bulge planetary nebulae, based on the observed line ratio of He II 1640/4686Å, shows that the ultraviolet reddening law towards the bulge is much steeper than in the solar neighbourhood. An analysis of the *optical* reddening law for 42 bulge PN, based upon observed Balmer line ratios and Hβ to radio free-free flux ratios, is presented. The optical reddening law towards the bulge is steeper than in the local ISM, and thus the ratio of total to selective extinction, $R_V = 2.29(\pm 0.50)$, is lower than the standard solar neighbourhood value of $R_V = 3.10$.

We present abundance determinations, in particular C/H and C/O ratios, for 11 Galactic bulge PN, based on spectrophotometry in the UV from IUE and in the optical from the Anglo-Australian Telescope. The derived abundances are compared with values for PN in the Galactic disk. The mean C/O ratio for bulge PN is significantly lower than that found for Galactic disk PNs. Additionally we present an abundance analysis of the very metal-poor halo population PN M2-29, which is located in the bulge.

Background & Method: The standard Galactic reddening law is generally adopted in studies of objects towards the bulge. But it is known that the value of R_V, the ratio of the total to selective extinction, can vary in certain directions. Reddening laws are normally derived from differential measurements of reddened and unreddened luminous early-type stars. But such stars are absent in the Galactic Bulge, so the reddening law towards the bulge has never been previously determined. However, an analysis of the differential optical extinction towards bulge planetary nebulae (PN) (based on a comparison of the observed Balmer line ratios for Hα, Hβ and Hγ with Case B recombination theory) allows a direct determination of R_V towards the bulge to be made. The steepness of the ultraviolet reddening law in the bulge can be diagnosed from a comparison with Case B recombination theory of the observed flux ratios of He II λ1640 to λ4686 from high-excitation PN.

The optical reddening law was derived using data from the literature: reliable *total* Hβ fluxes were taken mainly from Webster (1983, PASP, 95, 610) and Shaw & Kaler (1989, ApJS, 69, 495). Balmer line flux ratios were taken from Acker et al (1991, A&AS, 89, 237) and Aller & Keyes (1987, ApJS, 65, 405). 5 GHz radio fluxes were taken mainly from Zijlstra (1989, Thesis, University of Groningen) and Gathier et al (1983, A&A, 128, 325).

The 11 PN predicted to be bright enough in the UV were observed with the IUE telescope in 1990 and 1992. Optical spectrophotometry was obtained for these objects at the 3.9m Anglo-Australian Telescope. The carbon abundances were determined by analysis of the UV C III] and C IV lines, using an ultraviolet reddening law that was interpolated from our mean measurements plotted in Fig. 1. The

H. Dejonghe and H. J. Habing (eds.), Galactic Bulges, 337–338.

abundances of the elements He, O, N, Ne, Ar and S were determined by analysis of the UV and optical emission lines.

	He	C $\times 10^4$	N $\times 10^4$	O $\times 10^4$	Ne $\times 10^4$	C/O	N/O
Bulge mean	0.118	3.49	2.01	4.62	1.10	0.603	0.487
±	0.029	4.10	1.14	1.83	0.590	0.532	0.435

Table 1. The mean abundances determined for the 11 bulge PNs.

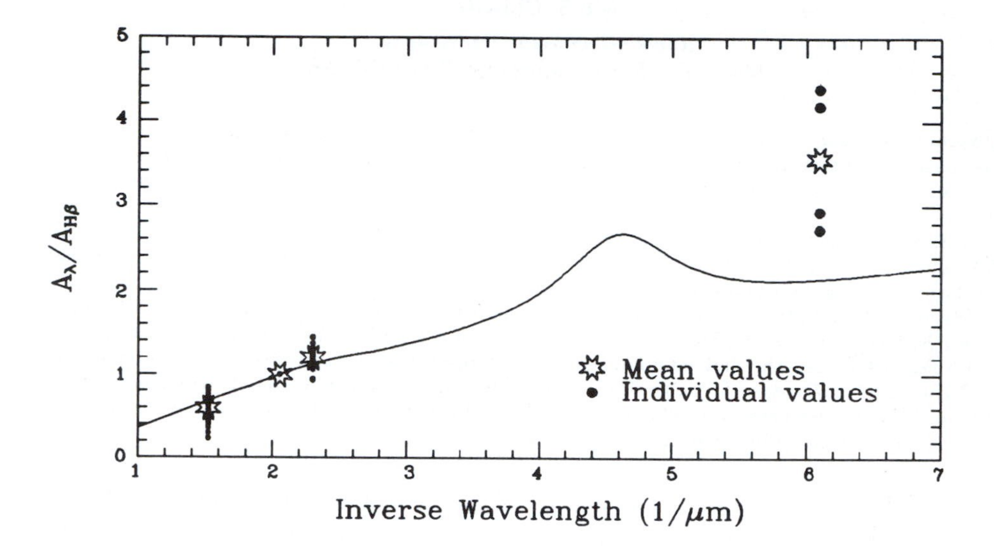

Fig. 1. The UV and optical reddening law. The standard reddening curve is that of Seaton (1979, MNRAS, 187, 79p) for R=3.10. The open circles show the $A_\lambda/A_{H\beta}$ values for the individual PNs at Hα, Hβ, Hγ and He II λ1640. The stars mark the mean values of $A_\lambda/A_{H\beta}$.

Conclusions: The optical and UV reddening law in the bulge is clearly steeper than the normal Galactic law (e.g. Seaton, 1979). The optical analysis of 42 galactic bulge PN, shows that R_V is 2.29 (±.50), significantly lower than the value of R_V=3.1 found for the Galactic diffuse interstellar medium in the solar neighbourhood. The optical data supports the conclusion, obtained from the UV data, that the reddening law is steeper towards the bulge.

Only 2 of the 11 PN have C/O>1, compared with 60% for Solar neighbourhood PN (Zuckerman & Aller, 1986, ApJ, 301, 772). The only bulge PN with C/O clearly in excess of unity (Cn 1-5) has a WC4 Wolf-Rayet central star. The mean O/H ratio for the bulge PN is similar to that found for local PN, a result in agreement with that of Ratag et al (1992, A&A, 255, 255). One of the bulge PN (M 1-42) is of Type I (i.e. it has high He/H and N/O ratios).

M 2-29, probably a halo population object currently passing through the bulge, has very low metallicity and in addition is extremely carbon-poor: we derive $O/H = 2.8\times10^{-5}$, He/H = 0.093 and C/O = 0.04.

DO THE OH/IR STARS WITHIN 100 PC OF THE GALACTIC CENTER BELONG TO THE DISK POPULATION?

A. E. Whitford
Lick Observatory, University of California
Santa Cruz, California 95064 USA

The 135 OH/IR stars within 0.8 deg (~ 100 pc) of the Galactic Center (GC) identified in a search with the VLA (Lindqvist et al. 1991) are in rapid rotation about the center with a regression line showing a velocity gradient of 1.2±0.1 km s^{-1}, equivalent to 178 km s^{-1} deg^{-1} for R_o = 8.5 kpc (Lindqvist et al. 1992). This is of the same order as the inward extrapolation of the Galactic rotation curve derived from the terminal velocities of H_I for $l < 2$ deg (Burton and Gordon 1978) and the general trend of the velocities of the CO clouds in the equatorial disk mapped by Dame et al. (1987) (Fig. 1b). These OH/IR stars are located along the ridge of highest projected CO density. This spatial location and the flat configuration argue for assignment to the disk population, a possibility suggested by Feast (1989). The VLA objects are not in general supergiants (Blommaert et al. 1992). On a K-band map of an area 1×2 deg around the GC, Catchpole et al. (1990) found a number of stars brighter than $(m_K)_o = 5$. On the assumption that $BC_K = 3.3$, (Frogel and Whitford 1987), and R_o = 8.5 kpc, $M_{bol} < -6.3$, i.e., in the supergiant range.

Lindqvist et al. (1992) interpreted the greater scale height and velocity dispersion of the less luminous fraction of the VLA sample (judged by smaller average shell expansion velocity V_{exp}) as an age difference; the longer life of these stars allows time for a disk heating process (Wielen et al. 1992) to operate. This suggests an age spread like that in the classical disk population.

In order to compare the kinematics of OH/IR stars near the GC with other OH/IR stars in the inner Galaxy, a sample was taken from the Effelsberg Parkes survey (te Lintel Hekkert et al.

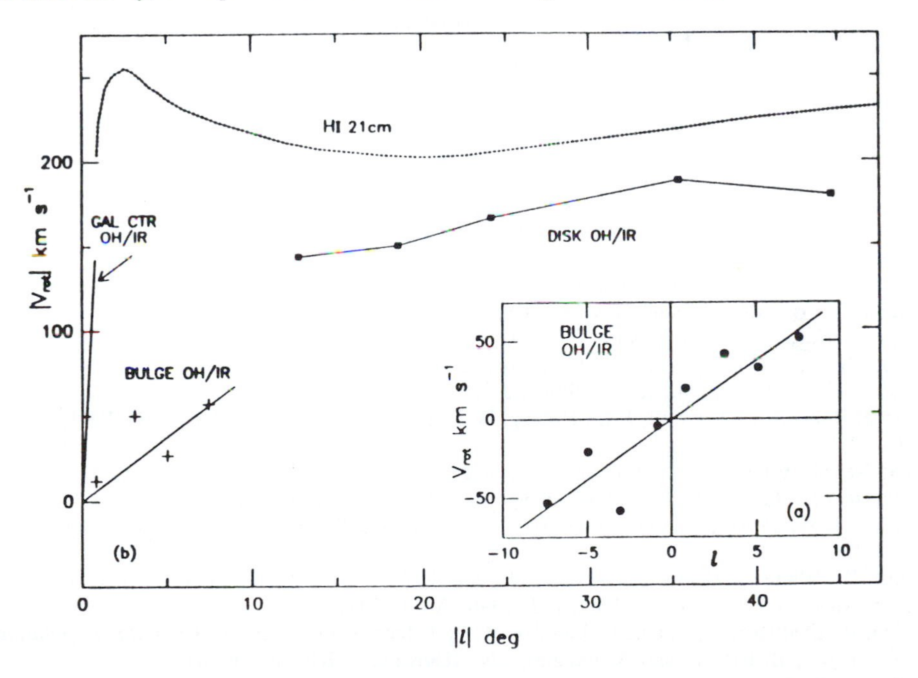

Figure 1. (a) Group means for bulge OH/IR stars, $|l| < 10$, and regression line. (b) Regression lines showing rotation velocity gradient for OH/IR stars in Galactic center and bulge,and group means for disk OH/IR stars. Dotted line: Galactic rotation curve from H_I clouds.

H. Dejonghe and H. J. Habing (eds.), Galactic Bulges, 339–340.

1991). Objects with thick circumstellar shells having IRAS colors IIIb were selected. Van der Veen and Habing (1990) showed that for this class of objects the survey is very nearly complete to the distance of the GC.

The 59 stars in the sample within the range $10 < | l | < 50$ show Galactocentric velocities, V_{rot}, that mark them as disk objects. Group means are shown in Figure 1b. The points fall below the terminal velocities for H_I because the stars are in general not at the tangent point, and because of the asymmetric drift found by Lewis and Freeman (1989).

The 80 OH/IR objects in the selected sample with $| l | < 10$ define the bulge kinematics. Group means for 8 bins show a slow average rotation with a regression line $V_{rot} = 7.6 \pm 1.4 - 0.6 \pm 6.0$ km s^{-1} deg^{-1} (Fig. 1a; reflected means in Fig. 1b). The dispersion is 83 km s^{-1}, larger than the mean rotational velocity for any bin; there are some retrograde orbits. The velocity gradient of the OH/IR stars is similar to that of planetaries and Miras in the bulge region (Menzies 1990). Minitti et al. (1992) found a similar gradient for two samples of K giants. These gradients are also similar to those observed in the unresolved bulge light of edge-on galaxies such as NGC 4565 (Kormendy and Illingworth 1982). The rotation in these bulges was interpreted as furnishing a fraction of the total support sufficient to account for the oblateness ratio c/a.

The large difference in the rotation speed of the bulge objects and of the gas clouds that define the Galactic rotation curve at the same longitudes presents no problem if the bulge is related to the halo, as advocated by Carney et al. (1990). These authors note that the very slow average rotation of halo objects at the solar radius R_o carries an angular momentum that, after contraction to a tenfold smaller radius in the bulge, is adequate to explain the average rotation of bulge objects.

It seems likely that there is a disk population embedded in the bulge that connects the OH/IR stars near the GC with disk objects outside the bulge. Its members should show rotational velocities, maximum luminosities, and latitude distribution that set them apart from bulge members. Identifying them may require a considerable effort.

REFERENCES

Blommaert, J.A.D.L., van Langenvelde, H.J., Habing, H. J., van der Veen, W.E.C.J., and Epchtein, N. 1992, *Feast Symposium*, preprint.

Burton, W. B., and Gordon, M. A. 1978, A&A, 63, 7.

Carney, B.W., Latham, D.W., and Laird, J.B. 1990, in *Bulges of Galaxies*, B. Jarvis and D. Terndrup, eds. (Garching: ESO), p. 127.

Catchpole, R.M., Whitelock, P.A., and Glass, I.S., 1990, MNRAS, 247, 479.

Dame, T. M., Ungerechts, H., Cohen, R. S., de Geus, E. J., Grenier, I. A., May, J., Murphy, D. C., Nyman, L.-A., and Thaddeus, P. 1987, ApJ, 322, 706.

Feast, M. W. 1989, in *The Gravitational Forces Perpendicular to the Galactic Plane*, A. G. Davis, and P. K. Lu eds. (Schenectady: L. Davis Press), p. 210.

Frogel, J. A., and Whitford, A.E., 1987, ApJ, 320, 199.

Kormendy, J., and Illingworth, G. 1982, ApJ, 256, 460.

Lewis, J. R., and Freeman, K. C. 1989, AJ, 97, 139.

te Lintel Hekkert, P., Caswell, J. L., Habing, H. J., Haynes, R. F. and Norris, R. F. 1991, A&AS, 90, 327.

Lindqvist, M., Winnberg, A., Habing, H. J., and Matthews, H. E. 1992, A&AS, 92, 43.

Lindqvist, M., Habing, H. J., and Winnberg, A. 1992, A&A, 259, 118.

Menzies, J. W. 1990, in *Bulges of Galaxies*, B. Jarvis and D. Terndrup, eds. (Garching: ESO), p. 115.

Minitti, D., White, S. D.M., Olzewski, E. W., and Hill, J. M. 1992, ApJ Lett., 393, L47.

van der Veen, W.E.C.J., and Habing, H.J. 1990, A&A, 231, 404.

Wielen, R., Dettbarn, C., Fuchs, B. Jahreiss, H., and Radons, G. 1992, in *The Stellar Populations of Galaxies*, B. Barbuy and A. Renzini, eds., (Dordrecht: Kluwer), p. 81

Dynamics of OH/IR Stars in the Inner Galactic Bulge

ANDERS WINNBERG and MICHAEL LINDQVIST
Onsala Space Observatory, S-439 92 Onsala, Sweden
and
HARM J. HABING
Sterrewacht Leiden, P.O. Box 9513, NL-2300 RA Leiden, The Netherlands

Abstract. Using the VLA at 1612 MHz Lindqvist et al. (1992a) have found 134 OH/IR stars close to the Galactic Centre (GC). These stars plus 15 from Habing et al. (1983) have been used as probes of the gravitational potential to derive the mass distribution in the inner galactic bulge between ≈ 5 to ≈ 100 pc from the GC (Lindqvist et al., 1992b). In this paper we present a progress report of a dynamical model which we have applied to the data.

Key words: Galaxy (the): kinematics and dynamics of – Galaxy (the): center of – stars: long-period variables – stars: OH/IR

1. Introduction

Estimates of the mass distribution within the inner regions of the Galaxy are crucial for our understanding of the dynamics in the inner few hundred pc and as a test of the existence of a massive central object. The gas dynamics in the inner region is complex and show non-circular motions (Genzel & Townes, 1987). Studies of stellar kinematics within this region is essential since the mass distribution derived from the gas kinematics is uncertain and so is also the mass distribution estimated from the 2 μm light distribution (Genzel & Townes, 1987).

2. Mass distribution

We calculate the enclosed mass, $M(r)$, using Jeans' equations under the assumptions given in Lindqvist et al. (1992b). The resulting enclosed mass connects well with the values from McGinn et al. (1989) based on the 2.3 μm CO bandhead for galactocentric distances of ≈ 0.3 to ≈ 5 pc (Fig. 1a). The latter work used the same method (Jeans' equations) under the same assumptions.

We have applied a model of the potential to our results and those of McGinn et al. (1989). We assume that the mass distribution is the sum of contributions from a point mass, M_c, and a stellar cluster with a density dependence $\rho(r) = \rho_0$ for $r < r_c$ (core radius), $\rho(r) = \rho_0 r^{-\alpha_1}$ for $r_c < r < r_1$, and $\rho(r) = \rho_0 r^{-\alpha_2}$ for $r > r_1$. The "free parameters" are M_c, ρ_0, α_1, α_2, r_c, and r_1. Fig 1a shows an example of a reasonably good fit to the data. However, we emphasize that this is only one of several possible solutions. McGinn et al. (1989) did a similar analysis to their set of data. The circular velocity curve is shown in Fig. 1b.

We have created $l - b$ and $l - v$ diagrams by Monte Carlo calculations of stars, with certain orbital eccentricity and inclination distributions, injected into the potential. This is represented by gray-scale maps with the VLA sample stars superposed as comparison (Fig. 1c and 1d). We plan to proceed with similar modelling work.

H. Dejonghe and H. J. Habing (eds.), Galactic Bulges, 341–342.

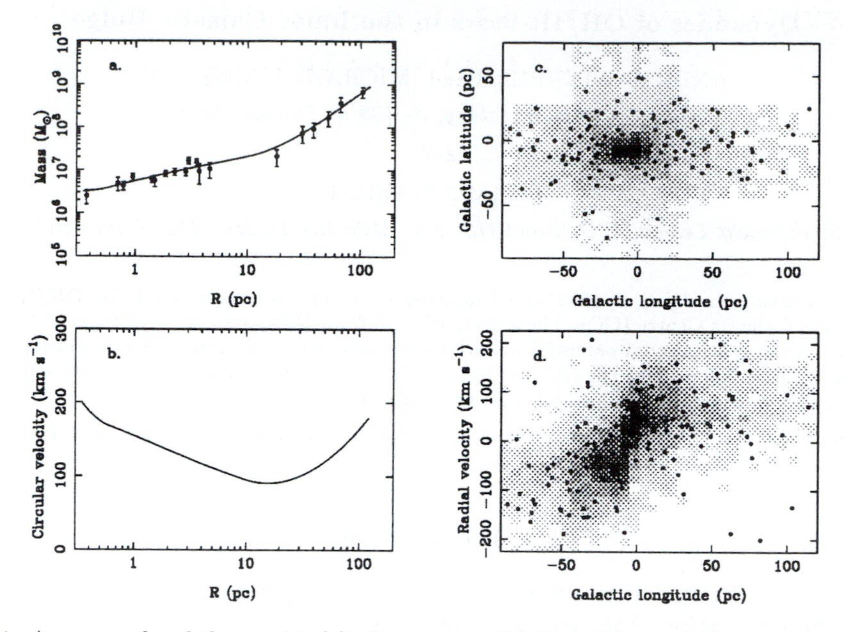

Fig. 1. An example of the model (a) given as a solid line with the following parameters $M_c = 3 \times 10^6\ M_\odot$, $\rho_0 = 10^6\ M_\odot\ pc^{-3}$, $\alpha_1 = -2.5$, $\alpha_2 = -1.0$, $r_c \approx 0.6$ pc, $r_1 \approx 10$ pc. The circular velocity curve is shown in (b). The gray-scale plots represent distributions from Monte Carlo simulations (c and d). The VLA sample is denoted by dots.

It has become clear in the course of this survey that we have not got a statistically significant number of stars within a few parsecs (≈ 5 pc) from the Galactic Centre. We therefore plan to carry out a more sensitive search in this area using the VLA. We also plan to extend the survey in galactic longitude and latitude.

In this work we have information on 3 (the position and the radial velocity) of the 6 variables of the problem. Thus the problem is heavily underdetermined. We therefore plan to search for H_2O masers in these OH/IR stars. Later we will try to determine their proper motions thus increasing the number of known variables to 5.

Acknowledgements

This project is supported by the Swedish Natural Science Research Council (NFR).

References

Genzel R., Townes C.H., 1987, ARA&A, 25, 377
Habing H.J., Olnon F.M., Winnberg A., Matthews H.E., Baud B., 1983, A&A, 128, 230
Lindqvist M., Winnberg A., Habing H.J., Matthews H.E., 1992a, A&AS, 92, 43
Lindqvist M., Habing H.J., Winnberg A., 1992b, A&A, 259, 118
McGinn M.T., Sellgren K., Becklin E.E., Hall D.N.B., 1989, ApJ, 338, 824

SYNTHETIC PHOTOMETRIC INDICES FOR GALACTIC GLOBULAR CLUSTERS

M.L. MALAGNINI

Dipartimento di Astronomia, Universita' degli Studi di Trieste, Italy

L.E. PASINETTI FRACASSINI and S. COVINO

Dipartimento di Fisica, Universita' degli Studi di Milano, Italy

and

A. BUZZONI

Osservatorio Astronomico di Merate, Milano, Italy

Abstract. A grid of synthetic spectral energy distributions, representative of old stellar populations, has been used to derive colors in different photometric systems, and to compare the theoretical predictions with observational data for galactic globular clusters.

1. Photometric indices

Evolutionary synthesis of stellar populations provides a powerful tool for understanding the role of individual stellar components in the integrated radiation of stellar systems, such as clusters and galaxies, once the models have been proved to be consistent with the observations. In order to investigate this point, we undertook a systematic comparison between synthetic and observed photometric indices of galactic globular clusters. The synthetic indices were obtained from the integrated spectral energy distributions (SED's) for simple stellar populations (Buzzoni 1989), which have been computed by taking into account a set of relevant physical parameters, namely metal and Helium abundances (Z,Y); slope of the initial mass function (s); mass loss parameter (η); and morphology of the horizontal branch (HB). A total number of 47 SED's have been considered; they were computed for ages τ = 10.0, 12.5, 15.0, and 18.0 Gyrs, and the following values for the input parameters:

- **log Z = -4.0 (Y=0.23), -3.0 (Y=0.23), -2.0 (Y=0.25);**
- **s = 1.35, 2.35;**
- **η = 0.30, 0.50.**

Out of 47 models, 44 refer to synthetic clusters having, at τ = 15.0 Gyrs, an intermediate type HB morphology, as in M3, while the remaining three models have HB morphologies Blue (NGC 6752 like; log Z = -4.0) or Red (47 Tuc like; log Z = -3.0, -2.0), and fixed values s = 2.35 and η = 0.30.

The photometric indices are those of Johnson UBVRIJK, Thuan and Gunn uvgr, and Washington Observatory CMT_1T_2 systems. We ought to limit our analyses to photometric bands for which synthetic values could be computed from theoretical SED's, by taking into account their wavelength sampling.

From the observational point of view, a total number of 130 galactic globular clusters have been considered; the sample includes globular clusters with different morphologies of their HR diagrams. For each cluster, the following data have been collected, when available, the colors in the above-mentioned photometric systems, the E(B-V) color excess, the integrated spectral type, the total visual magnitude,

H. Dejonghe and H. J. Habing (eds.), Galactic Bulges, 343–344.

and the parameter B/B+R. In order to make comparable the synthetic colors with the observed ones, we corrected the observational data for interstellar reddening.

First of all, the role of physical input parameters on the derived SED's has been considered, then synthetic colors have been computed, and their behavior in two-colors diagrams has been analyzed. Then comparisons have been made by using all available photometric data with the whole set of synthetic indices.

Here we present some results, in particular for the UBV system, referring to model predictions and to comparisons with observational data.

Acknowledgements

Partial support from MURST 40%-60%, and CNR bilateral grants is acknowledged.

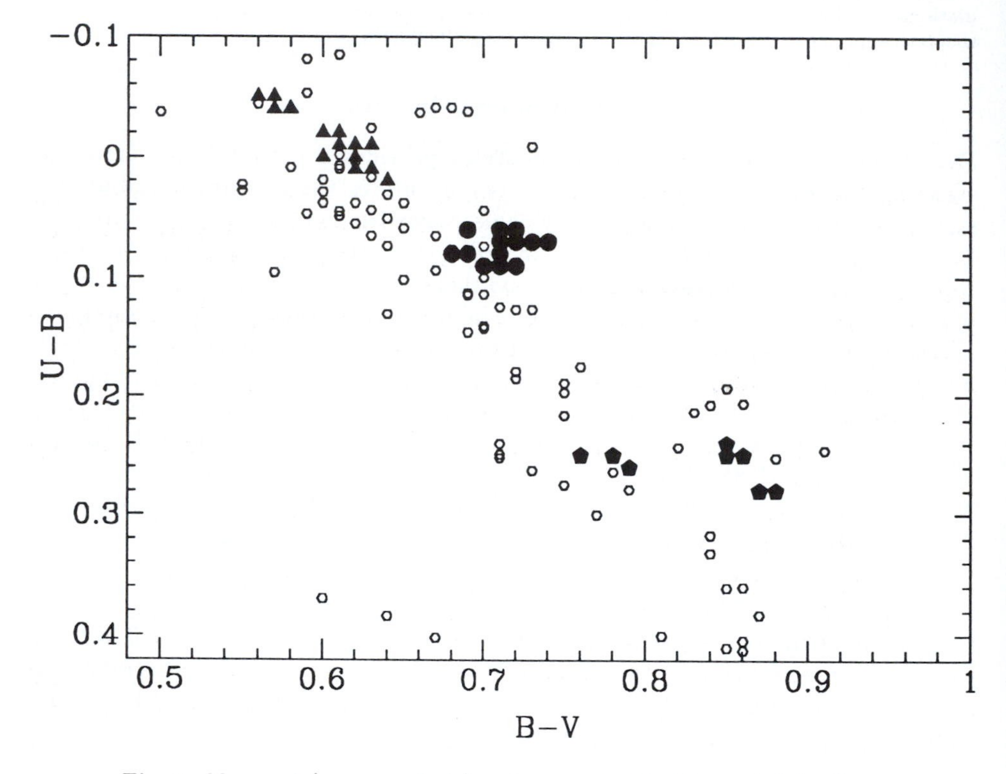

Fig. 1. Observed (open symbols) and computed (solid symbols) colors

References

Buzzoni, A., 1989, *Astrophys. J. Suppl. Ser.*,71, 817

HB MORPHOLOGY AND INTEGRATED SPECTRA OF GLOBULAR CLUSTERS: A THEORETICAL APPROACH

M.L. MALAGNINI
Dipartimento di Astronomia, Universita' degli Studi di Trieste, Italy
G. BONO, C. MOROSSI and L. PULONE
Osservatorio Astronomico di Trieste, Italy
and
E. BROCATO
Osservatorio Astronomico di Teramo, Italy

Abstract.
The aim of the project is to investigate the role of different horizontal branch morphologies on the integrated light of globular clusters.

1. Preliminary results

The evolutionary phase of HB stars is a fundamental testing ground for evaluating the parameters of population II globular clusters. In particular, it plays a relevant role in the determination of the contributions to the synthetic integrated spectral energy distribution (SED's) of different Horizontal Branch (HB) morphologies. Indeed, dating back to the HB classification of Dickens (1972), moving from the blue to the red types, the physical parameter governing the HB morphology is the metal content. This theoretical explanation does not reproduce properly the HB morphologies of the intermediate metallicity globular clusters, for which a second parameter is advocated to explain the non-monotonic behaviour of Dickens's classification. Therefore the appropriate HB morphology should be taken into account to obtain reliable evaluations of globular cluster metallicities and reddenings, based on observational integrated spectral types.

To build up a synthetic HB, a set of HB evolutionary tracks (Castellani et al. 1991) has been computed. The grid of models is available for different metallicities (Z= .0001, .0004, .001, .003, .006, .02) and a helium content Y=.23, so that it can be usefully adopted for population synthesis ranging from low to high metallicity stellar systems. The core masses and the extra-helium contents due to the first dredge-up event have been obtained from H-burning models reaching the tip of the red giant branch in 15 Gyr.

A synthetic HB, computed for Z = 0.02, has been populated with 1000 objects; a linear mass distribution in the range 0.55 - 0.80 $M\odot$ has been adopted. From the theoretical HB, the synthetic integrated spectral energy distributions are computed, by using a newly available set of theoretical fluxes computed from stellar model atmosphere by Kurucz (1992).

Here we present some preliminary results, computed for the case of solar chemical composition, for a synthetic horizontal branch of the intermediate type. Figure 1 shows the integrated SED for this case.

H. Dejonghe and H. J. Habing (eds.), Galactic Bulges, 345–346.

Acknowledgements

MURST 40%-60%, MFD, and CNR bilateral grants are acknowledged.

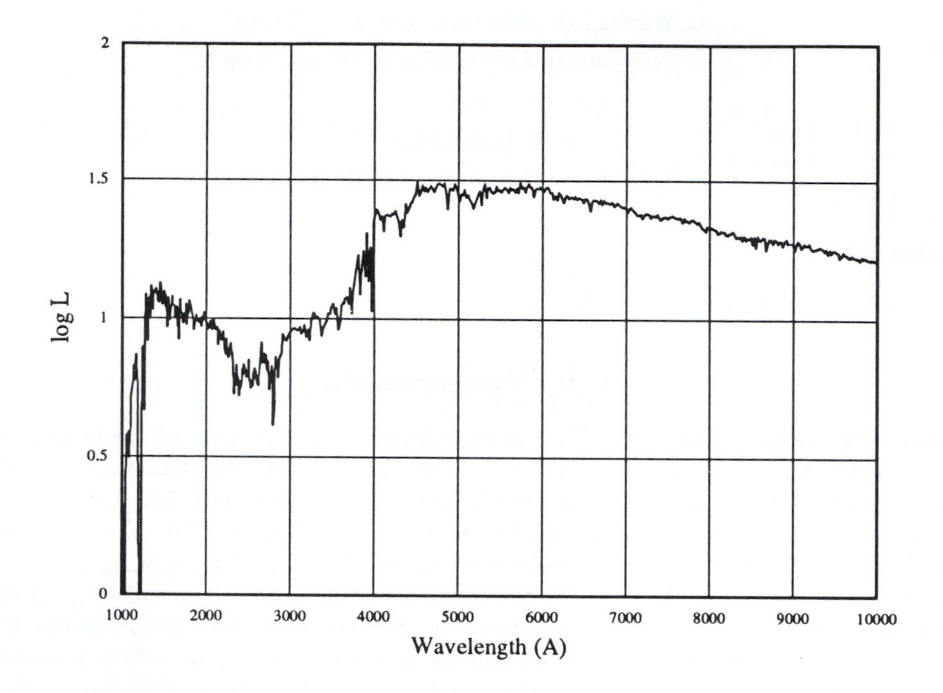

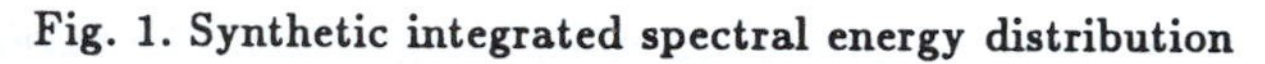

Fig. 1. Synthetic integrated spectral energy distribution

References

Castellani, V., Chieffi, A., Pulone, L., 1991, *Astrophys. J. Suppl. Ser.*,**76**, 911

Dickens, R.J., 1972, *Mon. Not. Roy. Astron. Soc.*, **157**, 281

Kurucz, R.L., 1992, in *The Stellar Populations of Galaxies*, eds. Barbuy and Renzini, Kluver Academic Publishers, 225

PROPER MOTIONS: WHAT DYNAMICAL INFORMATION CAN WE EASILY EXTRACT FROM THEM?

M. WYBO and H. DEJONGHE
Universiteit Gent, Belgium

Abstract. We explore the proper motion distributions of anisotropic Plummer models, and show that information on the orbital structure can be obtained, even if no distance information is available.

We may well be at the dawn of a new era, where proper motions will become commonly available. This is already the case for data on the internal kinematics of globular clusters, but other data sets, most notably samples of populations in our own Galaxy and the Galactic Bulge, are being produced (Cudworth (1986), Minniti (this symposium)).

Clearly, proper motions are very valuable, because they involve 2 components of the velocity. On the other hand, unlike radial velocities, their magnitudes depend on the distance, which, especially in Galactic work, is hard to get. Therefore, we concentrate on two applications which are not so sensitive to an uncertain distance: data in the Galactic Bulge, and data in globulars.

The distribution of proper motions (hereafter pm) is a 2-dimensional probability distribution. Is is defined *at every point in the sky* with coordinates (l, b), and gives the probability of finding a star with proper motion components v_c and v_t in the point (l, b). The v_c-component is directed towards the center of the system. In this contribution, we assume spherical symmetry, and it suffices to simply substitute (l, b) by the projected radius r_p (to the Galactic Center or the center of the globular).

We use the simplest dynamical equilibrium model for spherical star systems which is undoubtedly the Plummer model. We consider a one-parameter family of models, which all have the same Plummer law in the mass density, but have different anisotropic 2-integral distribution functions $F(E, L)$, and hence different orbital structures (Dejonghe 1986). This orbital structure is determined by one parameter q, which is related to Binney's anisotropy parameter β by

$$\beta = 1 - \frac{\sigma_\varphi^2}{\sigma_r^2} = 1 - \frac{\sigma_\vartheta^2}{\sigma_r^2} = \frac{q}{2}\frac{r^2}{1+r^2},$$

in standard notations. Note that q and β have the same sign.

We consider three examples. First we have the isotropic systems, with $q = \beta = 0$. This case is what is normally called a Plummer model. Then we have the radial systems, with $0 < q \leq 2$, and $\beta > 0$. Radial orbits are prevalent, especially in the outer parts. For $q = 2$ only radial orbits are present at large radii. At last we have the tangential system, with $q < 0$, and $\beta < 0$. Tangential orbits are prevalent, especially in the outer parts. For $q \to -\infty$ only circular orbits are present.

To carry out the calculations, it is convenient to normalise the velocities such that the escape velocity is set to unity (the normalisation factor is of course different at every point). In addition the pm-distribution is normalised to one with the projected mass density.

H. Dejonghe and H. J. Habing (eds.), Galactic Bulges, 347–348.

Since the distribution functions are known, one simply calculates a double integral: one over the velocity component along the line of sight and secondly through the stellar system. This can be done analytically for the isotropic case and for $q = -2n$.

We determined the pm's by calculating all the moments of the 2-dimensional pm-distribution, then calculate the probability of finding a star at r_p without proper motion, i. e. $\mathrm{pm}_{r_p}(0,0)$ (the "top value"). Assuming that the distribution can be well approximated by $\mathrm{pm}_{r_p}(v_c, v_t) = \sum_{i,j} a_{ij}(1 - v_c^2 - v_t^2)^{\alpha+i} v_c^{2j}$, with $\alpha = 4.5 - q$ from theoretical arguments, we can determine the a_{ij} by matching the moments of the approximation with the true moments and its top value.

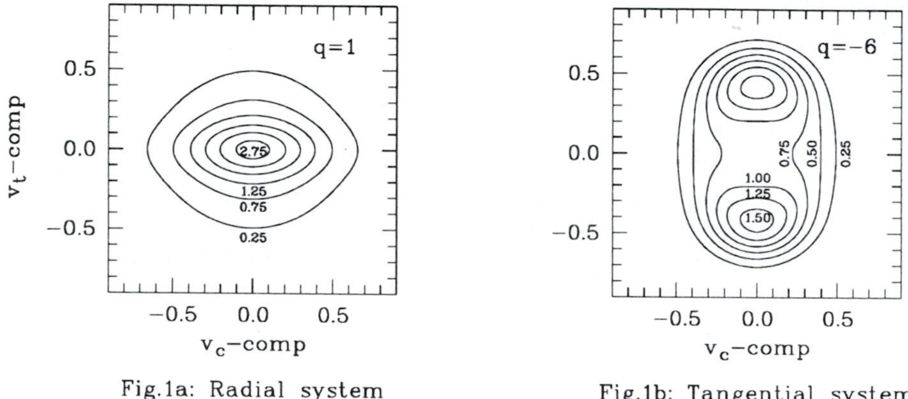

Fig.1a: Radial system

Fig.1b: Tangential system

As an illustration, we show in fig 1 the contour lines of the two-dimensional distribution, taken at a distance of 5 times the core radius. In the tangential system, the distribution function is bimodal along the v_t-axis, as can be expected due to the motion of the stars. However in the radial system the v_c velocity-component is predominant. The isotropic case is not plotted, since the contours (trivially) are circles.

Conclusions: The distribution of proper motions provides valuable information on the orbital structure of the system, *independently of the distance*, by simple inspection of the contour lines. As is well known, line profiles provide important information on the kinematics of the system which is intrinsically distance independent, but the determination of the orbital structure needs inversion of the data, which has its own problems. On the other hand, comparison of the line profiles (velocity dispersions) with the velocity dispersion components of the proper motions will enable us to *determine the distance* to the system. However, an accuracy problem remains; we still need a lot of data to be able to determine the pm distribution accurately.

References

Dejonghe, H., 1987. *Mon. Not. R. astr. Soc.*, **224**, 13
Cudworth, 1986, *AJ*, **92**, 348
Minniti, 1992, this symposium

KINEMATICAL FEATURES OF THE MAIN BULGE IN A MULTI-COMPONENT GALACTIC MODEL

RAFAEL CUBARSI and MANUEL HERNANDEZ-PAJARES

Departament de Matemàtica Aplicada i Telemàtica, UPC
P.O. Box 30002, E-08080 BARCELONA, Spain

The present state of the Galaxy can be idealized as a stellar system with several components: inner bulge, main bulge, thin disk, thick disk, stellar halo, corona, etc. In the regions where the diffusion phenomena are negligible, the dynamics can be determined from a conservative dynamic system, according to the superposition principle with a common potential. Here we describe the kinematical behavior of the main bulge, thin and thick disks, and stellar halo, depending on the cylindrical galactocentric coordinates r and z, under axisymmetric conditions. The angular dependence is treated in the work by Sanz-Subirana & Juan-Zornoza (1992).

Our model is based in the following two hypotheses: (*a*) The stars are moving under a gravitational potential $U(t,\mathbf{r})$ and, since the model is not applied to the inner bulge, gas or corona, the collisionless Boltzmann equation is satisfied for all the components. (*b*) According to Oort's approximation, the velocity distribution of one stellar component is idealized by a Gaussian function of the peculiar velocities.

For a time-depending axisymmetric model this situation leads us to Stäckel potentials (Sala, 1990). However, in the case of superposition of several stellar components, the more general potential function has the following form (Cubarsí, 1990)

$$U(\mathrm{t},\mathbf{r}) = A(t)(r^2+z^2) + k^{-1}U_1(k^{-1}(r^2+z^2)) + (r^2+z^2)^{-1}U_2(z^2r^{-2})$$

where U_1, U_2, $k(t)$ and A are arbitrary functions of the specified arguments.

For each stellar component the mean velocity field and the velocity distribution are completely determined. In particular, the rotational mean velocity and the diagonal central moments for the i-th component are

$$\Theta_0^{(i)} = -\Omega_i r\,(1+a_i r^2+b_i z^2)^{-1}; \qquad \mu_{rr}^{(i)} = k_i^{-1}(1+b_i r^2)(1+b_i(r^2+z^2))^{-1}$$

$$\mu_{\theta\theta}^{(i)} = k_i^{-1}(1+a_i r^2+b_i z^2)^{-1}; \qquad \mu_{zz}^{(i)} = k_i^{-1}(1+b_i z^2)(1+b_i(r^2+z^2))^{-1}$$

being a_i, b_i, k_i, and Ω_i specific parameters of the components.

The mass distribution of one component is modeled by the the potential and by the particular kinematical parameters of the component. Thus we have,

$$N(t,\mathbf{r}) \propto \frac{\exp\left(\frac{1}{2}k_i\Omega_i^2r^2(1+a_ir^2+b_iz^2)^{-1}-k_i(U(t,\mathbf{r})-A(t)(r^2+z^2))-k_ib_iU_2(z^2r^{-2})\right)}{(k_i^3(1+a_ir^2+b_iz^2)(1+b_i(r^2+z^2)))^{1/2}}$$

where $k_i = \alpha_i k(t)$, with α_i constant.

The imprecise information about numerical values of the Galactic potential, stellar density or velocity distribution for different regions of the Galaxy, in particular for the bulge, moves us to use an schematic version of the potential with $U_1 = ck^{-1}(r^2+z^2)$ and $U_2 = q+sz^2r^{-2}$ (q, s and c constants). The resulting potential has one term corresponding to an harmonic function, significant at large distances of the Galactic center,

H. Dejonghe and H. J. Habing (eds.), Galactic Bulges, 349–350.

and another term generating a short-range force. The corresponding stellar density is dominated by the exponential factor with three differentiated contributions: (1) A term due to the rotation of the stellar component. If the component is fast rotating, like the thin disk and the bulge, it can be more important than the other terms. (2) A term due to the harmonic part of the potential and to the specific kinematics of the component. If $c>0$, this term fixes the scale length of the Galactic components. (3) A term depending on the potential and also on the particular kinematics of the component, which is significant near the center. If $q<0$, the short-range force is attractive and this term produces an exponential increasing of the mass density towards the center.

According to this model, we can compare the kinematics and mass distribution of the main bulge, disk and stellar halo. But since there are not definitive data about the potential, we focus in the Galactic plane. For the parameters of the potential, and in order to fix the characteristic length of the thick disk in 4.5 kpc (Kruit, 1987), we take ck^{-2}=850 $kpc^{-2}(km/s)^2$, consistent with the accepted local potential values (Ninkovic, 1990). In order to obtain decreasing mass density functions in r for the bulge, we take q=-3000 $kpc^2(km/s)^2$ (case a), otherwise q can vanish (case b) since this value only modifies the stellar density within r<0.75 kpc. The kinematical features of the thin disk, thick disk and stellar halo at the solar neighborhood (Gilmore & Wyse, 1987; Sandage & Fouts, 1987; Cubarsí, 1992; Cubarsí *et al.* 1992) are shown in the table bellow. For the main bulge we take the velocity dispersions similarly to the stellar halo (Rich, 1992), in particular the first component. In the table, their values are given at 0.5 kpc and the tentative rotational velocity has a maximum of 250 km/s at 0.3 kpc. The comparative parameters are: h the characteristic length of the component, R the distance which satisfies $N(R)=N(h)e^{-1}$ (N is the normalized stellar density), H the distance with maximum stellar density, d the distance with maximum rotational velocity, and V the maximum rotational velocity value. The distances are given in kpc and the velocities in km/s. This method has been used in the synthetic approach in order to detect bulge stars in the solar neighborhood (Hernández-Pajares *et al.* 1992).

	$(\sigma_r:\sigma_\theta:\sigma_z)$	h	R	H	d	V
bulge	(130:65:65)	2	6	0 (a)/0.75 (b)	0.3	250
thin-d	(28:17:17)	4.5-6.5	4-7	5.5	6.5	230
thick-d	(70:50:45)	4.5	8	0	9	185
halo	(130:105:85)	4.5	10	0	11.5	140

REFERENCES

Cubarsí, R. (1990) Astron. J. **99**, 1558.
Cubarsí, R. (1992) Astron. J. **103**, 1608.
Cubarsí *et al.* (1992) *Astronomy from Large Databases II*. ESO. Haguenau.
Gilmore, G.; Wise, R.F.G. (1987) *The Galaxy*. Reidel:Dordrecht, p247.
Hernández-Pajares, M. *et al.*(1992) This Symposium.
Kruit, P.C. van der (1987) *The Galaxy*. Reidel:Dordrecht, p27.
Ninkovic, S. (1990) Bull. Astron. Inst. Czechosl. **41**, 236.
Rich, R.M. (1992) *The Stellar Population of Galaxies*. Reidel:Dordrecht, p29.
Sala, F. (1990) Astron. Astrophys. **235**, 85.
Sandage, A.; Fouts, G. (1987) Astron. J. **93**, 74.
Sanz-Subirana, J.; Juan-Zornoza, J.M. (1992) This Symposium.

Acknowledgements: Work supported by the spanish grant PB90-0478 from DGICIT.

POINT-AXIAL MASS DISTRIBUTION IN THE EXTERNAL GALACTIC BULGE

J. SANZ-SUBIRANA[1] and J.M. JUAN-ZORNOZA[2]

[1]Dept. de Matemàtica Aplicada i Telemàtica. [2]Dept. de Física Aplicada
Universitat Politècnica de Catalunya. Apdo. 30.002 - Barcelona. SPAIN
EAN: MATJSS@MAT.UPC.ES

1. Introduction

The main purpose in this work is to present some analytical results about the spatial distribution of the stars in the Galactic Bulge. These results have been obtained by considering that the stellar system verifies the collisionless Boltzman equation and the ellipsoidal hypothesis (non-axisymmetrical) for the distribution of peculiar velocities of the stars.
A qualitative study of the kinematical behavior and mass distribution for an axisymmetric multi-component galactic model is presented by Cubarsí & Hernández-Pajares in this Symposium [1]. The angular dependence for a stellar system characteristic of the external galactic bulge is studied in our work by considering a point-axial velocity distribution function.
The general solution for the potential in a stationary point-axial system model can be found in [2], also, the separable potentials in the time-depending models can be found in [3]. These potential functions are compatible with triaxial spatial distribution of stars when the hydrodymamical equations are fulfilled.

2. The galactic model

We adopt the galactic model based on the collisionless Boltzman equation:

$$\frac{d\Psi}{dt} = \frac{\partial\Psi}{\partial t} + \underline{V}\cdot\nabla_{\underline{r}}\Psi - \nabla_{\underline{r}}U\cdot\nabla_{\underline{v}}\Psi = 0 \tag{1}$$

and an ellipsoidal hypothesis for the distribution of peculiar velocities of stars:

$$\Psi(\underline{r},\underline{v},t) \equiv \Psi(Q+\sigma)= e^{-1/2\ (Q+\sigma)} \tag{2}$$

where: $Q = \underline{v}^t\cdot A\cdot\underline{v}$ and A and σ are functions of position and time.

Equation (1) under hypothesis (2) gives the well known Chadrasekhar equations [4], which leads to the elements of the velocity ellipsoid. A more precise description of the model is presented in [5]. The model admits the following potential function separable in ω and Z:

$$U = \frac{D_1}{2\ k_3^2}(\omega^2 + Z^2) + \text{ctt.} \tag{3}$$

The scalar function σ is invariant along of the local centroid trajectories [5], which is related to the mass distribution by :

$$\nu = \iiint_v \Psi(Q+\sigma)\ dv = \frac{2\pi}{|A|^{1/2}}\ e^{-\sigma/2} \tag{4}$$

This work has been supported by the DGICYT of Spain under Grand No. PB90-0478

H. Dejonghe and H. J. Habing (eds.), Galactic Bulges, 351–352.

The determinant $|\mathbf{A}|$ is a fourth degree polinomy in ω and z, which coefficients are functions of 2θ, and

$$\sigma= \frac{D_1}{k_3^2}\left(a(2\theta)\,\omega^2+ k_3 z^2\right) - \frac{\gamma^2}{k_2}\,\frac{a(2\theta)\,\omega^2}{(1 + pz^2)\kappa^2 + a(2\theta)\,\omega^2} \qquad (5)$$

where

$$\kappa^2= (K_1^2- Q^2)/k_2 \quad ; \quad a(2\theta)= K_1+ Q \sin(2\theta+\phi), \qquad (6)$$

being D_1, k_2, γ constants, and the others parameters are arbitrary functions of time.
Taking into account (5), the stellar density (4) in the Galactic Plane can be written as:

$$\nu= \frac{\exp\left(- \frac{D_1}{2k_3^2}\, a(2\theta)\,\omega^2 + \frac{\gamma^2}{2\,k_2}\,\frac{a(2\theta)\,\omega^2}{\kappa^2 + a(2\theta)\,\omega^2}\right)}{\left(k_2\,(\kappa^2+ a(2\theta)\,\omega^2)\,(k_3+ p\,a(2\theta)\,\omega^2)\right)^{1/2}} \qquad (7)$$

3.- Discussion

1) The stellar density (7) is characterized by two differentiated contributions in the exponential factor, both functions of the azimuthal angle θ:
- The first term is due for the harmonic potential and takes into account the anisotropic distribution of velocities. Being D_1 positive, this term fixes the scale length of the bulge.
- The second term is associated to the rotation of the stellar system. It produces a displacement forward of the mass density maximum. If the system is fast rotating, it can predominate in front of the other.
2) For a non-rotating system (γ=0), the isodensity surfaces are ellipsoids which axis-ratios are determined from the parameters of the velocity distribution. Nevertheless, in a rotating system the isodensity surfaces shows bulk shapes and also elliptic-toroidal shapes. The isodensity curves in a fixed meridian section are Cassini ovals.
For the numerical application, we have taken the values that corresponds to the last estimations publishes in the literature [6,7,8,9].

BIBLIOGRAPHY

[1] Cubarsí et al.: 1992, This Symposium.
[2] J.M. Juan-Zornoza et al.:1991, Astrophys. Space Sci. **185**, 95
[3] J. Sanz et al.:1987, Proc. IAU Twenty European Regional Astronomy Meeting **4**, 227.
[4] Chandrasekhar, S.: 1942, Principles of Stellar Dynamics, The University of Chicago Press, Chicago.
[5] Sanz et al.: 1989, Astrophys. and Space Sci. **156**, 19.
[6] T.D. Kinman et al.: 1989, Astroph. J. **95**, 804.
[7] R.M. Rich: 1992, Proc. IAU Symposium n°149, 29.
[8] R. Sharples et al.: 1990, M.N.R.S.A. **246**, 54.
[9] S. Ninkovich: 1987, Astrophys. and Space Sci. **136**, 299.

THE CONTRIBUTION OF THE GALACTIC BULGE TO THE GALACTIC ROTATION CURVE

I.V. PETROVSKAYA
Astronomical Institute of St. Petersburg University, Bibliotechnaya pl. 2, St. Petersburg, 198904, Russia

and

S. NINKOVIĆ
Astronomical Observatory, Volgina 7, 11050 Belgrade, Yugoslavia

It is not always clear what the bulge of the Galaxy is: a region close to the centre, a subsystem formed by a distinct population, or a mixture of populations but characterised by its own mass distribution. We consider the bulge of the Milky Way as a subsystem and thus contributing to the galactic gravitation field. We want to estimate the contribution of the galactic bulge to the rotation curve.

The rotation curve was obtained from the HII data and HI data (tangential points, 21cm line profile). We also added the HI results by Merrifield obtained from the hydrogen layer thickness. The distances to HII regions are known in kiloparsecs whereas the distances to the HI clouds are obtained as a ratio to R_0 - the distance of the sun from the galactic centre. The best agreement of HII and HI data was found for $R_0 = 7.9$ kpc (Nikiforov, I.I. & Petrovskaya, I.V., in prep.) and we adopted 208.6 km/s for the linear velocity of the LSR (Kerr, F.J. and Lynden-Bell, D.,(1986, *Mon.Not.R.Astron.Soc.*, **221**, 1023) recommend 26.4 km/s/kpc as the value of the LSR angular velocity).

In order to explain the circular velocity of the Galaxy we propose a model containing four subsystems: 1) the core; 2) the bulge; 3) the disk; 4) the corona. We try to explain some feature of the rotation curve, namely, the sharp maximum near the galactic centre followed by a corresponding minimum somewhat father from the centre (core) and the flat part of the curve at the larger distances (corona). For the core we assume a model of homogeneous spheroid. Its semiaxis major is equal to the distance at which the maximum occurs, the axial ratio is between 0.4 and 1 depending on the fit to the rotation curve. For the disk we adopted the truncated exponential model (Casertano, S., 1983, *Mon.Not.R.Astron.Soc.*, **203**, 735) with the radius $R_1 = 9$ kpc and the effective semithickness 0.25 kpc. We took the model of the corona by Ninković, S., (1992, *Astron. Nachr.*, **313**, 83) where the parameters were established through the fitting.

From the rotation curve (using the points for $R/R_0 > 0.64$) we found the mass of the disk $4.6 \times 10^{10} M_\odot$, the mass of the corona (up to 68 kpc) $1.1 \times 10^{12} M_\odot$ and the mass of the core+bulge $1.9 \times 10^{10} M_\odot$. We assume that the bulge is a spheroid of $0.6 - 0.8$ axial ratio. Using the rotation curve of Haud, U. (1979, *Pis'ma v Astron. zh.*, **5**, 124) for the central part of the Galaxy we found that the semimajor axis of the bulge is equal to 5 kpc, the central density $0.25\, M_\odot/pc^3$, the density being constant up to 1.9 kpc from the centre. For the density exponent of the outer part of the bulge we obtained the value -3, for the bulge mass $14.6 \times 10^9 M_\odot$.

H. Dejonghe and H. J. Habing (eds.), Galactic Bulges, 353.

Frank JANSEN

The Construction of Stäckel Potentials for Galactic Dynamical Modelling

P. BATSLEER and H. DEJONGHE
Sterrenkundig Observatorium, Gent, Belgium

Abstract. We present a set of axisymmetric Stäckel potentials which can be used for Galactic dynamical modelling. Each of them has a halo–disk structure with a flat rotation curve.

1. Introduction

Dynamical modelling of stellar populations in the bulge of our Galaxy necessitates a potential that approximates as much as possible our idea of the true potential. As is well known, of all requirements that we must impose on a potential, a flat rotation curve and the presence of a third integral are foremost. Therefore, we construct Stäckel potentials with flat rotation curves. Besides, the rotation curve must remain flat over a large distance (see Rubin et al. 1985).

To adapt such a potential to the true potential of our own Galaxy, one can use the force component K_{ϖ} and K_z, which can estimated from velocity measurements in the solar neighbourhood, as is well known.

2. Method

An axisymmetric Stäckel potential separates the Hamilton–Jacobi equation in spheroidal coordinates (λ, φ, ν) (see Dejonghe and de Zeeuw 1988a). Those coordinates are defined by two constants a and c which are related to the oblate or prolate shape of the system. For simplicity, we assign to both components a Kuzmin–Kutuzov potential (Kuzmin 1962), which is written in spheroidal coordinates as follows:

$$V(\lambda, \nu) = -\frac{GM}{\sqrt{\lambda} + \sqrt{\nu}}$$

with G the gravitational constant and M the total mass. To handle the different shape of disk and halo and to preserve the Stäckel formalism, two spheroidal coordinate systems are considered which are connected by $\lambda_{halo} = \lambda_{disk} - q$, $\nu_{halo} = \nu_{disk} - q$, $q \geq 0$. If k measures the relative contribution of the disk mass versus the total mass, the resulting potential can be written as:

$$V(\lambda, \nu) = -GM\left(\frac{k}{\sqrt{\lambda} + \sqrt{\nu}} + \frac{1-k}{\sqrt{\lambda - q} + \sqrt{\nu - q}}\right)$$

To obtain potentials which are compatible with what we know about the Galactic potential, we calculate for the solar neighbourhood the spatial and projected surface density and the circular velocity for each potential from the selected set and compare the values with the Bahcall–Schmidt–Soneira model in a Stäckel version (see Dejonghe and de Zeeuw 1988b).

H. Dejonghe and H. J. Habing (eds.), Galactic Bulges, 355–356.

3. Results

A survey of the parameter–space consisting of the disk and halo flatness, disk and halo mass produces a set of potentials with a flat rotation curve. The large variety in shape of the rotation curves led us to select the flat curves by simple inspection (see Fig.1). The following results, we should emphasize, are results for Stäckel potentials and the axis ratio mentioned is the ratio of the surfaces of constant density near the center:

- The requirement of flat rotation curves implies flat disks, with axis ratio larger than 1.5
- The halo must be nearly spherical: axis ratio smaller than 1.009
- The disk cannot contribute more than 0.15% to the total mass.

The Galactic potential is best fit with a disk which is rather flattened (axis ratio ≈ 4.4), while the halo is almost spherical. The total mass can be estimated to be 4×10^{11} $M_{\odot}$ and about 90% of that mass is contained in the halo. The spatial density in the solar vicinity equals 0.07 $M_{\odot}pc^{-3}$ while the projected surface density is 43 $M_{\odot}pc^{-2}$.

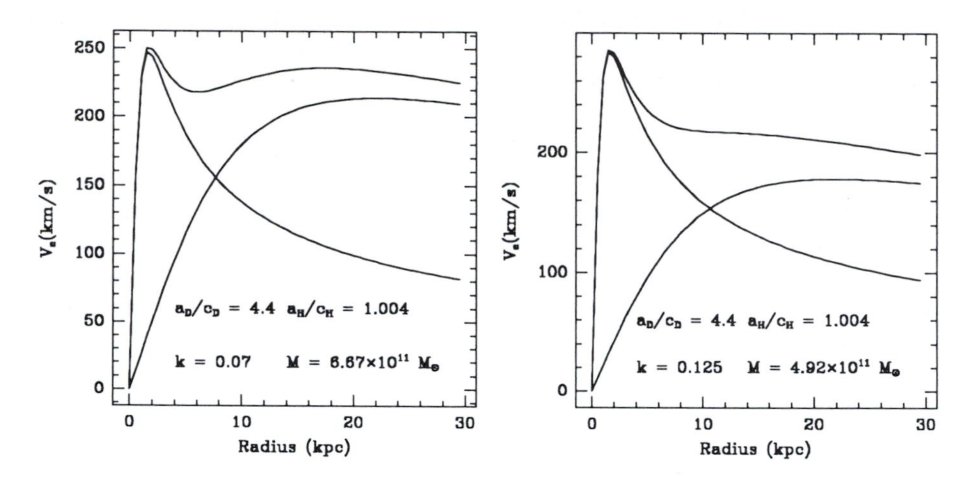

Fig. 1. Rotation curves of some of the retained potentials. Symbols are explained in the text.

References

Bahcall, J.N., Schmidt, M., and Soneira, R. M., 1982, ApJ L23, 258.

Dejonghe H. and de Zeeuw, P.T. 1988a, ApJ, 333, 90.

Dejonghe H. and de Zeeuw, P.T. 1988b, ApJ, 329, 720.

Gilmore, G., King, I., Van der Kruit, P. 1989, The Milky way as a Galaxy, ed. Buser. R, and King I. (Geneva Observatory).

Kuzmin, G.G., and Kutuzov, S.A. 1962, Bull. Abastumani Ap. Obs., 27, 82.

Rubin, V.C., Burstein, D. Ford, W.K., and Thonnard, N. 1985, ApJ, 289, 81.

Two-integral distribution functions for axisymmetric galaxies

C. HUNTER[1,2], E. QIAN[1]
[1] *Florida State University* & [2] *Sterrewacht Leiden*

Abstract. We present a new method for finding a distribution function f, which depends only on the two classical integrals of energy E and angular momentum J about the axis of symmetry, for a stellar system with a known axisymmetric potential and density.

Our method requires an analytical axisymmetric potential Ψ. This potential, together with the radial distance R from the axis of symmetry, define a coordinate system in terms of which we express the analytic density as $\rho(\Psi, R^2)$. The part of f that is even in J is then given by the contour integral

$$f(E, J^2) = \frac{1}{4\pi^2 i\sqrt{2}} \frac{\partial}{\partial E} \oint \frac{d\Psi}{(\Psi - E)^{\frac{1}{2}}} \rho_1 \left[\Psi, \frac{J^2}{2(\Psi - E)} \right].$$

The path of this integral starts and ends at the value of Ψ at large distances, and is determined from properties of the potential. The subscript 1 on ρ denotes a partial derivative with respect to its first argument. Our formula is the analogue for the axisymmetric case of Eddington's (1916) classical solution for the isotropic distribution function $f(E)$ for a known spherical density, and reduces to his solution when ρ has no R-dependence.

A numerical quadrature is generally required to evaluate this solution. Contour integrals can be computed simply and accurately by numerical quadrature even for complicated densities. This is a simpler and much more accurate procedure than direct solution of the integral equation that determines f. It may also be preferable to an explicit evaluation of f when the latter is an infinite or doubly infinite series obtained using Fricke's (1952) expansion method.

The figures display contours of the density and distribution function of one of Satoh's (1980) models. In both cases, only the region below the solid curve is physically relevant. The density of Fig.1 is in fact infinite on the dashed curve, so that the Laplace transform of it required for Lynden-Bell's (1962) method of determining f can not be taken. Generally, contours of f reflect the trends of those of ρ in exaggerated form.

Acknowledgements

CH acknowledges support from the NSF, and a visitor's grant in the Netherlands from the NWO.

References

Eddington, A.S., 1916, *Mon. Not. Roy. Astr. Soc.*, **76** 572
Fricke, W., 1952, *Astr. Nachr.*, **280** 193

H. Dejonghe and H. J. Habing (eds.), Galactic Bulges, 357–358.

Lynden-Bell, D., 1962, *Mon. Not. Roy. Astr. Soc.*, **123 447**
Satoh, C., 1980, *Publ. Astr. Soc. Japan*, **32 41**

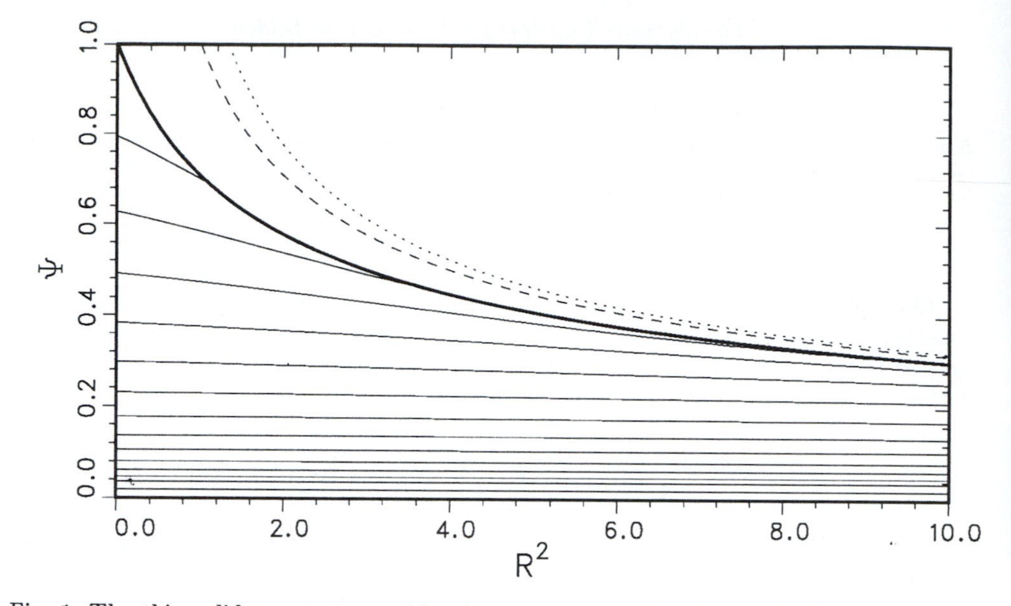

Fig. 1. The thin solid curves are equidensity contours, within the physically relevant domain, of a Satoh model with parameters $b/a = 1$. Outside this domain, $\rho(\Psi, R^2)$ is infinite on the dashed curve, and complex above the dotted curve.

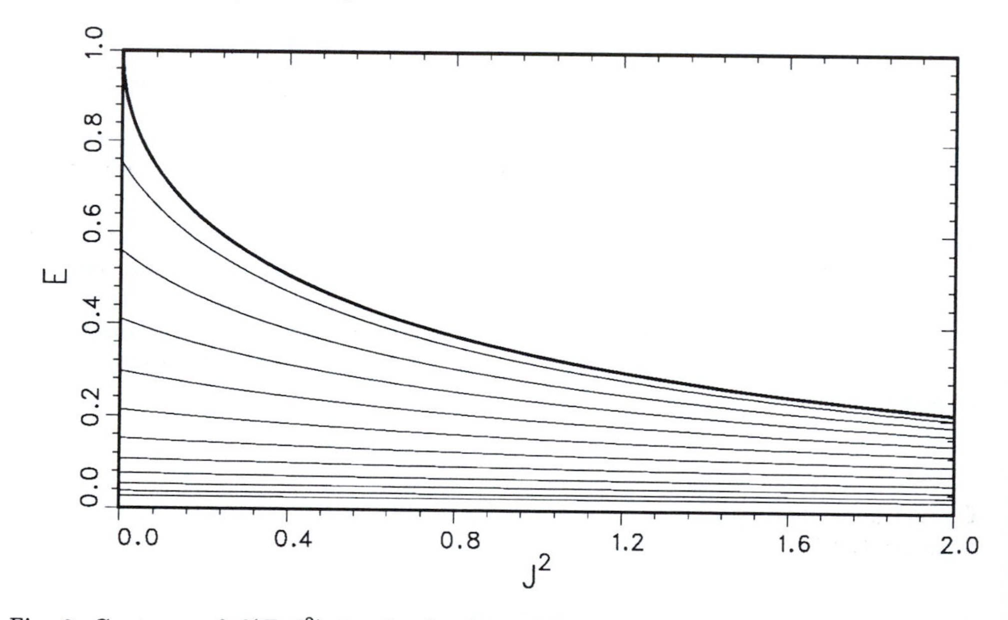

Fig. 2. Contours of $f(E, J^2)$ for the Satoh model of Fig.1. Successive contour levels of both figures differ by factors of 0.2.

Two–Integral Models with Almost Circular Orbits

P. BATSLEER and H. DEJONGHE
Sterrenkundig Observatorium, Gent, Belgium

Abstract. We present global dynamical models for bulge–disk systems. The phase–space structure is examined by means of two integral models. The models indicate that in the formation merger argument for ellipticals one cannot always assume that the maximum phase–space density occurs in the center of the system.

1. Introduction

In a disk, far from the center, stars travel on nearly circular orbits, while closer to the center the very nature of the bulge bring them out of the equatorial plane. As a consequence, one can distinguish two regions with different dynamics in a spiral galaxy. An adequate description of the dynamics of those two regions would require a three integral model. Yet, the construction of such a model is not well defined, because the distribution function is not unique. Therefore we will restrict the analysis to the (unique) two integral distribution function. We argue that a two integral approach is useful because a flat disk has a small extent in the z–direction and in the plane of the disk stars travel on nearly circular orbits and thus the velocity dispersion in the ϖ and z direction are both small and from a dynamical point of view where velocities of the order of the circular velocities are relevant, almost equal.

2. Method

The orbits the stars travel on in a disk can bring them out of the disk up to a certain height. For a given angular momentum L_z, any orbit can be populated with binding energy less than the binding energy of the circular orbit with angular momentum L_z and corresponding radius r. However, we look for bulge–disk systems with a finite extent in the z–direction, say z_0. Hence, for any L_z one must introduce a lower limit S for the binding energy below which no motion is allowed. S is the binding energy of the orbit with angular momentum L_z that just can reach the height z_0. Obviously, the farther from the center, the more the stars are confined to the equatorial plane and thus the more the orbits resemble nearly circular orbits, so the less S will differ from the binding energy of the circular orbit at that distance. Closer to the center, the stars must populate those orbits which can produce a bulge structure in configuration space. Hence, orbits different from circular orbits are populated. The energy $S(L_z)$ is given by $S(L_z) = \psi(r, z_0) - L_z^2/(2r^2(L_z))$. The two integral distribution function we propose has the following form:

$$F(E, L_z) = \sum_{\alpha_1,\alpha_2,\beta,\gamma} c_{\alpha_1,\alpha_2,\beta,\gamma}\, S^{\alpha_1}(L_z) e^{-\frac{\alpha_2}{S(L_z)}} [2S(L_z)L_z^2]^{\beta} [E - S(L_z)]^{\gamma} \qquad (1)$$

We introduce the exponential factor to obtain an exponential behaviour for the spatial density, as is observed in the light curves of most disks. To construct the distribution function (1) for a given bulge–disk system, we first select a large number

H. Dejonghe and H. J. Habing (eds.), Galactic Bulges, 359–360.

of those components which reveal roughly the features of a bulge–disk system. Those components are then fed to a Quadratic Programming routine (see Dejonghe 1989, hereafter QP), which will select in an iterative way a subset of components whose combined density will fit the Van der Kruit–density for that bulge–disk system. The latter density is given by

$$\rho(R,z) = \exp(-\frac{R}{R_0})\mathrm{sech}^2(\frac{z}{z_0}). \qquad (2)$$

Obviously, to obtain a physical model, one must add the constraint $F \geq 0$ to the QP–program.

3. Results

First of all we found that axisymmetric disks with a large variety of scale parameters R_0 and z_0, can be described by two integral models as given in (1). The density (2) is fitted extremely well. As an example, a model for the Galaxy was constructed (see Fig.1, with $R_0 = 2, z_0 = 0.25$). Moreover, the QP–method which determines the best components for the distribution function is suitable for this kind of work. The number of components is always small, at most 20.

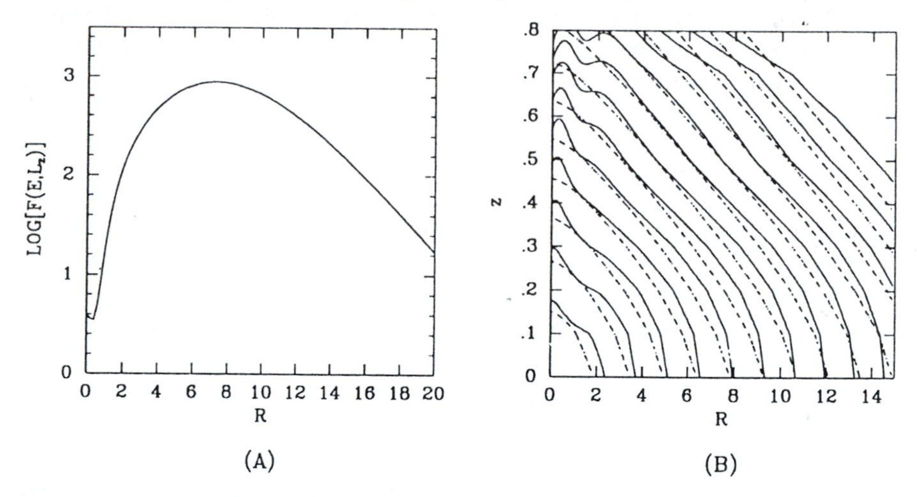

Fig. 1. (A): Plot of $\log[F(E,L_z)]$ vs. R along the locus of circular orbit. Notice the maximum phase–space density is not attained in the center and the 3 order of magnitude difference between the distribution function in the center and at its maximum. Notice also the exponential decline! (B): Contourmap of spatial density for the Galaxy model, where the solid line is the two integral density and the dashed line is the van der Kruit density with $R_0 = 2, z_0 = 0.25$

References

Dejonghe, H., 1989, ApJ, 343, 113.

Gilmore, G., King, I., van der Kruit, P. in *The Milky as a Galaxy*, Geneva Observatory.

Kruit, P.C. van der, and Searle, L., 1982, AA, 110, 61.

SIMPLE MODELS OF GALACTIC BULGES

N.W. EVANS
Institute of Astronomy
Madingley Rd
Cambridge
CB3 0HA

ABSTRACT. We present a simple axisymmetric model with an elementary distribution function capable of representing galactic bulges. The gravity field of the galaxy is based on the axisymmetric logarithmic potential, which has a flat rotation curve. Bulges are built as isothermal distributions of stars embedded within the potential.

1. Introduction

Using cylindrical polar coordinates (R, ϕ, z), the axisymmetric logarithmic potential is (Binney, 1981)

$$\psi = -\frac{1}{2}v_0^2 \log\left(R_c^2 + R^2 + \frac{z^2}{q^2}\right), \tag{1}$$

where R_c is the core radius and q is the axis ratio of the spheroidal equipotentials. We shall exploit the simplicity of the axisymmetric logarithmic potential to build bulge models. Suppose the bulge density embedded within the galaxy with gravity field (1) is

$$\rho_{bulge} = \frac{\rho_0 R_c^p}{(R_c^2 + R^2 + z^2 q^{-2})^{p/2}}, \tag{2}$$

where ρ_0 is the central density and $p \geq 2$ prescribes the asymptotic density decay. The strength of our model is that the (even part of the) bulge distribution function is just an isothermal, namely

$$F_{bulge} = D \exp(pE/v_0^2), \tag{3}$$

where D is a constant

$$D = \rho_0 R_c^p \left(\frac{p}{2\pi v_0^2}\right)^{3/2} \tag{4}$$

H. Dejonghe and H. J. Habing (eds.), Galactic Bulges, 361–362.

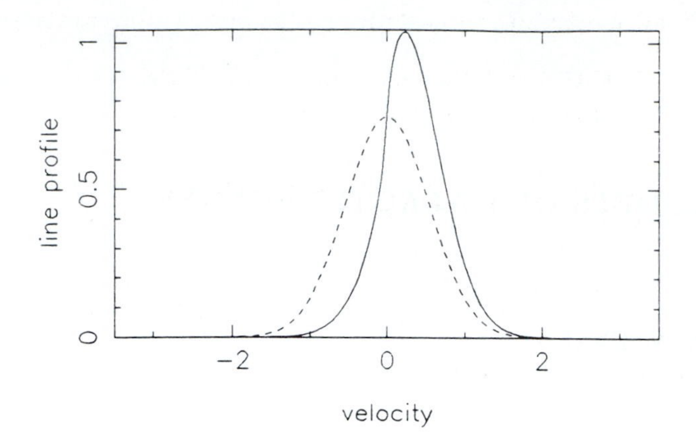

Fig. 1. Line profiles for rotating bulges

As an illustrative example, we shall take $p = 3.1$ as the asymptotic density fall-off. Fixing the effective radius of the bulge as 2.0kpc gives $R_c \sim 1.9$pc. The circular velocity of the halo v_0 is taken as $\sim 220\text{kms}^{-1}$ and the axis ratio q as 0.8, while the central bulge density $\rho_0 \sim 10^5 M_\odot \text{pc}^{-3}$. This gives an E2 bulge of total mass $\sim 7 \times 10^7 M_\odot$ embedded within the rest of the galaxy.

2. Kinematic Properties

All the properties of the bulge can be calculated very simply. For example, the velocity dispersion tensor is isotropic with all the second moments everywhere equal to the same value, i.e..

$$\sigma_R^2 = \sigma_z^2 = \sigma_\phi^2 = \frac{v_0^2}{p}. \tag{5}$$

With the parameters we have chosen, the velocity dispersion of bulge stars is $\sim 125\text{kms}^{-1}$. There is valuable kinematic information in the line profiles – that is, the distribution of line of sight velocities – derived from stellar absorption spectra. Generally, the line profile is very hard to calculate for models of bulges, but here the simplicity of our distribution function can be marvellously exploited. Without rotation, the line profile of our bulge is always exactly Gaussian, as shown in the dotted line in figure 1. Significant distortions from Gaussian only occur for rotating bulges – for example, the bold line is the line profile for the maximum streaming model. Further details are given in Evans (1992).

References

BINNEY, J.J., 1986 *M. N. R. A. S.*, **196**,
EVANS, N.W., 1992 *M. N. R. A. S.*, in press.

Generalized isochrone models for spherical stellar systems *

G.G. KUZMIN and Ü.-I.K. VELTMANN
Tartu Astrophysical Observatory, Estonia

The space density $\rho(r)$ of a spherical system with potential $\Phi(r)$ is related to the phase density $\Psi(x,\xi)$ as a function of specific energy x and angular momentum 2ξ by a well-known integral equation. If the function $\rho(r)$ is given (and thus also $\Phi(r)$), this integral equation has no unique solution. In order to transform this problem into one with a unique solution we suggest the following form for $\Psi(x,\xi)$:

$$\Psi(x,\xi) = \Theta^{-\mu}\psi(u), \quad u = \frac{x-\lambda\xi}{\Theta}, \quad \Theta = 1 + p\xi, \tag{1}$$

where μ (dimensionless), λ and p are parameters satisfying $p \geq 0$ and $\lambda \geq r_{max}^{-2}$, with r_{max} the limiting radius of the star system. The additive constant in the potential is chosen so that $\Phi(r_{max}) = 0$. To make the parameters λ and p dimensionless and give them some meaning we adopt the central potential $\Phi(0)$ as the unit of Φ, x and u, and the equivalent radius of the stellar system r_e as the unit of length. Then $\Phi(0)r_e^2$ will be the unit of ξ. We obtain the integral equation

$$g(\Phi) = 4\sqrt{2}\pi \int_0^\Phi (\Phi-u)^{\frac{1}{2}} F(1, 5/2-\mu, 3/2; ps)\psi(u)\, du \tag{2}$$

for the unknown function $\psi(u)$ which we call the partial density. The function F is the Gauss hypergeometric function,

$$g(\Phi) = [1 + (\lambda + \rho\Phi)r^2(\Phi)]\rho \tag{3}$$

and

$$s(\Phi,u) = \frac{(\Phi-u)r^2(\Phi)}{1+(\lambda+\rho\Phi)r^2(\Phi)}. \tag{4}$$

If we put $\mu = 5/2$ and $\lambda = p = 0$, then the equation (2) is of the well-known Eddington (1916) form:

$$g(\Phi) = 4\sqrt{2}\pi \int_0^\Phi (\Phi-u)^{\frac{1}{2}} \Psi(u) du. \tag{5}$$

In this case $g(\Phi) = \rho(\Phi)$. If $g(\Phi)$ is represented as a power series in Φ, then it is easy to obtain the solution in the form of a power series in u, the term with Φ^n corresponding to the term $u^{n-\frac{3}{2}}$. Let us suppose we have an expansion (finite of infinite)

$$g(\Phi) = \sum_{m,n} g_{m,n} \frac{\Phi^n}{(1-q\Phi)^n} \tag{6}$$

* Originally published in: T.B. Omarov (ed.), Dynamics of Galaxies and star clusters, "Nauka" Kaz.SSR, Alma–Ata, 1973, p.82–87, Translated by L.P. Ossipkov.

H. Dejonghe and H. J. Habing (eds.), Galactic Bulges, 363–366.

with q a parameter. Then we obtain the solution in the form

$$\psi(u) = \sum_{m,n} \psi_{m,n} F(m, n+1, n-\frac{1}{2}; qu) u^{n-\frac{3}{2}}, \tag{7}$$

where

$$\psi_{m,n} = g_{m,n}[4\sqrt{2}\pi B(n-1/2, 3/2)]^{-1}. \tag{8}$$

In the simplest case $g(\Phi)$ is proportional to Φ^n. If moreover $\lambda = p = 0$ then ρ is proportional to Φ^n and we recover the isotropic polytropes. If, on the other hand, $\lambda \neq 0$ but $p = 0$ then we find the quasi–polytropes studied by Camm (1952) and Idlis (1961).

Now let us consider generalised isochrone models. The potential of these models is

$$\Phi(r) = a(b+\zeta)^{-1}, \quad \zeta(r) = (1+a^2r^2)^{\frac{1}{2}}, \quad a = b+1. \tag{9}$$

Adopting $M/\frac{4}{3}r_e^3$ as the unit of the density, where M is the mass of the stellar system, we find the following density

$$\rho(r) = \frac{1}{3}a^3(b+3\zeta+2b\zeta^2)(b+\zeta)^{-3}\zeta^{-3}. \tag{10}$$

It is evident that the function $g(\Phi)$ can be represented in the form of the expansion (6), and $q = b/a$. In the simplest case of the Schuster model $b = 0 = q = 0$) the expansion (6) is reduced to a polynomial. We obviously recover a polynomial for $\psi(u)$:

$$\psi(u) = \frac{2\sqrt{2}}{\pi^2}[\lambda + \frac{8}{5}pu + \frac{16}{7}(1-\lambda)u^2 - \frac{64}{21}pu^3]u^{\frac{3}{2}}. \tag{11}$$

If $\lambda = 1, p = 0$ or $\lambda = 0, p > 0$ we obtain models considered in KK68b, resp KK68a.

In general the expansion (6) consists of six terms. Let us restrict ourselves for sake of simplicity to $\lambda = 0$. Then there will be only five terms in the expansion (6) and the coefficients will have a more simple form. The expansion is not unique of course. If we fix $n = 3$ then the coefficients of the expansion (2) will be the following:

$$\begin{aligned} g_{-1,3} &= \tfrac{1}{3}(1-b^{-2})pq, \quad g_{0,3} = \tfrac{1}{3}(1-b^{-2})(-2+5b^{-2}pq), \\ g_{1,3} &= \tfrac{1}{3}[2-3b^{-2}-b^{-2}(1-3b^{-2})pq], \quad g_{2,3} = \tfrac{1}{3}b^{-4}pq, \\ g_{3,3} &= \tfrac{1}{3}b^{-2}(1-b^{-2}pq). \end{aligned} \tag{12}$$

These coefficients are to be multiplied by the same factor $2\sqrt{2}\pi^{-2}$ for the transition to the expansion (7). It is evident that the first two hypergeometrical functions will yield only linear expression in u, so only three non–trivial hypergeometrical functions remain.

For the limiting model $b = \infty$, ($q = 1$) only one non–trivial hypergeometrical function will appear. We have

$$\psi(u) = \frac{4\sqrt{2}}{3\pi^2}[F(1,4,5/2;u) - 1 - p(1 - \tfrac{8}{5}u)]. \quad (13)$$

For the isochroneous model $b = 1$, ($q = \frac{1}{2}$) the solution is reduced to the formula (13.11) of KK68a. Three hypergeometrical functions can be found by recursion from

$$F(1,1,3/2,z) = \frac{\arcsin(\sqrt{z})}{\sqrt{z(1-z)}} = f(z), \quad z = qu \quad (14)$$

We find for $m = 1$, ($n = 3$)

$$F(1,4,5/2,z) = (1-z)^{-2}[\tfrac{3}{16}f(z)z^{-1} + \tfrac{7}{8} - \tfrac{3}{16}z] \quad (15)$$

and somewhat more complicated expressions for $m = 2$ and $m = 3$. The expressions for the full $\psi(u)$ has a similar structure. For the isochrone model, see KK68a. The complications for arbitrary q are not very essential.

The parameter q, which characterizes central concentration, also affects the velocity distribution of the models. For moderate values of q, the velocity distribution function in the central part of the model is close to Maxwellian. For larger values of q it becomes steeper. The parameter p, on the other hand, is a purely kinematic one. The radial elongation of the velocity distribution becomes larger when p increases and its central concentration becomes smaller.

The case $\mu = \frac{7}{2}$ is of some interest too. The integral equation for the function $\psi(u)$ is more complicated but the expression for the relation of the radial to transversal velocity dispersions is found to be very simple:

$$\frac{\sigma_r^2}{\sigma_t^2} = 1 + (\lambda + \rho\Phi)r^2. \quad (16)$$

Translator notes.

The above study was further developed in a series of papers by Veltmann (1979b, 1981, 1983). The distribution functions for generalized isochronous models (9), the two-parameter generalized isochronous models (see below) and the spherical models with the density law $\rho(r) = \rho_0[1 + (r/r_0)^2]^{-\beta}$ were studied there. In particular the limit values for $\xi = 0$ and asymptotics for $\xi \to \infty$ were found for the distribution function of the form $\Psi(x,\xi) = (x - \lambda\xi)^{\alpha}\psi(\xi)$ (suggested earlier by Veltmann) and the explicit expressions for the coefficients $\psi_{m,n}$, $g_{m,n}$ in (6), (7) were given there (if $n = 3$, $m = 0.5$) (Veltmann 1979b, 1981). Also the distribution function in the form

$$\Psi(\kappa,\xi) = \sum_{m,n} a_{m,n}\kappa^{n-\frac{3}{2}}(1+2\xi)^{-1-m-n}, \quad n \leq 4 \quad (17)$$

was considered by Veltmann (1981). If $\lambda = 0$ then the distribution function of the generalized isochronous models is expressed in elementary functions (Veltmann, 1983).

The case $p = 0$ corresponds to systems with an ellipsoidal velocity distribution (Merrit, 1985; Ossipkov, 1979; Turakulov, 1983). The explicit expression of $\psi(u)$ in this case was given by Malasidze (1987) for generalized isochronous models. The radial velocity dispersion was also found by him for $a \to 0$ (Parenago's potential), $a = 1$ (Schuster-Plummer's sphere), and $a \to \infty$.

Generalized isochronous models with the potential (9) were described by Kuzmin & Veltmann (1973) (see also Veltmann, 1979a) with ore details. Earlier the similar potential was used by Kuzmin & Malasidze (1969) for the study of plane galactic orbits. Some properties of these spherical models were also described by Ossipkov (1978), who has shown the stability of such models with the spherical velocity distribution (basing on Antonov's condition $d^3\rho/d\Phi^3 > 0$. The model $a \to \infty$ was also discovered by Hernquist (1990) who showed that it can approximate de Vaucouleurs' density profile. Again based on the potential (9), Kuzmin & Malasidze (1987) have constructed non-spherical models admitting a third and quadratic integral.

Veltann (1979a,b) has considered the two-parameter generalization of the isochronous model with the potential

$$\Phi(r) = \Phi_0 a(b^c + \zeta^c)^{-\frac{1}{c}}, \quad a/b < 1, \qquad \Phi(r) = \Phi_0[1 + (r/r_0)^c]^{-\frac{1}{c}}, \quad a/b = 1. \tag{18}$$

Malasidze (1981) has suggested another generalization. His potential consists of two terms: equation (9) and the square of (9) taken with various weights. Malasidze (1984) constructed the corresponding spherical models and showed their stability (according to Antonov's criterion).

At last we have to mention that Kuzmin *et.al.* (1986) studied the spherical models of mass distribution with the potential $\Phi(r) = \Phi_0 q^{-1} \ln(1 + b/\zeta)$ where $\zeta(r)$ is taken according to (9). The limiting case $q = 1$ coincides with Jaffe's model (1983).

References

Camm, G. L., 1952 *MNRAS*, **112**, 155.
Eddington, A. S., 1916, *MN*, **76**, 572.
Herquist, L., 1990, *Ap. J.*, **356**, 359-364.
Idlis, G.M., 1961, *Structurai dinamika zwiozdnykhsistem.* Izd. Akad. Nauk Kaz.SSR, Alma-Ata (Trudy Astrofyz. Inst. **1**.
Jaffe, W., 1983, *MN*, **202**, 995-999.
Kuzmin, G. G., Malasidze, G. A., 1969, *Tartu Publ.*, **38**, 181-250.
Kuzmin, G. G., Malasidze, G. A., 1987, *Tartu Publ.*, **52**, 48-63.
Kuzmin, G. G., Veltmann, Ü.-I. K., 1968a, *Tartu Publ.*, **36**, 470.
Kuzmin, G. G., Veltmann, Ü.-I. K. , 1968b, *Tartu Publ.*, **36**, 3.
Kuzmin, G. G., Veltmann, Ü.-I. K., 1973, *Tartu Publ.*, **40**, 281-323.
Kuzmin, G. G., Veltmann, Ü.-I. K., Tenjes, P. L., 1986, *Tartu Publ.*, **51**, 232-242.
Malasidze, G. A., 1981 *Bull. Acad. Sci. Georgian SSR*, **102**, 333-336.
Malasidze, G. A., 1984 *Proc. Georgian Politech. inst.*, **9**, no 279, 118-124.
Malasidze, G. A., 1997 *Abastumani Bull.*, **63**, 115-162.
Merrit, D., 1985, *AJ*, **50**, 1027-1037.
Ossipkov, L. P., 1978, *Astrofiz.*, **14**, 225-237.
Ossipkov, L. P., 1979, *Pis'ma v Astron. Zh.*, **5**, 77-80.
Turakulov, Z. Ya., 1983, *Astrofiz*, **19**, 791-801.
Veltmann Ü.-I. K., 1979a, *Astron. Zh.* **56**, 976-980.
Veltmann Ü.-I. K., 1979b,in *Star clusters* (ed. K. A. Barkhatova), Ural Univ. Press, Sverdlovsk, p. 50-71.
Veltmann Ü.-I. K., 1981, *Tartu Publ.* **48**, 232-261 (in English!).
Veltmann Ü.-I. K., 1983, *Astron. Zh.* **60**, 223-226.

BULGE AND DISK: A SIMPLE SELF-GRAVITATING MODEL

S.A. KUTUZOV and L.P. OSSIPKOV
St. Petersburg University, Russia

A method of finding the distribution function of some steady-state axially symmetrical mass models was suggested by Ossipkov, Kutuzov (1987). The models consist of a disk embedded in a bulge (halo). The total potential $\phi(R, z) = \phi_0\varphi(\xi)$, with $\phi_0 = \phi(0,0)$, $\varphi(\xi)$ arbitrary,

$$\xi^2 = \rho^2 + 2\mu|\zeta| + \zeta^2, \qquad \mu = const \geq 0 \tag{1}$$

and $(R; z) resp. (\rho, \zeta)$ dimensional resp. dimensionless cylindrical coordinates respectively. The total dimensionless configuration density has the following form:

$$\nu(\rho, \zeta) = \nu_b(\xi) + \delta(\zeta)\sigma_d(\rho) \tag{2}$$

where ν_b is the bulge density, σ_d is the disk surface density and $\delta(\zeta)$ is the Dirac delta function. The bulge lenslike equidensity surfaces coincide with equipotential ones. The parameter μ determines their flattening: they are spherical if $\mu = 0$ and flat if $\mu = \infty$. The following expressions are found for bulge and disk densities:

$$\nu_b(\xi) = 3\omega^2(\xi) + 2(\mu^2 + \xi^2)\frac{d\omega^2}{d(\xi^2)}, \qquad \sigma_d(\rho) = 2\mu\omega^2(\rho) \tag{3}$$

where $\omega^2(\xi) = -2d\varphi(\xi)/d(\xi^2)$, $\omega(\rho)$ is a dimensionless circular frequency. Both densities are connected with each other by means of the equations (3).

The distribution function has the following form:

$$\Psi(E, e, h) = \Psi_b(E) + \delta(\zeta)\delta(\mathcal{U}_\zeta)\Psi_d(e, h) \tag{4}$$

where $E = \varphi(\xi) - (\mathcal{U}_\rho^2 + \mathcal{U}_\theta^2 + \mathcal{U}_\zeta^2)/2$ and $e = \varphi(\rho) - (\mathcal{U}_\rho^2 + \mathcal{U}_\theta^2)/2$ are the energy integrals of spatial and flat motion respectively, $h = \rho\mathcal{U}_\theta$ is an integral of angular momentum, $\mathcal{U}_\rho, \mathcal{U}_\theta, \mathcal{U}_\zeta$ are dimensionless velocity components in cylindrical coordinates. The bulge does not rotate and its velocity distribution is isotropic. The bulge distribution function can be found as a solution of the integral equation:

$$\sqrt{8}\pi^2\Psi_b(E) = \frac{d^2}{dE^2}\left(\int_0^E (E - \varphi)^{\frac{1}{2}} G(\varphi) d\varphi\right) \tag{5}$$

Here $G(\varphi) = \nu_b(\xi(\varphi))$ is an augmented density (Dejonghe 1987) but the inverted potential law $\xi(\varphi)$ is supposed to be a single-valued function.
The disk phase density is decomposed into even and odd components with respect to the azimuthal velocity (and h):

$$\Psi_d(e, h) = \Psi_+(e, h) + \Psi_-(e, h) \tag{6}$$

The even component can be found on the basis of the surface density by known methods (Kalnajs 1976, Ossipkov 1978). In particular if it depends on the e only then (Dekker 1976)

H. Dejonghe and H. J. Habing (eds.), Galactic Bulges, 367–368.

$$2\pi\Psi_{+}(e) = \frac{dg(e)}{de} \tag{7}$$

where $g(\varphi) = \sigma_d(\rho(\varphi))$ is an augmented surface density. The odd component is determined by a rotation law $\langle \mathcal{U}_\theta \rangle = \rho\,\Omega(\rho)$. It is assumed to be separable in e and h, that gives rise to

$$\Psi_{-}(e,h) = u(e)h, \qquad 2\pi u(e) = \frac{d^2 f(e)}{de^2} \tag{8}$$

where $f(\varphi) = g(\varphi)\Omega(\rho(\varphi))$.

As an example we consider the particular cases of the Kuzmin-Malasidze (1969) potential law

$$\varphi(\xi) = \alpha\left(\alpha - 1 + (1+\kappa\xi^2)^{\frac{1}{2}}\right)^{-1}, \qquad \alpha,\ \kappa = const > 0 \tag{9}$$

and suggest the rotation law

$$\Omega(\rho(\varphi)) = a[1 - b\varphi^p(1-\varphi)^q]\varphi^\tau, \qquad a, b, p, q, \tau > 0 \tag{10}$$

which allows a rotation curve with one minimum and flat outer part. Some restrictions on the functions and the parameters are established from the condition of non-negativity of the $\Psi_d(e,h)$. The rotation velocity $\langle \mathcal{U}_\theta \rangle$ has to be considerably smaller everywhere than the circular one $\mathcal{U} = \rho\,\omega(\rho)$. In the case of $\alpha = \kappa\mu^2 = 1$ there is a pure Kuzmin-Toomre disk (Kuzmin 1953, Toomre 1963) without bulge. Then $\tau > 3/2$ and $\alpha \leq 0.30$ for $\tau = 1.6$, $b = 0$.

References

Dejonghe, H., 1987, *Inst.Adv.Studies Preprint*, IASSNS-AST 87/10
Dekker, E., 1976, *Phys.Rep.*, **24c**, 315
Kalnajs, A., 1976, *Ap.J.*, **205**, 751
Kuzmin, G.G., 1953, *Isv.Akad.Nauk Est.SSR*, **5**, 369 (Tartu Teated, No 1)
Kuzmin, G.G., Malasidze, G.A. 1969 (1970), *Tartu Publ.*, **38**, 181
Ossipkov, L.P. 1978, *Pis'ma Astr.Zh.*, **4**, 70
Ossipkov, L.P., Kutuzov, S.A. 1987, *Astrofizika*, **27**, 523
Toomre, A. 1963. *Ap.J.*, **138**, 385

ORBITAL ELEMENTS OF DIFFERENT GALACTIC POPULATION OBJECTS

L.P. OSSIPKOV and S.A. KUTUZOV
St. Petersburg University, Russia

1. Introduction

The study of prevalent orbits in galactic subsystems can help us understand galactic structure and clarify its history. The classical analysis of flat orbits and metallicities of old stars led Eggen *et al.* (1962) to formulate the rapid collapse of the primordial Galaxy. On the other side Yoshii & Saio (1979) studied three-dimensional orbits that separate in spherical coordinates. They found the Galaxy contracted quasi-stationary after the formation of halo objects. Here we shall briefly discuss the results of numerical orbit calculations (with Merson's method) for selected galactic subsystems. The axially symmetrical two-component model of the Galaxy (Kutuzov, Ossipkov 1989) was adopted. One-component models (Barkhatova et al. 1987, Kutuzov 1988) were used also but no significant difference in orbit elements was found (Kutuzov & Ossipkov 1992). Pericenter and apocenter distances, R_p and R_a, and the maximal height of objects over the galactic plane, z_m, were used as orbit elements as well as dimensionless quantities $e = (R_a - R_p)/(R_a + R_p)$ (eccentricity) and $c = 2z_m/(R_a - R_p)$ (the flatness of box filled by orbit projection on the meridional plane).

2. Results

Open clusters: This part of the work was initiated by late Prof. Barkhatova and some preliminary results were already published (Barkhatova et al. 1987). Orbits of 82 clusters were calculated. Heliocentric distances of clusters were taken according to Barkhatova's short "scale" (based on Kholopov's ZAMS), Hagen's "long" scale and some others. Corresponding changes of dimensionless elements (mainly c) were found to be small. All orbits were box ones. The average values of e, c are equal to 0.07, 0.50 respectively. Of all possible correlations of orbital elements and physical characteristics of clusters (mass, age, metallicity), the only significant turned out to be the correlation of e and [Fe/H]; the corresponding coefficient was equal to 0.65.

Orbits of globular clusters: The necessary data only for 13 cluster are available. Proper motions are taken from Cudworth (1974), earlier works of Meurers and Hallermann, Gamaley and others. Cluster distances and radial velocities were taken from Webbink and Hesser et al. As a rule the orbits were found to be almost hyperbolic (the cluster energy was positive for 5 from 11 clusters studied by Meurers and Hallermann). Most of the finite orbits are probably not box orbits and resemble the complicated figures found by Hayli (1965) and Innanen (1966). Possible ergodicity has no significance during the galactic lifetime. It seems that there is a week tendency of R_p to increase with [Fe/H] in the interval from -1 to 0.

H. Dejonghe and H. J. Habing (eds.), Galactic Bulges, 369–370.

Planetary nebulae: Data for the 49 most studied nebulae were used for the first orbit calculations. Proper motions and radial velocities ware taken from Cudworth (1974) and Perek & Kohoutek (1967) respectively. They used the "short" distance scale of Khromov (1985) with distances close to Sklovsky's ones (all orbits were found to be box orbits) and the "long" kinematical distance scale of Cudworth (1974) (one orbit was hyperbolic, two were probably tube orbits). The distributions of orbital elements e, z_m are multimodal. Maybe this shows that our sample is not homogeneous. The average values of e, c, z_m are equal to $(0.1 - 0.2)$, $(0.7 - 0.9)$ and $(0.7 - 0.9)$kpc respectively.

Short-periodic cepheids: We compiled data on proper motions and radial velocities of ca. 200 stars but the systematization of the material is necessary. At present we found orbits for 76 stars. Proper motions and radial velocities were taken according to Hemenway (1975); data of Clube et al. (1971) and some others were also used. Heliocentric distances were taken from Woolley et al. (1965).
We found that not only stars with retrogade motion but stars with hyperbolic or practically hyperbolic (the order of R_a is some hundred kiloparsecs) orbits are not rare. Many orbits are probably tube orbits and it is difficult to find elements e, c for them. The distributions of e, z_m, R_p were studied. The presence of some maxima of the two last distributions is to be mentioned. The relations between orbital elements and type and period (for RR Lyrae stars) were studied. The portions of hyperbolic, retrogade and probable tube orbits increase with the period as well as average values of e and z_m (for stars with $P < 0^d.4$ average e, z_m are 0.4 and 0.6 kpc and if $P > 0^d.6$ then the corresponding values are 0.6 and 0.7 kpc). Average e, z_m are 0.7, 7 kpc for RR Lyrae stars. As for four "shortest" RR Lyrae stars we have found $e = 0.4$, $z_m = 1$ kpc.
The investigation of relation between elements and the Preston metallicity index ΔS show that stars with $\Delta S = 5.6$ have the largest eccentricity 0.6 and the largest average z_m equal to 12 kpc.

References

Barkhatova, K.A., Kutuzov, S.A., Ossipkov, L.P., 1987, *Astr.Zh.*, **64**, 956
Clube, S.V.M., Aslan, Z., Russo, T.W., Clements, E.D., 1971, *ROB, n°*, **166**, 175
Cudworth, K.M. 1974, *A.J.*, **79**, 1384
Eggen, O.J., Lynden-Bell, D., Sandage, A.R., 1962, *A.J.*, **136**, 748
Hayli, A., 1965, *An.d'Ap.*, **28**, 49
Hemenway, M.K., 1975, *A.J.*, **80**, 199
Innanen, K.A., 1966, *Z.f.Ap.*, **80**, 445
Khromov, G.S., 1985, *Planetarnye tumannosti. Nauka*, Moscow
Kutuzov, S.A., 1988, *Kin.fiz.neb.tel*, **4**,39
Kutuzov, S.A., Ossipkov, L.P., 1989, *Astrono.Zh.*, **66**, 965
Kutuzov, S.A., Ossipkov, L.P., 1992, *In:Astron.-geodez.issl.,Ural Univ.Press, Ekaterinburg* (in press)
Perek, I., Kohoutek, L. 1967, *Catalogue of galactic planetary nebulae*, Academia, Prague
Woolley, R., Harding, G.A., Cassels, I. 1965, *ROB, n°*, **97**, E3
Yoshii, Y., Saio, H. 1979, *PASP*, **31**, 339

ORBITAL STRUCTURE OF TRIAXIAL EQUILIBRIUM MODELS OF VARIOUS SHAPES

S. UDRY and L. MARTINET
Geneva Observatory, CH-1290 Sauverny, Switzerland

1. Aim

To bring out the link between the observed shapes of triaxial dynamical systems (ellipticals, bulges, bars) and their orbital populations inferred from models.

2. Methods

1) We consider a series of triaxial systems of various shapes (elliptic, boxy, peanut, disky) built from N-body simulations of collisionless gravitational collapses. The used code is characterized by: a particle mesh algorithm; a polar grid exponentially spaced in the radial direction insuring a good central resolution; 200 000 particles; for details see Udry 1992 and Pfenniger & Friedli 1992.

2) We randomly select in each N-body configuration a set of about 2 000 particles and integrate the motion over more than $5 \cdot 10^9$ y in the fixed stabilized potentials.

3) We define parameters n_i as the numbers of sign changes of any coordinate i ($i = x$, y or z) along the orbit. We build histograms of n_i/n_j, normalized to the number of considered particles, and corresponding to the ratio ω_i/ω_j of orbital oscillation frequencies along the i- and j-axes.

4) We estimate the percentage of orbits with frequency ratios in various intervals corresponding to tube-, round-, boxy- or peanut-shaped orbits. The more populated intervals of frequency ratios and consequently the corresponding orbits may be related to the global shape of the models.

5) To establish a *shape – orbital population* relation, we calculate for each model the normalized fourth coefficient of Fourier analysis of the isophote departures from pure ellipses, a_4/a, which gives the degree of boxiness of the isodensity contours (Bender & Möllenhoff 1987). It is also compared to the model flattenings.

6) By means of Floquet theory, we have estimated ω_z/ω_x, the ratio of the normal frequency in the z direction to the orbital frequency of the x-axial orbit, in a peanut-shaped model and a boxy shaped model.

3. Results

1) Their is a clear correlation between the model shapes and their orbital populations parametrized by oscillation frequency ratios along the figure's main axes (n_i/n_j; Figure 1). Peanut-shaped models present a significant proportion of orbits with $n_i/n_j > 3/2$, whereas orbits in elliptical-shaped systems are rather characterized by $1 \lesssim n_i/n_j \lesssim 3/2$. This result implies a relation between Bender's

H. Dejonghe and H. J. Habing (eds.), Galactic Bulges, 371–372.

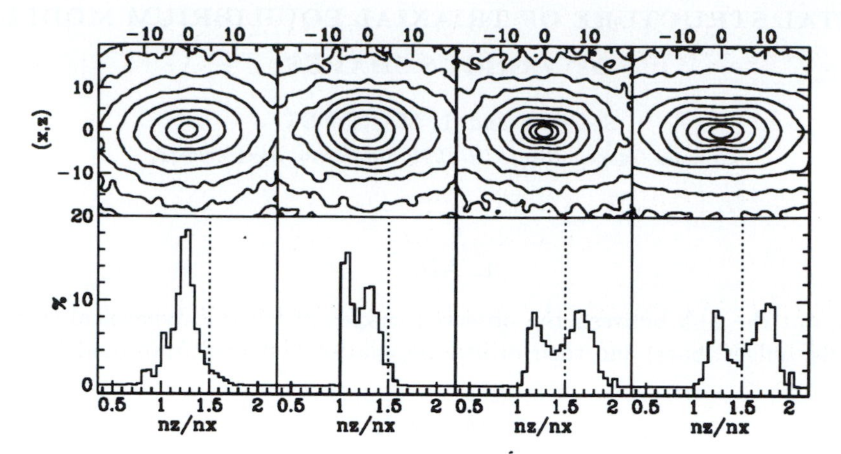

Fig. 1. Projected density contours in the (x, z) plane for an *elliptic-boxy-peanut* sequence of obtained N-body equilibrium models with corresponding normalized histograms of n_z/n_x values. Parameters n_i are the numbers of sign changes of the coordinate i ($i = x$ or z) along the orbit ($n_z/n_x \simeq \omega_z/\omega_x$).

a_4/a as a shape parameter and the percentage of orbits whose n_i/n_j values are in a particular interval, as an orbital population parameter.

2) We also observe a well defined relation between the degree of boxiness of the models and the configuration flattenings in the sense that the more flattened a model is, the more boxy- or peanut-shaped it looks like. This relation is due to the individual shape of resonant orbits whose vertical oscillation frequencies increase (n_z/n_x goes up) as their z-amplitudes go down.

3) The global model configuration depends on the sum of the effective populations of the existing orbital classes, and thus is related to the characteristics of the more numerous ones. In models with the particular oblate or prolate symmetries, the circular shape is induced by tube orbits in the corresponding planes. On the contrary, in triaxial models, tube orbits are not numerous and round shapes are rather given by resonant orbits with low n_i/n_j ($\sim$1).

4) The approximate percentage of box orbits is between 60 and 70% in strongly triaxial systems and less than 30% in prolate or oblate non rotating ones.

5) Finally, the vertical instability on the x-axial orbit, induced by the bifurcation with vertical banana orbit families in boxy- and peanut-shaped models, contributes to push motions out from the equatorial plane and thus to populate orbits determining the global shape of the system.

Details will be published elsewhere (Astronomy & Astrophysics).

References

Bender R., Möllenhoff C., 1987, A&A, 177, 71
Pfenniger D., Friedli D., 1992, submitted
Udry S., 1992, A&A, in press

GLOBALLY STABLE EQUILIBRIA OF COLLISIONLESS SELF-GRAVITATING MATTER

HEINZ WIECHEN, HARALD J. ZIEGLER
Ruhr-Universitaet Bochum, Theoretische Physik 4
D-4630 Bochum, Germany

ABSTRACT. We discuss the problem how to determine globally stable equilibrium states of collisionless self-gravitating matter with non-vanishing total angular momentum.

1. Introduction

We summarize the essentials of a techniques to calculate globally stable equilibrium states of collisionless matter with non-vanishing total angular momentum, that are gravitationally bound in all parts. The distribution functions of the resulting equilibria depend on Jacoby's integral. It can be seen that globally stable states cannot be rotationally symmetric for arbitrary fixed total angular momentum. With the help of a suitable relaxation model this techniques can be used to determine globally stable equilibria on macroscopic scales, which can be interpreted as final states of collisionless ("violent") relaxation (for details see Wiechen and Ziegler 1992).

2. Construction of globally stable equilibria

First we define the class of test functions. Let Φ_0 be the class of all distribution functions $f(\vec{r}, \vec{v})$ in 6-dimensional phase space meeting the following four constraints:

(i) All distribution functions $f \in \Phi_0$ correspond to the arbitrary fixed total angular momentum $\vec{L}_0 = L_0 \vec{e}_z$.

(ii) All states $f \in \Phi_0$ are accessible in collisionless, i.e. phase space density conserving dynamics. Thus all distribution functions $f \in \Phi_0$ correspond to the same arbitrary fixed memory function $g(\varphi)$ ($\Theta(x)$ denotes the unit step function)

$$\int \Theta(f(\vec{r}, \vec{v}) - g) d^3r\, d^3v = \varphi(g) \Rightarrow g(\varphi) \qquad \forall f \in \Phi_0. \qquad (1)$$

H. Dejonghe and H. J. Habing (eds.), Galactic Bulges, 373–374.

(iii) For simplicity and clearness we restrict the space of test functions to rotationally symmetric states in configuration space.
(iv) The distribution functions $f \in \Phi_0$ have to describe states that are gravitationally bound in all parts.

To determine the globally stable equilibrium out of the class Φ_0 we have to calculate the global minimum of the energy functional $E_{f,\psi}$

$$E_{f,\psi} = \int hf(\vec{r}, \vec{v}, t) d^3r d^3v + \frac{1}{8\pi G} \int (\nabla\psi)^2 d^3r \tag{2}$$

with $h = \frac{v^2}{2} + \psi$ (ψ: gravitational potential). We minimize the energy functional $E_{f,\psi}$ in two steps. In the first step we minimize $E_{f,\psi}$ with respect to f for arbitrary fixed ψ.

We construct the minimizing distribution functionf_ψ^* as a decreasing function of $\tilde{h} = v^2/2 + \psi - \lambda^2 r^2/2$ by an incompressible map in phase space (the fixed total angular momentum is considered by the Lagrangian multiplier λ, in equilibrium $\tilde{h}$ is equivalent to the Jacobi-integral). This yields

$$f_\psi^*(\tilde{h}) = g(\varphi_{\psi,\lambda}(\tilde{h})) \tag{3}$$

$(\varphi_{\psi,\lambda}(\tilde{h}) = \int \Theta(R_{\psi,\lambda}(z) - r)\, \Theta(\tilde{h} - (v^2/2 + \psi - \lambda^2 r^2/2))\, d^3r d^3v)$. $R_{\psi,\lambda}(z)$ denotes the boundary of the domain in configuration space where the system is gravitationally bound.

In the second step of the minimizing procedure we minimize with respect to ψ by variation (now with $f = f_\psi^*(\tilde{h})$). The result of that variational calculus is the following Euler-Lagrange equation.

$$\Delta\psi = 4\pi G\, \Theta(R_\psi(z) - r) \int f_\psi^*(\tilde{h}(\xi)) d^3v. \tag{4}$$

The globally stable equilibrium is given by that solution $\psi^*, f_{\psi^*}^*(\tilde{h})$ of (4) which corresponds to the global minimum of W_ψ.

It can be shown that the constraints (i)-(iv) may be so restrictive, that Φ_0 becomes the empty set. Then at least no rotationally symmetric, gravitationally bound stable equilibria corresponding to the fixed $g(\varphi)$ exist with the given total angular momentum. In that case, the globally stable equilibrium either has to be 3-dimensional (with uniform figure rotation) or it does not exist at all.

3. References

Wiechen, H. & Ziegler, H.J., 1992. *Astrophys. J.*, submitted.

MIXING IN COLLISIONFREE SYSTEMS

HARALD J. ZIEGLER, HEINZ WIECHEN
Ruhr-Universitaet Bochum, Theoretische Physik 4
D-4630 Bochum, Germany

ABSTRACT. Results of dynamical, numerical simulations of self-gravitating systems are used to show the existence of non-uniform mixing in phase-space.

The systems under consideration (3-D dynamics , vanishing total angular momentum) are collisionfree, so only phase space conserving states are accessible within the exact dynamics. This class of states Φ_0 can be described to be the set of all one particle distribution- functions $f(\vec{r}, \vec{v})$ leaving $g(\varphi)$ (the memory function) invariant.

$$\int \Theta(f(\vec{r}, \vec{v}) - g) d^3r \, d^3v = \varphi(g) \Rightarrow g(\varphi) \qquad \forall f \in \Phi_0. \tag{1}$$

Θ denoting the unit step function. A globally stable equilibrium state within collisionless dynamics is the minimum energy state of all $f \in \Phi_0$. For given Φ_0 non-equilibrium states are separated from globally stable states by an energy difference ΔE (Wiechen et al., 1988).
Due to filamentation on microscopic scales equilibrium states may be reachable on macroscopic scales. Microscopic filamentation and coarse-graining of f on macroscopic scales effects the "memory"-function $g(\varphi)$ in the following way (Wehrl, 1978)

$$\overline{M}(\varphi) = \int_0^\varphi \overline{g}(\varphi') d\varphi' \leq \int_0^\varphi g(\varphi') d\varphi' \quad \forall \varphi \tag{2}$$

with $\overline{g}(\varphi)$ being calculated from the averaged version of the distribution function $\overline{\varphi}(g) = \int \Theta(\overline{f}(\vec{r}, \vec{v}) - g) d^3r d^3v$. Because this "mixing" transformation of $g(\varphi)$ increases the energy of the corresponding lowest energy state (Ziegler and Wiechen, 1989) a globally stable state on macroscopic scales may be attainable and one can try to predict it by transforming the memory function $g(\varphi)$ obeying the mixing condition (2) to that extend, that the energy of the corresponding lowest energy state equals E_0. In that way the overall effect of mixing is calculated

H. Dejonghe and H. J. Habing (eds.), Galactic Bulges, 375–376.

self-consistently, but the special form of transformation (2) generally is unknown and additional assumptions are needed.
In the case of a strong and uniform filamentation of $f(\vec{r}, \vec{v}, t)$ with peaks of small width δ with typical distances $\tilde{\delta}$ and a macroscopic scale Δ obeying $\delta \ll \Delta < \tilde{\delta}$ the transformation of g corresponding to the coarse-graining of f reads

$$\overline{g}(\varphi) = \frac{1}{\alpha} g(\varphi/\alpha) \quad \alpha = \Delta/\delta \quad \delta, \Delta \to 0 \tag{3}$$

For systems initially being not to far away from an equilibrium state ($\Delta E/|E| < O(1)$) the theoretical predictions using assumption (3) are in very good agreement with the simulational results (see e.g. Ziegler, Wiechen 1989). This situation changes significantly if we go over to initial data very far from a stable equilibrium state (see also Nozakura, 1992). To show this and to clarify the reason we have performed several simulational runs using the TREECODE (Hernquist 1990).
$\overline{g}(\varphi)$ and $\overline{M}(\varphi)$ reconstructed from the particle data show that the final states are lowest energy states, being more mixed than the initial (exact) states according to relation (2). But the mixing turns out to be non-uniform. Especially for systems being initially very far away from an equilibrium state mixing is less efficient for small φ and more efficient for big φ compared to the assumption of strong and uniform mixing (equation (3)). For a lowest energy state (globally stable state) small φ correspond to high values of f and low values of h. So we can state, that the low energy part of the distribution function is less mixed and the high energy part is more mixed compared to the uniform case. Note that low energies typically correspond to particles in the core of the system while high energy particles are likely to be in the halo. Certainly the prediction for the final states can be improved significantly replacing (3) by a more appropriate operator. For the calculated examples a convolution of $g(\varphi)$ with e.g. a gaussian curve would help, but a sound motivation going beyond a good fitting reason is still missing.

References

Hernquist, L, 1990. *J. Comp. Phys.*, **87**, 137.
Nozakura, T., 1992 . *MNRAS* , **257**, 455.
Wehrl, A., 1978. *Rev. Mod. Phys.*, **50**, 221.
Wiechen, H., Ziegler, H.J. & Schindler, K., 1988 . *MNRAS* , **232**, 623.
Ziegler, H.J. & Wiechen, H., 1989. *MNRAS* , **238**, 1261.

ON THE FORMATION MECHANISMS FOR ELLIPTICALS AND BULGES

A.M. FRIDMAN and V.L. POLYACHENKO
Institute of Astronomy, Moscow, Russia

Abstract. We study the formation of elliptical galaxies and the bulges of disk galaxies as a result of the collisionless collapse of a rotating star cloud. At small amounts of rotation, this process is accompanied by the bar mode of the radial orbit instability slightly modified by rotation. We refer this case to (giant) ellipticals. For moderate rotation, when the radial orbit instability is suppressed, another mode takes over, which is the direct continuation of a strongly damping mode at the limit of almost radial orbits; it turns into a practically non-damping and long-lived mode (for many revolutions), and even a slowly rotating bar may eventually be formed. It is natural to refer this case to bulges and dwarf elipticals. Then spirals could be formed from the clouds with large amounts of rotation.

1. Introduction

There exists a very natural question: to what extent may the principal differences between galaxies (or their components) be attributed only to rotation. We drew attention to this problem a long time ago, see Polyachenko, Synakh, Fridman (1972) where we suggested to classify all galaxies according to their specific angular momentum. The general scheme might be like this:

giant ellipticals	dwarf ellipticals	spirals
slow bars	bulges	fast bars

$\longrightarrow$

specific momentum (L/M)

In this scheme, slow and fast bars are Lynden-Bell's bars and those in "standard" N-body simulations, respectively.
As the first step to solving the problem, it is useful to consider the possible eigenmodes in rotating gravitational systems. Here we study the eigenmodes of the simplest model of a rotating collisionless sphere with elongated orbits. In particular by studying the properties of modes in such a model, one may hope to understand what modes and instabilities can lead to formation of giant elliptical galaxies (practically, non-rotating) and bulges of spiral galaxies (with moderate rotation).

2. Modes of rotating spheres

Within the frame of our model (Polyachenko, 1989, 1992a), which represents strongly elongated orbits as rotating "needles", the simplest equilibrium distribution function for a rotating sphere is

$$f_0(\Omega, a) = F_0(\Omega, a)(1 + \mu \frac{J_z}{J}) \qquad (1)$$

where the function F_0 is the distribution function of a nonrotating sphere, Ω the precession velocity of an orbit-needle, a its size, J and J_z the orbital angular momentum and its z-component, respectively, μ a dimensionless parameter $(0 < \mu < 1)$

H. Dejonghe and H. J. Habing (eds.), Galactic Bulges, 377–379.

responsible for the rotation. Note that a distribution function of the same form (1) was first used by Synakh, Fridman, Shukhman (1971) in their study of a rotating sphere with circular orbits (but the sense of the distribution function in the case under consideration is different). It we also assume that the radial energy of each star is fixed (a =const), then one can easily derive the following dispersion relation for a small barlike perturbation of (1) (Polyachenko, 1992a,b)

$$\frac{3}{2}\left[Q^2 - 1 + x^2 g_-(x)\right] + \mu\left[-\sqrt{\pi}(x + \frac{1}{4x}) + (x^2 - \frac{1}{4})g_+(x)\right] = 0 \tag{2}$$

where we assumed for $F_0(\Omega, a)$ a Maxwellian distribution

$$F_0(\Omega) = exp(-\Omega^2/\Omega_T^2)/\pi\Omega_T^2, \tag{3}$$

and denoted by $x = \omega/2\Omega_T$ the dimensionless frequency (the time dependence of perturbation $\sim exp(-i\omega t)$, ω is the usual frequency). Further we have $g_\pm = f_- \pm f_+$, with

$$f_\mp = \int_0^\infty dt\ exp(-t^2)/(x \mp t), \tag{4}$$

with $Q = 2\Omega_T/\omega_G$ the dimensionless precession velocity dispersion, ω_G the growth rate in the limit of a cold ($\Omega_T = 0$) non-rotating ($\mu = 0$) sphere. The quantity Q is so defined that $Q = 1$ corresponds to a marginally stable model at $\Omega_T = 0$, $\mu = 0$; it is similar to Toomre's (1964) stability parameter Q in the stability theory of rotating disks. Table 1 shows the growth of critical values of Q (at the onset of the radial orbit instability) when the rotation of the sphere increases. Fig.1 shows real and imaginary parts of the frequency x against Q, for $\mu = 0, 0.1, 0.5$.

Table 1

μ	0.0	0.1	0.2	0.3	0.4	
Q_{cr}	1.000	1.145	1.220	1.280	1.335	
μ	0.5	0.6	0.7	0.8	0.9	1.0
Q_{cr}	1.385	1.430	1.475	1.520	1.560	1.600

3. Discussion

As is seen from Fig.1a and Table 1, the growing mode finishes at some critical value of Q: $Q = Q_{cr}(\mu)$, the latter being dependent on the amount of rotation (μ). In the region $Q < Q_{cr}$, the radial orbit instability (somewhat modified by rotation) dominates. Rather many facts are accumulated to date which support this scenario for the formation of elliptical galaxies (Polyachenko, 1981) as a dissipationless collapse accompanied by the radial orbit instability. All oscillations at $Q > Q_{cr}(\mu)$ are damping (see Fig.1b). But we would like to draw attention to

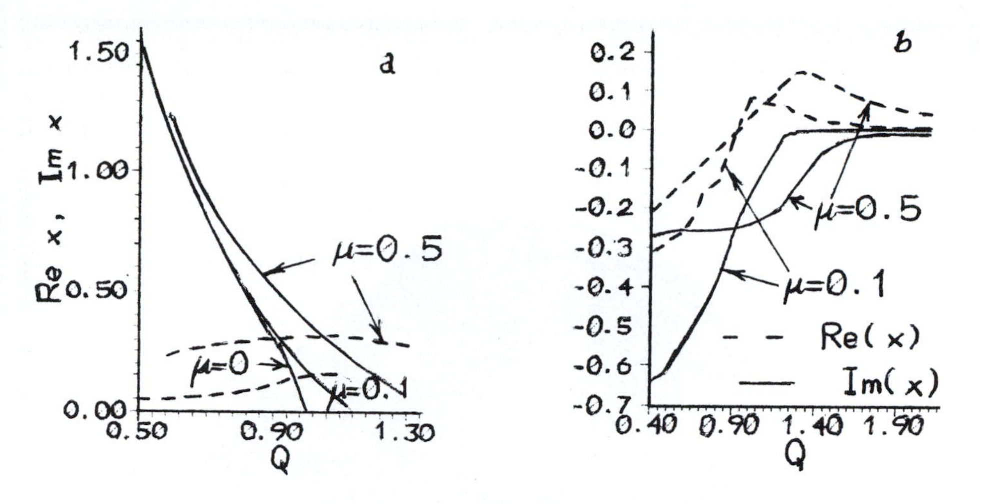

Fig. 1. Eigen frequencies (x) of the rotating sphere (1) against Q (a) growing mode, (b) damping mode

the fact that strongly damping eigenmodes at small Q's turn into practically non-damping modes at larger Q's when the radial orbit instability is suppressed. The decrement of the mode γ goes to values much less than the pattern's angular velocity Ω_p: at $Q > Q_{cr}$, $\gamma/\Omega_p = 2\sqrt{\pi}\Omega_p/\Omega_T \ll 1$. So in this limit we have a slowly rotating, long-lived bar. Once such a bar is formed due to suitable initial conditions, then it can exist for a long time. Possibly, such conditions occur when bulges are formed.

References

Polyachenko, V.L., Synakh,V.S., Fridman,A.M., 1972, *Sov.Astron.*, **48**, 1174

Polyachenko, V.L., 1981, *Sov.Astron.Lett.*, **7**, 142

Polyachenko, V.L., 1989, *Sov.Astron.Lett.*, **15**, 385

Polyachenko, V.L., 1992a, *Sov.Astron.*, **69**, to be published

Polyachenko, V.L., 1992b, *Astron.Tsirk*, to be published

Synakh, V.S., Fridman, A.M., Shukhman, I.G., 1971, *Reports of USSR Acad.Sci.*, **201**, 827

Toomre, A., 1964, *Astrophys.J.*, **139**, 1217

Valery POLYACHENKO

ANALYTICAL MODELS FOR ROTATING AND FLATTENED PERTURBATIONS IN BULGES

H. DEJONGHE and P. VAUTERIN
Sterrenkundig Observatorium, Gent, Belgium

Abstract. A power series approach for solving the linearized transport equation for perturbations in the central part of flat disks is presented. The application of this method is in principle independent of the mathematical complexity of the unperturbed distribution. As an illustration, this method was used to solve the transport equation in the case of Kalnajs' Omega models.

In this contribution, we want to show that solutions for the linearized Boltzmann equation for two-dimensional disks can be generated by substituting a series expansion form of the perturbed distribution. In combination with Poisson's equation, one can produce self-consistent models for rotating perturbations in the central part of a galaxy, such as bulges. More specifically, the velocity coordinates and the radius are expanded in a power series form, while the angular coordinate and the time appear in a uniformly rotating, harmonic perturbation.

In many strongly flattened rotating galaxies, most stars move on more or less circular orbits, so it is convenient to use these orbits as the zero point of the velocity expansions. The power expansion will then pertain to the description of small deflections from circular orbits. Moreover, since the radius is also expanded in a power series, one can easily see that the resulting distribution will be best suited for the central part of the galaxy. This is in contrast to other perturbation analysis techniques, such as the WKBJ approximation.

The distribution function and the equations of motion are first rewritten using a new coordinate $v'_\theta = v_\theta - r\Omega(r)$ (with $\Omega(r)$ the angular velocity for stars moving on circular orbits). The perturbing potential is has the form

$$V^m(r,\theta,t) = \sum_i a_i^m e^{i(m\theta-\omega t)}. \tag{1}$$

The integer parameter m indicates the order of the perturbing harmonic, while ω stands for the rotation speed. Our central assumption concerns the form of the perturbed distribution

$$df'(r,v_r,v'_\theta,t) = \sum_{i,j,k} p_{i,j,k} r^i v_r^j {v'_\theta}^k e^{i(m\theta-\omega t)}. \tag{2}$$

Since the equations will be linearized by neglecting second order terms in the perturbation, we are allowed to consider every order separately.

Filling in this form in the linearized transport equation generates a set of linear equations by collecting the terms with the same order. The coefficients of the perturbations are easily calculated by solving a subset of these equation recursively. However, the full set of equations is overdetermined. This is acceptable since general solutions of the transport equation can have an (unphysical) $\frac{1}{r}$ singularity in the centre, which can never be generated by a power series. It is possible to show that this limitation on the structure of the solution implies that the full set of equations only have a solution for a restricted set of perturbing potentials, i.e. if $a_i^m = 0$ for all

H. Dejonghe and H. J. Habing (eds.), Galactic Bulges, 381–382.

$i < |m|$. It is interesting to note that exactly the same restriction on the potential results from solving the Poisson equation for flat disks (Hunter 1963)

As a testcase, we used the previously described method to generate solutions for the linearized transport equation for perturbations in the case of the Omega models (Kalnajs, 1972), for which the full mode analysis has already been done analytically (ibid.). The proposed method should be equivalent to it, since the perturbing potentials are also polynomial. The thus calculated series expansion for the perturbed mass density turns out to be exactly the expansion of the analytic solution. Moreover, we checked that other important aspects of the distribution function are also identical.

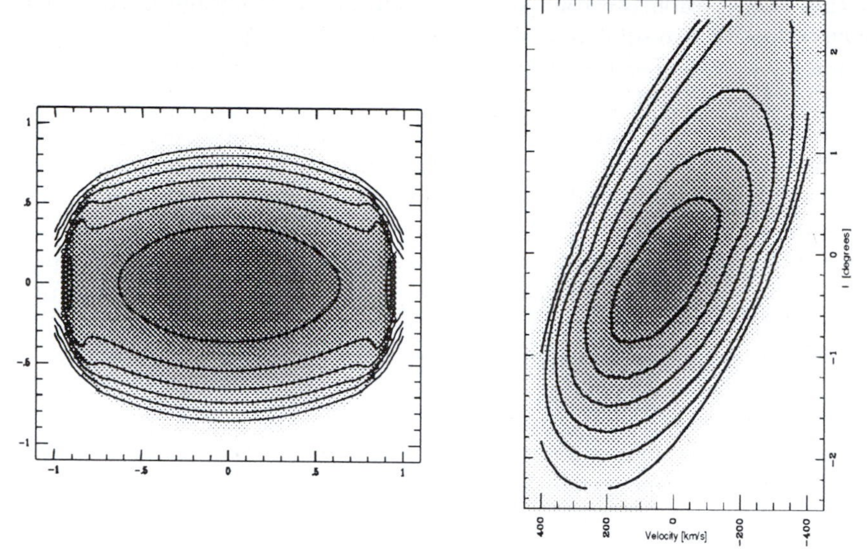

Fig. 1. Left panel: the surface density of a barlike perturbation (m=2). Right panel: the (l, v) diagram based on the barlike perturbation at the Galactic Centre, as seen from the sun. Notice that the rotation speed of the bar is different from the rotation of the bulge.

This series expansion strategy allows in principle to generate solutions for the linearized transport equation for flat perturbations in disks, regardless the mathematical complexity of the underlying unperturbed distribution. Moreover, the algorithm is easy to program and turns out to be very fast.

However, there is no guarantee that these series are convergent to the exact solution, although one can reasonably expect that there should be at least a small region with very small perturbed velocities for which this is the case. The fact that only a series expansion of the solution is generated is, in our opinion, a minor disadvantage, since most analytic solutions are so complex that the only way to handle them is to generate a series expansion anyway.

References

Dejonghe, H., 1984, *Ph. D. thesis (unpublished)*
Hunter, C., 1963, *MNRAS*, **126**, 299.
Kalnajs, A., 1972, *Astrophys. J.*, **175**, 63

TUMBLING INSTABILITY IN OBLATE ROTATING STELLAR SYSTEMS

E. SZUSZKIEWICZ, J.C.B. PAPALOIZOU, A. J. ALLEN and P.L. PALMER

School of Mathematical Sciences, Queen Mary & Westfield College, London

ABSTRACT. We present evolutionary calculations which show the tumbling instability found in spherical stellar systems by Allen *et al.*(1992) also occurs in some initially oblate rotating systems.

1. Introduction

Allen *et al.* (1992) have recently presented calculations for spherical systems with some degree of rotation obtained by reversing the z component of angular momentum, J_z, of some fraction of the stars. These showed a non-axisymmetric instability which resulted in a slowly tumbling triaxial bar with ratio of major to minor axis being about 1.5. In view of the potential importance of this instability for explaining the structure of elliptical and the bulges of spiral galaxies, we have extended these calculations to a variety of initially axisymmetric oblate rotating models to study the range of parameter space for which the instability occurs as well as the range of possible end states.

2. Initial models

The distribution function of the initial models was taken to be the sum of three components in the form

$$f = c_1(E_{01} - E)^{n_1} + c_2(E_{02} - E + \omega J_z)^{n_2} + c_3|J_z|^k(E_{03} - E)^{n_3+(k+3)/2}.$$

Here $k, E_{0i}, c_i, n_i (i = 1, 2, 3)$ and ω are constants. This leads to a mass density in cylindrical coordinates (r, z) given by

$$\rho = d_1(E_{01} - \phi)^{n_1+3/2} + d_2\left(E_{02} - \phi - \frac{1}{2}\omega^2 r^2\right)^{n_2+3/2} + d_3 r^k(E_{03} - \phi)^{n_3+(k+3)/2},$$

where $d_i, i = 1, 2, 3$ are new constants proportional to the c_i and ϕ is the gravitational potential. The first contribution is that of an isotropic polytrope, and the second corresponds to a polytrope rotating with constant angular velocity ω. The E_{0i} are chosen to limit the extent of any contribution to a fixed fraction of the surface equatorial radius and the d_i are chosen to fix the relative masses in the various contributions. The models which are invariant to reversing J_z for any number of stars are constructed by an iterative procedure and more details will be published elsewhere. As an illustrative example, we present evolutionary calculations for a model which contained $\frac{1}{3}$ of its mass in the first component with the remainder in the third with $k = 1$, the first extending up to $\frac{1}{2}$ of the full equatorial radius. The z component of angular momentum of $\frac{3}{4}$ of the particles was then reversed.

3. Numerical Results

The initial axisymmetric model had an axial ratio of 1.35 and was followed for 550 crossing times using the low noise numerical technique described in Allen *et al.* (1990). Here the basic time unit of one crossing time $\equiv 0.41(GMR^{-3})^{-1/2}$, M, R, and G being the total mass, equatorial radius and gravitational constant respectively. After about 150 crossing times the model developed a strongly triaxial

H. Dejonghe and H. J. Habing (eds.), Galactic Bulges, 383–384.

structure with ratio of major to minor axis of about 1.5. Figure 1 shows the axial ratios as a function of time. The rotating bar-like pattern had a period of 13 crossing times initially but this increased to about 35 crossing times by the end of the run. The rotational velocity curve at the end of the run is shown in figure 2. At this stage the model had ratio of maximum rotational velocity to dispersion velocity of about 0.5 and ellipticity of about 0.3, values which are typical of those found for bulges of spirals (Kormendy, 1981). There was some increase in the central condensation of the model the final density being $\propto r^{-2}$ over about two orders of magnitude.

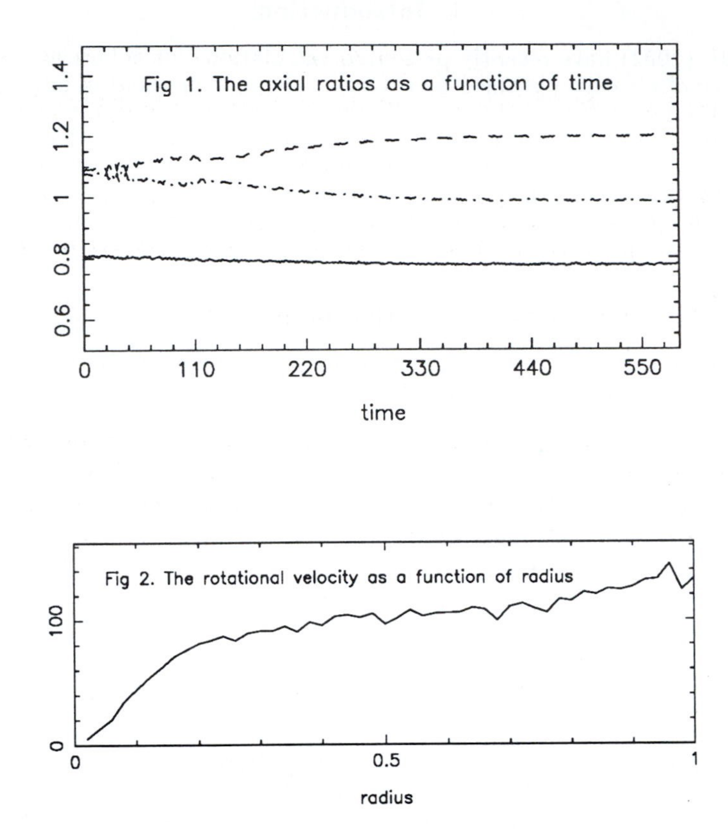

Although this model exhibited a tumbling instability similar to that found in Allen *et al.* (1992) not all oblate rotating models show this instability. For example uniformly rotating isotropic polytropes seem to be stable. More details of this work will be published elsewhere.

References

Allen, A.J., Palmer, P.L. & Papaloizou, J., 1990. Mon Not R astr Soc,**242**, 576.
Allen, A.J., Palmer, P.L. & Papaloizou, J., 1992. Mon Not R astr Soc,**256**, 695.
Kormendy, J., 1981. The Structure and Evolution of Normal Galaxies ed. Fall, S. M. & Lynden-Bell, D. , Cambridge University Press.
Papaloizou, J., Palmer, P.L. & Allen, A.J., 1991. Mon Not R astr Soc, **253**, 129 .

Investigating a self-consistent galactic potential with central mass concentration

H. HASAN

Space Telescope Science Institute, Baltimore, MD 21218, U.S.A.

J. A. SELLWOOD

Department of Physics and Astronomy, Rutgers University, P O Box 849, Piscataway, NJ 08855-0849, U.S.A.

and

C. A. NORMAN

Space Telescope Science Institute and Johns Hopkins University, Baltimore, MD 21218, U.S.A.

Abstract. The evolution of a barred galactic potential containing a central mass concentration is examined by means of a self-consistent 2-D N-body simulation. It is found that the bar weakens as the central mass grows and eventually dissolves, in agreement with earlier orbital studies of this problem.

Key words: barred galaxies, central mass concentration, orbits

1. Introduction

We pursue an earlier idea (Hasan and Norman, 1990; Hasan, Pfenniger and Norman, 1992) that the growth of a central mass in a rapidly rotating barred galaxy causes the bar to dissolve. This happens because some stellar orbits which support the bar become stochastic, while others are converted to orbits which are anti-aligned to the bar by the appearance and outward motion of an Inner Lindblad Resonance. Stable z orbits, which take stars outside the plane and could perhaps help in heating the disk stars into the bulge, were also found to exist. As a first step towards constructing fully self-consistent 3-D models of this phenomenon, we have used Sellwood's (1981) N-body code for simulations in two dimensions.

2. 2-D Simulations

All our simulations begin with 75% of the total mass represented by 50K particles in an axisymmetric Kuz'min/Toomre disk. The remaining mass is in the form of two concentric, unresponsive, Plummer spheres representing the bulge and core; initially both have a scale radius equal to one half that of the disk. The system is allowed to evolve self-consistently for 100 dynamical times, during which a bar forms and settles into steady rotation. Then during the next 4 bar tumbling periods (from $t = 100$ to $t = 150$), the scale radius of the core is gradually reduced to a value of 0.01 disk scale lengths after which it is held steady at its new value. The behaviour of the bar and disk are followed to at least $t = 200$. Simulations are performed for five different cases in which the mass of the core component is varied so that it contains a fraction x of the total mass, leaving a fraction $0.25 - x$ in the unchanging bulge.

H. Dejonghe and H. J. Habing (eds.), Galactic Bulges, 385–386.

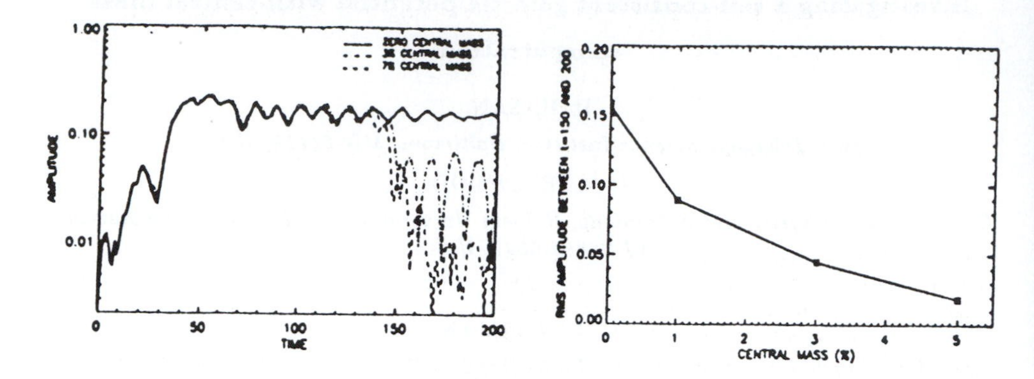

Fig. 1. Variation with central mass of (a) amplitude (b) rms amplitude of the $m = 2, \gamma = 0$ Fourier component. The weakening and destruction of the bar with increasing central mass is evidenced by a reduction in the amplitude.

3. Results

We find that when there is no central mass a steady bar remains till the end of the simulation. The bar weakens as the central mass is introduced, but thereafter it survives at a lower amplitude. The larger the fraction of mass in the core, the weaker the bar at late times. Figure 1(a) shows the time variation of the amplitude of the $m = 2, \gamma = 0$ Fourier component of the potential (which represents the bar) and Figure 1(b) the rms amplitude (between $t = 150$ and $t = 200$) for all the cases we have run. (The amplitude variations are caused by beats between the bar and weak spiral patterns in the outer disk which rotate at different rates.)

Surfaces of section were computed for various Hamiltonian values in a static potential approximated by the average of the potential computed for the period $t = 150$ to $t = 200$, for (a) no central mass and (b) 3% central mass. In the first case, invariant curves (IC) corresponding to B (or x_1) orbits which support a bar were found for all Hamiltonian values considered, while in the second case it was found that the ICs have dissolved for Hamiltonian values corresponding to tightly bound orbits, indicating a weakening of the bar in the presence of a central mass. We also see that some B orbits have been replaced by A orbits.

The present study supports our earlier results that a concentrated central mass ($\sim 7\%$ or greater than the total mass) would cause a bar to dissolve because some orbits become irregular and others switch from B orbits to A orbits which can no longer support the bar. It would be challenging to find observational evidence of this phenomenon as well as to study implications for the structure and stability of barred and elliptical galaxies as such central mass concentrations form.

References

Hasan, H. and Norman, C., 1990, *ApJ* **361**, 69.
Hasan, H., Pfenniger, D. and Norman, C., 1992, *ApJ*, submitted.
Sellwood, J., 1981, *Astron. Astrophys.* **99**, 362.

DELAYED FORMATION OF BULGES BY DYNAMICAL PROCESSES

DANIEL PFENNIGER
Geneva Observatory, CH-1290 Sauverny, Switzerland

September 29, 1992

Abstract. Two mechanisms involving purely dynamical processes can lead to the formation of a bulge after its disc: **1)** small bulges (1 – 2 kpc), including box-shaped bulges and mildly triaxial bulges, can result from the formation and destruction of a bar; **2)** big bulges (> 2 kpc) à la Sombrero can grow following the accretion of small satellites. Fully consistent N-body simulations show that the fraction of galaxy mass accreted in this way needs to be larger than about 5%. Less accretion does not create smaller bulges, but heats the whole disc. These dynamical effects transforming Hubble types from SB to SA and vice-versa over $\approx 1 - 2$ Gyr also indicate, by the secular growth of bulges, a general sense of galactic evolution from Sd to Sa.

Key words: Bulge formation - Galaxy evolution - N-body simulation

1. Constraints from Observations and Models

The existence of a large fraction of metal rich stars in the Galactic Bulge, discussed many times in the recent years (cf. Rich, this conference) suggests that the classical scenario of galaxy formation, in which bulges form before their disc, is not the unique possibility. Scenarios in which bulges grow from disc material have to be considered. Also a number of other facts concerning galaxies that were not clearly perceived before the 70's should now be taken into account:

1) Numerous bulgeless galaxies (Sm, Sd) do exist. The percentage of low surface brightness galaxies is probably underestimated due to sky light contamination. These galaxies are example of discs, sometimes extremely thin, able to form without bulge. Such discs should fit in the galaxy formation scenario. Late type galaxies typically lack of symmetry, contrary to early type ones. The earlier the galaxies, the more wound are the spiral arms, suggesting that late type galaxies are dynamically "young"; N-body models show that large scale irregularities imply rapid morphological changes with a time-scale of order of 0.2 Gyr. After several rotations an unpertubed disc increases its symmetry by winding up its spiral arms.

2) Small bulge galaxies also exist (Sc's). Such small bulges have a radius smaller than or similar to the disc scale-length. In several cases Sc bulges are bluish and rapidly rotating (Kormendy, this conference) suggesting a causal link with their disc. Yet small bulges can hardly be made by accretion after the formation of the disc without heating the disc to a similar "temperature". Indeed for such a scenario, falling in material needs not only to have an unrealistically small amount of initial angular momentum in order to hit directly the central region, but also to lose all its kinetic energy in less than half a crossing time. Otherwise at the subsequent rebounds the disc would inevitably be traversed and heated.

3) Decades of N-body experiments show that galactic discs are extremely "live" objects. Small perturbations on collisionless discs, such as the flyby of Magellanic satellites[1], or the inclusion of gas effects[2], often produce large evanescent spiral arms. Basically the disc fragility results from the Q stability parameter being near

H. Dejonghe and H. J. Habing (eds.), Galactic Bulges, 387–390.

to critical over a wide range of radii. Marginal stability can be maintained by a feedback mechanism between the general energy dissipation, due to gas cooling, and the heating due to massive star formation triggered by large scale gravitational instabilities[3]. Disc instabilities easily produce rotating bars. Also the gas consumption rate is often short. All these mechanisms are able to induce a secular evolution over a time-scale much shorter than 10 Gyr. Assuming steady galactic potentials over such long time-intervals now appears as unrealistic.

4) Galaxy catalogues show that truly axisymmetric disc galaxies constitute a minority ((SA:SAB:SB) $\approx$ (1:1:1)). Non-axisymmetric deformations occur especially often within the bulge region. Furthermore, SA's include edge-on or dusty galaxies for which axisymmetry is assumed, but increasingly disproved by observations peering through dust (e.g. Zarinsky et al., this conference). Now, according to several studies about bar dynamics[4–5], even small or weak bars induce effects far from negligible. Indeed bars imply the existence of numerous radial and vertical orbital resonances, especially inside corotation[6]. Therefore their interaction and their causal connection with bulges must be investigated.

2. Formation of Small Bulges

In recent years, two dynamical mechanisms able to build small bulges from a disc have been described in the literature. Both mechanisms require a bar.

A number of N-body studies in 3D have shown that dynamically cold discs at first produce, through a well documented global instability, a classical rotating bar. Then, for a wide range of initial conditions, a 2/1 vertical orbital resonance in the bar triggers another collective bending instability *perpendicular* to the galactic plane[7–11], lifting stars moving close to the plane up to $1-2$ kpc. The resulting bar is made of disc material and appears as a peanut-shaped bulge when the bar is seen edge-on, and as a round bulge when seen end-on. When no dissipative effect is included, this boxy structure is stable over more than 10 Gyr[8]. Furthermore, insensitive to particular initial conditions, 1) the surrounding stellar disc tends toward an exponential shape due to the relaxation induced by the bar, and 2) the bar region develops a steeper nearly exponential bulge-like profile[12–13,8].

Studies including small dissipative effects lead to considering the effect of condensing mass in the central regions of a bar. As the specific angular momentum is no longer conserved inside a bar, accretion proceeds much faster than in a disc. Furthermore a central mass concentration inside a bar always creates a number of radial and vertical resonances, in particular low order 2/1 resonances. It turns out that in fully self-consistent models a mass concentration inside the ILR of $1-3\%$ of the total stellar disc mass is sufficient to dissolve a strong bar[14]. Any process secularly accumulating mass near the centre leads to the dissolution of the bar[15–16]. Among the specifically studied mechanisms are artificial slow drag of a fraction of the N-body particles[14], satellite accretion by dynamical friction[1], and gas accretion by hydrodynamic SPH simulations[2]. The dissolving bar first takes a hotter and triaxial shape, and later on a spheroidal shape. This mechanism can therefore produce small round bulges made from disc material. In addition, the exponential disc and the steeper bulge profiles follow naturally.

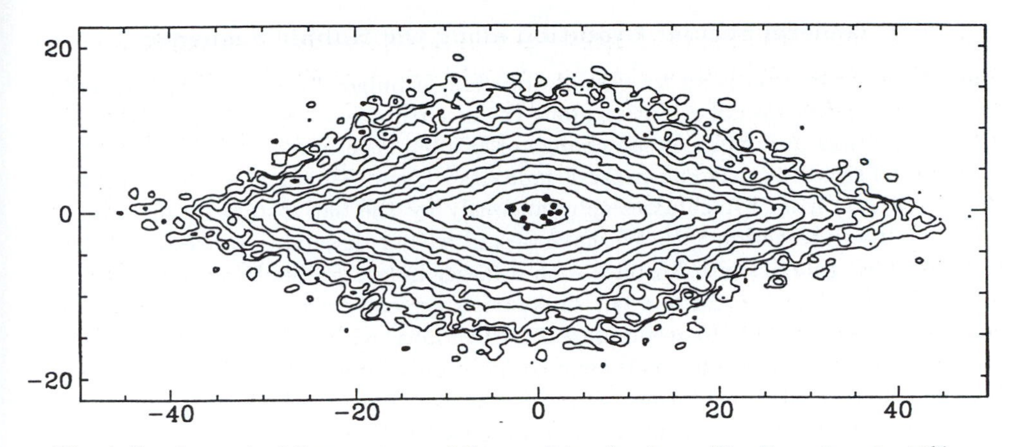

Fig. 1. Isophotes, in 0.5 mag steps, of the resulting Sombrero like disc when $f = 10\%$, seen edge-on. The ten satellites are marked by dots near the centre. The length unit is the kpc.

3. Formation of Big Bulges

The bulge class includes also big members, such as the Sombrero bulge. Such bulges, having a characteristic radius much larger than the disc scale-length can not result from disc material heated up in z by internal instabilities such as bars, since the required energy exceeds the total mechanical energy that the disc and a possible bar can supply. But mergers falling in from large distances are an interesting source of large energy per unit mass. We know that equal sized mergers destroy discs and typically produce an ellipsoidal object[17]. Therefore it is natural to ask whether the cumulated accretion of small satellites can result in a galaxy type intermediate between a spiral and an elliptical, i.e. an early type spiral.

To quantify this scenario, we have performed a series of 3D collisionless PM N-body simulations with a polar grid and $N = 2 \cdot 10^5$. Ten equal point mass satellites, without systematic rotation but supported by kinetic pressure start within a 30 kpc radius sphere, and slowly accrete by dynamical friction toward the centre of a disc. The initial Miyamoto disc has a scale-length of 4 kpc and a scale-height of 1 kpc, is in equilibrium, but is slightly bar unstable, as in [10]. Unperturbed it produces a peanut-shaped bulge-bar. The integration time amounts to 4 Gyr. The only varied parameter is the ratio f of satellite mass to galaxy mass.

Below $f = 2\%$, the overall effect is negligible, also because dynamical friction is insufficient to drag the ten satellites close to the center within 4 Gyr. The peanut-shaped bar is not significantly changed. Between $f = 3\%$ and $f = 5\%$, the satellites destroy the peanut shape of the bar, but not the bar itself; the bar becomes more oval. In this case, the persistence of the peanut shape traces the *absence* of merging events (and vice versa). Around $f = 10\%$, the bar is destroyed and the disc is significantly heated into a characteristic Sombrero like bulge with pointed isophotes along the disc (Fig. 1). Here all the mass in the big bulge comes from the disc. The bulge is hot and consequently rotates moderately. Above $f = 10\%$, the disc is more and more damaged and finally destroyed into an ellipsoidal object.

4. General Secular Evolution along the Hubble Sequence

Since bulge formation is an irreversible process, if bulges form progressively *after* their disc, the direction of evolution along the disc Hubble sequence is from Sd to Sa. The recognition that galaxies can change of Hubble type in less than 10 Gyr is not new. Equal sized mergers can make E galaxies[17], and SB galaxies can be formed (in ≈ 0.2 Gyr) from SA galaxies spontaneously by the bar instability or induced by a satellite interaction[18,1]. We have described bar dissolving processes able to transform SB galaxies into SA ones with a larger bulge. A dissolving bar produces a bulge of similar size. Such a bar dissolution typically occurs as quickly as the accreting process is able to accumulate a critical mass within the bar ILR. Directly, repeated satellite accretion events tend to make only large bulges. Indirectly, satellites can boost central evolution by first inducing a bar by tidal interaction which can be eventually destroyed into a small bulge a few Gyr later.

The other properties of the Hubble sequence also suggest qualitatively that the sense of evolution is from Sd to Sa. The rotation speed and surface luminosity increases can result from the virial theorem applied to dissipative systems (present kinetic energy = dissipated mechanical energy). The degrees of symmetry and spiral arm opening can be understood by slow dynamical effects, because discs need several rotation periods to organize themselves. The gas fraction, the metallicity, and the colour can be attributed to the irreversible stellar activity processes.

In this picture the Sm, Sd galaxies are closer to proto-galaxies than Sa's. Since late type galaxies are dynamically young, galaxies do not "form" necessarily only at an early brief epoch, but over an extended time interval, perhaps even up to now. Galaxy formation and galaxy evolution can not be separated, since late type galaxies can result from early type ones.

Several ideas in this work owe their origin to the influential lectures given by John Kormendy in the 1982 Saas Fee Course.

References

[1] Pfenniger, D.: 1991, in *Dynamics of Disc Galaxies*, ed. B. Sundelius, Göteborg, 191
[2] Friedli, D., Benz, W.: 1992, *A.A.* in press
[3] Kennicutt, R.C.: 1989, *Ap.J.* **344**, 685
[4] Athanassoula, E., Bienaymé, O., Martinet, L., Pfenniger, D.: 1983, *A.A.* **127**, 349
[5] Contopoulos, G., Papayannopoulos, Th.D.: 1980, *A.A.* **92**, 33
[6] Pfenniger, D.: 1984, *A.A.* **134**, 373, and 1985, *A.A.* **150**, 112
[7] Combes, F., Sanders, R.H.: 1981, *A.A.* **96**, 164
[8] Combes, F., Debbasch, F., Friedli, D., Pfenniger, D.: 1990, *A.A.* **233**, 82
[9] Friedli, D., Pfenniger. D.: 1990, in *Bulges of Galaxies*, ESO Conference and Workshop Proceedings No. 35, eds. B.J. Jarvis, D.M. Terndrup, 265
[10] Pfenniger, D., Friedli, D.: 1991, *A.A.* **252**, 75
[11] Raha, N., Sellwood, J.A., James, R.A., Kahn, F.D.: 1991, *Nature* **352**, 411
[12] Hohl, F.: 1971, *Ap.J.* **168**, 343
[13] Pfenniger, D.: 1990, *A.A.* **230**, 55
[14] Friedli, D., Pfenniger, D.: 1991, in *Dynamics of Galaxies and Molecular Clouds Distribution*, IAU Symp. No 146, eds. Casoli F., Combes, F., Dordrecht: Kluwer, 362
[15] Hasan, H., Norman, C.: 1990, *Ap.J.* **361**, 69
[16] Pfenniger, D., Norman, C.: 1990, *Ap.J.* **363**, 391
[17] Barnes, J.E.: 1992, *Ap.J.* **393**, 484
[18] Noguchi, M.: 1987, *M.N.R.A.S.* **228**, 635

PEANUT SHAPED BARS

J. A. SELLWOOD
Department of Physics and Astronomy, Rutgers University
P O Box 849, Piscataway NJ 08855-0849, U.S.A.

ABSTRACT. Recent 3-D N-body simulations have shown that the bars formed through the well-known global instability of discs later acquire a pronounced peanut shape when seen from the side. This paper reports a new calculation of the phenomenon using a finer grid and a slightly different initial model. The final peanut shaped bar, when viewed from within the disc, has an appearance remarkably similar to that of the Milky Way bulge. When viewed as a distant edge-on galaxy, an observer would see cylindrical rotation in the "bulge".

1. Model and Evolution

Combes & Sanders (1981), Combes *et al.* (1990) and Raha *et al.* (1991) have reported 3-D N-body simulations of barred disc galaxies in which the bar always acquired a peanut shape when viewed from the side. The new experiment described here started from a $Q = 1.2$, axisymmetric Kuz'min-Toomre disc represented by 50K particles containing 70% of the total mass; the remaining 30% is in a rigid Plummer sphere which has half the scale length of that of the disc. (The earlier experiments by Raha *et al.* showed that the behaviour studied here is not greatly affected by this rigid bulge/halo approximation.) The model was evolved using a 3-D Cartesian particle-mesh code having $127^2 \times 31$ cubic grid cells (double the resolution used by Raha *et al.*).

As expected, the model first formed a bar which then began to bend out of the plane through the fire-hose instability. After reaching a bend angle of some 20°, the bar abruptly regained symmetry about the disc plane, but was quite a bit fatter normal to the plane and noticeably shorter. The bar is thickest at its ends, giving it a pronounced peanut shape when viewed from the side, but the thickness of the disc at larger radii is not much affected by the bending of the bar.

2. "Observations" of the Final Model

The peanut shape of the object is visible from viewing angles in the plane which are greater than about 30° to the bar major axis, but at even smaller angles, the

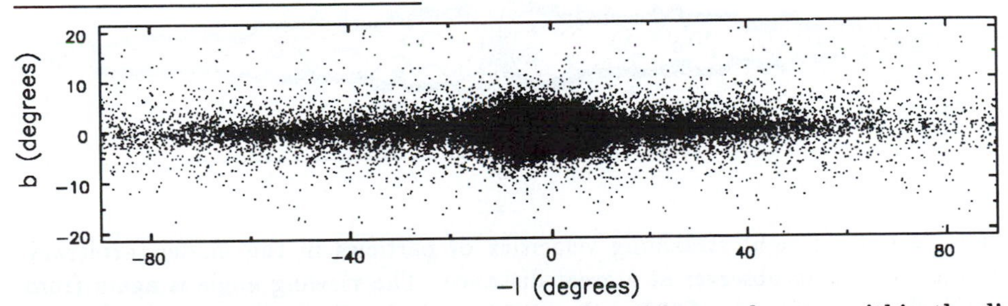

Figure 1 The sky projection of the model as "seen" by an observer within the disc looking towards the inner galaxy at an angle of 30° to the bar major axis.

H. Dejonghe and H. J. Habing (eds.), Galactic Bulges, 391–392.

object still appears boxy. Given the full 3-D spatial and velocity information, it is possible to "observe" the model from any arbitrary angle to deduce quantities to be compared with observational data.

2.1 AS A MODEL OF THE MILKY WAY

Figure 1 shows the sky projection of the model from a point at a distance from the centre approximately four times the bar semi-major axis, and at an angle of 30° to that axis. This galactocentric distance was chosen such that the two lobes of the peanut subtend a longitudinal angle of approximately 10°. If the observing point is moved closer to the bar major axis the bulge loses some of its peanut appearance, becoming more box-like, and I would have to move the observing point inwards for the box feature to subtend the same angle; this would also magnify the longitudinal asymmetry. The resemblance to the bulge of the Milky Way is very striking, even the asymmetry in the *b* extent of the bar between the near and far sides is comparable. Apart from obvious aspects such as the absence of obscuration in the model, the comparison could be improved still further by scaling the intensity of each point as the inverse square of its distance.

2.2 AS A MODEL OF A DISTANT GALAXY

Alternatively we can project the entire model to mimic the view from a great distance; the projected velocity field of the particles is shown in Figure 2. The absence of a dark halo does not affect the kinematics of the bulge region, but gives the model a slightly declining outer rotation curve which is somewhat unrealistic. One of the most striking features of Figure 2 is that the bulge-like object possesses apparent cylindrical rotation, as has been observed (*e.g.* Shaw *et al.* 1992).

References

Combes F & Sanders R H, 1981. *Astron. Astrophys.* **96**, 164.

Combes F, Debbasch F, Friedli D & Pfenniger D, 1990. *Astron. Astrophys.* **233**, 82.

Raha N, Sellwood J A, James R A & Kahn F D, 1991. *Nature* **352**, 411.

Shaw M, Wilkinson A & Carter D, 1992. Preprint

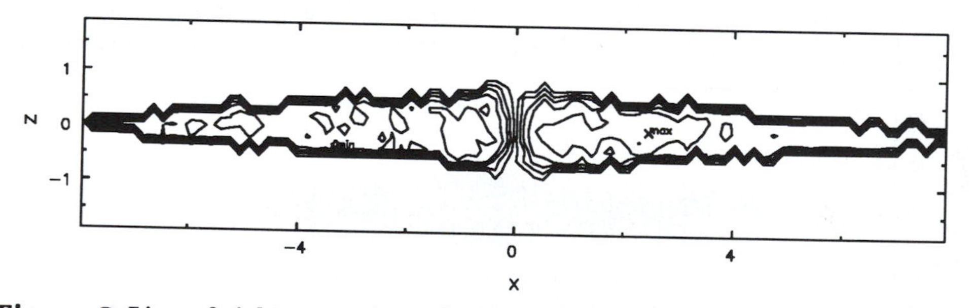

Figure 2 Line of sight streaming velocities of particles in the model (arbitrary scale) as seen by an observer at a great distance. The viewing angle is again from the mid-plane at an angle of 30° to the bar major axis. The value of the streaming velocity has been set to zero in any bin containing fewer than 10 particles.

The Influence of a Halo on the Evolution of Elliptical Configurations

F.THEEDE and K.O.THIELHEIM

Institut für Reine und Angewandte Kernphysik, Abt. Mathematische Physik
Christian-Albrechts-Universität zu Kiel, Otto-Hahn-Platz 3, Germany

Abstract. Results of n-body simulations of self-gravitating systems containing 100,000 particles are presented. A program based on the particle-mesh method is used to compute the evolution of a hot stellar-disk till an oval configuration at the center is formed. The system is evolved again from this point onwards a) with halo, b) without halo, and c) with the halo but with total mass of the system reduced to that of case b. Results are presented in form of a diagram showing the development of ellipticity.

1. The Program

The program is based on a two-dimensional particle-mesh method in a polar grid (r, ϕ) with exponentially increasing radial coordinates (Miller 1976, Sellwood 1981):

$$r = le^{\alpha u} \quad , \quad \phi = \alpha v \quad , \text{and} \quad \alpha = \frac{2\pi}{v_{max}} . \tag{1}$$

It has been vectorized for the CRAY XMP in Kiel. For calculating the potential we are using a FFT method. The softening parameter is chosen to be $s = 1.2\,kpc$.

The parameters of the grid are :

$$l = 0.2856\,kpc \quad , \quad v_{max} = 64 \quad , \text{and} \quad u_{max} = 60 . \tag{2}$$

2. The Initial Configuration

The simulation is started with 100,000 stars of total mass $M = 10^{11} M_\odot$ in a homogeneous disk of radius $10\,kpc$ where $M_\odot$ is the solar mass.

The angular velocity of a star is given by

$$\omega(r) = \kappa \sqrt{\frac{GM_r}{r^3}} , \tag{3}$$

where $0 \leq \kappa \leq 1$ is the relative reduced initial angular momentum (Thielheim 1990), G the gravitational constant and M_r the integral mass inside the radius r. The initial velocity dispersion is chosen to be isotropic–gaussian, with

$$\left| \frac{2K}{W} \right| = 0.0025 , \tag{4}$$

where W is the total potential energy and K the total kinetic energy.

3. The Simulations

For defining an ellipse and calculating the ellipticity a method based on determining the diagonalized tensor of inertia (I_{kl}) is used. The ellipticity in this case is defined as

$$e = \max \begin{cases} 1 - \sqrt{I_{11}/I_{22}} & : \quad I_{11} < I_{22} \\ 1 - \sqrt{I_{22}/I_{11}} & : \quad I_{22} < I_{11} \end{cases} \tag{5}$$

H. Dejonghe and H. J. Habing (eds.), Galactic Bulges, 393–394.

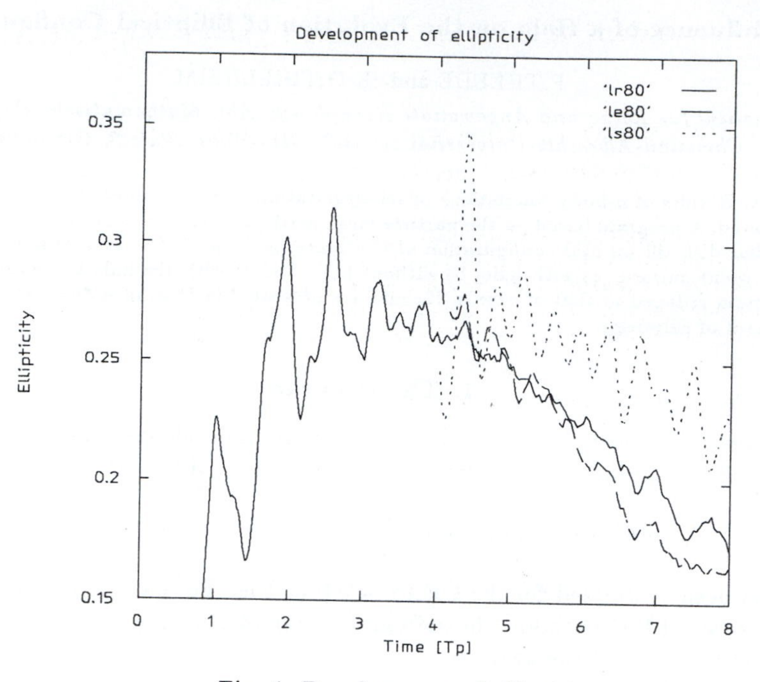

Fig. 1: Development of ellipticity

At the time $4\,T_p$ the elliptical configuration formed in $lr80$ is cut out and simulated as $le80$ till $8\,T_p$ both shown in fig. 1. If the corresponding number of stars is taken away at random from $lr80$ at $4\,T_p$ the subsequent evolution of $ls80$ exhibits a development of ellipticity which is also shown in fig. 1.

Comparing these results it is obvious that the influence of the halo on the evolution of the system is modest.

References

Miller, R.H., 1976, *J.Comp.Phys.*, **21**, 400
Sellwood, J., 1981, *Astron. Astrophys.*, **99**, 362
Thielheim, K.O., 1990, *Astrophys. Space Sci.*, **170**, 381

SPH Simulations of the Chemical and Dynamical Evolution of the Galactic Bulge

T. TSUJIMOTO, K. NOMOTO, T. SHIGEYAMA and Y. ISHIMARU
Department of Astronomy, University of Tokyo, Bunkyo-ku, Tokyo 113, Japan

Abstract. We simulate the chemical and dynamical evolution of the galactic bulge with the smoothed particle hydrodynamics (SPH) method. We calculate the early phase of galaxy formation in which the bulge is formed through a burst of star formation. The calculated abundance distribution function of stars in the bulge is consistent with the observations of bulge K giants, if the heavy element yields are three times larger than those expected from Salpeter's IMF.

Key words: galaxies: dynamics — abundances — supernovae

The observed abundance distribution function of bulge K giants (Rich 1990) shows the presence of stars with a wide range of metallicity ($-1 \leq$[Fe/H]$\leq +1$). Matteucci & Brocato (1990) constructed one zone chemical evolution models for the bulge and found that the observed distribution function is reproduced only if the timescale of galaxy collapse is as short as 10^7yr. This seems to be too short to be compatible with the free fall timescale, thus indicating the necessity of realistic hydrodynamical simulations. We construct three dimensional models for the chemical and dynamical evolution with SPH, and calculate the collapse of a protogalaxy until a burst of star formation ceases.

Our model has three components, i.e., gas, dark matter and stars, with the Jeans criterion of star formation. During collapse, the thermal and chemical evolutions of gases and stars are calculated by taking into account radiative cooling and supernova heating as well as enrichment of heavy elements due to supernova explosions.

The results are as follows. The core which consists of stars and gas are formed in the central part of dark matter at $\sim 2 \times 10^8$yr (Fig.1). A burst of star formation occurs in a dense gaseous core where radiative cooling is very effective. Massive stars then explode as type II supernovae and rapidly contaminate the core with heavy elements (Fig.2). The metallicity [Fe/H] at the peak of the abundance distribution function of stars in the core is lower than that of bulge K giants, if the yield expected from Salpeter's IMF is used. With the three times lager yield, the average metallicity of gas in the core becomes about twice the solar abundance. Some stars are formed from gases of more metal rich than the average due to the inhomogeneous metal contamination. To reproduce the abundance distribution function of bulge K giants as seen in Fig.3, it is very important to consider the inhomogenity which does not appear in one zone models. In addition, our model predicts that the heavy elements in the bulge stars are mostly produced from type II supernovae. This is consistent with the observations in which the abundance ratios between the α-elements and Fe in the bulge M giants are significantly higher than the solar ratios (Rich 1992).

References

Matteucci F., Brocato E., 1990, *ApJ*, **365**, 539.
Rich R.M., 1990, *ApJ*, **362**, 606.
Rich R.M., 1992, in *IAU Symposium 149, The Stellar Populations of Galaxies* (Kluwer), p. 29.

H. Dejonghe and H. J. Habing (eds.), Galactic Bulges, 395–396.

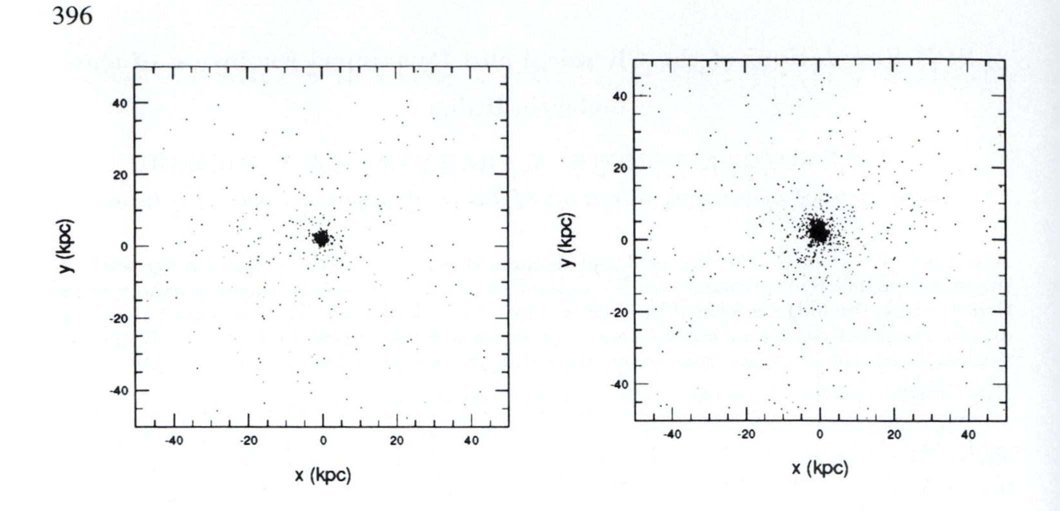

Fig. 1. Distributions of gas (left) and dark matter (right) at 2.3×10^8 yr.

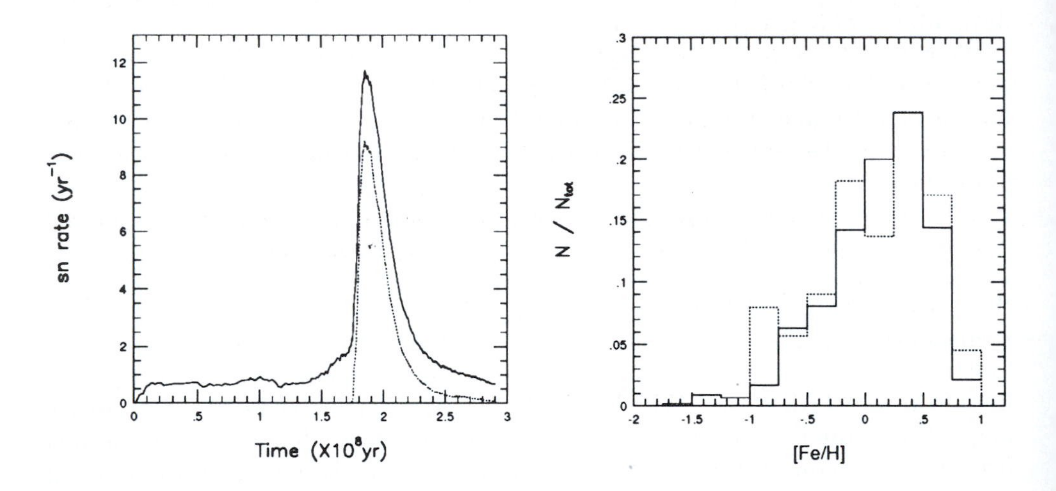

Fig. 2. Supernovae rate vs. time. The solid and dotted lines show the total supernova rate, and the rate in the core within 2 kpc, respectively (left).

Fig. 3. Abundance distribution function of stars in the core (solid) as compared with that of bulge K giants (dotted:Rich 1990) (right).

PRIMEVAL STARBURST AND BULGE FORMATION

Keiichi WADA *and* **Asao HABE**
Department of Physics, Hokkaido University, Sappro 060 Japan

1. Introduction

Starburst phenomena in galaxies should have been common in the epoch of galaxy formation. How did the primeval starbursts affect the structure or the properties of the host galaxies?

Recently, some high-redshift radio-galaxies with a superwind and a secondly formed stellar system along radio lobe axis is observed(McCarthy *et al.* 1992). It is suggested that the expanding super bubble in a gaseous halo triggers the formation of the stellar system(Heckman *et al.* 1990).

Many hydrodynamical simulations of superwind originated in multiple supernovae showed that the superwind interacts with the ambient gas and forms a hot cavity and a cooled dense gas shell (e.g. Tomisaka and Ikeuchi 1988). Star formation in the gas shell should occur, if the gas shell radiatively cools and becomes gravitationally unstable(McCray and Kafatos 1987). We can interpret that part of the stellar systems observed in high-redshift galaxies were secondly formed during these process.

According to our estimate, the stellar shell secondly formed by super bubble associated with the primeval starbursts could have several $10^{10} M_\odot$ and several *kpc* radius. We study the dynamical evolution and the effects of the relaxation of such stellar system on the structure of host galaxies. In the present paper we aim to clear basic physics of the evolution of the (shell + disk) stellar system by 3-D N-body simulations.

2. Method and Models

2.1. METHOD

The N-body calculations were performed on the workstation with GRAPE-3. GRAPE(GRAvity PipE) is a series of special-purpose computers to simulate many-body systems, developed at University of Tokyo (Sugimoto *et al.* 1990). GRAPE calculates directly the gravitational force between particles, and the host computer performs all other calculations such as time integration.

2.2. MODELS

The stellar shells were calculated under two type of environments; *(a)in a fixed external disky potential(Miyamoto and Nagai 1975): (b)in a thin stellar disk.* The initial shells have uniform surface density; no rotation; and

H. Dejonghe and H. J. Habing (eds.), Galactic Bulges, 397–398.

thin spherical shape. Parameters are mass of shell(M_s) and initial velocity dispersion of the shell stars(σ_s).

3. Summary of our results

1. The relaxation processes of the shells differ depending on M_s and σ_s and so do the finally formed bulge;
   (a) *Type I*: Relaxation by phase-mixing caused by the disk potential. A large and vague bulge is formed. The relaxation time is greater than $10^9 yr$.
   (b) *Type II*: Relaxation dominated by self-gravity of the shell under the influence of the disk potential.
   A compact core with a diffuse halo is formed in a following way($t <$ *several* $10^8 yr$);
      i. Cylindrical/torus structure appears in the early stage of the relaxation.
      ii. A few clumps are formed in this expanding torus at the disk plane by gravitational instability.
      iii. They merge and form the bulge.
   (c) *Type III*: Relaxation dominated by self-gravity. A triaxial bulge is formed.
2. The structure of the disk is changed during the relaxation of the shell. The disk thickens and non-axisymmetrical structure disappears due to the heating of the disk stars by the gravitational interaction between the disk and the shell stars.
3. The angular momentum of the disk is transferred to the shell stars during the relaxation.

Acknowledgments
We would like to thank Professors D.Sugimoto and T.Ebisuzaki, and Dr J.Makino for letting us use their GRAPE-3. We alse acknowledge the Nukazawa Memorial Foundation for our travel support.

References

Heckman,T.M. *et al.* 1990. *Ap. J. Supplement* **74**, 833.
McCarthy, P.J., Persson, S.E., & West, S.C. 1992. *Astrophys. J.*, **336**,52.
McCray, R. & Kafatos, M., 1987. *Astrophys. J.*, **317**, 190.
Sugimoto, D. *et al.* 1990. *Nature*, **345**, 33.
Tomisaka, K. & Ikeuchi, S. 1988. *Astrophys. J.*, **330**,695.

SECULAR EVOLUTIONARY TRENDS IN ELLIPTICALS AND BULGES DUE TO MERGERS

TAPAN K. CHATTERJEE
Facultad de Ciencias, Fisico-Matematicas, Universidad A. Puebla, (A.P.1316), Puebla, Mexico

Using basically the impulsive approximation and a modification of the method used by Alladin S.M., (1965, *Ap.J.*, **141**, 768), and described in detail in Chatterjee T.K., (1990, *IAU Col.*, **124**, 519, 569) we study the evolution of binary interacting galaxies, leading ultimately to mergers. Each collision is characterised by the initial separation between the galaxies and the relative velocity therein. In each case the orbital evolution and largescale structural changes in the galaxies are studied by taking into account the change in relative velocity due to dynamical friction, leading ultimately to mergers. The evolution is considered from a time when the gravitational interaction between the progenitor pairs is physically significant (Chatterjee, 1992, *Astroph. Sp.Sc.*, in press).

Results indicate that the expected frequency of merging galaxies, not considering the galaxies to be embedded in massive halos, is $10^{-2}\%$, under favourable environmental conditions. This value falls short of the observational value in the present epoch by an order of magnitude and falls short of the extrapolated past value by two orders of magnitudes. But in median environmental conditions, let alone unfavourable ones, the merger rates are drastically low (Chatterjee, 1992,a). The frequency of different types of progenitor pairs in merger is of the same order of magnitude. The majority of the mergers are achieved in several (2 to 3) orbital periods, and only about 10% of them are achieved in a single orbital period; and only about 1% of them are due to central impacts. During binary evolution substancial changes in the structures of the participating galaxies take place due to intense tidal effects. Several bursts of star formation will also take place during this time and the resultant dissipation is likely to cause a substancial change in the mass to luminosity ratio (M/L) of the system.

Binary evolution will have its reflection on the fundamental parameter plane for ellipticals (e.g. Kormendy and Djorgovski, 1989, *Ann. Rev. Astron. Astr.*, **27**, 235). The variation of the power coefficients of the fundamental plane variables from the pure virial coefficients are indicative of the dependence of the factor of proportionality in the fundamental plane solutions, upon the fundamental plane variables. This factor depends not only upon the formation but also upon the evolutionary history of galaxies. This implies that the structural changes and dissipation undergone during binary evolution and mergers is relevant in defining the finescale structure of the fundamental plane; in particular a curvature is expected to be induced. These secular evolutionary trends may also be applicable to bulges of spirals, as they seem to occupy almost the same fundamental plane as ellipticals.

I would like to thank all those with whom I had helpful discussions, especially Prof. A. Poveda.

H. Dejonghe and H. J. Habing (eds.), Galactic Bulges, 399.

Tapan Kumar
CHATTERJEE

Analytical models for ellipticals and bulges with rotation

S.N. NURITDINOV
Tashkent University, Physical Faculty, Astronomy Department, Tashkent, Uzbekistan

Abstract. The results of the analytical study of two disspationless non–linear models are generalized in case of rotation.

1. Non–Linear Models without Rotation

The following non-linear models are very useful for the investigation of dissipationless collapse and the conditions of the formation of ellipticals: model 1 (see V.A. Antonov & S.N. Nuritdinov, 1981, *Sov.Astr.Zh.*, **58**,1158)

$$\Psi_1 = \frac{\rho}{2\pi v_b}\delta(v_r - v_a)\delta(v_\perp - v_b)\chi(\Pi - r) \tag{1}$$

and model 2 (see S.N. Nuritdinov, 1983, *Sov.Astr.Zh.*, **60**, 40)

$$\Psi_2 = \frac{\rho\Pi^3}{\pi^2}[(\Pi^2 - r^2)(\Pi^{-2} - v_\perp^2) - \Pi^2(v_r - v_a)^2]^{-\frac{1}{2}}\chi(\Pi - r). \tag{2}$$

Here $\rho(t)$ is the density at the time t, v_r and $v_\perp$ are the radial and the transverse velocities, χ is the Heaviside step function, $\Pi = \frac{1+\lambda\cos(\psi)}{(1-\lambda^2)}$, $t = \frac{\psi+\lambda\sin(\psi)}{(1-\lambda^2)^{\frac{3}{2}}}$, $v_a = \frac{-\lambda r\sin(\psi)}{\sqrt{1-\lambda^2}\Pi^2}$, $v_b = \frac{r}{\Pi^2}$, $1-\lambda = (\frac{2T}{|U|})_0$. Antonov & Nuritdinov (1981) and Nuritdinov (1983) found the critical value $\Lambda_{cr} = \frac{\sqrt{21}}{5}$ for the ellipsoidal oscillations mode that corresponds to the value of $(\frac{2T}{|U|})_0 = 0.084$ and that is connected with the radial–orbit instability. The numerical experiments of L.Aguilar & D.Merritt (1991, *Ap.J.*, **345**, 33) confirm this result. Nuritdinov (1985, *Sov.Astr.Zh*, **62**, 506) noted an ellipsoidal instability zone $\lambda \in [0.611, 0.873]$, which has a resonance nature and corresponds to $(\frac{2T}{|U|})_0 \in [0.127, 0.389]$. Unfortunately up to now the accuracy of our numerical experiments does not allow to reveal the presence of this zone.

2. Models with Rotation

For the case with rotation we have (see Nuritdinov, 1992, in press)

$$\Psi_\Omega = [1 + \Omega(\frac{v_\perp}{v_b})\sin\theta\,\sin\gamma]\Psi_1, \quad \Psi_\Omega^* = (1 + \Omega r v_\perp \sin\theta\,\sin\gamma)\Psi_2, \tag{3}$$

where $\sin\theta = \frac{\sqrt{x^2+y^2}}{r}$, $\tan\gamma = \frac{v_\varphi}{\theta}$, and Ω characterizes the rotation ($0 \le \Omega \le 1$). The angular velocities of these models are $\omega_1 = \frac{\Omega}{2\Pi^2}$ and $\omega_2 = \frac{\Omega}{4\Pi^2}$. We found a critical dependence of $(\frac{2T}{|U|})_0$ on Ω. It is interesting that the area of the radial–orbit instability connects to the resonance instability area at lowest value of Ω. On the base of these models we can construct a new model $\Psi = (1-\nu)\Psi_{\Omega_1} + \nu\Psi_{\Omega_2}^*$, where ν, Ω_i, $i = 1, 2$ are free parameters.

H. Dejonghe and H. J. Habing (eds.), Galactic Bulges, 401–402.

Singers and dancers performing during the cultural event

Bulges and ellipticals: can formation mechanisms be the same?

S.N. NURITDINOV
Tashkent University, Physical Faculty, Astronomy Department, Tashkent, 700095, Republic Uzbekistan

We consider three mechanisms for ellipticals and bulge formation.

1. Mechanisms within a dissipationless collapse scenario: for example, the radial-orbits instability (for non-stationary models see V.A. Antonov & S.N. Nuritdinov, 1981, *Sov. Astr. Zh.*, **58**, 1158; L. Aguilar & D. Merrit, 1990, *Ap. J.*, **345**, 33; Nuritdinov, this issue).
2. Evolution of (proto)galaxy from an anisotropic sphere or a spheroidal model (A.M. Fridman & V.L. Polyachenko, 1984, *Physics of gravitating systems*, Springer). Here axisymmetric oscillations ($m = 0, N = 4$) correspond to the case of a galaxy with a bulge, and the ellipsoidal mode ($m = N = 2$) corresponds to the case of ellipticals. In principle these models can have a halo or a corona with a given mass (Nuritdinov, 1978, *Sov. Astr. Zh.*, **55**, 37).
3. Dissipation phenomena in non-stationary models.

Now we proceed to analyse the role of the "dome" instability during non-stationary evolution in order to check a relation of this instability to the bulge formation problem. Nuritdinov (1987, *Dinamica gravitiruyshchih system i metodi analyt. neb. meh.*, p65) has constructed two phase models of pulsating disk. One of these models is

$$\Psi = \frac{\sigma_0}{2\pi\Pi\sqrt{1-\Omega^2}}\left[\frac{1-\Omega^2}{\Pi^4}(\Pi^2 - r^2) - (v_r - v_a)^2 - (v_\perp - \frac{\Omega r}{\Pi^2})^2\right]^{-\frac{1}{2}}$$

where all notations are according Nuritdinov (this issue). Recently we have studied warps in this disk, assuming vertical displacements of the form $B(t)\frac{1}{\xi}P_N^m(\xi)e^{im\varphi}$, where B is a time function and $\xi = \sqrt{1 - r^2/\Pi^2}$. Here we suffice to give the result for the dome perturbation ($m = 0, N = 3$): the stability region in the $(2T/|\mathcal{U}|, \Omega)$ plane (see figure) shows some interesting narrow channels. Moreover, we have calculated the unstable modes as a function of $2T/|\mathcal{U}|$ and Ω. Our analysis shows that the dome instability can play a role in the formation of bulges.

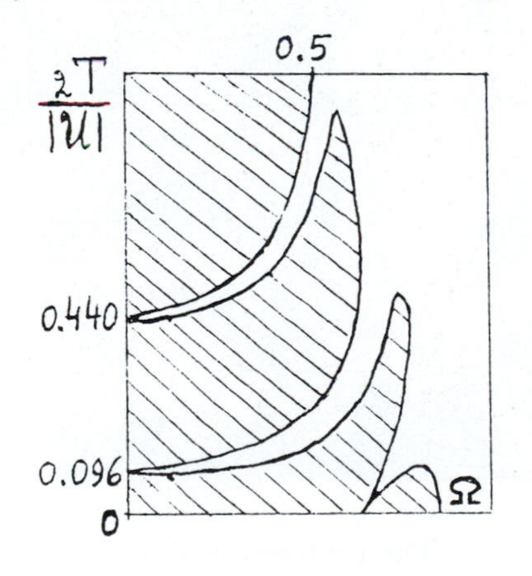

H. Dejonghe and H. J. Habing (eds.), Galactic Bulges, 403–404.

Reception at the town hall

The conference dinner

THE EFFECT OF A BAR ON THE KINEMATICS OF A CENTRAL RING STRUCTURE IN NGC4736

P. PIŞMIŞ & E. MORENO
Instituto de Astronomía, Apartado Postal 70-264
C.P. 04510, México, D.F.

ABSTRACT. We analyze the kinematics of the gaseous ring in the central region of NGC 4736 considering an ejection mechanism for the formation of the ring in a non-axisymmetric potential. A barred spiral of strength 10 or 25 per cent of the axisymmetric background is consistent with van der Kruit (1974) observational data.

1. The model

Extending a recent work (Pişmiş & Moreno 1992) we have taken a barred spiral galaxy with a composite potential, consisting of an overall axisymmetric potential and a barred one superposed on it, to investigate the motion of material ejected from the nuclear region, taking into account its interaction with the interstellar medium. This interaction leads an ejecta to follow a circular orbit in the axisymmetric case and a deformed circle when a bar plus a two-armed spiral pattern is introduced. This provides a way to form spiral and ring-like structures in galactic inner regions by means of an ejection mechanism. In NGC4736, showing a ring structure, the axisymmetric part of the potential is the one used by Pişmiş and Moreno (1992). The non-axysimmetric part is of the form proposed by Roberts et al. (1979), and we have tested a force strength of 10 and 25 per cent of this component with respect to the axisymmetric background. The velocity field of the interstellar gas in the overall potential is taken as that considered by Prendergast (1962), i.e. along the effective equipotential curves in the non-inertial frame defined by the bar-and-spiral pattern.

2. Results

In Fig. 1 we show four situations modeling the kinematics of the ring in NGC4736, with N = 1, 3 (see eq. 6 of Roberts et al. 1979) and a bar-spiral strength of 10 and 25 per cent for each value of N. This figure is to be compared with observational data given by van der Kruit (1974, Fig. 3). The ring is built in $\sim 10^8$ years, about an order of magnitude greater than in the axisymmetric potential, but the required energy is still of the order $10^{56} - 10^{57}$ ergs (Pişmiş & Moreno 1992).

H. Dejonghe and H. J. Habing (eds.), Galactic Bulges, 405–406.

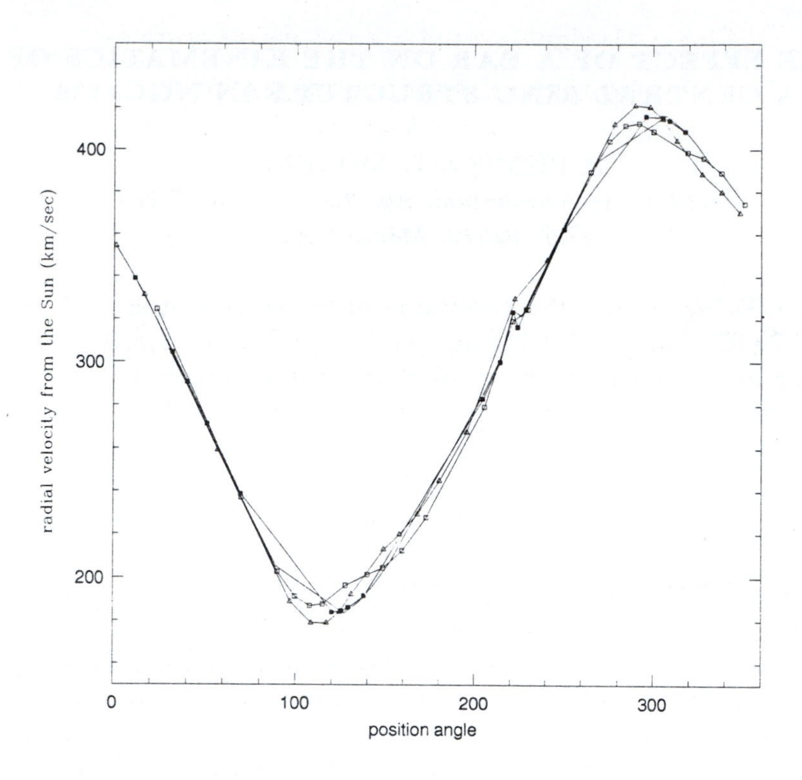

Fig. 1. Kinematics of the ring in NGC 4736 with a bar-and-spiral force strength of 10 per cent (filled symbols) and 25 per cent (open symbols) of the axisymmetric potential.

The large scatter in velocity around position angle 90 and 270 (Fig. 3 of van der Kruit, 1974) can be attached to the interaction of the ejecta with the barred spiral pattern. Sanders & Bania (1976) gave this interpretation in their single explosion model; here we propose a discrete ejection to form the ring, requiring less power and energy of a central engine.

REFERENCES

Pişmiş, P. & Moreno, E. 1992, in preparation.
Prendergast, K.H. 1962, in *Interstellar Matter in Galaxies*, ed. L. Woltjer, p. 217.
Roberts, W.W., Huntley, J.M., & van Albada, G.D. 1979, *Ap.J.*, 233, 67.
Sanders, R.H. & Bania, T.M. 1976, *Ap.J.*, 204, 341.
van der Kruit, P.C. 1974, *Ap.J.*, 188, 3.

Kinematic signatures of triaxial stellar systems

RICHARD ARNOLD, TIM DE ZEEUW
Sterrewacht Leiden
and
CHRIS HUNTER
Sterrwacht Leiden and Florida State University

Abstract. Analytic dynamic models of triaxial stellar systems, such as elliptical galaxies and galactic bulges, can be used to calculate the velocity fields of systems in a wide range of potentials without the need for orbit integrations. We present results from a first application of these models, in the form of velocity fields projected onto the sky. The appearance of the velocity field depends strongly on the viewing angle. Thin orbit models provide a theoretical upper limit to streaming in all possible kinematic models in a given potential.

In order to model the observed velocity field of a stellar system, one needs to know the mass and momentum densities of the stellar orbits in the gravitational potential. In an arbitrary potential these densities can only be determined by direct numerical integration of the motion of a test particle around each orbit.

Such numerical integrations can be avoided in certain special potentials where the orbital densities are known analytically. Hunter and de Zeeuw (1991) have found analytic descriptions of self-gravitating stellar systems built from 'thin' orbits – which have no radial thickness – in Stäckel potentials. In the cases where all stars rotate in the same sense, these thin orbit models represent the maximum streaming possible in the adopted potential. Any other model will consist of thicker orbits, or will have contra-rotating stellar populations, and will have less streaming.

We have investigated the maximum streaming models based on perfect ellipsoid potentials with a variety of axis ratios. We project the mass and momentum densities of the models onto the sky for a range of viewing angles, yielding in each case a two dimensional map of radial velocities. Two examples are shown in figure 1.

As can be seen from figure 1, and as has already been shown in detail by Statler (1991), the velocity field of a given triaxial stellar system can look radically different from different directions. For example, consider the difference in slope of the *zero velocity curve* (ZVC) between small and large radii in a velocity field. In figure 1(a) the difference in slope is very large (135°), and the velocity field shows a counter-rotating core. Figure 1(b), which is the same model viewed from a different direction, and which also has same projected ellipticity, shows a ZVC slope difference in slope of only 5°. Contour maps of ZVC slope difference as a function of viewing direction are shown in figure 2 for two different models. For a range of viewing angles – larger in model (a) than model (b) – a counter-rotating core is visible.

From the thin orbit models we learn for the first time how to put bounds on the streaming velocities in a triaxial potential. Using these bounds, and the further constraints provided by the equation of continuity, we create more realistic models with less streaming and thicker tube orbits. Because the mass and momentum densities have an analytic form, the resulting models provide a unique means to survey a wider range of galaxy shapes than would be feasible with methods relying on numerical orbit integration.

H. Dejonghe and H. J. Habing (eds.), Galactic Bulges, 407–408.

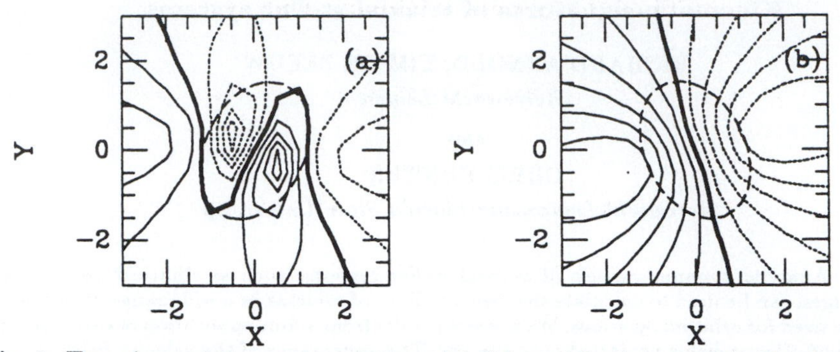

Fig. 1. Two views of the same prolate model ($b/a = 0.5, c/a = 0.375$). The heavy curve is the *zero velocity curve:* the locus of points of zero radial velocity. The ellipse is the projection of the ellipsoid which encloses half of the mass, and indicates the shape of the light distribution. [Viewing angles (θ, ϕ): (a) ($45°, 15°$), (b) ($45°, 165°$)]

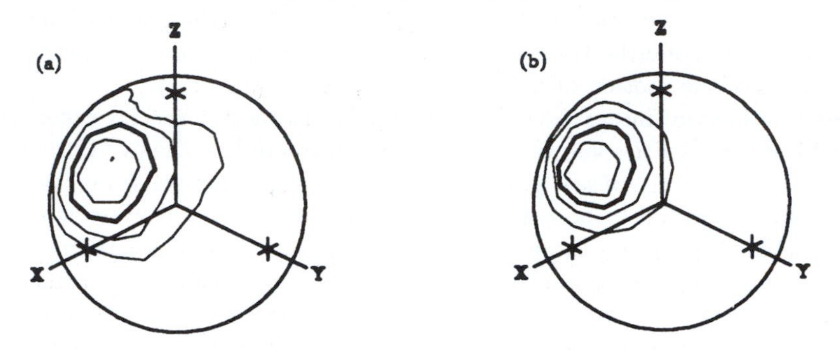

Fig. 2. The difference in slope of the zero velocity contour between small and large radii in the velocity field, drawn on the unit sphere of viewing angles. The results for two models are shown: (a) $b/a = 0.5$, $c/a = 0.375$; (b) $b/a = 0.75$, $c/a = 0.625$; The contours are spaced at intervals of 30°. For viewing directions inside the heavy contour the deviation is greater than 90° and the velocity field shows a distinct counterrotating core.

It is our intention to explore the probability distributions of the observables which can be derived from velocity fields (ellipticity, minor/major axis rotation etc.) in a variety of potentials using the thin orbit models. The ultimate aim is to constrain the possible shapes of stellar systems from the observed distributions (see for example Franx, Illingworth and de Zeeuw 1991).

Acknowledgements

CH acknowledges support from the NSF, and a visitor's grant in the Netherlands from NWO. RA is supported by an ESA external fellowship.

References

Franx, M., Illingworth, G. D., & de Zeeuw, P. T., 1991, *Astrophys. J.*, **383** 112
Hunter, C., & de Zeeuw, P. T., 1991, *Astrophys. J.*, **389** 79
Statler, T. S., 1991, *Astron. J.*, **102** 882

Colour Gradients in Galaxy Bulges

MARC BALCELLS
Kapteyn Lab, Postbus 800, 9700 AV Groningen, Netherland
and
R. F. PELETIER
E.S.O., Karl Schwarzschildstrasse, 2, D-8046 Garching bei München, Germany

November 27, 1992

1. Introduction

Colour gradients, when due to stellar populations rather than dust, are potential diagnostics on the formation history of galaxy bulges. We have carried out measurements of colour gradients on a sample of 30 bulges, the first such study using CCD detectors.

We have looked for two types of dependencies: 1) with total and bulge luminosities; 2) with disk parameters, such as central surface brightness and disk colours. We also compare bulges' gradients to colour gradients in elliptical galaxies.

2. Data, analysis

Our sample comprises galaxies of types S, S0–Sb from the UGC catalog with $B_T \leq 14$, with diameters above 2' and axial ratios above 2.1. Barred galaxies have been excluded.

Images in U, B, R, I were obtained at the prime focus of the Isaac Newton Telescope with a GEC CCD. Conditions were photometric, but there was some Sahara dust. Flat fielding was always better than 0.4%. Uncertainties in the absolute photometry were 0.05 mag in R and I, and 0.1 mag in U and B. Bulge-disk decomposition was carried out following the scheme used by Kent (1986).

Identifying the effects of dust is the main difficulty of this project. Galaxies were labeled "dusty" if patchy dust was obvious to the eye, or if the disk scale lengths in U and in R were very different, a consequence of dust extinction. Not surprisingly, the "dust-free" sample comprises fundamentally S0 and Sa galaxies.

3. Results

Colour gradients are present in most of the galaxies. While some positive gradients exist, negative gradients (bluer outward) are more common. Gradients in $B-R$ range from -0.25 to +0.15 magnitudes per decade in radius. Gradients in $U-R$ range from -0.4 to +0.1, in the same units. We find a correlation between colour gradients and absolute magnitude of the bulge: larger bulges show negative gradients, while small bulges have positive gradients. Such relation cannot be a dust effect; when dusty galaxies are not excluded the trend changes sign, a consequence of the strong inward reddening in small dusty bulges. Bulge colours correlate with the absolute magnitude of the galaxy and the absolute magnitude of the bulge. All bulges with

H. Dejonghe and H. J. Habing (eds.), Galactic Bulges, 409–410.

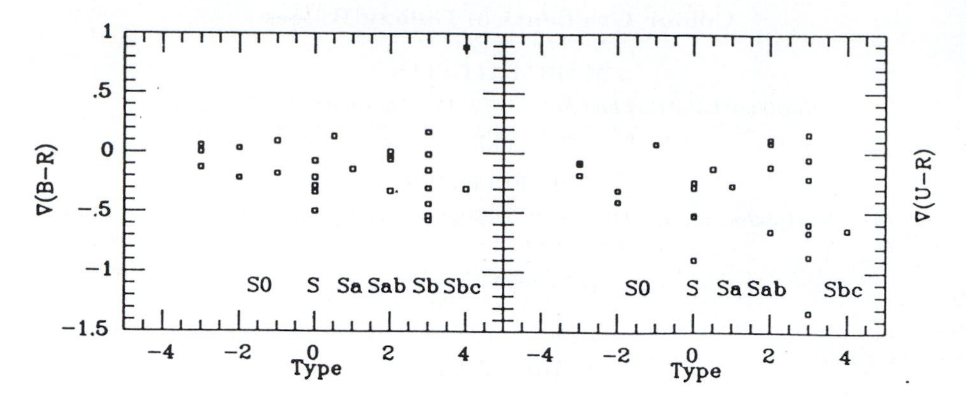

Fig. 1. *left:* $\nabla(B-R)$ as a function of galaxy type for the bulges sample. *right:* $\nabla(U-R)$ as a function of galaxy type. The increased dispersion in the gradients in small bulges reflects the effects of dust reddening and star formation.

$B - R \geq 1.7$ turn out to be signifficantly affected by dust. The dust-free sample shows colours (at 5") ranging from $B - R = 1.4$ to 1.65. The bulge colours correlate well with the disk central colours, both for dusty and dust-free galaxies. However, the bulge colour gradients do not correlate with any disk photometric parameter.

Colour gradients in bulges and in ellipticals are similar, though not identical. The mean values for our dust-free bulges sample and the elliptical sample of Peletier *et al.* (1990) are essentially identical ($< B - R >= -0.07$ and $< U - R >= -0.20$) For both ellipticals and bulges gradients become increasingly negative at absolute magnitudes M_B between -18.5 and -21. The relation gradient–luminosity is cleaner for bulges than it is for ellipticals. The main difference between ellipticals and bulges is the existence of positive colour gradients in small bulges, which is not seen in small ellipticals. Note however that the bulges sample extends to lower luminosities.

In summary, population gradients in early-type spiral bulges appear to be similar to population gradients in elliptical galaxies.

Acknowledgements

The Isaac Newton Telescope is operated on the island of La Palma by the Royal Greenwich Observatory in the Spanish Observatorio del Roque de los Muchachos of the Instituto de Astrofísica de Canarias.

References

Peletier, R. F., Davies, R. L., Illingworth, G. D., Davis, L. E. & Cawson, M. 1990, *Astron. J.*, **100**, 1091.

Kent, S. M. 1986, *Astron. J.*, **91**, 1301.

VELOCITY DISPERSIONS AND METALLICITIES OF BOX-SHAPED GALAXIES

R. E. DE SOUZA, S. DOS ANJOS, B. BARBUY
1 IAG-USP, CP 9638, 01065-970 São Paulo, Brazil

ABSTRACT. Velocity dispersions and Mg_2 indices were measured from CCD spectra of 13 S0 box-shaped galaxies and 5 ellipticals. Evidence for a different relation between velocity dispersion and metallicity, relative to ellipticals, was found for our S0 sample; we attribute this difference to the S0 disk component, affecting both the mean velocity dispersion and mean metallicity.

1. Introduction

The objects used in this work were selected from the catalog of box-shaped galaxies (de Souza & dos Anjos, 1987), classified from visual inspection of the ESO-B films and POSS prints. A sample constituted mainly of S0 edge-on galaxies, extracted from this list, was selected for photometric and spectroscopic observations. The photometric properties of 10 S0 box-shaped galaxies, 9 of which are in common with the present sample, is presented elsewhere by dos Anjos & de Souza (1992). The present work is concerned to the analysis of spectroscopic medium resolution data for 13 S0-Sa and 5 E galaxies, for which velocity dispersions and Mg_2 indices were measured. The spectroscopic observations were done with the Cassegrain spectrograph at the 1.6m Boller & Chivens telescope of the Laboratório Nacional de Astrofísica (LNA), Brazópolis, Brazil, in january/1991 and in september/1991. We use a grid of 900 tr/mm centered at 510 nm with a reciprocal dispersion of 45.9 Å/mm corresponding to 1.009 Å/pixel in the detector. The spectral coverage was 580 Å and the slit aperture was 400 μ corresponding to 4 arcsec projected in the sky and a spectral resolution of 4.1 Å, measured from the comparison spectrum of a HeAr lamp. The detector was a GEC CCD (400x600 pixels) with 22 μ/pixel. Each object was observed using three exposures of 20 minutes centered at the nucleus. The reductions were carried out with the VAX 8530 of the Astronomy Department of IAG/USP, using the eVe package routines developed at Meudon Observatory. The photometric calibration was done by observing standard spectrophotometric stars from the list of Stone and Baldwin (1983), and Taylor (1984). The extinction correction was estimated using broad band photoeletric photometry measurements observed at LNA during the period 89-91.

2. Radial velocities, velocity dispersions and Mg_2 indices

In the determination of velocity dispersion we basically adopt the Fourier Quotient algorithm described by Sargent et al. (1977), using as reference spectra a sample of K0III giant stars. After the observation of each galaxy we have observed at least one reference star in the same night and in the same sky region of the program objects. The radial velocities of the reference stars were determined by direct identification of prominent isolated spectral features: Hβ(λ4861.3), Fe(lambda4910.03), Fe(λ4920.51), Fe(λ5079.75), Mg(λ5183.30),

H. Dejonghe and H. J. Habing (eds.), Galactic Bulges, 411–412.

Fe(λ5226.87), Fe(λ5328.05) and Fe(λ5446.92). The mean square error in the final velocities of the stars is of the order of 10 km.s^{-1}.
The Mg_2 index was measured from the calibrated spectra in the way described by Burstein et al. (1984), where the Mg_2 bandpass is located at $\lambda\lambda$ 515.6 - 519.725 nm with blue continuum at 489.7-495.825 nm and red continuum at 530.3-536.675 nm.

As pointed out by Terlevich and Davies (1981) there is a relation good correlation between velocity dispersion and mettalicity. We confirm the presence of this correlation for our SO sample as can be seen in Fig. 1. However the observed correlation is steeper than the one observed in the sample of pure ellipticals of Faber et al. (1989). The presence of this relation is probably related to differences in the evolution of galaxies of different masses (Binney, 1989). In low mass galaxies the gravitational potential is not suficcient to bound the chemically enriched galactic wind driven by supernovae explosions, and as a result they can loose as much as 70% of their original mass (Arimoto, Yoshi, 1987). On the contrary larger mass galaxies retain the enriched material increasing therefore their mettalicities. In Fig 1. we show the trend present in their model

References

dos Anjos, S., de Souza, R.E.: 1992, this Symposium

Arimoto, N., Yoshii, Y.: 1987, A&A 173, 23

Binney, J., Tremaine, S.: 1987, *Galactic Dynamics*, Princeton Univ. Press

Burstein, D., Faber, S.M., Gaskell, C.M., Krumm, N.: 1984, ApJ, 287, 586

Faber, S.M., Wegner, G., Bustein, D., Davies, R.L., Dressler, A., Lynden-Bell, D., Terlevich, R.J.: 1989, ApJS, 69, 763

Terlevich, R., Davies, R.L., Faber, S.M., Burstein, D.: 1981, MNRAS 196, 381

Taylor, B.J.: 1984, ApJS 54, 259

Sargent, W.L.W., Schechter, P.L., Boksenberg, A., Shortridge, K.: 1977, ApJ 212, 326

de Souza, R.E., dos Anjos, S.: 1987, A&AS 70, 465

Stone, P.S., Baldwin, J.A. : 1983, MNRAS 204, 347ex

Figure 1 - Velocity dispersions (km s^{-1}) vs [M/H]

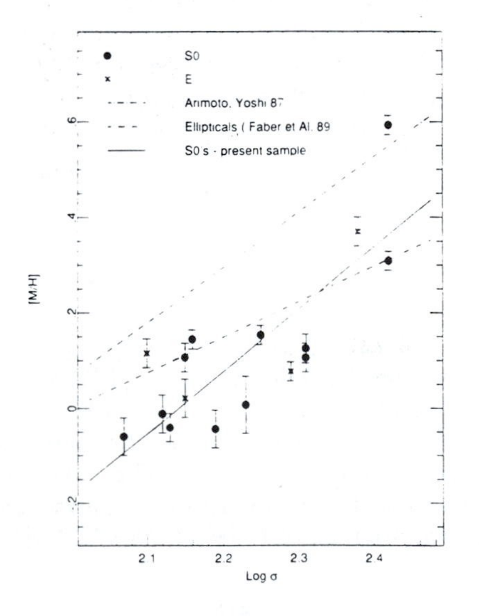

PHOTOMETRIC PROPERTIES OF BOX-SHAPED GALAXIES

S. DOS ANJOS[1], R. E. DE SOUZA[1]
1 IAG-USP, São Paulo, Brazil

ABSTRACT. A bulge/disk decomposition technique using a thin and a thick disk model is applied to a sample of 11 objects. In the thick disk modeling, the profiles are obtained in five differents radial directions. The results show that the model of thick disk is more realist.

1. Introduction

The photometric and dynamical properties of box-shaped galaxies are still poorly known. Recent studies about box-shaped bulges indicate that this morphology is a common phenomena in spiral galaxies and tends to occur in S0 and Sab-Sb galaxies (Jarvis, 1986). De Souza and dos Anjos (1987), published a complete sample ($m_{lim} < 13.2$) of this kind of objets with 74 galaxies and concluded that among lenticulars 33% shows this effect, indicating that these objects correspond probably to edge-on view of barred galaxies. They also statistically analysed if the environmental influence can affect the morphology of the galaxies by verifying if the galaxies are isolated or in groups or clusters. Information about the general structure of the environmental influence can be obtained by using the photometric technique of decomposition of brightness profile. In this present work our goal is to obtain the structural parameters of a sample of box-shaped galaxies using CCD photometry, and adopting two models (thin and thick disk) for the decomposition of the brightness profile into components. The objects were selected from de Souza and dos Anjos (1987) sample and observed with 1.60m Perkins-Elmer telescope of National Laboratory of Astrophysics (LNA-Brazil), at Pico dos Dias (lat.-23°). Each object was observed on B and V filters of Johnson system with mean exposure time of 1200 and 900 seconds, respectively. The detector used was a GEC CCD with 400x600 pixels (22μ), corresponding to 0.57 arcsec projected on the sky. The CCD frames were processed according to the standard rules : bias subtraction, flatfielding and sky subtraction, using the eVe package. The galaxies reduced are: NGC0128, NGC1381, NGC1596, NGC3115, NGC4565, NGC4594, NGC4958, NGC5253, NGC5864, NGC7041, A23-48. The comparison of the observed and estimated brightness profile indicate that both the thin and thick models fit the observational data. Because thick model provides a better fit of the small deviations of the observed profile than thin model, we present here the results concern to thick model.

2. Photometric Decomposition

In this model the disk is represented by an exponential law and has a transversal thickness (z_0) obtained from the solution of equation of hydrostatic equilibrium, $g(z/z_0) = sech^2(z/z_0)$, given by van der Kruit (1981). The exponential disk

H. Dejonghe and H. J. Habing (eds.), Galactic Bulges, 413–414.

presents a brightness distribution along the major axis given by

$$E(y) = \int_{-\infty}^{\infty} e^{-(x^2+y^2)^{1/2}/r_d} dx.$$

Considering the transversal thickness we obtain, $I(y, z) = I_{0d} g(z/z_0) E(y)$, or in terms of modified Bessel function K1(s), $E(s) = 2sK_1(s)$.

The profiles used to test this model were obtained along the 5 directions, ($\theta = 0°$, 23°, 45°, 67° and fitting parameters 90°), and μ_{0d}, μ_{0b}, r_d, r_b, z_0, q_{b23}, q_{b45}, q_{b67}, q_{b90} were simultaneously determinated. The results indicates that thick model fit the brightness profile of the observed galaxies except for NGC5864 for which no possible fitting was obtained with thick model. In the Figure below we can see, for example, the comparison of the observed (•) and calculated (solid curve) brightness profiles in several directions of the NGC1381 and NGC4958. The dotted and dashed curves represents the contribution of thick disk and bulge, respectively. We can see that the agreement of observed and calculated brightness profiles is very good, with exception of direction along major axis, where we can see an excess of luminosity between radius 25" to 60" for NGC1381 and 15" to 50" for NGC4958. It could indicate the presence of a subsystem with ring morphology, as suggested by Buta and de Vaucouleurs (1983) for NGC4958.

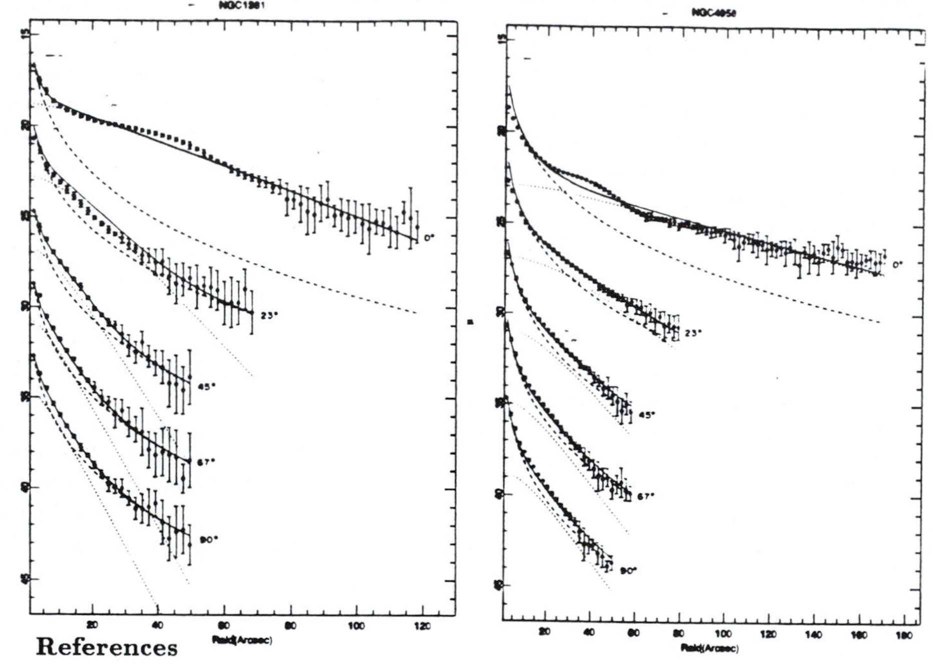

References

Buta, R., Vaucouleurs, G de 1983, ApJS **51**, 149
Jarvis, B.J. 1986, AJ **91**, 65
Kruit, P.C. van der, Searle, L. 1981, A&A **95**, 105
Souza, R.E. de, Anjos, S. dos 1987, A&AS **70**, 465

Hot Stellar Populations in the M31 Bulge

HENRY C. FERGUSON
University of Cambridge, Institute of Astronomy
and
ARTHUR F. DAVIDSEN
Center for Astrophysical Sciences, The Johns Hopkins University

Abstract. We investigate the possibility that the Far-UV spectral energy distributions (SED's) of M31 and NGC 1399 differ only in the relative contribution from classical PAGB stars and in the amount of extinction. A good fit to the M31 spectrum is obtained for E(B-V) = 0.11 if about 65% of the flux comes from PAGB stars and the rest from stars of the type producing the far-UV emission in NGC1399. We speculate that the UV continuum in both galaxies is dominated by stars in post-horizontal branch phases of evolution, with a distribution of post-HB masses governed primarily by the metallicity and metallicity spread of the population. This hypothesis can qualitatively explain both the relative fluxes of UV rising branches in NGC1399 and M31 and the shapes of their SED's.

While post-asymptotic-giant-branch (PAGB) stars now seem unlikely to be the dominant contributor to the UV flux from the most metal-rich galaxies (Greggio & Renzini 1990; Castellani & Tornambè 1991; Ferguson *et al.* 1991), PAGB stars are known to be present in old metal-rich populations (as evidenced by the existence of planetary nebulae), and it is possible that they contribute significantly to the UV flux in galaxies with weaker upturns. To investigate this possibility, we have considered a simple model for the Hopkins Ultraviolet Telescope spectrum of M31 (Ferguson & Davidsen 1992), in which a portion of the far-UV light comes from a single-mass population of PAGB stars, and a portion from the population (whatever it is) that is producing the far-UV emission in NGC 1399. We then adjust the relative proportions of the two populations, the PAGB mass, and the extinction to see whether we can achieve a good fit. This is by no means a unique or even realistic model, however it could indicate whether the *concept* of a composite population, with differing mixes of PAGB, PEAGB, EHB, and AGB Manqué stars (Greggio & Renzini 1990), is at least viable.

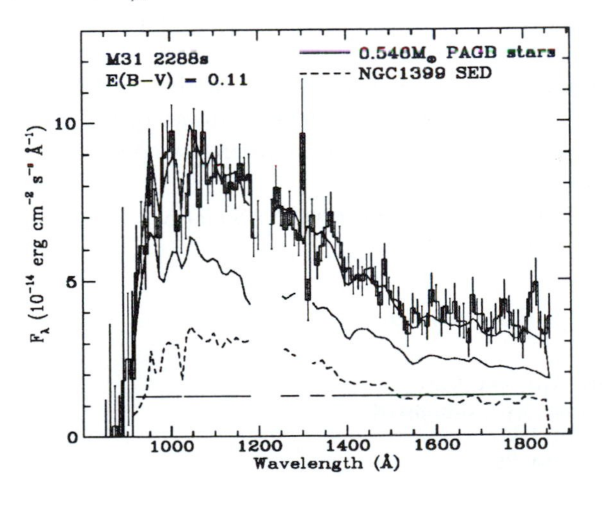

FIG. 1 — Best fit PAGB + NGC 1399 model for the M31 spectrum. The M31 spectrum is shown, corrected for extinction assuming $E(B-V) = 0.11$. The solid curve shows the best fit model for this extinction. The dashed curve shows the contribution from a NGC 1399-like stellar population, and the dotted curve shows the contribution from $0.546 M_\odot$ PAGB stars.

H. Dejonghe and H. J. Habing (eds.), Galactic Bulges, 415–416.

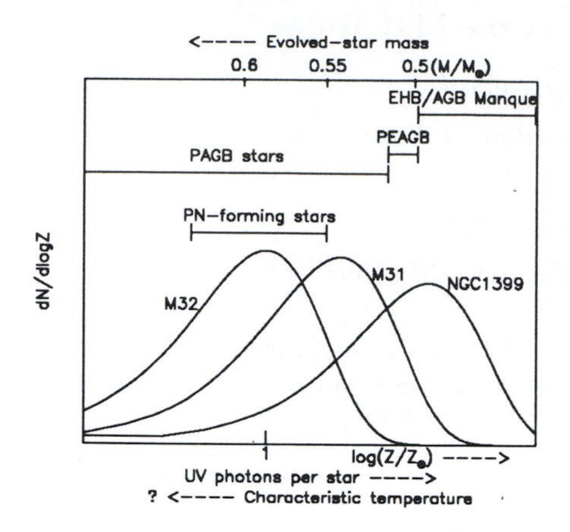

FIG.2 — Shows schematically the distribution of metallicities in M32, M31, and NGC 1399, galaxies with weak, moderate, and strong UV upturns, respectively. As metallicity increases, the core mass of each evolved star tends to decrease (Greggio & Renzini 1990), the number of UV photons emitted over the lifetime of each star tends to increase, and the temperature at which most of the photons are emitted tends to decrease. Because of the rapid variation of the rate of evolution as a function of mass, PN sample only a narrow range of PAGB-star mass.

The best fit is shown in Fig. 1. Details of the fitting are discussed in our paper (Ferguson & Davidsen 1992). The M31 SED *can* be fit with our simple model. If we hold $E(B-V)$ fixed, low-mass PAGB models are preferred, and contribute $\sim 65\%$ of the flux at 1425Å. Models with just an NGC 1399-like component are excluded with a high degree of confidence for any extinction. Our best estimate (Ferguson & Davidsen 1992) is that central stars of planetary nebulae could account for no more than 1% of the flux, while the point sources detected by the HST FOC (King *et al.* 1992; Bertola 1992) could account for about 15%. If the FOC calibration is correct, then the PAGB component in the M31 spectrum must come from very low mass ($< 0.55 M_\odot$) stars.

The result that M31 has a larger PAGB component than NGC 1399 is consistent with the hypothesis that the UV continuum in both galaxies is dominated by stars in HB and post-HB phases of evolution, with a distribution of masses governed by the metallicity and metallicity spread of the population. This is illustrated schematically in Fig. 2. The number of UV photons produced per star increases toward lower mass, while the characteristic temperature of the UV emitting population decreases, qualitatively explaining both the relative UV fluxes of NGC 1399 and M31 and the shapes of their SED's. Because planetary nebulae sample the high-mass end of the PAGB mass spectrum, we predict an anticorrelation in the number of PN per unit luminosity with $1550-V$ color, which we confirm with existing samples (Ferguson & Davidsen 1992).

The HUT project is supported by NASA contract NAS5-27000 to the Johns Hopkins University. HCF is supported by SERC.

References

Bertola, F. 1992, preprint.
Castellani, M. & Tornambè, A. 1991, ApJ, 381, 393
Ferguson, H. C. & Davidsen, A. F. 1992, ApJ, submitted.
Ferguson, H. C., *et al.* , 1991, ApJ, 382, L69.
Greggio, L. & Renzini, A. 1990, ApJ, 364, 35.
King, I., *et al.* , 1992, ApJ, 397, L35.

LINE PROFILES IN THE BULGE OF NGC7217

Konrad Kuijken[1] *and Michael R. Merrifield*[2]

ABSTRACT. Line-of-sight velocity distributions are becoming increasingly important for our understanding of galactic structure. We present results obtained for the bulge of NGC7217 with a new deconvolution technique.

1. The Need for Measuring Absorption Line Profiles in Galaxies

The stellar kinematics of three-dimensional systems are usually governed by three integrals of motion, and hence a three-dimensional data set is required to describe completely the intrinsically three-dimensional distribution function (DF). Even measurements of mean velocities and velocity dispersions over the entire face of a galaxy can never provide this much information: at best they yield several two-dimensional functions. Moreover, the rms velocity dispersion is never actually measured from galactic spectra: instead a measure of the line-broadening is obtained by assuming a particular (usually gaussian) line profile and finding the dispersion that best reproduces the spectrum. Since this may differ from the true rms dispersion of the stellar line-of-sight velocity distribution (LOSVD), it is not clear if use of these 'dispersions' in the Jeans equations is valid. Only more direct measurements of the LOSVD can help us decide. Measurement of the LOSVD over the face of a galaxy also provides the extra dimensionality in the data set which may eventually allow us to transcend the Jeans equations and get to the heart of the DF.

It has become increasingly feasible to measure the LOSVD directly, or at least to constrain it beyond measuring a best-fit gaussian (see van der Marel and Franx 1992 and refs. therein). Here we present a new method, based on quadratic programming (QP).

2. Line Profiles using Narrow Gaussians and QP

Any non-negative function which is smooth on scales below Δv can be approximated by a sum of gaussians with dispersion Δv and means v_i separated by $< 2\Delta v$ (the maximum separation for which there is no dip between the two peaks). If we write the LOSVD $f_{\rm los}$ as such a sum, a continuum-subtracted galaxy spectrum (assumed to be the doppler-shifted and -broadened spectrum S of a suitable late-type giant) should be a sum of copies of S with amplitude x_i, all broadened by the same gaussian dispersion Δv but shifted by different v_i. A least-squares fit of this model to the observed galaxy spectrum can then be formulated as a QP problem, and solved efficiently for x_i with a variation

[1] **Center for Astrophysics, 60 Garden St, Cambridge MA 02138**
[2] **Dept. of Physics, University of Southampton, Southampton SO9 5NH, UK**

H. Dejonghe and H. J. Habing (eds.), Galactic Bulges, 417–418.

of the simplex algorithm. Positivity constraints on f_{los}, or indeed any other constraints linear in the x_i, can be built in easily, and if we choose the v_i such that they correspond to shifts of whole numbers of pixels, we only need to compute one convolution of the template.

Since the smoothing scale Δv can be set explicitly, our method falls somewhere between the parametric approach of van der Marel and Franx (1992) and the essentially non-parametric fit of Rix and White (1992). We have typically set Δv to two pixels, with adjacent gaussians shifted by three pixels.

3. Line Profiles in the Bulge of NGC7217

We have applied our method to deep (4hrs with the 3.9m CFHT), long-slit spectra of the Sab disk galaxy NGC7217, with 28 km s^{-1} pixel^{-1}. Because of the moderate inclination of the galaxy (35°), all three velocity components project down the line of sight, raising hopes that a full solution for the three-dimensional kinematics of the system can be found.

The Figure shows our line profiles for the major axis of NGC7217. Note the large degree of consistency between opposite sides of the nucleus, the manifestly non-gaussian nature of almost all the major axis profiles, and the presence of around 30% of stars with apparently retrograde velocities in the outermost region of the bulge. The existence of retrograde stars in fact persists well out into the disk (Merrifield and Kuijken, in preparation).

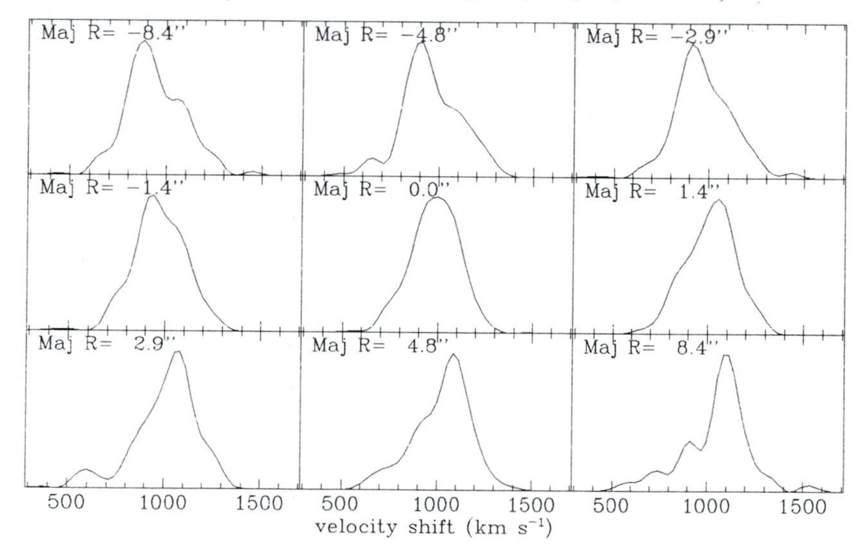

References

van der Marel, R. and Franx, M. 1992. ApJ, in press.
Rix, H.-W. and White, S.D.M. 1992. MNRAS 254, 389.

On the photometric characteristics of ellipticals and bulges of spirals in interacting systems

V.P.RESHETNIKOV

Astronomical Institute of St.Petersburg University, Special Astrophysical Observatory of Russian Academy Sciences

September 2, 1992

Abstract. The results of the study of the global photometric parameters of ellipticals and bulges of spiral galaxies are presented.

Key words: interacting galaxies - photometry

1. Introduction.

In order to gain additional information about the influence of tidal interactions on the photometric properties of the involved galaxies, a **R** band CCD photometric study was conducted for a sample of close interacting galaxies (Reshetnikov et al. 1993). Observations were made during 1990 October 8-11 and 1991 May 7-8 using 6-m telescope of Special Astrophysical Observatory and CCD detector with 512×512 pixels. The observations were made with the filter **R**, that realized a pass band close to $\mathbf{R_C}$ system of Cousins. Details of our observations and data reductions can be found in Reshetnikov et al. (1993).

For analysis of photometric structure of interacting galaxies we chose their *equivalent profiles.* For the decomposition of equivalent profiles in the contributions of the bulge and disk we used an interactive procedure similar to that described by Kormendy (1977).

2. Results and discussion

2.1. Bulges of spiral galaxies

Fig.1a plots the distribution of bulges of interacting spirals in the $\mu_e - \log R_e$ plane (values of μ_e are in $\mathbf{R_C}$ band and $\mathbf{H_0} = 75$ km/s Mpc). The solid lines represents the standard relations for the bulges of ordinary spiral galaxies according to the data of various authers. As one can see, the bulges of interacting spirals are placed in the plane $\mu_e - \log R_e$ along the average relation for the bulges of normal galaxies. This may means that even strong tidal interaction has a small influence on the inner regions of spiral galaxies and keep the general photometric properties of their bulges without large changes.

2.2. Ellipticals and SO bulges

Fig.1b plots the distribution of our sample ellipticals and SO bulges in the $\mu_e - \log R_e$ plane as *rectangles.* Elliptical galaxies belonging to close binary systems according to the data by Peletier et al. (1990) are shown as *rhombs.* The solid lines represents the standard relations for the normal galaxies according to different

H. Dejonghe and H. J. Habing (eds.), Galactic Bulges, 419–420.

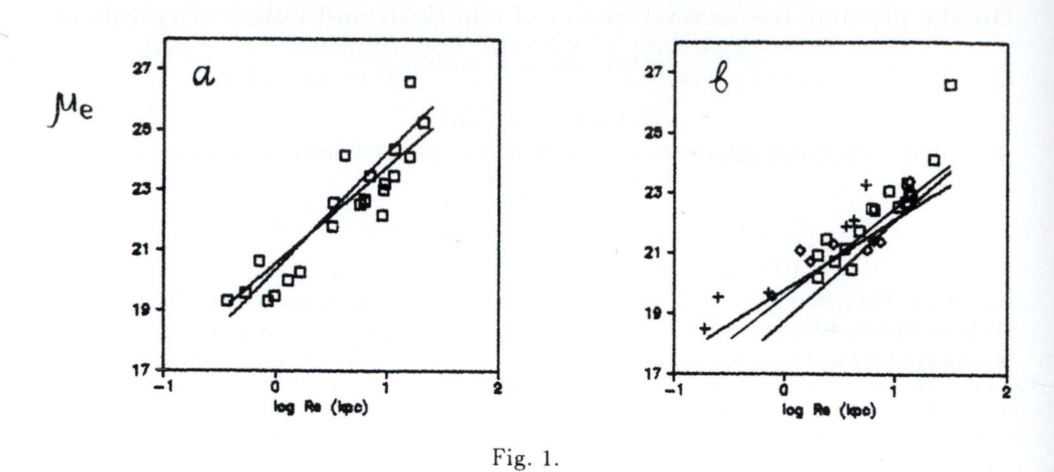

Fig. 1.

authers. ¿From the examination of Fig.1b, it is apparent that the interacting E-SO galaxies, on average, lie above the mean relation for normal ellipticals and SO galaxies. This shift in the location of interacting galaxies relative to normal E-SO galaxies is in opposite direction to that of for brightest cluster members (Schombert 1986). We find that the effective radii of interacting ellipticals and SO bulges are about 30% smaller at a given effective brighnesses than radii of normal galaxies. Therefore, interacting galaxies are more compact and shrinkaged with respect to the average structure of ordinary ellipticals and SO galaxies.

2.3. Bulges of polar-ring galaxies

Polar-ring galaxies (PRG) are among the most interesting examples of interactions between galaxies (Whitmore et al. 1990). Accretion, either from tidal capture of matter from a nearby system or the merger of a gas-rich companion, is generally considered as an explanation of the observed structure of PRG.

The *crosses* in Fig.1b are the bulges characteristics of six SO galaxies with polar rings. For three galaxies (NGC 4650A, ESO 415-G26, A0136-0801) we have taken values of R_e from Whitmore et al. (1987) and have calculated μ_e assuming $\mathbf{r}^{1/4}$ model with absolute magnitudes of bulges from this work. For the galaxies NGC 2685, IC 1689, and UGC 7576 we have obtained μ_e and R_e parameters by the decomposition of their major axis profiles published in Whitmore et al. (1990). In this figure one can see that PRG are placed in the plane of the effective parameters in the same region as an interacting galaxies. Therefore, the photometric structure of PRG give us a new surprising evidence of the interaction events in its recent history.

References

Kormendy J., 1977, *Astrophys. J.*, **217**, 406
Peletier R.F. et al., 1990, *Astron. J.*, **100**, 1091
Reshetnikov V.P., Hagen-Thorn V.A., Yakovleva V.A., 1993, *Astron. Astrophys. Suppl.* (in print)
Schombert J.M., 1986, *Astrophys. J. Suppl.*, **60**, 603
Whitmore B.C., McElroy D.B., Schweizer F., 1987, *Astrophys. J.*, **314**, 439
Whitmore B.C. et al., 1990, *Astron. J.*, **100**, 1489

The Counter-Rotating Twin Disks in NGC 4550

HANS-WALTER RIX

Institute for Advanced Study, Princeton, NJ, 08540, U.S.A.

MARIJN FRANX

Harvard-Smithsonian Center for Astrophysics, Cambridge, MA 02138

and

DAVID FISHER AND GARTH ILLINGWORTH

University of California, Santa Cruz, CA 95064

Abstract. We discuss the striking kinematic properties of the S0 galaxy NGC 4550. A detailed analysis of the line-of-sight velocity distribution (LOSVD) along the major axis shows that this galaxy contains two cold, cospatial, counterrotating disks with indistinguishable scale lengths and luminosities. The two disks, twins save the signs of their spin, are also found to have exponential luminosity profiles. We discuss qualitatively how this system might have formed.

Key words: galaxies: kinematics - galaxies: lenticulars - galaxies: individual: NGC 4550

The E7/S0 galaxy NGC 4550 was recently discovered by Rubin *et al.* (1992) to have counter-rotating, cospatial, stellar disk components. In this contribution we summarize the results of a detailed photometric and kinematic study by Rix *et al.* (1992) which quantified the striking properties of this object. The kinematics were studied by analyzing the line-of-sight velocity distribution (LOSVD) along the major axis which is reflected in the broadening of the stellar absorption line spectra (see e.g. Rix and White 1992). Outside the bulge, this velocity distribution is bi-modal, each component arising from the stars in one disk. The qualitative behavior is illustrated in Figure 1.

The Structure of NGC 4550

- The disks have identical luminosities and exponential profiles with indistinguishable scale lengths. Thus, the two disks appear photometrically to be one.
- The bulge contributes only $\sim 15\%$ to the total light, appears to be non-rotating and shows no sign of multiple components.
- Both disks are kinematically as "cold" ($v/\sigma \sim 2-3$) as normal S0 disks at one scale length.
- In both disks the velocity dispersion appears to be roughly constant over ~ 2 scale lengths, while for all other known disks the dispersion drops with radius.
- The stellar populations are old (probably at least a few Gyrs) and the galaxy appears to be in a steady state.

These properties make NGC 4550 the kinematically most unusual galaxy known!

Speculations about the Formation History

So far, no coherent, quantitative scenario of how to form a galaxy with the above properties exists.

All attempts to simulate the dissipationless formation of such a system through the merger of spirals with anti-parallel spin have failed (J. Dubinski *priv. comm.*, Heyl, Hernquist, Spergel *in prep.*) because the the simulations show that the two disks get heated too much.

H. Dejonghe and H. J. Habing (eds.), Galactic Bulges, 421–422.

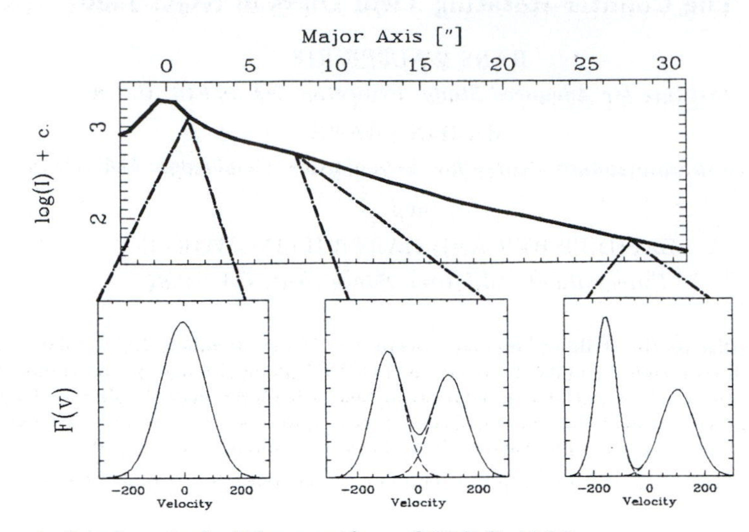

Figure 1: Major Axis Kinematics of NGC 4550

The top panel shows the major axis profiles of NGC 4550 along with the LOSVD at several distances from the center. At the center the LOSVD is close to a single Gaussian, while it is bi-modal throughout the disk, with comparable flux contributions at 7″ and 26″.

Consequently, we are led to assume that the formation of the second disk occurred through the accretion of gas, which subsequently formed stars. The advantages of this scenario are that no exact anti-alignement of the spin of the new material is required, because gas can settle dissipatively, and that an adiabatic addition of gas will not heat the pre-existing disk excessively.

Nonetheless, this scenario has a number of problems: in particular it is unclear, why the second disk also has an exponential luminosity profile and why the two scale lengths are the same. If the scale length of accreted material were determined by its initial angular momentum, such a coincidence cannot be expected.

It is likely that a massive halo is needed to prevent the adiabatic compression (and increase in v_{circ}) of NGC 4550 during the gas accretion, which would otherwise have moved the galaxy off the Tully-Fisher relation (Rubin *et al.* 1992, Bothun and Gregg 1990).

Acknowledgements

H.-W.R. and M.F. acknowledge support through Hubble Fellowships from STScI under contract to NASA (HF-1027.01-91A).

References

Bothun, G. and Gregg, M. (1990), ApJ, 350, 73.
Rubin, V. *et al.* 1992, ApJL, 394, L9.
Rix, H.- W. and White, S. D. M. (1992), *MNRAS*, **254**, 389.
Rix, H.-W., Franx, M., Fisher, D. and Illingworth, G. (1992) *ApJL*, **400**, L5.

Stellar Kinematic Evidence for Massive Black Holes Revisited

HANS-WALTER RIX
Institute for Advanced Study, Princeton, NJ, 08540, U.S.A.

Abstract. The existing stellar kinematic evidence for massive black holes at the centers of galaxies is based on $\lesssim 30\%$ discrepancies between constant M/L models and kinematic data. We show that asymmetries in the observed velocity distributions may introduce systematic errors in the data/model comparisons which are of the same order and have the same sign as these discrepancies. Existing modeling has so far failed to account properly for these asymmetries. Major axis data of M32 are analyzed as a specific example.

Key words: galaxies: kinematics - galaxies: dynamics - black holes

After a decade of work (see Kormendy 1992, for a summary) there are by now half a dozen nearby galaxies for which the presence of a central, massive dark object, presumably a black hole (BH), has been inferred, because dynamical models with constant mass-to-light ratio, herafter $(M/L)_{const.}$, failed to match the observed kinematics. Even though the inferred BH masses and the distances to the host galaxies vary by more than three orders of magnitude, the gravitational radius, $R_{grav.} \equiv GM_{BH}/\langle\sigma^2_{galaxy}\rangle$, which is the distance out to which the BH significantly impacts the stellar kinematics, subtends an angle of $\sim 1''$ in every case and is thus always close to the resolution limit (see Figure 1). Leaving all other potential implications aside, this apparent correlation implies that in *all* cases the discrepancy between the predicted kinematics of a M/L_c model and the observed kinematics is $\lesssim 30\%$ over the observed radial range. Consequently, to reject $(M/L)_{const.}$ models we must feel confident that we can compare a given dynamical model with the kinematic data to at least that accuracy.

For most of the BH candidates this data/model comparison is complicated by the fact the observed line-of-sight velocity distribution (LOSVD) is significantly asymmetrical. Such asymmetries may be due to (1) the "intrinsic" distribution function, (2) the line-of-sight integration in the presence of a steep velocity gradient, and (3) seeing. Existing modeling (e.g. Tonry 1987, Kormendy and Richstone 1992) takes effects (2) and (3) into account, but assumes that the velocity distribution is locally a Gaussian.

We have analyzed the LOSVD of M32, NGC 3377 and NGC 3115 (see Rix and White, 1992, for the technique) and have found them to be significantly asymmetrical at small projected radii $\lesssim 1.5''$. Specifically, we found for M32 that these asymmetries are much too strong to be attributed mainly to effects (2) and (3) and must reflect a highly anisotropic distribution function.

With such strong asymmetries the 1st and 2nd moment of the LOSVD (the quantities predicted, e.g., by Jeans equation models) differ from the velocity v and the dispersion σ estimated from the data by standard means, e.g. cross-correlation. Since existing models for most BH candidate galaxies predict much weaker asymmetries than are observed, this comparison of moments is necessarily flawed. Figure 2 shows that for M32 the difference between the moments and v and σ amounts to $\sim 20\%$, comparable to the discrepancy between the "conventionally" measured kinematics and the $(M/L)_{const.}$ models.

H. Dejonghe and H. J. Habing (eds.), Galactic Bulges, 423–424.

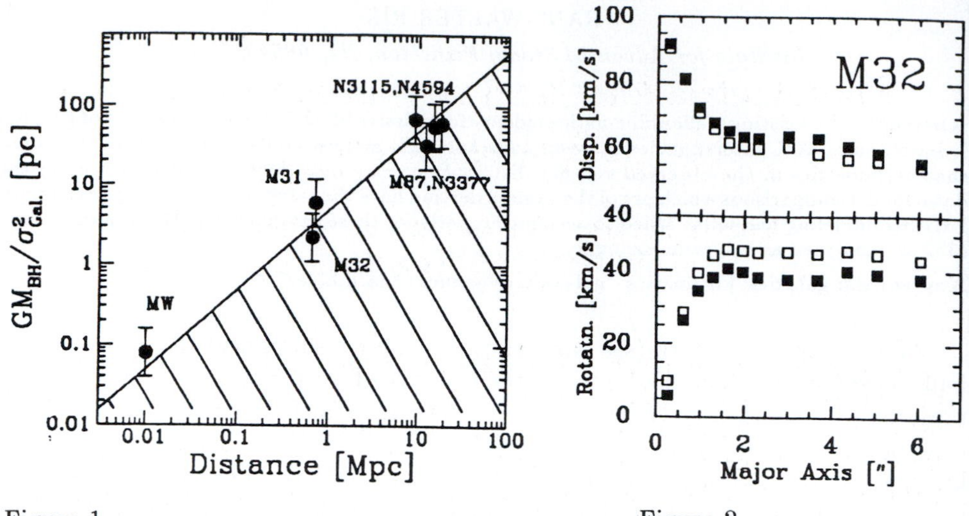

Figure 1 Figure 2

Figure 1: This figure compares the distances of the host galaxies with the gravitational radii of the claimed BH, taken from the literature (e.g. Kormendy 1992, and references therein). The shaded area represent the region of parameter space in which R_{grav} subtends less than one second of arc. The error bars represent a typically quoted uncertainty of a factor of 3 in the BH mass. In all cases the gravitational radius has an angular size of $\sim 1''$ and is thus very close to the observational resolution limit.

Figure 2: This figure compares the the first two moments (filled symbols) of the observed LOSVD with the best fitting Gaussian parameters, v and σ, in major axis spectra of M32. The LOSVD is sufficiently asymmetrical that the two quantities differ by $\sim$ 20%.

The existence of strong asymmetries in the LOSVDs of nearly all BH candidates casts doubt on one's ability to compare models with the data at the required accuracy level ($\sim$ 25%) with the methods employed so far. At present, it is an open question whether it will be possible to construct $(M/L)_{const.}$ models, since no self-consistent models producing sufficient LOSVD asymmetries exist. Nonetheless, it is clear that not all "plausible" $(M/L)_{const.}$ models have been explored and rejected – a necessary condition to conclusively infer the presence of unseen mass – and that indeed the models tried so far are incompatible with the observations merely on the basis of their LOSVDs.

The author acknowledges support through Hubble Fellowships from STScI under contract to NASA (HF-1027.01-91A) and is grateful to Dave Carter for permitting the presentation of his data on M32.

References

Tonry, J. (1987), *Ap.J.*, **332**, 932.
Rix, H.- W. and White, S. D. M. (1992), *MNRAS*, **254**, 389.
Kormendy, J. and Richstone, D. (1992), *Ap.J.*, **393**, 559.
Kormendy, J. (1992) in *High Energy Neutrino Astrophysics*, eds. V. Stenger *et al.*, World Scientific

Photometric and kinematic properties of disky elliptical galaxies

Cecilia Scorza. Landessternwarte-Königstuhl,
D-W6900 Heidelberg, Germany

New insights in the structure of elliptical galaxies have been obtained indicating the presence of faint stellar disks in some of these systems. Clear correlations between the isophotal shapes, the kinematic, radio and X-ray properties of disky E's were found, suggesting that these galaxies form together with SO's a continuous transition of D/B ratio in the Hubble sequence (Bender et al. 1989). The isophotal shapes of these galaxies have been quantified by the fourth cosine coefficient a_4 of the Fourier expansion of the deviations from perfect ellipses, which yields positive values when the isophotes are pointed along the major axis (Carter 1987, Lauer 1985, Jedrzewski 1987, Bender and Möllenhoff 1987).

Based on a photometric decomposition, we were able to disentangle, in a first aproximation, the kinematic curves of a sample of disky E's into the corresponding rotation and velocity dispersion curves of their disk and bulge components. This decomposition turns out to be an important task to test the continuity of physical properties towards lower D/B ratios in the Hubble sequence, and hence to investigate common formation processes with SO galaxies.

The photometric decomposition consists in the iteractive subtraction of a sum of exponential or/and ring-type thin disk models from the original frames. The disk-model parameters are free. They are successively varied until a bulge having perfectly elliptical isophotes remains. In contrast with other decomposition methods, in this new approach special emphasis has been made on the disk modelling, because in this way the inclination of the disk (and of the galaxy) can be constrained, which is of interest for dynamical models. A first version of this method was limited to single exponential thin disk-models (for details see Scorza and Bender 1990).

The kinematic decomposition applied here relies on the fact that the line-of-sight velocity distributions (LOSVDs) of many disky ellipticals are asymmetric. These are interpreted as the superposition of two distinct kinematical components: a rapidly rotating cold component (a disk) and a more slowly rotating hot component (a bulge). The developement of new methods for kinematic analysis had made possible the detection of these asymmetries (Franx and Illinworth 1988, Bender 1990, Rix and White 1992). In a first aproximation, the observed LOSVD can be described by assuming that the line profiles of both components are gaussians. A double gaussian fit to the LOSVD of the spectral lines is then performed under the constraint that the ratio of the integrated flux of the two gaussians is equal to the

H. Dejonghe and H. J. Habing (eds.), Galactic Bulges, 425–426.

local D/B ratio, as derived from the photometry.

The kinematic decomposition is ilustrated here for the case of NGC 3610. The photometric decomposition of this galaxy has been described with detail in Scorza and Bender (1990). Shown in Fig. 1.a is the LOSVD of the major axis spectra at 3.5, 5.5, 8, 11 arcsec from the center of the galaxy (solid line), the gaussians corresponding to the disk and bulge components (dotted lines) and the fit to the LOSVD (dashed line). The D/B ratio values used to constrain the fit are given in each panel. The quality of the fit indicates that the shape of the LOSVD of NGC 3610 can be undestood within the simple two-component picture. Once the fit has been made, the rotation and velocity dispersion curves of the disk and bulge can be obtained. These are shown in Fig. 1.b together with the kinematical curves derived in the conventinal way from a single-Gauss fit to the LOSVD (see small vs. big symbols). The v/σ=3.4 of the disk corresponds to a cold stellar component. This result is in good agreement with that obtained by Rix and White (1992). The $(v/\sigma)^*$=1.13 of the bulge indicates that it is rotationally flattened. From ten disky E's examined so far, eight have rotationally flattened bulges and cold inner disks. More details and discussion are given in Scorza 1992.

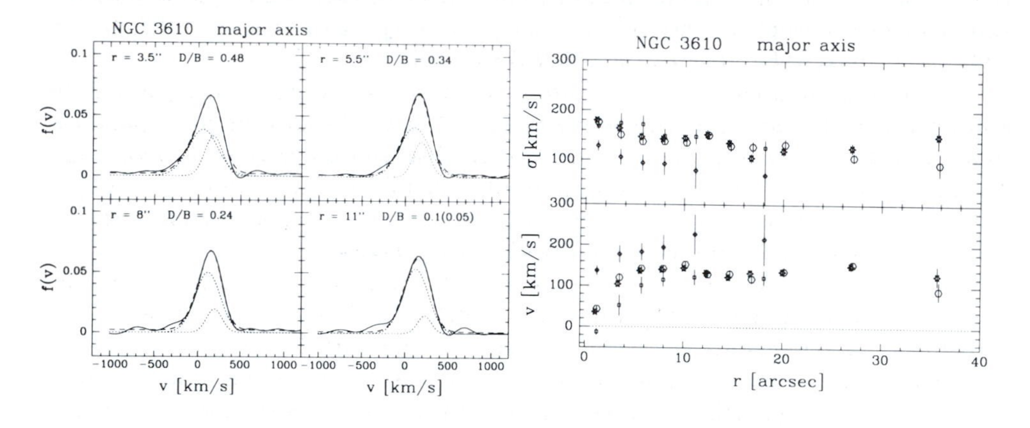

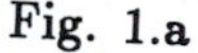
Fig. 1.a

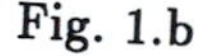
Fig. 1.b

Bender, R., Möllenhoff, C.: 1987, *Astron. Astrophys.* **177**, 71
Bender, R.: 1988, *Astron. Astrophys. Letters* **L193**, L7
Bender, R.: 1990, *Astron. Astrophys.* **229**, 441
Carter, D.: 1987, *Astrophys. J.* **312**, 514
Franx, M., Illingworth, G.D: *Astrophys. J.* , 344**L55**,
Jedrzewski, R.I.: 1987, *Monthly Notices Roy. Astron. Soc.* **226**, 747
Lauer, T.R.: 1985, *Monthly Notices Roy. Astron. Soc.* **216**, 429
Nieto, J-L., Bender, R., Surma, P. : 1991, *Astron. Astrophys. Letters* **L244**, L37
Rix, H-W., White, S.: 1992, *Monthly Notices Roy. Astron. Soc.* **254**, 389
Scorza, C., Bender, R.: 1990, *Astron. Astrophys.* **235**, 49
Scorza, C. : 1992. Ph.D. Thesis, Univ. of Heidelberg.

ABUNDANCE JUMP IN THE INNER BULGES OF GALAXIES

O.K.SIL'CHENKO
Sternberg Astronomical Institute, University av. 13, 119899 Moscow.

Abstract. The metallicity difference of about an order between the kinematically decoupled nuclei and the surrounding bulges is found in 3 spiral galaxies possessing an axisymmetric potential in their centres.

Key words: Galaxies - Metallicity - Nuclei of galaxies

It is known that some galaxies, spirals (M 31, NGC 4594) and ellipticals (IC 1459, NGC 5813), have kinematically decoupled nuclei. From our study of gas kinematics in the inner parts of spiral galaxies we have found that almost 50% of our sample galaxies have decoupled nuclei, and half of these have fast rigid-body circular gas rotation [1]. The interesting question is if kinematically decoupled nuclei are distinguished by their stellar population properties from neighbouring bulges.

In 1989–90 we have investigated 12 bright nearby galaxies of various morphological types with the two-dimensional Multi-Pupil Fiber Spectrograph (MPFS) of the 6-m telescope of Special Astrophysical Observatory of RAS. In the spectral range 4800–5400 A, with the spectral resolution 7–10 A, the spectra from a rectangular area of 10"x11" have been obtained for each galaxy (one spatial element is of 1.25"x1.25"). The direct image mode of MPFS has allowed us to construct isophotes for the central 30" of the galaxies under consideration, and the spectra have been used to derive line-strength profiles in the very centres of galaxies with rather high spatial resolution. The seeing quality during the observations was from 1" to 2".

The analysis of the isophotes for NGC 615, 7013, and 7331, previously known as possessing decoupled nuclei with circular gas rotation, has shown the coincidence of the major axis position angles for the very central isophotes and the outer disk ones. So these galaxies *may* have axisymmetric cores which are implied by their circular gas rotation. On the contrary, NGC 2655 and 7217, suspecting to have strong non-circular gas motions in the centres, show significant twisting of the isophotes together with the increasing ellipticity in their nuclei, so they *must* have triaxial central components.

The analysis of the line-strength profiles has revealed a jump of the $MgI\lambda5175$ absorption line equivalent width between the nuclei and the neighbouring bulges for 7 of 12 investigated galaxies, and among them there are distinct-nucleus galaxies NGC 615, 7013, 7331. Using the broad-band colour $B-V$ of the bulges obtained earlier [4], I have compared distributions of the bulges and globular clusters (the data being from [2]) on the diagram (Mg,$B-V$) where effects of age and metallicity are separated [3]. The bulges have mostly the same position as the reddest globular clusters with $[Fe/H]$ -0.7 – -1.0. So the jump of MgI-strength between the nuclei and the neighbouring bulges is a jump of metallicity reaching almost an order.

The behaviour of MgI-profiles in NGC 2655 and 7217 is rather complex, and the nuclei are not distinguished. It seems to be natural because a triaxial form of the central potential must result in strong radial gas flows which make impossible for a nucleus to possess an autonomous evolution.

H. Dejonghe and H. J. Habing (eds.), Galactic Bulges, 427–428.

Acknowledgements

I am grateful to the Local Committee for the financial support giving me the possibility to attend the Symposium.

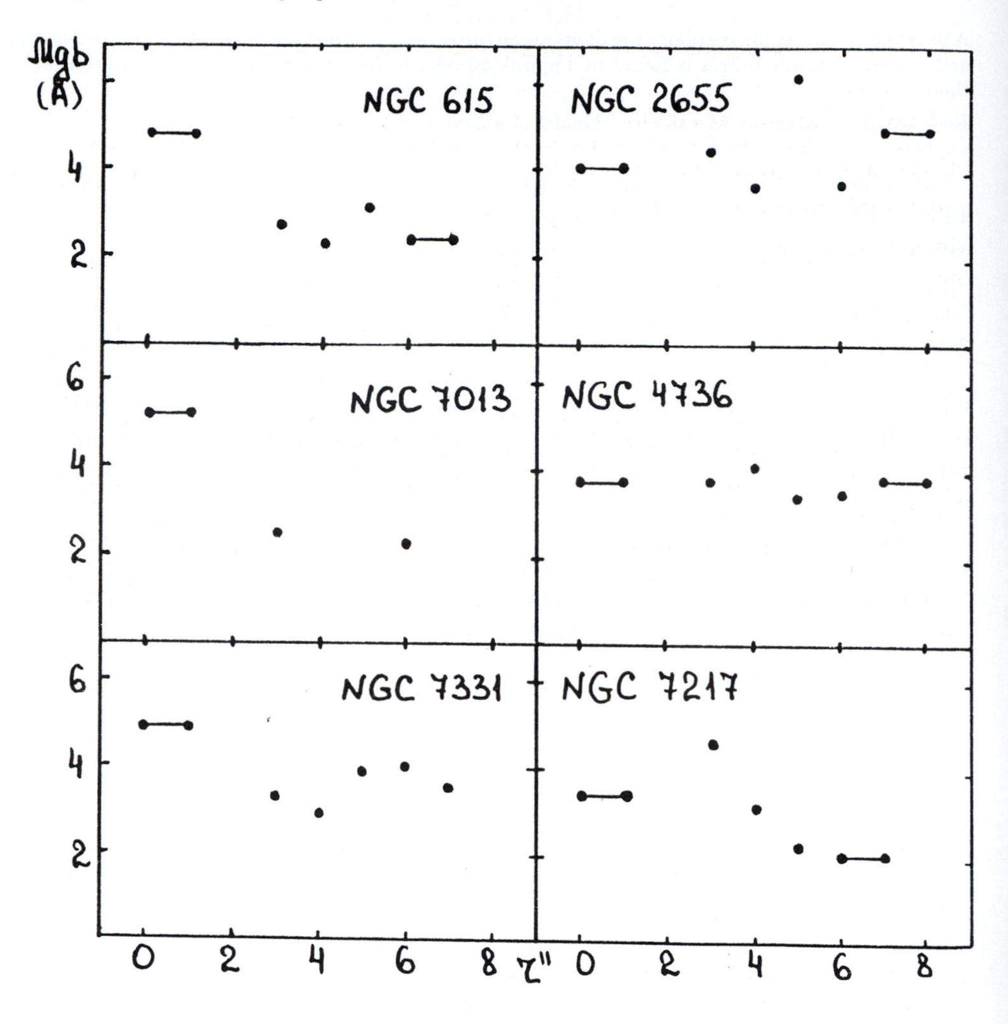

Fig. 1. The azimuthally-averaged magnesium line profiles for 6 spiral galaxies: NGC 615, 7013, and 7331 have an axisymmetric potential in the centres, NGC 2655, 4736, and 7217 have ovally distorted cores. The accuracy of equivalent widths is 0.3 A.

References

[1] Afanasiev, V.L., Sil'chenko, O.K., Zasov, A.V., 1989. *Astron. Astrophys.*, **213**, L9.
[2] Burstein, D., *et al.* 1984. *Astrophys. J.*, **287**, 586.
[3] Sil'chenko, O.K., 1984. *Pis'ma v Astron. Zh.*, **10**, 19.
[4] Sil'chenko, O.K., 1989. *Pis'ma v Astron. Zh.*, **15**, 493.

STELLAR POPULATION IN THE NUCLEI OF EARLY-TYPE GALAXIES

O.K.SIL'CHENKO

Sternberg Astronomical Institute, University av. 13, 119899 Moscow.

Abstract. The results of nuclear stellar population investigation are presented for 100 "normal" galaxies. The significant fraction of intermediate-age stars (T being about 1 billion years) is found in the nuclei of 50% of early-type disk galaxies.

Key words: Galaxies - Spectra of galaxies - Age of population

In 1986-1989 we have undertaken a spectral investigation of normal galaxies' nuclei at the 6-m telescope of Special Astrophysical Observatory of RAS to study their stellar populations. Using the 1000-channel skaner, we have obtained more than 120 spectra for the central parts of 100 nearby galaxies of different morphological types. The sample includes 29 E-galaxies, 21 S0's, 30 Sa-Sb's, and 20 Sc's; all the galaxies possess bright star-like nuclei. The spectral dispersion is 1.8 A/px (the spectral resolution 6-8 A), the spectral range is 3700-5500 AA. We have used two sets of aperture: rectangular one, with the height of 4" and the width from 0.6" to 2", and the round one, with the diameter from 1.5" to 2.5". The equivalent widths of 12 strong absorption lines have been measured (for the numerical data - see [2] and [3]). The intrinsic accuracy of equivalent width for lines of the moderate strength ($EW = 1 - 5A$) is estimated as 0.8 A. For the strong lines $CaIIH$ and K the accuracy is somewhat worse: 1.5-2.0 A.

All the absorption spectra have been divided by us into two groups:

- the "E"-type ones, having strong metallic lines (Mg,Ca,Fe) and some molecular bands, but only one Balmer line H_β,

- and the "H"-type ones, having, except the metallic lines, also the strong Balmer lines H_γ and H_δ in absorption.

Almost all elliptical galaxies have the "E"-type nuclear spectra; this spectral type can result from an old stellar population with nearly solar metallicity. Among the early-type disk galaxies (S0, Sa-Sb), 50% of nuclei have the "H"-type spectra which indicate the presence of A-stars (Fig.1). This fact may be the evidence of either the intermediate age of stellar population (T=1-2 billion years) or the low abundance in the nuclei of 50% early-type disk galaxies.

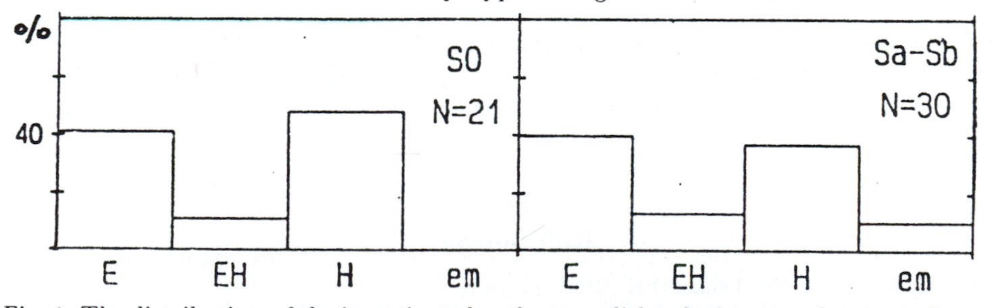

Fig. 1. The distribution of the investigated early-type disk galaxies over the spectral types: "E" and "H" - see the text,"EH" - the type is unclear, "em" - the HII-type emission.

H. Dejonghe and H. J. Habing (eds.), Galactic Bulges, 429–430.

To test these hypotheses we have calculated some evolutionary models with solar metallicity and varying star formation rate [1] and have compared the equivalent widths of absorption lines for the observational spectra and synthetic ones (Fig.2). It is quite visible that on the diagram ($< H >$, $CaIIK$) the "E"–nuclei follow the sequence of globular clusters so they may be old stellar systems with metallicities about from 0.1 $Z_{\odot}$ to $Z_{\odot}$. And the "H"–nuclei mostly follow the model sequence "SFR" (which is calculated with the exponentially decreasing star formation rate, characteristic decay time being from infinitely large to 1 billion years); so they must have nearly solar metallicity and star formation rate rather intense in some cases (among those there are S0–galaxies such as NGC 3412, 3489, 4150, 4262, and S0/a galaxies NGC 2681, 3166, etc.)

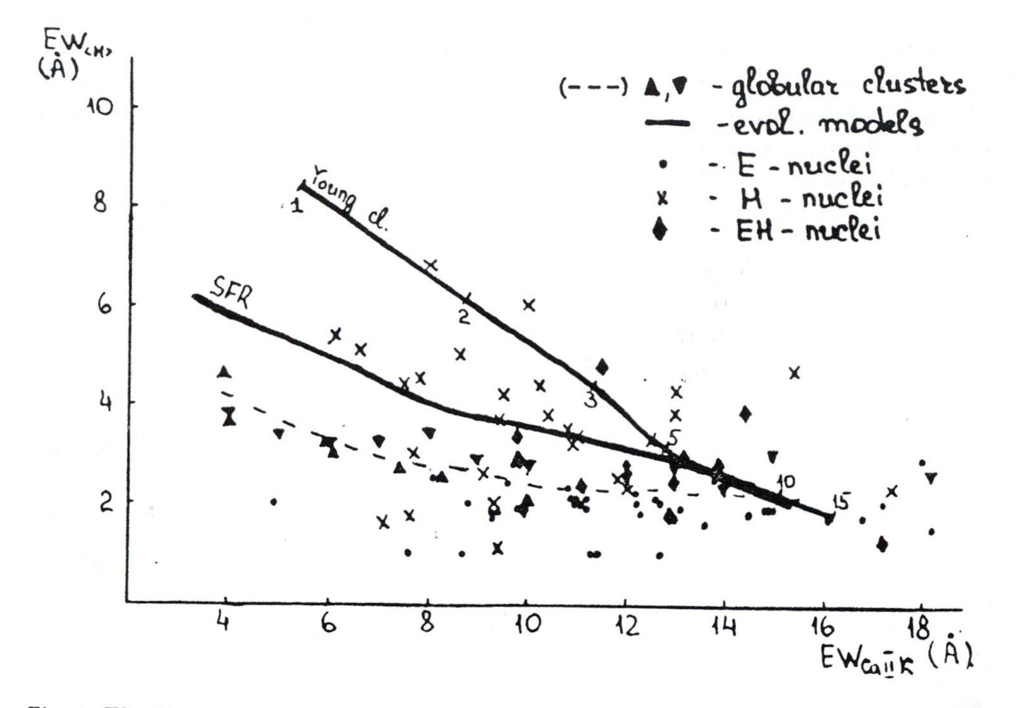

Fig. 2. The diagram ($< H >$,$CaIIK$) for the observed early–type disk galaxies compared with Galactic globular clusters and with model sequences calculated for the solar metallicity and different star formation histories. Numbers are the ages of model star clusters in billion years.

References

[1] Balinskaya, I.S., & Sil'chenko, O.K., 1992. *In press.*

[2] Sil'chenko, O.K., & Shapovalova, A.I., 1989. *Preprint SAO RAS* No.37.

[3] Sil'chenko, O.K., 1990. *Preprint SAO RAS* No.42.

ON THE DIFFERENCE IN FORMATION HISTORY BETWEEN BULGES AND ELLIPTICALS

B.ROCCA-VOLMERANGE[1], O.K.SIL'CHENKO[2]

[1] *Institut d'Astrophysique de Paris, 98bis Boulevard Arago, F-75014, Paris.*
[2] *Sternberg Astronomical Institute, University av. 13, 119899 Moscow.*

Abstract. On the diagram (Fe5270, Mg5175) ellipticals and bulges of disk galaxies (from S0 to Sc) maintain a very different position: E's seem to be overabundant in Mg, bulges seem to have a solar ratio Mg/Fe, except several ones who are overabundant in Fe.

Key words: Galaxies - Chemical evolution - Evolutionary synthesis

Recently Faber et al.[1] have claimed that on the diagram ($Fe5270$, Mg_2) evolutionary models with different ages and metallicities represent a common sequence, according with galactic globular clusters as well. But most ellipticals seem to be shifted to the right from this sequence so being enriched in magnesium relative to Fe.

We have checked this conclusion with our models [3] and observations [4,5]. Indeed, on the diagram ($Fe5270$, $Mg5175$) all the models with solar metallicities, but with different ages and star formation histories, represent an universal sequence which is in accordance with some galactic globular clusters; and some ellipticals of various luminosities (a compact dwarf NGC 4486B among them) are shifted to the right (Fig.1). So, we have confirmed the Mg overabundance in a significant part of ellipticals.

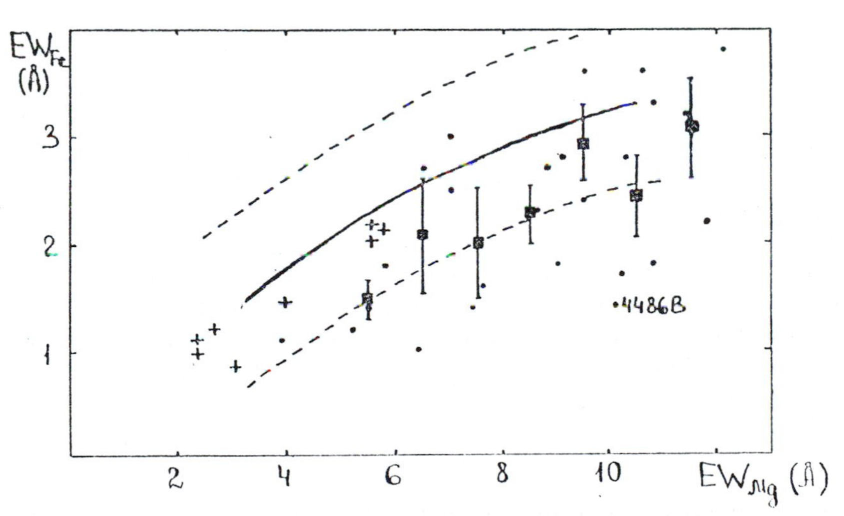

Fig. 1. The diagram (Fe5270, Mg5175) for the ellipticals: the solid line represent the theoretical sequence for varying age and star formation history, crosses – galactic globular clusters, dashed lines roughly mark the range in EW_{Fe} "theoretical sequence+/-one observational mean error", dots – observational data for galaxies, squares – the averaged values for each one–angstrom EW_{Mg} interval.

H. Dejonghe and H. J. Habing (eds.), Galactic Bulges, 431–432.

But when we compare the location of bulges of disk galaxies (from S0 to Sc) with the theoretical sequence, we see that the most bulges follow the sequence very well, with the dispersion fully explained by our observational equivalent width accuracy (Fig.2). Moreover, few galaxies (among them 4 S0–Sa galaxies from the nearby Leo group – NGC 3368, 3384, 3412, 3489) in a very narrow range of EW_{Mg} – from 5 to 6 A – have significantly stronger Fe–lines, than the bulk of galaxies, and not only $Fe5270$, but also $Fe4383$ and $Fe4528$. So, for some bulges we establish a Fe overabundance relative to Mg.

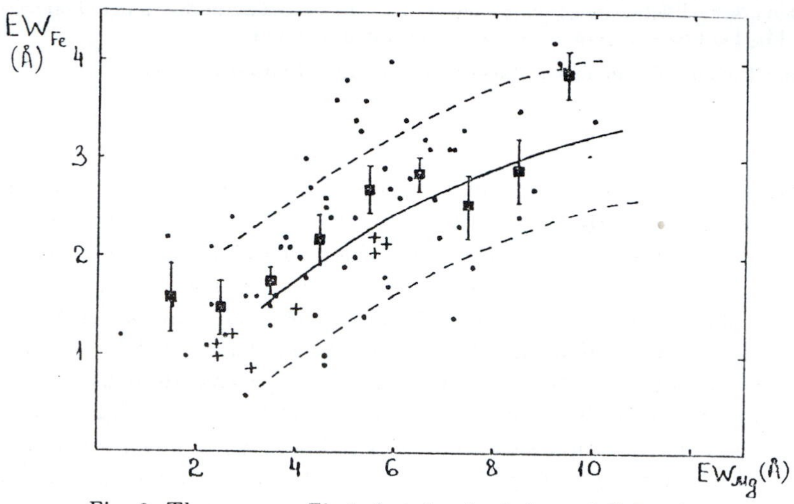

Fig. 2. The same as Fig.1, but for the bulges of disk galaxies.

The above mentioned difference between bulges and ellipticals may be explained as a difference of their star formation histories. For more detailed hypotheses on ellipticals – see [1]. Here we briefly summarize. There may be a difference in early Initial Mass Function (IMF) – in the sense that in ellipticals the IMF must be enriched in massive stars, and in Fe–overabundant bulges the IMF must have a lower upper limit. The second, even more attractive possibility is to utilize the time delay for the Fe production keeping in mind that Mg is produced by SNII, and Fe – by SNIa. If most ellipticals have only one initial star formation burst rather short to give stars with a lot of magnesium but without a bulk of Fe; and if some bulges have several star formation bursts with an interval of some billion years between them – we would obtain a Mg overabundance for the first [1] and Fe overabundance for the latter [2].

References

[1] Faber, S.M., Worthey, G., Gonzalez, J.J., 1991. In *Stellar Populations of Galaxies*, IAU Symposium 149, Eds. B.Barbuy & A.Renzini (Dordrecht: Kluwer), p.255.
[2] Gilmore, G., & Wyse, R.F.G., 1991. *Astrophys. J.*, **367**, L55.
[3] Guiderdoni, B., & Rocca–Volmerange, B., 1987. *Astron. Astrophys.*, **186**, 1.
[4] Sil'chenko, O.K., & Shapovalova, A.I., 1989. *Soobshch. Spets. Astrofiz. Obs.*, No.60, 44, or *Preprint SAO RAS* No.37.
[5] Sil'chenko, O.K., 1990. *Soobshch. Spets. Astrofiz. Obs.*, No.65, 75, or *Preprint SAO RAS* No.42.

THE VELOCITY FIELDS OF ELLIPTICAL GALAXIES: CONSTRAINTS ON INTRINSIC SHAPES

T. S. STATLER and A. M. FRY
University of North Carolina, Chapel Hill, NC, 27599-3255, USA

Abstract. The problem of determining the intrinsic shapes of elliptical galaxies cannot be solved using photometry alone. Measuring rotation on the apparent major and minor axes adds a kinematic constraint, but does not significantly improve the situation. We find that having two *more* spectra, at the $\pm 45°$ position angles, gives enough kinematic information that much tighter limits can be placed on the intrinsic axis ratios than are possible otherwise.

Key words: Elliptical galaxies - Radial velocities - Intrinsic shapes - Triaxiality

1. Introduction

One of the few surviving signatures, at low redshift, of the process of galaxy formation should be the distribution of shapes of elliptical galaxies. Yet the problem of inferring this distribution from the observed ellipticals is still unsolved, because insufficient use has been made of kinematic information. The kinematic data available for most ellipticals consists of only major and minor axis spectra; and Franx *et al.* (1991) find, using simple geometric models, that the addition of only one kinematic parameter (the ratio of minor axis to major axis rotation velocity) to the photometry is just not enough to finely constrain the intrinsic shape distribution. On the other hand, the more elaborate self-consistent models (*e.g.*, Levison and Richstone 1987, Statler 1987) have made only infrequent and model-dependent predictions of complicated velocity patterns, mostly at small radii, and have not discussed how they should vary with intrinsic shape.

2. Theoretical Models

We have developed (Statler 1993*a*, Paper I) an approximate but reliable method of computing the velocity field (hereafter VF) for a model elliptical of arbitrary shape. To avoid problems connected with strongly dissipative evolution in cores and the presence of central black holes, we take the view that the most useful VF features are to be found at large radii. We then assume (1) radial self-similarity at large r; (2) negligible rotation of the figure, which implies (3) intrinsic circulation (rotation) only around the long (x) and short (z) axes; (4) flow of the stellar "fluid" on spherical shells (or, with a slight added complication, on ellipsoidal shells), on which (5) the streamlines of the x and z circulations are given by coordinate lines in a confocal ellipsoidal system. This last assumption is suggested by the analytically tractable Stäckel potentials, in which the flow is exactly along those lines, but is *much less restrictive* than asserting the potential is separable. With the streamlines specified, each of the x and z flows is dictated by the equation of continuity, and the projected velocity follows with a little geometry.

A boundary condition is, of course, required to solve for the complete flow. An exact expression for the boundary condition for any one model would require knowing the complete distribution function for the tube orbits; however, we argue

H. Dejonghe and H. J. Habing (eds.), Galactic Bulges, 433–434.

on the grounds that the mean velocity should not be spatially discontinuous, and from the existing fully self-consistent models, that the boundary condition — a function of one variable giving the mean flow across the plane containing the long and short axes — should take only a limited variety of forms. (The details of this vague statement can be found in Paper II.) We are then able to calculate VFs at any intrinsic shape and projection for a small number of ***distinguishable classes*** of models — for instance, with intrinsic streaming about the long axis, about the short axis, or about both axes. These models are distinguishable from each other because we are able to look not merely at the apparent axis of rotation (*cf.* Franx *et al.* 1991), but at the *asymmetries* of the VF, which are the true signatures of triaxiality.

3. Model Fitting

Perfect mapping of the VF is not presently feasible for any galaxy, so we ask whether a reasonable extra observational effort can give greatly improved results. We define (Statler & Fry 1993, Paper II) three dimensionless kinematic parameters (V_1, V_2, V_3) that are combinations of the velocities ($v_{\rm maj}, v_{\rm min}, v_{+45}, v_{-45}$) measured at one radius, and then average over radial bins. V_1 is essentially the "kinematic misalignment angle" used by Franx *et al.* ; V_2 and V_3 are measures of the VF asymmetry. We find that a combination of apparent ellipticity and V_2 correlates well with axis ratio c/a, and a combination of V_1 and V_3 correlates with triaxiality, for all classes of models we have tried. A Monte-Carlo modeling procedure (Statler 1993*b*, Paper III) has also been developed to fit galaxies one by one, and a preliminary application to NGC 4589 (Möllenhoff & Bender 1989) indicates that this is a prolate-triaxial system.

The intrinsic shape distribution of the full population of ellipticals is likely to be formally consistent with most types of internal streaming models, though (we would hope) different for each type. Comparing with predictions from galaxy formation theories will be necessary to settle the problem decisively, and in turn to constrain those theories.

References

Franx, M., Illingworth, G. D., and de Zeeuw, T. 1991, *Ap. J.*, **383**, 112
Möllenhoff, C. and Bender, R. 1989, *Astron. Astrophys.*, **214**, 61
Levison, H. F. and Richstone, D. O. 1987, *Ap. J.*, **314**, 476
Statler, T. S. 1987, *Ap. J.*, **321**, 113
Statler, T. S. 1993*a*, *Ap. J.*, submitted (Paper I)
Statler, T. S. 1993*b*, *Ap. J.*, submitted (Paper III)
Statler, T. S. and Fry, A. M. 1993, *Ap. J.*, submitted (Paper II)

TRIAXIALITY IN THE BULGES OF SPIRALS: DYNAMICAL IMPLICATIONS

A.M. VARELA[1], E. SIMONNEAU[2], C. MUÑOZ–TUÑÓN[1]

1. Instituto de Astrofísica de Canarias, 38200, La Laguna, Tenerife, Spain.
2. Institut d'Astrophysique de Paris, 98 bis, Boul. Arago, 75014 Paris, France.

The presence of triaxial bulges has been suggested as a mechanism for sweeping out the interstellar gas into the nuclear zones of normal spiral galaxies, giving rise to star formation processes (Zaritsky and Lo, 1986). In an initial phase of the present study, we have analysed the bulge component of a sample of spiral galaxies, obtained using the 4.2m WHT at La Palma Observatory, selected from among those which exhibit moderate activity. The results obtained are: (a) There is evidence, from the misalignment between the semi–major axes of the isophotes of the bulge and disc components, to support the presence of triaxial bulges in these galaxies; (b) The high spatial resolution of the images allows us to sample the bulges, showing that the surface brightness distribution is not always represented by a family of homologous coaxial ellipses, but rather by a set of ellipses which precess around a common centre, and whose ellipticity varies (see Beckman *et al.*, 1991; Varela, 1992). It is possible to reproduce this structure with a family of concentric triaxial ellipsoidal shells. The presence of a disc offers the advantage of enabling us to interpret the geometry of the bulge correctly, but poses the disadvantage of requiring to interpret the disc emission in order to reconstruct that associated only with the bulge, *i.e.* to decontaminate the bulge from the disc light. To this end we carried out a photometric analysis, which enabled us to effect a valid bulge/disc decomposition. The preliminary results yielded by the analysis of the bulge component (after decontaminating the disc) show the presence of "transparent" discs in the galaxies analysed.

The absence of axisymmetry in the inferred bulge potential, in the context of galactic rotation, might be responsible for driving the gas into the central region of the bulge, and must therefore be considered when accounting for the formation of rings, the production of gas flows, etc., and in the evolutionary model of starbursts. In order to analyse quadrupole moments and, more generally, the dynamics of these triaxial bulges, we must determine the spatial distribution of the bulges in the spirals analysed, as derived from the observed isophotes, assuming a mass to light ratio. In some cases, the spatial source and the surface brightness distributions are related by the Abel integral transform (Stark, 1977; Simonneau, Varela, Muñoz–Tuñón, 1992), but if differential precession exists between isophotes we do not, in general, obtain an integral relation, nor is it even possible to specify analytically the surface brightness contours. In these cases we adopt a direct method to synthesize photometrically the 3–D structure of the bulge. It is possible to reproduce this morphology with a family of

H. Dejonghe and H. J. Habing (eds.), Galactic Bulges, 435–436.

ellipsoidal shells of equal source density which are concentric and coaxial, but with varying axial ratios. For each individual object under study, we intend to model a three–dimensional structure (see in bottom figure the spatial source distribution associated to different 3–D solutions for NGC 4736, corresponding to different axial ratios families) capable of reproducing the observations (pixel size is 27"). The distorsions detected could play a fundamental role in the dynamics of the circumnuclear gas (see Muñoz–Tuñón and Varela, 1991).

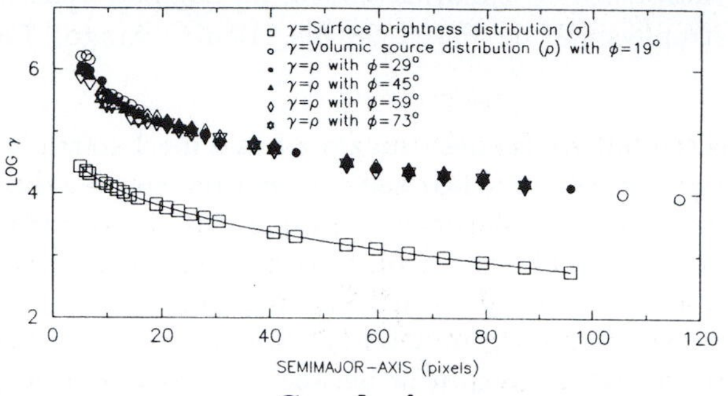

Conclusions

1. The surface brightness within the bulge is reproduced by a family of ellipses that precess around a common centre.
2. The spatial source distribution has been modelled and we have developed a spectral analytical method to inverse the 2–D structure.
3. It is possible to reproduce the surface brightness distribution with a family of concentric, coaxial ellipsoids with variable axial ratios.
4. In general, the concepts **projection** and **integration** along line of sight have different meanings, and one cannot be used in the context of the other.
5. Quadrupolar moments for the gravitational potential are present on the disc and are not negligible, depending on the azimuth.
6. The presence of a triaxial potential associated to the bulge component should be taken into account when modelling the stable inner annular structures. This is an alternative to the **classical** interpretation of explosive nuclear starburst.

References

Beckman, J.E.; Varela, A.M.; Muñoz–Tuñón, C.; Vílchez, J.M.; Cepa, J.: 1991, *Astron. Astrophys.*, **245**, 436.

Muñoz–Tuñón, C.; Varela, A.M.: 1991, Proc. Workshop *Relationships between* Active Galactic Nuclei and Starburst Galaxies, China (in press).

Simonneau, E.; Varela, A.M.; Muñoz–Tuñón, C.: 1992, *JQSRT* (in press).

Stark, A.A.: 1977, *Astrophys. J.*, **213**, 368.

Varela, A.M.: 1992, *Ph.D. Thesis*, Univ. La Laguna.

Zaritsky, D.; Lo, K.Y.: 1986, *Astrophys. J.*, **303**, 66.

EXTRACTING KINEMATICS FROM SPECTRA: A NEW METHOD AND NGC 4406

T. B. WILLIAMS
Dept. of Physics and Astronomy, Rutgers University,
Piscataway, NJ 08855-0849, USA.
P. SAHA
CITA, University of Toronto,
60 St. George Street, Toronto, Ontario M5S 1A7, Canada.

Introduction

The observed spectrum of a galaxy is the composite spectrum of its stellar population broadened by the stellar velocities. Our topic is the extraction of the broadening functions (i.e., line-of-sight velocity distributions) from spectra. In the past, observers tried to extract only mean velocities and dispersions. But recent work on deconvolution [Franx & Illingworth *ApJL* **327**, L55 (1989), Bender *A&A* **229**, 441 (1990), Rix & White *MNRAS* **254**, 389 (1992), van der Marel & Franx *ApJ*, to appear] shows that more information can be recovered. The general idea is to compare a galaxy spectrum with one or more stellar 'template' spectra; but various methods differ widely in the treatment of noise and the control of sources of systematic error.

We have developed a new deconvolution method, though drawing on previous work. Here we show rotation curves, dispersions, and deviations from Gaussian broadening functions for the major axis of the E3 NGC 4406.

Observations

The galaxies were observed using the Large Cassegrain Spectrograph and on the McDonald Observatory 2.7 meter telescope on February 8–9, 1992. The spectra cover the wavelength region 4800Å to 5500Å with a resolution of 1.8Å; the seeing ranged from 1.5 to 2.5 arcseconds. The galaxies NGC 1439, 1700, 3379, 4261, and 4406 were observed at 2–5 different position angles. Some preliminary results for NGC 4406 are presented here.

Analysis method

We have used the form of van der Marel and Franx for the broadening function, with four parameters V, σ, h_3, h_4—h_3 measures the skewness and h_4 the kurtosis (relative to a Gaussian). In other respects, our method is closest to Rix and White's—it does the fitting in pixel space rather than Fourier space (so pixels with higher S/N automatically get more weight), and includes a simple population synthesis. However, we introduce two improvements:

1. Our 'pixels' for the fit really are the pixels on which the galaxy spectrum is recorded, not a rebinned version. This avoids correlations in the noise introduced by rebinning.
2. We assume that the noise in the data is Gaussian, but we *don't* assume that the errors in the parameter fits are Gaussian. Previous approaches supply parameter values and formal Gaussian error estimates. We use a Bayesian

H. Dejonghe and H. J. Habing (eds.), Galactic Bulges, 437–438.

method to compute a probability distribution for the parameters. The estimates shown in here are the medians and 66% boundaries of such probability distributions.

Conclusions

Significantly non-Gaussian line-of-sight velocity distributions are the rule rather than the exception. The deviations from Gaussian provide new input for modelling the dynamics of galaxies. Ignoring these deviations is likely to produce systematic errors in estimates of dispersions.

The deconvolution method we have developed has the advantages that (i) the 'fitting' process is very transparent; (ii) prior information, if available, can be incorporated through the Bayesian prior probability; and (iii) the error analysis is more general (and, we believe, more reliable) than for previous methods.

Possible applications: (i) The search for central black-holes. (ii) It is now feasible to map potentials from discrete velocity measurements (see Merritt, preprint 1992); broadening functions (such as presented here) could probably also be used for this purpose.

Programs are available to anyone interested.

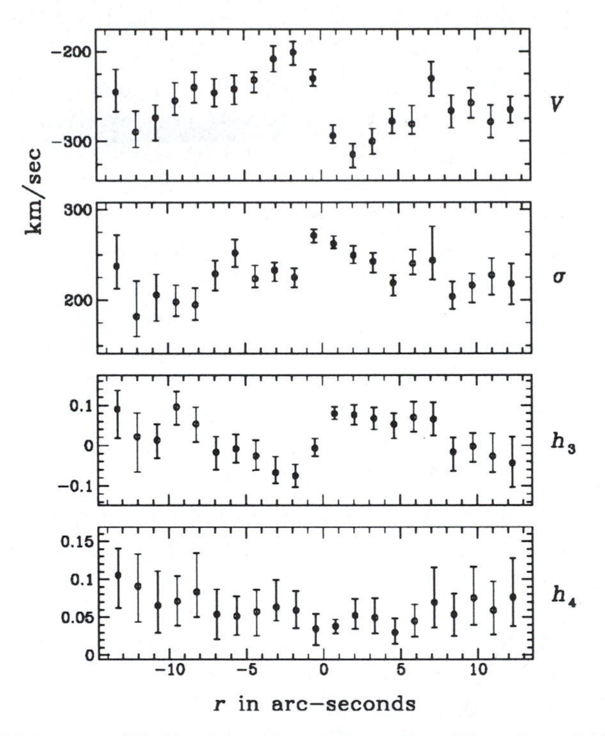

Figure 1. NGC 4406 (M86) along the major axis. The plot of V shows a counter-rotating core. Note also the antisymmetry of the skewness (h_3) plot as we move away from the centre. The latter seems to be a fairly common feature in ellipticals; it may result from a combination of velocity anisotropy and rotation or it may imply two kinematically distinct components—it is not yet clear which.

Nuclear Molecular Gas in the Virgo Cluster S0 Galaxy NGC 4710

J.M. WROBEL
NRAO, P.O. Box O, Socorro, NM 87801, USA

and

J.D.P. KENNEY
Dept. of Astronomy, Box 6666, Yale University, New Haven, CT 06511, USA

Abstract. The CO(J=1→0) emission from NGC 4710, a star–forming S0 galaxy in the Virgo Cluster, was synthesized with spatial and velocity resolutions of 7″ and 26 km s^{-1}, respectively. The CO shows a compact morphology and co–rotates with the galaxy's stars and nuclear optical emission line gas. Analysis of the CO distribution and kinematics indicates that the nuclear molecular gas is probably gravitationally unstable, and this may explain why the galaxy is presently forming stars. Four possible origins for the nuclear molecular gas are considered. An origin via bulge star ejecta being deposited into a residual interstellar medium is favored.

This work is fully described in the 1992 November 1 issue of *The Astrophysical Journal.* NRAO is operated by Associated Universities, Inc., under cooperative agreement with the National Science Foundation.

D. Pfenniger

H. Dejonghe and H. J. Habing (eds.), Galactic Bulges, 439.

PREDICTIONS
Henry FERGUSON

IR Imaging of Spiral Galaxy Bulges

DENNIS ZARITSKY
Carnegie Observatories, 813 Santa Barbara St., Pasadena, CA, 91101, U.S.A.
MARCIA RIEKE
Steward Observatory, Univ. of Arizona, Tucson, AZ, 85721, U.S.A.
and
HANS-WALTER RIX
Institute for Advanced Study, Princeton, NJ, 08540, U.S.A.

Abstract. Imaging in the infrared (2.2μ) minimizes the impact of dust obscuration and allows reliable mapping of the mass-tracing stellar population in spiral galaxies. We find dramatic differences compared to photometry at shorter wavelengths (e.g. 0.8μ). As an example, the observations of the mini-bar and inner spiral arms of M 51 are discussed.

Key words: galaxies: photometry - galaxies: spiral - galaxies: individual: M 51

The apparent shape of the disk and bulge components of spiral galaxies is best studied through infrared imaging because distortions of the light from the mass-tracing stellar population by light from young stars and by dust are minimized. Imaging face-on galaxies eliminates complications from projection effects. Although our project consists of a systematic study of about 25 galaxies with inclinations of less than 10°, in this paper we only discuss preliminary results from observations of M 51.

The data presented here were obtained with the Steward Observatory 2.3m telescope using either a CCD (B and I images, 0.45 and 0.8μ respectively) or a Nicmos 3 (256×256) array (K images, 2.2μ). Exposure times range from 2 min (CCD) to 15 min (IR) and the data were reduced using standard techniques.

The oval distortion seen at I by Zaritsky and Lo (1986) and Pierce (1986) is found to be much more prominent at 2.2μ. Whether this distortion is a prolate bulge or a bar is an observationally subtle distinction, which may, however, imply different origins. Based on the IR surface photometry, we argue that the oval distortion is a bar, because the spiral arms appear to begin at the edges of this feature (cf. Figure 1 in which a smooth model has been subtracted from the K-band image) and because a power-law fit to the intensity profile in the inner region overpredicts light at large radii (i.e. there must be a distinct truncation of the oval component). However, the distinction between bar and oval bulge is not unambiguous.

Pierce (1986) also claimed to have observed an inner ring of material at the edge of the bulge. We cannot comment on the possible presence of a gaseous ring, but we do not observe a stellar ring. Instead we see the continuous winding of spiral arms to the edge of the bar. Comparison of J and K images shows that these arms are primarily ridges of stellar emission, not contrast enhancements from dust lines. Since the arms are tightly wound these could have been mistaken for a ring. This distinction can only be made with the K-band images. Models of the interaction between M 51 and its companion suggest that the spirals arms, at least at large radii, are material arms generated by tidal interaction (Hernquist 1990). However, if the inner arms are material arms as well we can use the winding of the arms to estimate

H. Dejonghe and H. J. Habing (eds.), Galactic Bulges, 441–442.

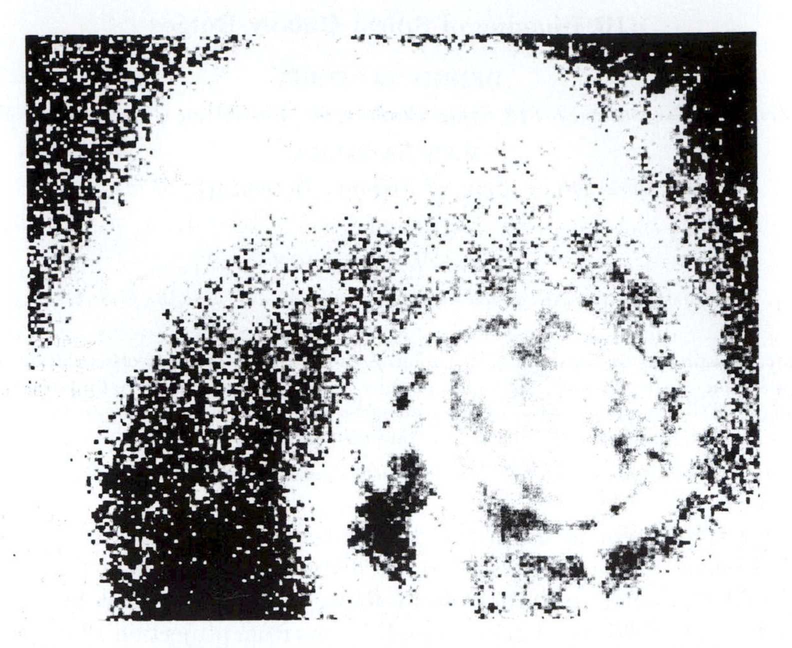

Fig. 1. Model-Subtracted K-Band Image of Inner Region of M 51

the time since their formation (cf. Binney and Tremaine 1987). We find that if the rotation curve is presumed to be flat with a value of 200 km sec^{-1} (Tully 1974, Goad et al. 1979) then the implied age (2.3×10^7 yrs) is inconsistent with the estimated time since the pericenter passage of NGC 5195 ($\sim 3 \times 10^8$ yrs; Hernquist 1990). Unfortunately, the shape of the rotation curve is not sufficiently well determined to place strong constraints. The continuity of these arms through the inner Lindblad resonance, their thinness, and their tight winding are remarkable.

Acknowledgements

D.Z. and H.-W.R. acknowledge support through Hubble Fellowships from STScI under contract to NASA (HF-1027.01-91A and HF-1027.01-91A respectively). D.Z. thanks the IAS for its hospitality during his visit. We gratefully acknowledge financial support for the infrared camera from the National Science Foundation and NASA for providing the Nicmos array.

References

Binney, J. and Tremaine, S. (1987), *Galactic Dynamics*, (Princeton University Press: Princeton).

Goad, J.W., DeVeny, J.B., and Goad, L.E. (1979), *Ap. J. Suppl.*, **39**, 439.

Hernquist, L. (1990), in *Dynamics and Interactions of Galaxies*, ed. R. Wielen, (Heidelberg: Springer-Verlag), p. 108.

Pierce, M.J. (1986), *A.J.*, **92**, 285.

Tully, R.B. (1974), *Ap. J. Suppl.*, **27**, 449.

Zaritsky, D., and Lo, K.-Y. (1986), *Ap. J.*, **303**, 66.

A triaxial model for the bulge of NGC 4697

W.W. ZEILINGER
European Southern Observatory, Garching bei München, Germany D-8046
and
M.L. WINSALL and H. DEJONGHE
Sterrenkundig Observatorium Gent, Krijgslaan 281 (S9), Gent, Belgium B-9000

There is now mounting evidence that the intrinsic shape of elliptical galaxies and bulges of disk galaxies may generally be triaxial. There are different signatures of non–axisymmetric structures observed: velocity gradients along the minor axis (e.g. Davies & Birkinshaw 1986), twisting of the isophotes (e.g. Williams & Schwarzschild 1979) and misalignment between bulge and disk major axes (e.g. Bertola, Vietri & Zeilinger 1991). Therefore, a framework of reliable algorithms for the dynamical modelling of such triaxial potentials is needed.

Of all triaxial potentials, Stäckel potentials have the advantage that all orbits are regular. Moreover, the three integrals of the motion are simple quadratic functions of the velocities. Therefore one can construct simple components, representing families of stars, which are in (not necessarily self–consistent) equilibrium in a Stäckel potential.

Very flattened elliptical galaxies were selected as observational prerequisite for such models. In the past most of the effort was concentrated on round "bona–fide" ellipticals, and there are no explicit dynamical models for flattened ellipticals. If kinetic energy (pressure or rotational) is responsible for their flattenings, their most advantageous viewing angle is one that is very nearly edge-on, since in that direction the large kinetic energy will appear as motion along the line-of-sight, which is detectable.

The E6 galaxy NGC 4697 is considered to be one of the "prototypes" of non–rotationally supported systems. The observations were carried out at the ESO 3.5-m NTT with EMMI in the Red Medium Spectroscopy Mode using a FA 2048 × 2048 CCD (15 μm pixelsize). Long–slit Spectra with a spectral resolution of 20 km/s were obtained along the optical major axis, 45° intermediate axis and at an offset of 5″ parallel to the major axis. The calibrated spectra were rebinned to have $S/N > 30$ in the centre and 15 in the outer parts. The spectra were analyzed with the Fourier Quotient technique yielding radial velocities and velocity dispersions as a function of radius.

Unlike the spherical case, we must make a number of assumptions when modelling triaxial galaxies from their projected quantities: First of all, there are no direct indicators for the determination if the viewing angles, such as a gaseous disk (as in NGC 5077). Surface photometry of NGC 4697 reveals however a stellar disk, seen almost edge–on (being the most favourable inclination for detection). Hence, we place the z-axis in the plane of the sky. There is no obvious choice for the orientation of the x-axis. We arbitrarily place it in the plane of the sky. Further, we assume a Kuzmin–Kutuzov potential for simplicity. Using ellipsoidal coordinates with axis ratios of $a/b = 1.1$, $b/c = 1.5$, and a scale length $b + c = 1.75$ kpc, we

H. Dejonghe and H. J. Habing (eds.), Galactic Bulges, 443–444.

were able to produce a projected surface density which is a surprisingly good fit to the photometry.

The dynamical model has components of Abel type (Dejonghe & Laurent 1991), and has a distribution function $F(E, I_2, I_3) = F(E + wI_2 + uI_3)$, where E, I_2 and I_3 are the Stäckel integrals of the motion. In order to produce rotation, we added new components which are constant on similar planes in phase space, but are restricted to the region in phase space occupied by short axis tubes. The calculations for the moments (mass density, mean rotation and velocity dispersion) are only approximate at this point.

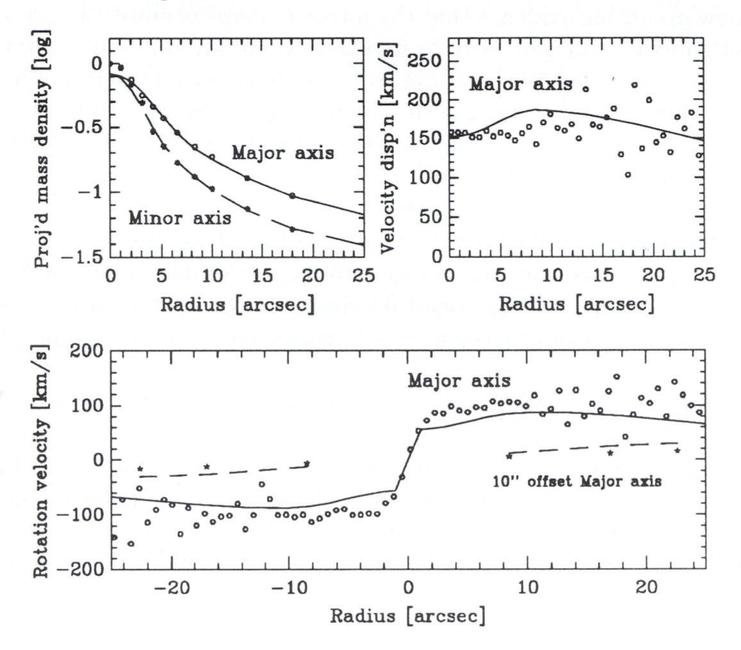

Fig. 1. Projected mass density (top left), projected velocity dispersion (top right) and rotation velocity (bottom). Lines (solid and dashed) show the model.

As is clear from the figure, the model produces an excellent fit to the photometry. The velocity dispersion too is fitted within the observational error (suggested by the spread of the data points). Clearly no hidden matter is needed on these scales. The rotational velocity is, if anything, a bit low. In the model, the different kinds of short axis tubes are populated fairly evenly. This suggest that there is more structure in the short axis orbit population, with more emphasis on circular orbits.

References

Bertola, F., Vietri, M. & Zeilinger, W.W., 1991, *Astrophys. J. (Lett.)*, **374**, L13.
Binney, J.J., Davies, R.L. & Illingworth, G.D. 1990, *Astrophys. J.*, **361**, 78.
Davies, R.L. & Birkinshaw, M. 1986, *Astrophys. J. (Lett.)*, **303**, L45.
Dejonghe, H. & Laurent, D. 1991, *Mon. Not. R. astr. Soc.*, **252**, 606.
Williams & Schwarzschild, M., 1979, *Astrophys. J.*, **227**, 56.

The technical staff: P. Vauterin and P. De Groote

P. Batsleer and the photographer: D. Laurent

Subject Index

'Micro' assistants during discussions : L. Sjouwerman and M. Sevenster

If you develop plans to visit Gent in the future ...

Object Index

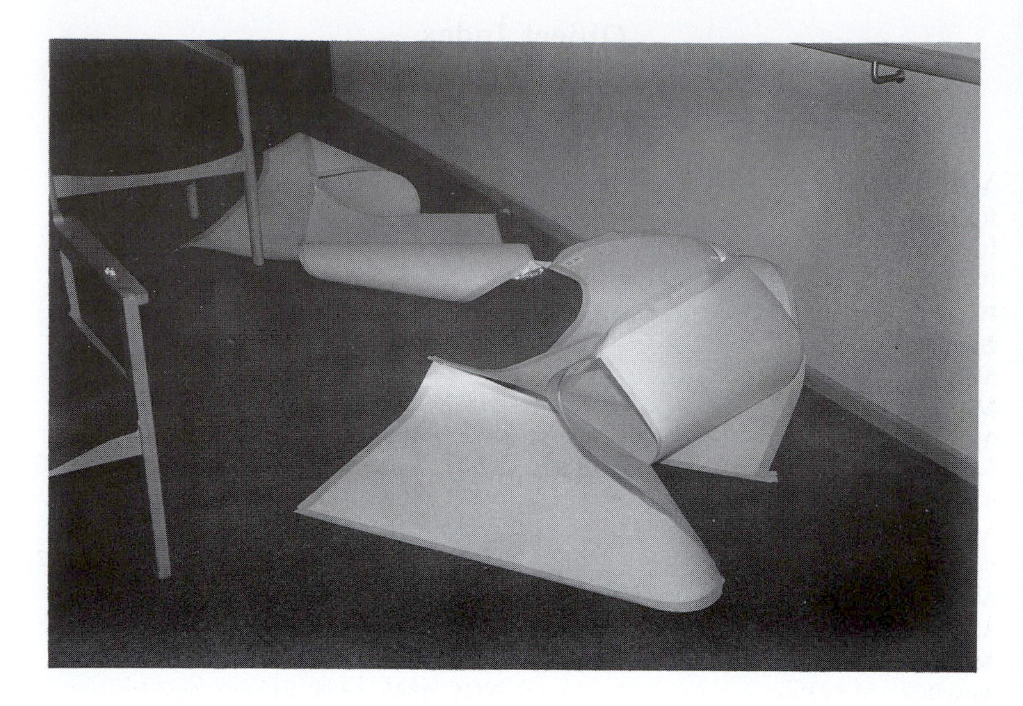

The day after...

Author Index